HANDBOOK
OF INFRARED
OPTICAL MATERIALS

OPTICAL ENGINEERING

Series Editor

Brian J. Thompson

Provost
University of Rochester
Rochester, New York

Other Volumes in Preparation

HANDBOOK OF INFRARED OPTICAL MATERIALS

**EDITED BY
PAUL KLOCEK**

*Texas Instruments, Inc.
Dallas, Texas*

CRC Press
Taylor & Francis Group
Boca Raton London New York

CRC Press is an imprint of the
Taylor & Francis Group, an **informa** business

First published 1991 by Marcel Dekker, Inc.

Published 2019 by CRC Press
Taylor & Francis Group
6000 Broken Sound Parkway NW, Suite 300
Boca Raton, FL 33487-2742

First issued in paperback 2019

No claim to original U.S. Government works

ISBN 13: 978-0-367-45057-1 (pbk)
ISBN 13: 978-0-8247-8468-3 (hbk)

Library of Congress Cataloging-in-Publication Data

Handbook of infrared optical materials / edited by Paul Klocek.
        p.   cm. – (Optical engineering ; vol. 30)
    Includes bibliographical references and index.
    ISBN 0-8247-8468-5
    1. Infrared technology--Materials.   2. Optical materials.
I. Klocek, Paul.   II. Series.
TA1570.H36   1991
    621.36'2 - - dc20
                                                        91-17507
                                                        CIP

About the Series

The series came of age with the publication of our twenty-first volume in 1989. The twenty-first volume was entitled *Laser-Induced Plasmas and Applications* and was a multi-authored work involving some twenty contributors and two editors: as such it represents one end of the spectrum of books that range from single-authored texts to multi-authored volumes. However, the philosophy of the series has remained the same: to discuss topics in optical engineering at the level that will be useful to those working in the field or attempting to design subsystems that are based on optical techniques or that have significant optical subsystems. The concept is not to provide detailed monographs on narrow subject areas but to deal with the material at a level that makes it immediately useful to the practicing scientist and engineer. These are not research monographs, although we expect that workers in optical research will find them extremely valuable.

There is no doubt that optical engineering is now established as an important discipline in its own right. The range of topics that can and should be included continues to grow. In the ''About the Series'' that I wrote for earlier volumes, I noted that the series covers ''the topics that have been part of the rapid expansion of optical engineering.'' I then followed this with a list of such topics which we have already outgrown. I will not repeat that mistake this time! Since the series now exists, the topics that are appropriate are best exemplified by the titles of the volumes listed in the front of this book. More topics and volumes are forthcoming.

Brian J. Thompson
University of Rochester
Rochester, New York

Preface

Advances in science and technology have led to the exploitation of various parts of the electromagnetic spectrum, and new uses continue to be developed and refined. This is particularly true in the field of optics that concentrates on the ultraviolet, visible, and infrared regions of the electromagnetic spectrum, as can be seen in the exploding number of optical and electro-optical systems being used and developed for applications in telecommunications, medicine, industry, the military, consumer products, science, and art. Fiber optics, light emitting diodes (LEDs), lasers, sensors, optical computing, data/voice/image storage and retrieval, visual displays, and holography are some of the optics technologies these systems are based on.

One of the common denominators of the various technologies in optics, and for that matter in most fields, is the materials they rely on. In the case of optics and electro-optics, these are materials that transmit, reflect, refract and diffract light in the wavelength region of interest. Because of these different regions of interest, there is not one set of optical materials to rely on but several, often overlapping sets. This handbook is designed to describe the set of optical materials used for infrared optical applications.

The *Handbook of Infrared Optical Materials* has been designed to be useful to a wide range of readers. It is intended to aid scientists, engineers, and other professionals involved in the research, development, design, and production of infrared optics and infrared optical and electro-optical materials, devices, and systems. It is also intended to serve those in related or overlapping fields.

This handbook includes a comprehensive presentation of infrared optical materials. The fundamental physics of optical matter, the definition of material physical properties, the listing and comparison of the physical properties of infrared optical materials, and the theory, design, and survey of infrared optical coatings are thoroughly presented. Physical property data for over 100 crystalline and glass infrared optical materials and over 50 infrared optical coatings are compiled in the handbook. Along with the supporting text and tabulature, the handbook contains over 136 additional tables, 300 figures, 300 equations and 500 references. The objective is to provide as comprehensive a source as possible in a format allowing the reader to easily and quickly answer a range of questions, from the basic physics to measured properties.

This handbook is divided into eight chapters. Each chapter is intended to support the others so that questions arising in one are answered in another. The first three chapters describe the fundamental physics of the physical properties of materials, particularly infrared optical materials, the interaction of light and matter, and methods of material physical property determination. These chapters are intended to provide fundamental understanding of the physical properties of the materials. They are also intended of allow the reader to make better use of the material data in other chapters in terms of recognizing their limitations and determining how they may be modified, controlled and extrapolated, or inter-polated if measured data are not available.

Chapters 4 through 7 contain the physical property data of infrared optical materials. These chapters are formatted to allow the reader quick and easy access to a substantial database. Chapter 4 discusses and defines the physical properties listed for each material presented in later chapters. Chapter 5 provides a comprehensive comparison, physical property by physical property, of all the infrared optical materials. Chapters 6 and 7 contain data sheets, tables, and figures for each of over 100 crystalline and glass materials. The materials listed include both well-developed commercially available materials as well as some that have just recently become available or are still in development. The appearance of a material in this handbook, therefore, does not imply general availability. The intention is to provide the reader with some insight into new materials that may be of interest or need. The physical property data in these chapters was obtained from literature or producer sources. Since development work continues on both existing and new materials in the area of infrared optical materials, the reader is encouraged to allow for updating by following the technical literature and producer information regarding the physical properties of the materials in this handbook.

Chapter 8 describes the theory, design, and properties of infrared optical coatings. Various design methods, coating types, materials, and coating properties are detailed. This chapter was included since infrared optical coatings not only represent another set of infrared optical materials but are usually inextricably involved with the application of the material set described in Chapters 6 and 7. The properties of over 50 coatings are tabulated. As in other chapters, the data were obtained from literature sources and the reader should allow for updates as development work produces improvements in the listed coatings as well as new coatings.

The editor and authors would like to acknowledge the financial support of Texas Instruments in the preparation of this book, particularly in terms of illustration and manuscript preparation and permission to use various photographs. We also wish to extend our appreciation to the Technical Publications Department of Texas Instruments.

For Chapters 1, 2 and 3 we wish to acknowledge the many discussions and interactions with colleagues at AT&T Bell Laboratories that, over the years, have assisted us immensely in developing an understanding of the theoretical aspects discussed. We particularly thank K. Nassau and A. M. Glass; AT&T Bell Laboratories administration.

The authors of Chapters 4, 5, 6, and 7 wish to thank several former students of the Jacksonville University Physics Department for their diverse contributions. The most significant contribution came from Mark Grant (now a graduate student at the University of Florida). He produced most of the figures used in these four chapters and assisted throughout the entire project. Vitally important contributions in collating data and organizing the comparison tables of Chapter 5 and the data sheets of Chapter 6 were provided by Amanda Philips Estep and Sean Seeba. Special recognition is given for the important contributions of Margaret Smith, Douglas Sherman, Visarath In and Richard DeSalvo. The project also benefited from the assistance of Steve Batey and Paula Hilliker. We sincerely thank each student for their special contributions. The authors extend a special acknowledgement to C. A. Klein and R. L. Gentilman both at Raytheon Company for their contributions to Chapter 6.

Acknowledgement in also extended to Katharyn Hunter of Texas Instruments for manuscript preparation and to Jo Satloff for editing. Finally, we thank all the members of our families for their understanding and patience during the copious hours spent away from them during this project.

Paul Klocek

Contents

Contents

Contributors

Stanley S. Ballard *University of Florida, Gainesville, Florida*

James Steve Browder *Jacksonville University, Jacksonville, Florida*

Paul Klocek *Texas Instruments, Inc., Dallas, Texas*

Malcolm E. Lines *AT&T Bell Laboratories, Murray Hill, New Jersey*

Dale E. Morton *Optic-Electronic Corporation, Dallas, Texas*

1

Physical Properties of Materials: Theoretical Overview

Malcolm E. Lines *AT&T Bell Laboratories, Murray Hill, New Jersey*

1.1. DEFINITIONS, SYMBOLS, AND UNITS

Although there is now an international agreement to adopt the Système International (SI) system of units throughout science and technology, and presumably future generations of research workers and technicians will use it exclusively, present-day spectroscopists still commonly use the cgs system together with such other officially deprecated units as calories, decibels, and microns. In the United States it remains a common practice to use SI in undergraduate instruction but to revert to cgs (or else atomic units) in graduate courses and research papers. In Europe, on the other hand, the use of SI is, in most countries, directed by law.

In this handbook SI units will be preferred except in situations where the literature cited overwhelmingly uses a different unit. One such example would be the use of cm^{-1} (rather than m^{-1}) as a measure of wavenumber in the context of infrared optics. One other reality that must be accepted is the still very common use in the theoretical research literature of the cgs-Gaussian system of units for electromagnetic equations. Although we shall cast the primary electromagnetic outline of this handbook in SI, we shall, in deference to this fact, also include the cgs-Gaussian forms (where they differ) in an effort to make parallel usage simple and natural. Such dual-unit pairs of equations (unprimed equation numbers for SI, primed equivalents for cgs-Gaussian) occur mainly in Section 1.5 and Chapter 2.

The theoretical overview is affected by this duality only in those areas where electric or magnetic quantities enter in a direct fashion. Other equations are completely independent of any system of units. For example, the equation $\rho = M/V$, relating density ρ to mass M and volume V, needs no further elaboration. The reason the electromagnetic equations take on a different appearance in the SI and cgs-Gaussian formats is the decision to express the former in terms of the practical unit of current, the ampere. Now the defining equations that relate electrical and magnetic quantitites to mechanical properties do not require the

1

introduction of an extra unit. In other words, it is perfectly possible to define current in terms of the three fundamental dimensions (mass, length, and time) of mechanics, although it does take on a curious-looking dimension $(\text{mass})^{1/2}$ $(\text{length})^{3/2}(\text{time})^{-2}$. The cgs-Gaussian system adopts this course while the SI system retains the freedom of choosing one of the electrical units arbitrarily and assigning to it a dimension. The advantage is the retention of familiar practical electrical units, but at the expense of introducing a dimensionality to previously dimensionless proportionality constants that relate electrical and mechanical properties.

More specifically, the Coulomb inverse square law, relating the force F of interaction between two electric charges e_1 and e_2 in vacuo to their separation distance r $(F = Ce_1e_2/r^2)$ can be used with C set equal to 1 to define electrostatic (or esu) units of charge in terms of mass, length, and time. If, on the other hand, charges e are measured in the SI units of coulombs (i.e., ampere-seconds), then C can be written as $1/4\pi\epsilon_o$, where ϵ_o is called the permittivity of free space, and has the value of $10^7/4\pi c^2$ farad/meter, where $c = 2.9979 \times 10^8$ (m/s) is the velocity of light in vacuo. Similarly, for the magnetostatic inverse square equivalent $F = C'm_1m_2/r^2$ involving the force between magnetic "poles" m_i, setting $C' = 1$ defines magnetostatic (or emu) units of m in terms of mass, length, and time. By contrast, if m is measured in SI units (joules/ampere), then C' can be cast in the form $1/4\pi\mu_o$ in which μ_o is called the permeability of free space and has the value of $4\pi \times 10^{-7}$ henry/meter (H/m). The factors $1/4\pi$ introduced into these SI relationships are inserted solely to simplify (that is, remove π from) other more frequently used equations of electromagnetic theory.

The cgs-Gaussian system therefore adopts mixed esu/emu units and is cast only in the dimensions of mass, length, and time. The SI system accepts the additional dimension of electric current. With it the practical units of charge (coulomb), potential (volt), resistance (ohm), and capacitance (farad) are retained, but at the expense of introducing ϵ_o and μ_o as set out above. In this chapter the essential differences between SI and cgs-Gaussian units appear in the Maxwell equations of Section 1.5.

In the wider context of materials science in general, the SI units used in this handbook are based on the dimensions of length (meters, m), mass (kilogram, kg), time (seconds, s), current (amperes, A), and temperature (kelvins, K). The fundamental SI units with special names, together with their symbols and dimensions, are shown in Table 1.1, and a tabulation of SI prefixes (i.e., decimal multiples and submultiples) in Table 1.2. A full list of the optical and material properties discussed in this theoretical overview, together with the symbols used to represent them, is given in Tables 1.3 and 1.4. The material compliances shown in Table 1.4 are also accompanied by their defining equations in terms of the relevant stimulus and response variables taken from Table 1.3. Table 1.4 gives the full tensor notation for the compliances concerned; subscripts i,j,k, etc.

Table 1.1 SI Derived Units with Special Names

Quantity	Name	Symbol	Dimension
Frequency	hertz	Hz	s^{-1}
Force	newton	N	$m \cdot kg \cdot s^{-2}$
Pressure	pascal	Pa $(= N/m^2)$	$m^{-1} \cdot kg \cdot s^{-2}$
Energy, work	joule	J $(= N \cdot m)$	$m^2 \cdot kg \cdot s^{-2}$
Power	watt	W $(= J/s)$	$m^2 \cdot kg \cdot s^{-3}$
Electric charge	coulomb	C	sA
Electric potential	volt	V $(= W/A)$	$m^2 \cdot kg \cdot s^{-3} \cdot A^{-1}$
Capacitance	farad	F $(= C/V)$	$m^{-2} \cdot kg^{-1} \cdot s^4 \cdot A^2$
Electrical resistance	ohm	Ω $(= V/A)$	$m^2 \cdot kg \cdot s^{-3} \cdot A^{-2}$
Conductance	siemens	S $(= A/V)$	$m^{-2} \cdot kg^{-1} \cdot s^3 \cdot A^2$
Magnetic flux	weber	Wb $(= V \cdot s)$	$m^2 \cdot kg \cdot s^{-2} \cdot A^{-1}$
Magnetic induction	tesla	T $(= Wb/m^2)$	$kg \cdot s^{-2} \cdot A^{-1}$
Inductance	henry	H $(= Wb/A)$	$m^2 \cdot kg \cdot s^{-2} \cdot A^{-2}$

Table 1.2 Decimal Multiples and Submultiples of SI units

Factor	Prefix	Symbol	Factor	Prefix	Symbol
10^{-1}	deci	d	10	deca	da
10^{-2}	centi	c	10^2	hecto	h
10^{-3}	milli	m	10^3	kilo	k
10^{-6}	micro	μ	10^6	mega	M
10^{-9}	nano	n	10^9	giga	G
10^{-12}	pico	p	10^{12}	tera	T
10^{-15}	femto	f	10^{15}	peta	P
10^{-18}	atto	a	10^{18}	exa	E

depict Cartesian components, which can each consequently assume three different values, for example, $i = 1, 2, 3$ corresponding to the three Cartesian axes. Commonly used contracted matrix notations are discussed fully in the sections in which each specific compliance is introduced. Finally, Table 1.5 presents some relevant fundamental constants and their numerical values.

1.2. TENSOR NATURE OF THE PHYSICAL PROPERTIES OF CRYSTALS

Physical properties **T** of crystals, which result from the measurement of a response ΔR to a small stimulus ΔS according to the equation

$$\Delta R = T \cdot \Delta S \qquad (1)$$

Table 1.3 Names, Symbols, and SI Units of Quantities Appearing in the Overview Chapters (Excluding compliances, which are shown in Table 1.4)

Quantity	Symbol	SI units	Quantity	Symbol	SI units
Angular frequency	ω	Hz	Energy density:		
Mass	M	kg	Internal	U	J/m^3
Volume	$V(v)$	m^3	Heat	Q	J/m^3
Pressure	p	Pa	Helmholtz	A	J/m^3
Density	ρ	kg/m^3	Gibbs	G	J/m^3
Light velocity	c	m/s	Enthalpy	H	J/m^3
Electronic charge	e	C	Entropy density	S	J/(m$^3\cdot$K)
Current density	$\mathbf{J}$	A/m^2	Permittivity of free space	ϵ_0	F/m
Wavenumber	q (k)	m^{-1}			
Vacuum wavelength	λ	m	Permeability of free space	μ_0	H/m
Light intensity	I	W/m^2	Specific heat at constant pressure	c_p	J/(kg·K)
Electric field	$\mathbf{E}$	V/m	Latent heat	L	J/kg
Magnetic field	$\mathbf{H}$	A/m	Stress	σ	Pa
Electric polarization	$\mathbf{P}$	C/m^2	Strain	η	—
Electric displacement	$\mathbf{D}$	C/m^2	Thermal expansion (linear)	α	K^{-1}
Magnetic induction	$\mathbf{B}$	T	Thermal expansion (volume)	β	K^{-1}
Magnetization	$\mathbf{M}$	T			
Time	t	s	Sound velocity (longitudinal)	v_l	m/s
Temperature	T	K	Sound velocity (transverse)	v_t	m/s
Dilatation	θ	—			
Melting point	T_M	K	Yield strength	σ_Y	Pa
Coherence length	ℓ_c	m	Tensile strength	σ_T	Pa
Stefan's constant	s	W/(m$^2\cdot$K^4)	Shear strength	σ_S	Pa
Relaxation time	τ	s	Resilience	W_R	Pa
Radiance	R	W/m^2	Toughness	W_T	Pa
Emissivity	e	—	Ductility	η_F	—
Absorptivity	a	—	Anelastic decrement	δ	—

(continued)

Table 1.3 Names, Symbols, and SI Units of Quantities Appearing in the Overview Chapters (Excluding compliances, which are shown in Table 1.4) (Continued)

Quantity	Symbol	SI units	Quantity	Symbol	SI units
Radiance loss coefficient	α_R	m^{-1}	Creep activation energy	E_c	J
Optical thickness	L_o	m	Band energy gap	E_g	J
Loss coefficient	α	m^{-1}	Skin depth	δ	m
Oscillator strength	S_j	—	Plasma frequency	ω_p	Hz
Electromagnetic vector potential	$\mathbf{A}$	T·m	Momentum	$\mathbf{p}$	N·s
Electromagnetic scalar potential	ϕ	V	Force	$\mathbf{F}$	N
Electron mass	m	kg	Optic mode frequency (longitudinal)	ω_{LO}	Hz
Effective electron (hole) mass	m_e^* (m_h^*)	kg	Optic mode frequency (transverse)	ω_{TO}	Hz
Electric dipole	$\mathbf{p}$	C·m	Mode damping parameter	γ_j	Hz
Power	P	W	Solubility	s	—
Enthalpy of formation	ΔHf_T	J/mole	Solubility product	K_{sp}	—
Gibbs energy of formation	ΔGf_T	J/mole	Fermi level for conduction electrons	F_c	J
Thermal entropy	S_T^o	J/(mole·K)	Fermi level for valence electrons	F_v	J
Specific Faraday rotation	F	rad/m	Verdet constant	V	rad/(T·m)

are tensors. Their detailed nature and the precise form of the implied multiplication $\mathbf{T} \cdot \Delta \mathbf{S}$ depends upon the degree of complexity of the particular stimulus and response in question. The simplest case of all occurs when $\Delta \mathbf{R}$ and $\Delta \mathbf{S}$ are both scalars, that is, quantities that possess only magnitude and can therefore be uniquely defined by single numbers (e.g., temperature T, mass M, and volume

Table 1.4 Material Compliances—Symbol, Defining Equation, and SI Units

Quantity	Symbol	Defining equation	SI units
Pyroelectric coeff.	Π_i	$\Delta P_i = \Pi_i\,\Delta T$	C/(m^2·K)
Dielectric constant	ϵ_{ij}	$D_i = \epsilon_o\epsilon_{ij}E_j$	—
Polarizability	χ_{ij}	$P_i = \epsilon_o\chi_{ij}E_j$	—
Permeability	μ_{ij}	$B_i = \mu_o\mu_{ij}H_j$	—
Magnetic polarizability	$(\chi_m)_{ij}$	$M_i = \mu_o(\chi_m)_{ij}H_j$	—
Refractive index	n_{ij} ⎱		
Extinction coefficient or absorption index	k_{ij} ⎰	$(n-ik)^2_{ij} = \epsilon_{ij}$	—
Impermeability	B_{ij}	$B_{ij} = (\epsilon_{ij})^{-1}$	—
Piezoelectric strain components	e_{ijk}	$P_i = e_{ijk}\eta_{jk}$	C/m^2
Piezoelectric stress components	d_{ijk}	$P_i = d_{ijk}\sigma_{jk}$	m/V
Linear electro-optic ⎱	r_{ijk}	$\Delta B_{ij} = r_{ijk}E_k$	m/V
(Pockels) coefficients ⎰	f_{ijk}	$\Delta B_{ij} = f_{ijk}P_k$	m^2/C
Quadratic electro-optic (Kerr) coefficients	$K_{ijk\ell}$	$\Delta B_{ij} = K_{ijk\ell}E_kE_\ell$	m^2/V^2
Piezo-optic coefficients	$q_{ijk\ell}$	$\Delta B_{ij} = q_{ijk\ell}\sigma_{k\ell}$	Pa^{-1}
Elasto-optic coefficients	$p_{ijk\ell}$	$\Delta B_{ij} = p_{ijk\ell}\eta_{k\ell}$	—
Elastic stiffness	$c_{ijk\ell}$	$\sigma_{ij} = c_{ijk\ell}\eta_{k\ell}$	Pa
Elastic modulus (or compliance)	$s_{ijk\ell}$	$\eta_{ij} = s_{ijk\ell}\sigma_{k\ell}$	Pa^{-1}
Electrical conductivity	σ_{ij}	$J_i = \sigma_{ij}E_j$	S/m
Second harmonic compliance	d_{ijk}	$P_i^{(2\omega)} = \epsilon_o d_{ijk}E_j^\omega E_k^\omega$	m/V
Miller's delta	δ_{ijk}	$E_i^{(2\omega)} = \epsilon_o^{-1}\delta_{ijk}P_j^\omega P_k^\omega$	m^2/C
Third-order harmonic compliance	$C_{ijk\ell}$	$P_i^{(3\omega)} = \epsilon_o C_{ijk\ell}E_j^\omega E_k^\omega E_\ell^\omega$	m^2/V^2
For isotropic solids:			
Lamé constant	λ	$\sigma_{11} = \lambda\eta_{22}$	Pa
Shear modulus	G	$\sigma_{ij} = 2G\eta_{ij}\ (i \neq j)$	Pa
Bulk modulus	K	$\Delta p = -K\,\Delta\theta$	Pa
Young's modulus	E	$\sigma_{11} = E\eta_{11}$	Pa
Longitudinal modulus	M	$M = \rho v_\ell^2$	Pa
Poisson ratio	μ	$\eta_{22} = -\mu\eta_{11}$	—

V). The associated tensors **T** are then referred to as being of rank zero. An example is density ρ, which is defined by

$$\Delta M = \rho\Delta V \qquad (2)$$

In this relationship, ΔM and ΔV are small, but macroscopic, quantities (i.e., large compared to molecular mass and molecular volume) and consequently define a macroscopic density that is independent of atomic position in crystals but that may fluctuate on the scale of 10–100 nm in glasses. Evidently, from their definition, rank zero tensors are also scalars.

Table 1.5 Some Fundamental Constants and Their Numerical Values

Quantity	Symbol	Value
Velocity of light	c	2.9979246×10^8 m/s
Permittivity of free space	ϵ_o	$8.8541878 \times 10^{-12}$ F/m
Planck's constant	h	6.6262×10^{-34} J/Hz
Electronic charge	e	1.60219×10^{-19} C
Electronic mass	m_e	9.1095×10^{-31} kg
Proton mass	m_p	1.67265×10^{-27} kg
Neutron mass	m_N	1.67495×10^{-27} kg
Boltzmann's constant	k	1.3806×10^{-23} J/K
Avogadro's number	N_o	6.0220×10^{23} mol^{-1}
Molar gas constant	R	8.314 J/(K·mol)
Stefan's constant	s	5.670×10^{-8} W/(m^2·K^4)
Permeability of free space	μ_o	1.256637×10^{-6} H/m
Gravitational constant	G	6.67×10^{-11} N·m^2/kg^2
Bohr radius	a_o	0.529177×10^{-10} m
Electron g factor	g	2.0023193
Bohr magneton	μ_B	9.2741×10^{-24} J/T
Atomic mass unit	amu	1.66056×10^{-27} kg

If the response $\Delta\mathbf{R}$ is a vector, possessing direction as well as magnitude, then, in a Cartesian reference frame x_1, x_2, x_3, it can be completely defined by its three components along the Cartesian axes. The response of such a quantity to a scalar stimulus then defines a tensor of the first rank. An example would be the response of electrostatic polarization $\mathbf{P}$ (or dipole moment per unit volume) to a small change of temperature T in a spontaneously polarized, or pyroelectric, crystal. The relevant equation

$$\Delta\mathbf{P} = \Pi\,\Delta T \qquad (3)$$

or

$$\Delta P_i = \Pi_i\,\Delta T \qquad (4)$$

in Cartesian component form (with $i = 1, 2, 3$), defines the pyroelectric coefficient Π. Requiring three scalars for its complete description, Π is a vector.

If both stimulus and response are in vector form, then the set of Cartesian equations that define the tensor property $\mathbf{T}$ are of the form

$$\Delta R_i = T_{ij}\,\Delta S_j \qquad i,j = 1, 2, 3 \qquad (5)$$

where here (and throughout this chapter) a repeated index implies summation; for example, $T_{ij}\,\Delta S_j = \Sigma_j T_{ij}\Delta S_j$. With nine components, the tensor T_{ij} is said to be of the second rank (i.e., has two component subscripts). With response equal to the polarization **P** induced by an electric field stimulus **E**, we can generate the example

$$\mathbf{P} = \epsilon_o \, \chi \cdot \mathbf{E} \tag{6}$$

or

$$P_i = \epsilon_0 \chi_{ij} \, E_j; \qquad i,j = 1, 2, 3 \tag{7}$$

defining the nine components χ_{ij} of the polarizability tensor. Other well-known physical properties that are second-rank tensors are stress, strain, electrical conductivity, and dielectric constant.

It is evident that as the tensor rank of the response and/or stimulus is increased, ever higher ranked physical properties can be defined. A typical third-rank tensor results, for example, when the response of the (second-rank) dielectric constant ϵ is probed by a (vector) applied electric field **E**. This particular relationship is normally expressed in terms of the reciprocal dielectric constant $\mathbf{B} = \epsilon^{-1}$ (often called the impermeability) and adopts the Cartesian form

$$\Delta B_{ij} = r_{ijk} E_k \tag{8}$$

which defines the linear electro-optic (or Pockels) tensor **r** with 27 components, i, j, $k = 1$, 2, 3. Another important physical third-rank tensor measures the response of polarization **P** to stress σ via the relationship

$$P_i = d_{ijk} \sigma_{jk}, \qquad i,j,k = 1, 2, 3 \tag{9}$$

Also with 27 components, **d** is referred to as the piezoelectric stress tensor.

The elastic properties of crystals, which relate stress σ to strain η (both second-rank tensor quantities), are perhaps the most common examples of fourth-rank tensors in solid-state physics. The defining relationship

$$\sigma_{ij} = c_{ijk\ell} \, \eta_{k\ell} \tag{10}$$

determines the 81 components (i, j, k, $\ell = 1$, 2, 3) of the elastic stiffness tensor (also commonly referred to as the elastic constant tensor). The inverse equation

$$\eta_{ij} = s_{ijk\ell} \, \sigma_{k\ell} \tag{11}$$

defines the analogous components $s_{ijk\ell}$ of the elastic compliance tensor. The components $c_{ijk\ell}$ and $s_{ijk\ell}$ are obviously interrelated via these two sets of equations. Other common examples of fourth-rank tensor properties relevant for infrared optics are the piezo-optic and elasto-optic coefficients $q_{ijk\ell}$ and $p_{ijk\ell}$, which relate dielectric impermeability B_{ij} to stress $\sigma_{k\ell}$ and strain $\eta_{k\ell}$, respectively, and the quadratic electro-optic (or Kerr) coefficients $K_{ijk\ell}$, which relate this same response to the quadratic electric field $E_k E_\ell$.

Tensor physical properties beyond the fourth rank are not commonly encountered in materials science, although, quite obviously, it is very simple to define them. The 81 independent scalars necessary to define fourth-rank tensors in their most general form are challenge enough. Fortunately, it is nearly always possible to reduce this number to more manageable proportions by use of symmetry arguments and particularly by choosing crystalline axes of symmetry for the Cartesian axes in terms of which the tensor components are to be expressed.

Tensors defining physical properties are, of course, entities that exist independently of any particular coordinate framework in which we choose to express them. Their scalar components (or elements), on the other hand, are dependent upon this choice. As previously stated, we shall restrict ourselves to Cartesian (i.e., rectilinear rectangular) coordinates in this book. Since this is an almost universal choice in the present context, no mention will be made of other frameworks; in particular we shall have no need to distinguish between covariant and contravariant tensors (Margenau and Murphy 1956).

1.2.1. Transformation Laws for Cartesian Tensor Elements

A tensor property exists independently of the system of coordinate axes in terms of which its components are represented. These components, on the other hand, change whenever there is a change of (in our case Cartesian) coordinates. This change takes place according to well-defined transformation laws, which, however, vary with the rank of the tensor (Boardman et al. 1973). Let us consider a linear orthogonal transformation in three-dimensional space from one Cartesian reference frame, x_j ($j = 1, 2, 3$), to another, x_i' ($i = 1, 2, 3$). Such a transformation can be written using a transformation matrix $\mathbf{A} = [a_{ij}]$ in the linear form

$$x_i' = a_{ij}x_j \tag{12}$$

where (for example), for real rotations, the coefficients a_{ij} are direction cosines. The inverse transformation from the primed to unprimed frame can be expressed as

$$x_i = a_{ji}x_j' \tag{13}$$

from which it follows that successive transformations between primed and unprimed coordinates generate the relationships

$$x_i = a_{ji}a_{jk}x_k, \qquad x_i' = a_{ij}a_{kj}x_k' \tag{14}$$

from which follow the orthogonality conditions

$$a_{ji}a_{jk} = a_{ij}a_{kj} = \delta_{ik} \tag{15}$$

where δ_{ik} is the Kronecker delta for which $\delta_{ik} = 1$ ($i = k$) and $\delta_{ik} = 0$ ($i \neq k$).

The determinant $|\,a_{ij}\,|$ of the transformation matrix elements is equal to ± 1. The positive sign applies if the original and transformed axes have the same (left- or right-handed) sense, the negative sign if the transformation involves reflections or inversions that change the handedness.

A tensor of rank zero (that is, a scalar) is defined by a single numerical value and remains unaltered under the Cartesian transformation **A**. A tensor of the first rank (that is, a vector) is defined by three components T_i ($i = 1, 2, 3$). Under the coordinate transformation **A**, the tensor, of course, remains invariant; its components, on the other hand, change according to

$$T_i' = a_{ij}T_j \tag{16}$$

the inverse of which is

$$T_i = a_{ji}T_j' \tag{17}$$

As a trivial example we can consider the transformation from an unprimed right-handed coordinate system x_i to a primed equivalent by rotation through a right angle about the x_3 coordinate axis such that $x_1' = x_2$, $x_2' = -x_1$. The relevant transformation matrix is

$$\mathbf{A} = \begin{bmatrix} 0 & 1 & 0 \\ -1 & 0 & 0 \\ 0 & 0 & 1 \end{bmatrix} \tag{18}$$

from which we can relate the above first-rank tensor components, using Eq. (16), in the manner

$$\begin{bmatrix} T_1' \\ T_2' \\ T_3' \end{bmatrix} = \begin{bmatrix} 0 & 1 & 0 \\ -1 & 0 & 0 \\ 0 & 0 & 1 \end{bmatrix} \begin{bmatrix} T_1 \\ T_2 \\ T_3 \end{bmatrix} = \begin{bmatrix} T_2 \\ -T_1 \\ T_3 \end{bmatrix} \tag{19}$$

One word of caution is necessary. In a practical application of tensor transformation we must always ensure that the relevant transformation is admissible. For example, the magnetic field vector **H** is not an "ordinary" vector, because its sign does not change upon reflection. This is easily seen by considering a horizontal loop of wire carrying a current and situated above a horizontal plane. On reflection in the plane, the current direction is not reversed, implying that the induced electromagnetic field has not reversed either. The magnetic field **H** is therefore a vector when restricted to transformations that do not include reflection. It is called a pseudovector or sometimes an axial vector to distinguish it from a normal or "polar" vector (Boardman et al. 1973). In general, therefore, we can transform from one reference frame to another only in a manner commensurate with the character of the problem under investigation.

To determine the equations of component transformation for a second-rank tensor T_{ij} under a coordinate transformation **A** we turn to its defining equation $r_i = T_{ij}s_j$ in terms of response and stimulus vectors **r** and **s**. In the primed coordinate system this equation can be recast as $r'_i = T'_{ij}s'_j$. Using rules (16) and (17) for vector transformation, it follows that

$$r'_k = a_{ki}r_i = a_{ki}T_{ij}s_j = a_{ki}T_{ij}a_{mj}s'_m \tag{20}$$

for which, by direct comparison with $r'_k = T'_{km}s'_m$, we obtain

$$T'_{km} = a_{ki}a_{mj}T_{ij} \tag{21}$$

which is the required transformation. That is, the second-rank tensor **T**, with elements T_{ij} in the unprimed system, can be recast with elements T'_{km} in the primed system. The inverse relationship follows from parallel arguments in the form

$$T_{ij} = a_{ki}a_{mj}T'_{km} \tag{22}$$

An extension to tensor elements for higher rank tensors is straightforward and leads to the transformation equations

$$T'_{ijk} = a_{im}a_{jn}a_{kp}T_{mnp} \tag{23}$$

$$T_{ijk} = a_{mi}a_{nj}a_{pk}T'_{mnp} \tag{24}$$

for third-rank tensors, and

$$T'_{ijk\ell} = a_{in}a_{jp}a_{kq}a_{\ell r}T_{npqr} \tag{25}$$

$$T_{ijk\ell} = a_{ni}a_{pj}a_{qk}a_{r\ell}T'_{npqr} \tag{26}$$

for fourth-rank tensors, with an obvious extension to tensors of higher rank. In the context of this handbook such transformations are important because the choice of axes x_ℓ ($\ell = 1, 2, 3$) is not universally standard for all crystal symmetries.

1.2.2. The Role of Crystal Symmetry

When we refer to the group of symmetry operations of a crystal structure, we strictly mean space-group operations (Boardman et al. 1973). However, if we are concerned only with a measured macroscopic property, then we can restrict our attention to the point groups alone, since other symmetry operations (such as glide and screw transformations) are detectable only by microscopic probes. Within this restriction, the general assumption is that the symmetry elements of the group of operators under which a particular macroscopic physical property of a crystal is unchanged include at least all the elements of the point group. This assumption, known as Neumann's principle, implies that no such property can ever be of *lower* symmetry than that of the crystal. On the other hand, physical restrictions of the property itself can often result in a higher symmetry.

An nth-rank tensor of the lowest symmetry point group (Class 1) can, in general, have as many as 3^n independent components. This number decreases, depending on any symmetry restrictions inherent in the physical property itself, and as the point-group symmetry increases. A property like thermal expansion, for example, is physically a *symmetric* tensor of the second rank, with $T_{ij} = T_{ji}$ by definition. It therefore possesses a maximum of six (instead of nine) independent elements, namely, $T_{11}, T_{22}, T_{33}, T_{12}, T_{23}, T_{31}$. The same is true for the dielectric constant. In similar fashion, the elastic stiffness, which is a fourth-rank tensor, is physically restricted to have $T_{ijkm} = T_{kmij}$. This reduces its maximum possible number of independent components from $3^4 = 81$ to 21. For photoelastic tensors (e.g., piezo-optic and elasto-optic coefficients), the physical restriction is less stringent, leading to a maximum of 36 independent components.

As the point-group symmetry of the crystal structure increases, an increasing number of these components can be related or shown to be necessarily zero. If the components of a particular physical tensor property are measured for one position (using *any* Cartesian frame), and we perform a symmetry operation **A** using the transformation rules of Section 1.2.1, then after transformation we must obtain exactly the same values, since the point-group operations leave the tensor components unchanged. This procedure provides the necessary point-group restrictions for the tensor in question (Narasimhamurty 1981). The final standard Cartesian form for any particular tensor property will be set out for each of the relevant point groups in the section of this chapter under which that particular property is discussed. We reiterate finally that a tensor is a physical property that is independent of reference frame—it is only its representation (in terms of its tensor elements or components) that is modified by changes of coordinate system.

1.2.3. Compliances and Thermodynamic Constraints

It is normally assumed that the thermal, elastic, and dielectric properties of any homogeneous nonmagnetic material can be fully described by the six variables temperature T, entropy S, stress σ, strain η_o, electric field **E**, and electric displacement **D**, where the last is related to polarization **P** by the equation

$$D = \epsilon_o E + P \tag{27}$$

$$\text{cgs:} \quad D = E + 4\pi P \tag{27'}$$

and ϵ_o is the permittivity of empty space. Since stress and strain are each determined by six variables (symmetric second-rank tensors), **E** and **D** by three variables (vectors), and T and S by 1 (scalars), this gives a total of 20 thermodynamic coordinates with which to describe the system. For spontaneously magnetic systems (which we shall not dwell upon here), a further six coordinates, corresponding to magnetic field **H** and induction **B** are required.

According to the first law of thermodynamics, the change in internal energy U per unit volume when an infinitesimal quantity of heat dQ is received by a unit volume of dielectric is given by $dU = dQ + dW$, where dW is the work done on this same volume (by electrical and mechanical forces) during the resulting quasistatic transformation. Noting that $dQ = T\,dS$ (second law of thermodynamics) and $dW = \mathbf{E} \cdot d\mathbf{D} + \boldsymbol{\sigma} \cdot d\boldsymbol{\eta}$, we can now rewrite the internal energy change as (Lines and Glass 1977)

$$dU = T\,dS + \sigma_i d\eta_i + E_i\,dD_i \qquad (28)$$

where stress and strain have been cast in a single-subscript notation σ_i, η_i, ($i = 1, \ldots ,6$), according to the (Voigt) convention

$$
\begin{aligned}
\sigma_1 &= \sigma_{11}, & \sigma_2 &= \sigma_{22}, & \sigma_3 &= \sigma_{33} \\
\sigma_4 &= \sigma_{23}, & \sigma_5 &= \sigma_{31}, & \sigma_6 &= \sigma_{12}
\end{aligned} \qquad (29)
$$

$$
\begin{aligned}
\eta_1 &= \eta_{11}, & \eta_2 &= \eta_{22}, & \eta_3 &= \eta_{33} \\
\eta_4 &= 2\eta_{23}, & \eta_5 &= 2\eta_{31}, & \eta_6 &= 2\eta_{12}
\end{aligned} \qquad (30)
$$

From Eq. (28) we can derive the three relationships

$$T = \left[\frac{\partial U}{\partial S}\right]_{\eta,\mathbf{D}}, \qquad \sigma_i = \left[\frac{\partial U}{\partial \eta_i}\right]_{S,\mathbf{D}}, \qquad E_i = \left[\frac{\partial U}{\partial D_i}\right]_{S,\eta} \qquad (31)$$

which are respectively the calorimetric, elastic, and dielectric equations of state. If we now write the differential form of these equations, we have

$$dT = \left[\frac{\partial T}{\partial S}\right]_{\eta,\mathbf{D}} dS + \left[\frac{\partial T}{\partial \eta_j}\right]_{S,\mathbf{D}} d\eta_j + \left[\frac{\partial T}{\partial D_j}\right]_{S,\eta} dD_j \qquad (32)$$

$$d\sigma_i = \left[\frac{\partial \sigma_i}{\partial S}\right]_{\eta,\mathbf{D}} dS + \left[\frac{\partial \sigma_i}{\partial \eta_j}\right]_{S,\mathbf{D}} d\eta_j + \left[\frac{\partial \sigma_i}{\partial D_j}\right]_{S,\eta} dD_j \qquad (33)$$

$$dE_i = \left[\frac{\partial E_i}{\partial S}\right]_{\eta,\mathbf{D}} dS + \left[\frac{\partial E_i}{\partial \eta_j}\right]_{S,\mathbf{D}} d\eta_j + \left[\frac{\partial E_i}{\partial D_j}\right]_{S,\eta} dD_j \qquad (34)$$

The coefficients are called the linear *compliances* and provide a measure of the coupling between state variables. Together they form a square array of 100 compliance matrix elements, 55 of which are independent [the others being related via the second derivatives of internal energy U in Eqs. (31)]. Only a few of these are the subject of common experimental investigation, the best known from the above set perhaps being the elastic stiffness

$$c_{ij} = (\partial \sigma_i / \partial \eta_j), \qquad i,j = 1, \ldots ,6 \qquad (35)$$

the contracted notational form of Eq. (10), and the inverse dielectric constant (or inverse permittivity)

$$B_{ij} = (\partial E_i / \partial D_j), \qquad i,j = 1, 2, 3 \qquad (36)$$

To be specific, Eq. (33) defines the elastic stiffness measured at constant entropy (i.e., adiabatic) and constant displacement, whereas Eq. (34) defines adiabatic inverse permittivity at constant strain (clamped). If other constraints are preferred (e.g., constant temperature T or stress σ), then the corresponding compliances can be derived by considering other thermodynamic functions in terms of which the equations of state can be written in an equally concise manner.

Since three independent variables can be chosen in eight different ways from among the conjugate pairs (T, S), (σ_i, η_i), and (E_i, D_i), eight thermodynamic potentials can be defined. In addition to the internal energy U, the others are

$$
\begin{array}{lll}
\text{Helmholtz free energy} & A = U - TS & \\
\text{Enthalpy} & H = U - \sigma_i\eta_i - E_iD_i & \\
\text{Elastic enthalpy} & H_1 = U - \sigma_i\eta_i & \\
\text{Electric enthalpy} & H_2 = U - E_iD_i & (37) \\
\text{Gibbs free energy} & G = U - TS - \sigma_i\eta_i - E_iD_i & \\
\text{Elastic Gibbs energy} & G_1 = U - TS - \sigma_i\eta_i & \\
\text{Electric Gibbs energy} & G_2 = U - TS - E_iD_i &
\end{array}
$$

By differentiation and use of Eq. (28) we readily derive the following differential forms describing infinitesimal changes in these potentials:

$$
\begin{aligned}
dA &= -S\,dT + \sigma_i\,d\eta_i + E_i\,dD_i \\
dH &= T\,dS - \eta_i\,d\sigma_i - D_i\,dE_i \\
dH_1 &= T\,dS - \eta_i\,d\sigma_i + E_i\,dD_i \\
dH_2 &= T\,dS + \sigma_i\,d\eta_i - D_i\,dE_i \\
dG &= -S\,dT - \eta_i\,d\sigma_i - D_i\,dE_i \\
dG_1 &= -S\,dT - \eta_i\,d\sigma_i + E_i\,dD_i \\
dG_2 &= -S\,dT + \sigma_i\,d\eta_i - D_i\,dE_i \qquad (38)
\end{aligned}
$$

It is apparent that the most general situation can now involve as many as 55 compliance elements, each measurable in any of four different conditions of thermodynamic constraint. Experimentally static or low-frequency measurements are most commonly made at constant temperature, stress, or field. At optical frequencies, on the other hand, measurements are more usually effectively clamped (constant strain) and adiabatic (constant entropy), since both thermal and elastic relaxation times are likely to be long on the scale ($\approx 10^{-14}$s) of a period of optical vibration.

As mentioned above, the demands of material symmetry in general impose conditions on the elements of the compliance tensors regardless of the nature of the thermodynamic constraint. Three subclassifications of point groups are of special interest. First there are the centrosymmetric groups. For them an application of stress cannot produce an electric polarity, because charges are displaced in a manner that is relatively compensating about the center of symmetry. An applied electric field does produce strain (electrostriction), but the sign of this strain is not changed when the field is reversed.

Eleven of the 32 point groups are centrosymmetric. Of the remaining 21, all except one (the cubic group 432) exhibit a polarization (or change of polarization) when subjected to stress. The effect is linear (i.e., reverses when the sign of the stress is reversed) and is called the piezoelectric effect. In a similar fashion, application of an electric field produces a strain that is linearly proportional to the field in lowest order. These groups are called the piezoelectric groups. Ten of them are characterized by the fact that they possess a unique polar axis. They are therefore spontaneously polarized in the absence of stress and are referred to as the polar or pyroelectric groups. We list the 32 crystallographic point groups with their international and Schoenflies symbols and their centrosymmetric, piezoelectric, or polar classifications in Table 1.6.

1.3. MECHANICAL PROPERTIES

The application of stress to a solid causes a change in its dimensions. If the stress is small the solid regains its former shape on removal of the stress, and the deformation is said to be elastic. In this regime the relationships between stress and strain are linear, and the resulting proportionality ''constants,'' which are the tensor compliances of the system, are independent of stress. For certain crystallographic symmetries, an applied stress also linearly produces an induced electric polarization. This ''piezoelectric'' effect is also completely reversible if the stress is not too large.

Larger stresses, however, can cause permanent changes in the solid, the most evident of which are permanent deformations that do not disappear upon removal of the stress. This type of behavior reflects a nonlinear relationship of stress to strain and introduces the concept of plasticity. The capability of metals to undergo extensive plastic deformation without fracture is well known, but nearly all crystalline materials can be plastically deformed to some degree, particularly at higher temperatures.

In this section we discuss both the linear reversible mechanical properties of solids and the onset of plasticity. In the former domain we define the tensor compliances as a function of crystal symmetry and give particular attention to the

Table 1.6   The 32 Crystallographic Point Groups Arranged by Crystal System

Crystal system	Symbol Inter-national	Symbol Schoen-flies	Pyro-electric	Piezo-electric	Centro-symmetric
Triclinic	1	C_1	✓	✓	
	$\bar{1}$	C_i			✓
Monoclinic	2	C_2	✓	✓	
	m	C_s	✓	✓	
	2/m	C_{2h}			✓
Orthorhombic	222	D_2		✓	
	$mm2$	C_{2v}	✓	✓	
	mmm	D_{2h}			✓
Trigonal	3	C_3	✓	✓	
	$\bar{3}$	S_6			✓
	32	D_3		✓	
	$3m$	C_{3v}	✓	✓	
	$\bar{3}m$	D_{3d}			✓
Tetragonal	4	C_4	✓	✓	
	$\bar{4}$	S_4		✓	
	4/m	C_{4h}			✓
	422	D_4		✓	
	4 mm	C_{4v}	✓	✓	
	$\bar{4}2m$	D_{2d}		✓	
	4/mmm	D_{4h}			✓
Hexagonal	6	C_6	✓	✓	
	$\bar{6}$	C_{3h}		✓	
	6/m	C_{6h}			✓
	622	D_6		✓	
	6 mm	C_{6v}	✓	✓	
	$\bar{6}m2$	D_{3h}		✓	
	6/mmm	D_{6h}			✓
Cubic	23	T		✓	
	$m3$	T_h			✓
	432	O			
	$\bar{4}3m$	T_d		✓	
	$m3m$	O_h			✓

A tick (✓) indicates that the point group is pyroelectric, piezoelectric, or centrosymmetric, as the case may be many scalar moduli definable in the isotropic limit. At the present level of sophistication, nonlinear effects are discussed only in a scalar formalism.

1.3.1. Elastic Stiffness and Compliance

Macroscopic theories of dielectrics treat the material in question as a continuum. In general, fields are postulated in sufficient number to describe the properties under study, and the laws of thermodynamics and classical mechanics are used to obtain relationships between them. For nonmagnetic materials it is usually assumed to be possible to describe a dielectric by three independent variables chosen from the pairs (temperature T, entropy S), (stress σ, strain η), and macroscopic electric field $\mathbf{E}$, macroscopic electric displacement $\mathbf{D}$) as discussed in Section 1.2.3. In this section we will consider in detail the response of strain to stress or vice versa in lowest order, remembering always that such a response can be studied with any two of the field variables chosen from the pairs (T, S) and $(\mathbf{E}, \mathbf{D})$ held constant.

Stress

Stress is defined broadly as force $\mathbf{F}$ (a vector) divided by the area $\mathbf{A}$ (also a vector) upon which the force acts. It is therefore a second-rank tensor. To define the state of stress at a point in a continuous body, it is customary to examine the forces acting on the faces of an infinitesimal rectangular parallelopiped surrounding the point. The faces of this parallelopiped are chosen normal to each of the three Cartesian reference coordinate axes x_i ($i = 1, 2, 3$). There are three components of force F_i acting on each of the faces A_j leading to nine stress components $\sigma_{ij} = F_i/A_j$. Thus, in the double-subscript notation for σ, the first subscript denotes the direction of the force component and the second the direction of the normal to the plane across which the force component acts (Fig. 1.1). The diagonal stress components $i = j$ are called the *normal* components of stress, while the off-diagonal components with $i \neq j$ are termed *shear* components. By taking moments about one of the axes x_i, one can readily demonstrate that $\sigma_{ij} = \sigma_{ji}$. The implication is that the stress tensor is symmetrical about its principal diagonal; accordingly it has only six independent components. Hydrostatic pressure p, a scalar, represents the special case $\sigma_{11} = \sigma_{22} = \sigma_{33} = -p$, $\sigma_{12} = \sigma_{23} = \sigma_{31} = 0$.

Strain

If a point in a continuum body is expressed by the position coordinate R_i ($i = 1, 2, 3$) and the body is deformed such that $R_i \rightarrow r_i$, then an elastic displacement vector $\mathbf{u}$, with components $u_i = r_i - R_i$, can be defined. Consider a line element dR_i at R_i in the undeformed body. After deformation it becomes dr_i at r_i, where

$$dr_i = dR_i + (\partial u_i/\partial R_j)\, dR_j \qquad (39)$$

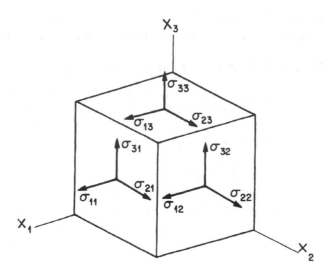

Figure 1.1 The stress components acting on each face of an elementary parallelepiped when a uniform force acts on the body.

It follows immediately that $du_i = dr_i - dR_i = (\partial u_i/\partial R_j) \, dR_j$. However, it is customary to separate this finding into two terms (Pollard 1977),

$$du_i = \eta_{ij} \, dR_j - \omega_{ij} \, dR_j \qquad (40)$$

in which

$$\eta_{ij} = (1/2)[(\partial u_i/\partial R_j) + (\partial u_j/\partial R_i)] \qquad (41)$$

is called the strain tensor, and

$$\omega_{ij} = (1/2)[(\partial u_j/\partial R_i) - (\partial u_i/\partial R_j)] \qquad (42)$$

The above decomposition is made because ω_{ij} can be identified as a pure rotation (Nye 1960) and therefore does not qualify as a component of deformation in the normally accepted sense. The strain tensor η_{ij} is seen to be symmetric ($\eta_{ij} = \eta_{ji}$) from its definition in Eq. (41) and therefore, like stress, is a second-rank tensor with only six independent components. The diagonal components are referred to as normal or *longitudinal* strains, and the off-diagonal components, as shear strains. Physically η_{ii} is the fractional elongation along the axis x_i while $\eta_{ij} + \eta_{ji}$ ($i \neq j$) is the decrease in angle between the axes x_i and x_j upon distortion. In general, both stress and strain can be functions of the position vector **R** in a continuum solid, but the most common circumstance has them constant throughout the material, in which case we refer to *uniform* stress and strain, respectively.

Table 1.7 Relation Between c_{ij} and $c_{ijk\ell}$

$c_{11} = c_{1111}$

$c_{12} = c_{1122} = c_{2211}$

$c_{13} = c_{1133} = c_{3311}$

$c_{14} = c_{1123} = c_{1132} = c_{2311} = c_{3211}$

$c_{15} = c_{1131} = c_{1113} = c_{3111} = c_{1311}$

$c_{16} = c_{1112} = c_{1121} = c_{1211} = c_{2111}$

$c_{22} = c_{2222}$

$c_{23} = c_{2233} = c_{3322}$

$c_{24} = c_{2223} = c_{2232} = c_{2322} = c_{3222}$

$c_{25} = c_{2231} = c_{2213} = c_{3122} = c_{1322}$

$c_{26} = c_{2212} = c_{2221} = c_{1222} = c_{2122}$

$c_{33} = c_{3333}$

$c_{34} = c_{3323} = c_{3332} = c_{2333} = c_{3233}$

$c_{35} = c_{3331} = c_{3313} = c_{3133} = c_{1333}$

$c_{36} = c_{3312} = c_{3321} = c_{1233} = c_{2133}$

$c_{44} = c_{2323} = c_{2332} = c_{3223} = c_{3232}$

$c_{45} = c_{2331} = c_{2313} = c_{3231} = c_{3213} = c_{3123} = c_{3132} = c_{1323} = c_{1332}$

$c_{46} = c_{2312} = c_{2321} = c_{3212} = c_{3221} = c_{1223} = c_{1232} = c_{2123} = c_{2132}$

$c_{55} = c_{3131} = c_{3113} = c_{1331} = c_{1313}$

$c_{56} = c_{3112} = c_{3121} = c_{1312} = c_{1321} = c_{1231} = c_{1213} = c_{2131} = c_{2113}$

$c_{66} = c_{1212} = c_{1221} = c_{2112} = c_{2121}$

Since both stress and strain tensors possess six independent elements, they are commonly expressed in the single-subscript notation σ_i, η_i ($i = 1, \ldots ,6$) as set out in Eqs. (29) and (30).

Hooke's Law

A perfectly elastic body is defined as one for which stress is linearly related to strain. In full tensor notation this relationship (called Hooke's law) defines the elastic stiffness constants $c_{ijk\ell}$ and the elastic compliances (or moduli) $s_{ijk\ell}$ via the equations

$$\sigma_{ij} = c_{ijk\ell}\, \eta_{k\ell} , \qquad (43)$$

$$\eta_{ij} = s_{ijk\ell}\, \sigma_{k\ell}, \qquad (44)$$

in which $i, j, k, l = 1, 2, 3$. In contracted notation these equations reduce to

$$\sigma_i = c_{ij}\, \eta_j, \qquad (i,j = 1, \ldots ,6) \qquad (45)$$

$$\eta_i = s_{ij}\, \sigma_j, \qquad (i,j = 1, \ldots ,6) \qquad (46)$$

where the detailed relationships between c_{ij} and $c_{ijk\ell}$ (and between s_{ij} and $s_{ijk\ell}$) are shown in Tables 1.7 and 1.8, respectively. The two sets of elements c_{ij} and

Table 1.8   Relation Between s_{ij} and $s_{ijk\ell}$

$$s_{11} = s_{1111}$$
$$s_{12} = s_{1122} = s_{2211}$$
$$s_{13} = s_{1133} = s_{3311}$$
$$(1/2)s_{14} = s_{1123} = s_{1132} = s_{2311} = s_{3211}$$
$$(1/2)s_{15} = s_{1131} = s_{1113} = s_{3111} = s_{1311}$$
$$(1/2)s_{16} = s_{1112} = s_{1121} = s_{1211} = s_{2111}$$
$$s_{22} = s_{2222}$$
$$s_{23} = s_{2233} = s_{3322}$$
$$(1/2)s_{24} = s_{2223} = s_{2232} = s_{2322} = s_{3222}$$
$$(1/2)s_{25} = s_{2231} = s_{2213} = s_{3122} = s_{1322}$$
$$(1/2)s_{26} = s_{2212} = s_{2221} = s_{1222} = s_{2122}$$
$$s_{33} = s_{3333}$$
$$(1/2)s_{34} = s_{3323} = s_{3332} = s_{2333} = s_{3233}$$
$$(1/2)s_{35} = s_{3331} = s_{3313} = s_{3133} = s_{1333}$$
$$(1/2)s_{36} = s_{3312} = s_{3321} = s_{1233} = s_{2133}$$
$$(1/4)s_{44} = s_{2323} = s_{2332} = s_{3223} = s_{3232}$$
$$(1/4)s_{45} = s_{2331} = s_{2313} = s_{2331} = s_{3213} = s_{3123} = s_{3132} = s_{1323} = s_{1332}$$
$$(1/4)s_{46} = s_{2312} = s_{2321} = s_{3212} = s_{3221} = s_{1223} = s_{1232} = s_{2123} = s_{2132}$$
$$(1/4)s_{55} = s_{3131} = s_{3113} = s_{1331} = s_{1313}$$
$$(1/4)s_{56} = s_{3112} = s_{3121} = s_{1312} = s_{1321} = s_{1231} = s_{1213} = s_{2131} = s_{2113}$$
$$(1/4)s_{66} = s_{1212} = s_{1221} = s_{2112} = s_{2121}$$

s_{ij} (or equivalently $c_{ijk\ell}$ and $s_{ijk\ell}$) can readily be related by use of the above equations defining them.

The 36 coefficients c_{ij} (or s_{ij}) are not all independent. The condition that the elastic energy be a single-valued function of strain requires that $c_{ij} = c_{ji}$ and $s_{ij} = s_{ji}$, reducing the number of independent elements for the most general anisotropic solid to 21 for each. Although all of these 21 elements are required for a triclinic crystal, the number of independent elastic stiffness (or compliance) elements is dramatically reduced as the crystal symmetry increases. A detailed derivation has been given by Nye (1960). The independent stiffness elements c_{ij} for the different crystal classes are shown in Table 1.9; the full matrices are all symmetric about the principal diagonal. The pattern of elements is identical for the corresponding compliance matrices ($c_{ij} \rightarrow s_{ij}$), except for the trigonal, hexagonal, and isotropic classes.

It is important to remember that the two-subscript forms c_{ij} and s_{ij} are *not* the components of second-rank tensors. In particular, they do not transform as in Eqs. (21) and (22) under change of coordinate axes. Axis transformations must always be accomplished using the full fourth-rank tensor formalism $c_{ijk\ell}$ ($s_{ijk\ell}$) via Eqs. (25) and (26).

Table 1.9 Elastic Stiffness and Compliance Matrices

Triclinic

$$
\begin{bmatrix}
c_{11} & c_{12} & c_{13} & c_{14} & c_{15} & c_{16} \\
c_{12} & c_{22} & c_{23} & c_{24} & c_{25} & c_{26} \\
c_{13} & c_{23} & c_{33} & c_{34} & c_{35} & c_{36} \\
c_{14} & c_{24} & c_{34} & c_{44} & c_{45} & c_{46} \\
c_{15} & c_{25} & c_{35} & c_{45} & c_{55} & c_{56} \\
c_{16} & c_{26} & c_{36} & c_{46} & c_{56} & c_{66}
\end{bmatrix}
\qquad
\begin{bmatrix}
s_{11} & s_{12} & s_{13} & s_{14} & s_{15} & s_{16} \\
s_{12} & s_{22} & s_{23} & s_{24} & s_{25} & s_{26} \\
s_{13} & s_{23} & s_{33} & s_{34} & s_{35} & s_{36} \\
s_{14} & s_{24} & s_{34} & s_{44} & s_{45} & s_{46} \\
s_{15} & s_{25} & s_{35} & s_{45} & s_{55} & s_{56} \\
s_{16} & s_{26} & s_{36} & s_{46} & s_{56} & s_{66}
\end{bmatrix}
$$

Monoclinic (all classes) (diad $\parallel x_3$)

$$
\begin{bmatrix}
c_{11} & c_{12} & c_{13} & 0 & 0 & c_{16} \\
c_{12} & c_{22} & c_{23} & 0 & 0 & c_{26} \\
c_{13} & c_{23} & c_{33} & 0 & 0 & c_{36} \\
0 & 0 & 0 & c_{44} & c_{45} & 0 \\
0 & 0 & 0 & c_{45} & c_{55} & 0 \\
c_{16} & c_{26} & c_{36} & 0 & 0 & c_{66}
\end{bmatrix}
\qquad
\begin{bmatrix}
s_{11} & s_{12} & s_{13} & 0 & 0 & s_{16} \\
s_{12} & s_{22} & s_{23} & 0 & 0 & s_{26} \\
s_{12} & s_{23} & s_{33} & 0 & 0 & s_{36} \\
0 & 0 & 0 & s_{44} & s_{45} & 0 \\
0 & 0 & 0 & s_{45} & s_{55} & 0 \\
s_{16} & s_{26} & s_{36} & 0 & 0 & s_{66}
\end{bmatrix}
$$

Orthorhombic (all classes)

$$
\begin{bmatrix}
c_{11} & c_{12} & c_{13} & 0 & 0 & 0 \\
c_{12} & c_{22} & c_{23} & 0 & 0 & 0 \\
c_{13} & c_{23} & c_{33} & 0 & 0 & 0 \\
0 & 0 & 0 & c_{44} & 0 & 0 \\
0 & 0 & 0 & 0 & c_{55} & 0 \\
0 & 0 & 0 & 0 & 0 & c_{66}
\end{bmatrix}
\qquad
\begin{bmatrix}
s_{11} & s_{12} & s_{13} & 0 & 0 & 0 \\
s_{12} & s_{22} & s_{23} & 0 & 0 & 0 \\
s_{13} & s_{23} & s_{33} & 0 & 0 & 0 \\
0 & 0 & 0 & s_{44} & 0 & 0 \\
0 & 0 & 0 & 0 & s_{55} & 0 \\
0 & 0 & 0 & 0 & 0 & s_{66}
\end{bmatrix}
$$

Tetragonal ($4\,mm$, $\bar{4}2m$, 422, $4/mmm$)

$$
\begin{bmatrix}
c_{11} & c_{12} & c_{13} & 0 & 0 & 0 \\
c_{12} & c_{11} & c_{13} & 0 & 0 & 0 \\
c_{13} & c_{13} & c_{33} & 0 & 0 & 0 \\
0 & 0 & 0 & c_{44} & 0 & 0 \\
0 & 0 & 0 & 0 & c_{44} & 0 \\
0 & 0 & 0 & 0 & 0 & c_{66}
\end{bmatrix}
\qquad
\begin{bmatrix}
s_{11} & s_{12} & s_{13} & 0 & 0 & 0 \\
s_{12} & s_{11} & s_{13} & 0 & 0 & 0 \\
s_{13} & s_{13} & s_{33} & 0 & 0 & 0 \\
0 & 0 & 0 & s_{44} & 0 & 0 \\
0 & 0 & 0 & 0 & s_{44} & 0 \\
0 & 0 & 0 & 0 & 0 & s_{66}
\end{bmatrix}
$$

Tetragonal (4, $\bar{4}$, $4/m$)

$$
\begin{bmatrix}
c_{11} & c_{12} & c_{13} & 0 & 0 & c_{16} \\
c_{12} & c_{11} & c_{13} & 0 & 0 & -c_{16} \\
c_{13} & c_{13} & c_{33} & 0 & 0 & 0 \\
0 & 0 & 0 & c_{44} & 0 & 0 \\
0 & 0 & 0 & 0 & c_{44} & 0 \\
c_{16} & -c_{16} & 0 & 0 & 0 & c_{66}
\end{bmatrix}
\qquad
\begin{bmatrix}
s_{11} & s_{12} & s_{13} & 0 & 0 & s_{16} \\
s_{12} & s_{11} & s_{13} & 0 & 0 & -s_{16} \\
s_{13} & s_{13} & s_{33} & 0 & 0 & 0 \\
0 & 0 & 0 & s_{44} & 0 & 0 \\
0 & 0 & 0 & 0 & s_{44} & 0 \\
s_{16} & -s_{16} & 0 & 0 & 0 & s_{66}
\end{bmatrix}
$$

Trigonal (classes 3, $\bar{3}$)

$$
\begin{bmatrix}
c_{11} & c_{12} & c_{13} & c_{14} & -c_{25} & 0 \\
c_{12} & c_{11} & c_{13} & -c_{14} & c_{25} & 0 \\
c_{13} & c_{13} & c_{33} & 0 & 0 & 0 \\
c_{14} & -c_{14} & 0 & c_{44} & 0 & c_{25} \\
-c_{25} & c_{25} & 0 & 0 & c_{44} & c_{14} \\
0 & 0 & 0 & c_{25} & c_{14} & \tfrac{1}{2}(c_{11}-c_{12})
\end{bmatrix}
\qquad
\begin{bmatrix}
s_{11} & s_{12} & s_{13} & s_{14} & -s_{25} & 0 \\
s_{12} & s_{11} & s_{13} & -s_{14} & s_{25} & 0 \\
s_{13} & s_{13} & s_{33} & 0 & 0 & 0 \\
s_{14} & -s_{14} & 0 & s_{44} & 0 & 2s_{25} \\
-s_{25} & s_{25} & 0 & 0 & s_{44} & 2s_{14} \\
0 & 0 & 0 & 2s_{25} & 2s_{14} & 2(s_{11}-s_{12})
\end{bmatrix}
$$

(continued)

Table 1.9   Elastic Stiffness and Compliance Matrices (Continued)

Trigonal (classes 32, $\bar{3}m$, 3m)

$$
\begin{bmatrix}
c_{11} & c_{12} & c_{13} & c_{14} & 0 & 0 \\
c_{12} & c_{11} & c_{13} & -c_{14} & 0 & 0 \\
c_{13} & c_{13} & c_{33} & 0 & 0 & 0 \\
c_{14} & -c_{14} & 0 & c_{44} & 0 & 0 \\
0 & 0 & 0 & 0 & c_{44} & c_{14} \\
0 & 0 & 0 & 0 & c_{14} & \tfrac{1}{2}(c_{11}-c_{12})
\end{bmatrix}
\qquad
\begin{bmatrix}
s_{11} & s_{12} & s_{13} & s_{14} & 0 & 0 \\
s_{12} & s_{11} & s_{13} & -s_{14} & 0 & 0 \\
s_{13} & s_{13} & s_{33} & 0 & 0 & 0 \\
s_{14} & -s_{14} & 0 & s_{44} & 0 & 0 \\
0 & 0 & 0 & 0 & s_{44} & 2s_{14} \\
0 & 0 & 0 & 0 & 2s_{14} & 2(s_{11}-s_{12})
\end{bmatrix}
$$

Hexagonal (all classes)

$$
\begin{bmatrix}
c_{11} & c_{12} & c_{13} & 0 & 0 & 0 \\
c_{12} & c_{11} & c_{13} & 0 & 0 & 0 \\
c_{13} & c_{13} & c_{33} & 0 & 0 & 0 \\
0 & 0 & 0 & c_{44} & 0 & 0 \\
0 & 0 & 0 & 0 & c_{44} & 0 \\
0 & 0 & 0 & 0 & 0 & \tfrac{1}{2}(c_{11}-c_{12})
\end{bmatrix}
\qquad
\begin{bmatrix}
s_{11} & s_{12} & s_{13} & 0 & 0 & 0 \\
s_{12} & s_{11} & s_{13} & 0 & 0 & 0 \\
s_{13} & s_{13} & s_{33} & 0 & 0 & 0 \\
0 & 0 & 0 & s_{44} & 0 & 0 \\
0 & 0 & 0 & 0 & s_{44} & 0 \\
0 & 0 & 0 & 0 & 0 & 2(s_{11}-s_{12})
\end{bmatrix}
$$

Cubic (all classes)

$$
\begin{bmatrix}
c_{11} & c_{12} & c_{12} & 0 & 0 & 0 \\
c_{12} & c_{11} & c_{12} & 0 & 0 & 0 \\
c_{12} & c_{12} & c_{11} & 0 & 0 & 0 \\
0 & 0 & 0 & c_{44} & 0 & 0 \\
0 & 0 & 0 & 0 & c_{44} & 0 \\
0 & 0 & 0 & 0 & 0 & c_{44}
\end{bmatrix}
\qquad
\begin{bmatrix}
s_{11} & s_{12} & s_{12} & 0 & 0 & 0 \\
s_{12} & s_{11} & s_{12} & 0 & 0 & 0 \\
s_{12} & s_{12} & s_{11} & 0 & 0 & 0 \\
0 & 0 & 0 & s_{44} & 0 & 0 \\
0 & 0 & 0 & 0 & s_{44} & 0 \\
0 & 0 & 0 & 0 & 0 & s_{44}
\end{bmatrix}
$$

Isotropic

As cubic, but with

$$c_{44} = (\tfrac{1}{2})(c_{11}-c_{12})$$
$$s_{44} = 2(s_{11}-s_{12})$$

1.3.2. The Isotropic Limit

As can be seen from Table 1.9, there are only two independent elastic stiffness constants, c_{11} and c_{12}, in an isotropic body. Historically, theories of stress and strain response in this isotropic limit have focused on the two quantities c_{12} and $(1/2)\,(c_{11} - c_{12})$. They, in turn, are traditionally symbolized as λ and G, respectively, and are referred to as Lamé constants. In terms of these Lamé constants it follows that Hooke's law of Eq. (43) can be cast in the form

$$\sigma_{ij} = 2G\eta_{ij} + \lambda\theta\delta_{ij} \tag{47}$$

in which δ_{ij} is the Kronecker delta and $\theta = \Sigma_i \eta_{ii}$ is referred to as the *dilatation*. Physically it is the fractional change in volume.

In terms of the Lamé constants, a number of related elastic moduli can be defined for isotropic bodies. Some of the more commonly used are discussed in the following paragraphs.

Shear Modulus or Modulus of Rigidity G. This is the Lamé constant $G = (1/2) (c_{11} - c_{12})$ itself. It is a direct measure of the ratio of shear stress to shear strain η_{ij}, $i \neq j$, in the form

$$\sigma_{ij} = 2G\eta_{ij}, \qquad i \neq j \tag{48}$$

It is also, as we shall see below, directly related to the transverse velocity of sound waves propagating in the solid.

Bulk Modulus K. Defined as the ratio of hydrostatic pressure p to fractional decrease in volume (i.e., $K = -dp/d\theta$). Substituting the definition $\sigma_{ij} = -p\,\delta_{ij}$ into Eq. (47) enables us to derive the equation

$$K = \lambda + (2/3)G \tag{49}$$

which relates bulk modulus to the fundamental Lamé constants.

Young's Modulus E and Poisson's Ratio μ. If an isotropic body is subjected to a longitudinal stress σ_{11} (say), the resulting strains follow directly from Eqs. (47) in the form

$$\begin{aligned}
\sigma_{11} &= 2G\,\eta_{11} + \lambda\theta \\
0 &= 2G\,\eta_{22} + \lambda\theta \\
0 &= 2G\,\eta_{33} + \lambda\theta \\
0 &= \eta_{ij}, \qquad i \neq j
\end{aligned} \tag{50}$$

Young's modulus E is defined as the ratio σ_{11}/η_{11} of longitudinal stress to longitudinal strain and, from Eq. (50), is readily seen to be given by

$$E = \frac{G(3\lambda + 2G)}{\lambda + G} \tag{51}$$

In the same experimental circumstances, the ratio of the magnitude of lateral strain $|\eta_{22}| = |\eta_{33}|$ to longitudinal strain η_{11} is called the Poisson ratio and, again using Eqs. (50), can be related to the Lamé constants in the form

$$\mu = \frac{\lambda}{2(\lambda + G)} \tag{52}$$

Longitudinal Modulus M. In general, two types of elastic waves can propagate in an isotropic solid (sound waves). One is a dilatational wave involving volume fluctuations, while the other is a rotational wave that does not perturb volume. In any given direction x_1, plane wave solutions exist, respectively, in the form of a longitudinal wave and two degenerate transverse waves. The former has the equation (Pollard 1977)

$$\frac{\partial^2 u_1}{\partial t^2} = \frac{\lambda + 2G}{\rho} \frac{\partial^2 u_1}{\partial x_1^2} = v_\ell^2 \frac{\partial^2 u_1}{\partial x_1^2} \qquad (53)$$

in which u_1 is the elastic displacement in the direction of propagation x_1, t is time, and ρ is density, and travels with the (so-called longitudinal) velocity $v_\ell = (M/\rho)^{1/2}$, where

$$M = \lambda + 2G \qquad (54)$$

is the longitudinal modulus of elasticity. In an ideal fluid, for which by definition $G = 0$, the longitudinal modulus M, bulk modulus K, and Lamé constant $\lambda = c_{12}$ are all equal.

The two degenerate transverse wave equations are

$$\frac{\partial^2 u_2}{\partial t^2} = \frac{G}{\rho} \frac{\partial^2 u_2}{\partial x_1^2} = v_t^2 \frac{\partial^2 u_2}{\partial x_1^2} \qquad (55)$$

$$\frac{\partial^2 u_3}{\partial t^2} = \frac{G}{\rho} \frac{\partial^2 u_3}{\partial x_1^2} = v_t^2 \frac{\partial^2 u_3}{\partial x_1^2} \qquad (56)$$

with elastic displacements u_2 and u_3 perpendicular to the direction x_1 of propagation, and possess (transverse) velocities $v_t = (G/\rho)^{1/2}$. Note that in the ideal fluid ($G=0$), transverse, or shear, waves cannot propagate.

In an anisotropic crystal there are (in general) no simple relationships between the directions of propagation and the elastic displacements for elastic waves. In other words, the sound waves are not purely longitudinal or transverse. There are, however, certain high-symmetry directions for which pure longitudinal and transverse waves do exist.

1.3.3. The Piezoelectric Effect

Piezoelectricity is the phenomenon in which a crystal under the action of a mechanical stress σ_{ij} becomes electrically charged, with opposite charges appearing at the two ends of the crystal. The effect does not occur in centrosymmetric crystals, since for them a uniform stress redistributes charges in a manner that is relatively compensating with respect to the center of symmetry. The absence of piezoelectricity is also manifested in the absence of any linear relationship between induced strain η_{ij} and applied electric field E_i. For centrosymmetric systems an applied field produces a strain (called electrostriction) that is quadratic in that field.

In piezoelectric crystals (which include all noncentrosymmetric point groups except the cubic 432 group; see Table 1.6), a mechanical stress induces a polarization P_i according to the linear relationship

$$P_i = d_{ijk} \, \sigma_{jk} \qquad (57)$$

defining a 27-component piezoelectric stress tensor of the third rank. Mathematically it can be shown that this tensor must also linearly relate the strain induced in a crystal by application of an electric field via the relationship

$$\eta_{jk} = d_{ijk} E_i \qquad (58)$$

a property usually referred to as the converse piezoelectric effect. Electrostriction terms quadratic in E_i are also still present but, for piezoelectric systems, represent a higher order and a usually negligibly small perturbation of the linear piezoelectric effect.

Note that crystals that are inversion images of each other have opposite signs of d_{ijk}. The absolute sign of d_{ijk} therefore depends on the definition of the direction of the polar axis. The standard piezoelectric convention is that the positive end of the polar axis is that which develops a negative charge upon compression along that axis.

Since elastic strain is proportional to stress, it is also possible to linearly relate induced polarization to strain in the manner

$$P_i = e_{ijk} \, \eta_{jk} \qquad (59)$$

which defines the 27-component piezoelectric strain tensor, and to define a corresponding converse relationship involving the stress induced by an applied field:

$$\sigma_{jk} = e_{ijk} E_i \qquad (60)$$

In tabulations it is customary to use contracted notations for the direct piezoelectric elements according to

$$P_i = d_{ij} \, \sigma_j, \qquad i = 1, 2, 3; j = 1, \ldots , 6 \qquad (61)$$

$$P_i = e_{ij} \, \eta_j, \qquad i = 1, 2, 3; j = 1, \ldots , 6 \qquad (62)$$

where the reduced stress σ_j ($j = 1, \ldots , 6$) and strain η_j ($j = 1, \ldots , 6$) components are defined in Eqs. (29) and (30). In a parallel fashion one may write reduced converse piezoelectric forms

$$\eta_j = d_{ij} E_i, \qquad i = 1, 2, 3; j = 1, \ldots , 6 \qquad (63)$$

$$\sigma_j = e_{ij} E_i \qquad i = 1, 2, 3; j = 1, \ldots , 6 \qquad (64)$$

There are consequently a maximum of 18 independent piezoelectric components d_{ij} (or e_{ij}) that are readily related to the 27 tensor elements d_{ijk} (e_{ijk}) using the

Table 1.10 The Nonzero Piezoelectric Matrices

Triclinic
Class 1, C_1

e_{11}	e_{12}	e_{13}	e_{14}	e_{15}	e_{16}	d_{11}	d_{12}	d_{13}	d_{14}	d_{15}	d_{16}
e_{21}	e_{22}	e_{23}	e_{24}	e_{25}	e_{26}	d_{21}	d_{22}	d_{23}	d_{24}	d_{25}	d_{26}
e_{31}	e_{32}	e_{33}	e_{34}	e_{35}	e_{36}	d_{31}	d_{32}	d_{33}	d_{34}	d_{35}	d_{36}

Monoclinic (diad $\parallel x_3$)
Class 2, C_2

0	0	0	e_{14}	e_{15}	0	0	0	0	d_{14}	d_{15}	0
0	0	0	e_{24}	e_{25}	0	0	0	0	d_{24}	d_{25}	0
e_{31}	e_{32}	e_{33}	0	0	e_{36}	d_{31}	d_{32}	d_{33}	0	0	d_{36}

Class m, C_s ($m \perp x_3$)

e_{11}	e_{12}	e_{13}	0	0	e_{16}	d_{11}	d_{12}	d_{13}	0	0	d_{16}
e_{21}	e_{22}	e_{23}	0	0	e_{26}	d_{21}	d_{22}	d_{23}	0	0	d_{26}
0	0	0	e_{34}	e_{35}	0	0	0	0	d_{34}	d_{35}	0

Orthorhombic
Class 222, D_2

0	0	0	e_{14}	0	0	0	0	0	d_{14}	0	0
0	0	0	0	e_{25}	0	0	0	0	0	d_{25}	0
0	0	0	0	0	e_{36}	0	0	0	0	0	d_{36}

Class mm2, C_{2v}

0	0	0	0	e_{15}	0	0	0	0	0	d_{15}	0
0	0	0	e_{24}	0	0	0	0	0	d_{24}	0	0
e_{31}	e_{32}	e_{33}	0	0	0	d_{31}	d_{32}	d_{33}	0	0	0

Tetragonal
Class $\bar{4}$, S_4

0	0	0	e_{14}	e_{15}	0	0	0	0	d_{14}	d_{15}	0
0	0	0	$-e_{15}$	e_{14}	0	0	0	0	$-d_{15}$	d_{14}	0
e_{31}	$-e_{31}$	0	0	0	e_{36}	d_{31}	$-d_{31}$	0	0	0	d_{36}

Class 4, C_4

0	0	0	e_{14}	e_{15}	0	0	0	0	d_{14}	d_{15}	0
0	0	0	e_{15}	$-e_{14}$	0	0	0	0	d_{15}	$-d_{14}$	0
e_{31}	e_{31}	e_{33}	0	0	0	d_{31}	d_{31}	d_{33}	0	0	0

Class $\bar{4}2m$, D_{2d} (diad $\parallel x_1$)

0	0	0	e_{14}	0	0	0	0	0	d_{14}	0	0
0	0	0	0	e_{14}	0	0	0	0	0	d_{14}	0
0	0	0	0	0	e_{36}	0	0	0	0	0	d_{36}

Class 422, D_4

0	0	0	e_{14}	0	0	0	0	0	d_{14}	0	0
0	0	0	0	$-e_{14}$	0	0	0	0	0	$-d_{14}$	0
0	0	0	0	0	0	0	0	0	0	0	0

Class 4mm, C_{4v}

0	0	0	0	e_{15}	0	0	0	0	0	d_{15}	0
0	0	0	e_{15}	0	0	0	0	0	d_{15}	0	0
e_{31}	e_{31}	e_{33}	0	0	0	d_{31}	d_{31}	d_{33}	0	0	0

(continued)

Table 1.10　The Nonzero Piezoelectric Matrices (Continued)

Trigonal
Class 3, C_3

e_{11}	$-e_{11}$	0	e_{14}	e_{15}	$-e_{22}$	d_{11}	$-d_{11}$	0	d_{14}	d_{15}	$-2d_{22}$
$-e_{22}$	e_{22}	0	e_{15}	$-e_{14}$	$-e_{11}$	$-d_{22}$	d_{22}	0	d_{15}	$-d_{14}$	$-2d_{11}$
e_{31}	e_{31}	e_{33}	0	0	0	d_{31}	d_{31}	d_{33}	0	0	0

Class 32, D_3

e_{11}	$-e_{11}$	0	e_{14}	0	0	d_{11}	$-d_{11}$	0	d_{14}	0	0
0	0	0	0	$-e_{14}$	$-e_{11}$	0	0	0	0	$-d_{14}$	$-2d_{11}$
0	0	0	0	0	0	0	0	0	0	0	0

Class 3m, C_{3v} $(m \perp x_1)$

0	0	0	0	e_{15}	$-e_{22}$	0	0	0	0	d_{15}	$-2d_{22}$
$-e_{22}$	e_{22}	0	e_{15}	0	0	$-d_{22}$	d_{22}	0	d_{15}	0	0
e_{31}	e_{31}	e_{33}	0	0	0	d_{31}	d_{31}	d_{33}	0	0	0

Hexagonal
Class $\bar{6}$, C_{3h}

e_{11}	$-e_{11}$	0	0	0	$-e_{22}$	d_{11}	$-d_{11}$	0	0	0	$-2d_{22}$
$-e_{22}$	e_{22}	0	0	0	$-e_{11}$	$-d_{22}$	d_{22}	0	0	0	$-2d_{11}$
0	0	0	0	0	0	0	0	0	0	0	0

Class $\bar{6}m2$, D_{3h} $(m \perp x_2)$

e_{11}	$-e_{11}$	0	0	0	0	d_{11}	$-d_{11}$	0	0	0	0
0	0	0	0	0	$-e_{11}$	0	0	0	0	0	$-2d_{11}$
0	0	0	0	0	0	0	0	0	0	0	0

Class 6, C_6

0	0	0	e_{14}	e_{15}	0	0	0	0	d_{14}	d_{15}	0
0	0	0	e_{15}	$-e_{14}$	0	0	0	0	d_{15}	$-d_{14}$	0
e_{31}	e_{31}	e_{33}	0	0	0	d_{31}	d_{31}	d_{33}	0	0	0

Class 622, D_6

0	0	0	e_{14}	0	0	0	0	0	d_{14}	0	0
0	0	0	0	$-e_{14}$	0	0	0	0	0	$-d_{14}$	0
0	0	0	0	0	0	0	0	0	0	0	0

Class 6mm, C_{6v}

0	0	0	0	e_{15}	0	0	0	0	0	d_{15}	0
0	0	0	e_{15}	0	0	0	0	0	d_{15}	0	0
e_{31}	e_{31}	e_{33}	0	0	0	d_{31}	d_{31}	d_{33}	0	0	0

Cubic
Class 23, T and $\bar{4}3m$, T_d

0	0	0	e_{14}	0	0	0	0	0	d_{14}	0	0
0	0	0	0	e_{14}	0	0	0	0	0	d_{14}	0
0	0	0	0	0	e_{14}	0	0	0	0	0	d_{14}

above equations if required. Because of the various symmetry elements associated with the crystal classes, the number of independent piezoelectric constants for most point groups is much less. The full d_{ij} and e_{ij} matrix forms for the 20 piezoelectric crystal classes are given in Table 1.10.

The special case of a crystal subjected to a hydrostatic pressure p is readily investigated by putting $-p = \sigma_{11} = \sigma_{22} = \sigma_{33}$. Direct substitution into Eq. (57) now leads to the contracted result

$$-P_i = \sum_{j=1}^{3} d_{ij}p \tag{65}$$

The remaining d_{ij} with $j = 4, 5, 6$ are absent, indicating the fact that uniform pressure induces no shear stresses.

The converse piezoelectric effect of Eq. (58) underlies the operation of transducers, which are devices to convert oscillatory electrical input into mechanical energy. The usual mode of operation is at acoustical frequencies (Bradfield 1970) with the applied electric field tuned to a natural mechanical resonance frequency of the particular crystal specimen involved in the device.

1.3.4. Thermal Expansion

All solids change their linear dimensions and volume with change of temperature T, the vast majority expanding when heated and contracting when cooled. For a crystal of length L_i in coordinate direction x_i, we may define a coefficient α_i of linear expansion as the fractional change in L_i per unit of temperature increment at constant pressure p, that is,

$$\alpha_i = (1/L_i) \, (\partial L_i/\partial T)_p, \qquad i = 1, 2, 3 \tag{66}$$

From it one can also define a related coefficient β of volume expansion in the form

$$\beta = (1/V) \, (\partial V/\partial T)_p \tag{67}$$

for which $\beta = \Sigma_i \alpha_i$.

Thermal expansion, like many other thermal properties, has it origin in lattice vibrational motion, the amplitude of which increases as the temperature rises. A qualitative theoretical understanding can be given in terms of classical oscillators and, in particular, in terms of the effect of anharmonic contributions to the potential energy U on the mean separation of pairs of atoms at temperature T (Kittel 1976). Let us write the potential energy of the atoms in a crystal as a function of their displacement x from equilibrium separation at absolute zero. Assuming it to be of anharmonic form,

$$U(x) = ax^2 - bx^3 + \text{higher order terms} \tag{68}$$

the thermal average separation $\langle x \rangle$ can be calculated by use of the Boltzmann distribution

$$\langle x \rangle = \frac{\int_{-\infty}^{\infty} x \exp[-U(x)/kT]\, dx}{\int_{-\infty}^{\infty} \exp[-U(x)/kT]\, dx} \tag{69}$$

which weights the possible values of x according to their thermodynamic probability. In the limit of small displacements, the anharmonic terms in the integrand exponentials can be expanded, reducing Eq. (69), in lowest order, to

$$\langle x \rangle = \frac{\int_{0}^{\infty} bx^4 \exp[-ax^2/kT]\, dx}{kT \int_{0}^{\infty} \exp[-ax^2/kT]\, dx} \tag{70}$$

or, since

$$\int_{0}^{\infty} x^4 e^{-\lambda x^2}\, dx = (3/8)(\pi/\lambda^5)^{1/2} \tag{71}$$

and

$$\int_{0}^{\infty} e^{-\lambda x^2}\, dx = (1/2)(\pi/\lambda)^{1/2} \tag{72}$$

to

$$\langle x \rangle = (3b/4a^2)kT \tag{73}$$

In this classical (or high-temperature) approximation, it follows that the thermal expansion coefficients are independent of temperature and proportional to the lowest order anharmonicity.

More generally, for quantum oscillators the expansion coefficients are related to specific heat and consequently approach zero (in a Debye T^3 fashion for crystals) as $T \to 0$. Empirically, data also show a correlation with melting point T_M, the total fractional volume change between $T=0$ and $T=T_M$ tending to be approximately constant. This would indicate that materials with low melting points tend to have larger thermal expansion coefficients, and those with high melting points, smaller ones. On the other hand, abrupt changes in α_i may occur as a function of T if any crystallographic transitions take place. Other, more gradual, anomalies may occur in materials that experience thermodynamic cooperative ordering processes (e.g., ferromagnets, ferroelectrics, ferroelastics), particularly near their Curie temperatures. These can even induce negative or near-zero (invar) expansion coefficients over certain restricted ranges of temperature.

1.3.5. Hardness

Hardness is one of the most investigated properties of solids. It is also one of the most difficult to represent theoretically in any elementary fashion. It is measured by resistance to indentation or scratching, but, primarily because of the absence of any simple fundamental theory, the various scales of hardness in use are rather arbitrary. The hardness of ceramics and minerals is still often measured simply by scratching the surface with different types of minerals. In the well-known Mohs hardness test, the scratch resistance is matched with that of one of ten minerals ranked in order of their increasing hardness from 1 (for talc) to 10 (for diamond).

Most present-day hardness tests are more quantitative and consist of pushing a penetrator into the material under test and measuring the effect in some way. The geometric form of the penetrator and the load mechanism vary with the different tests, as does the technique for measuring the magnitude of the indentation. Unfortunately, hardness numbers measured by one such scheme cannot be quantitatively converted to comparable hardness measured on another scale by use of any well-defined formalism. However, an approximate empirical conversion can be made between scales, and such a comparison for some of the more commonly used scales is shown in Fig. 1.2. (Jastrzebski 1976). Hardness correlates well with Young's modulus for many materials, particularly glasses. Its measurement helps to assess the vulnerability of a material to surface damage and is also an indicator of the ease with which a surface can be polished.

1.3.6. Nonlinear Elastic Properties

Linear elastic relationships, such as those discussed in Section 1.3.1 and typified by Hooke's law of Eq. (43), hold only for relatively small values of elastic deformation. For large deformations, nonlinear effects become important, so compliances and moduli then become functions of the magnitude of the stimulus or response. In such cases it is possible to define tangential elastic compliances as the slope of the response versus stimulus curve (which now deviates, at large values of response and stimulus, from a straight line). In the linear, or elastic, regime, these tangential compliances are equal to the conventional compliances as defined in Sections 1.3.1 and 1.3.2.

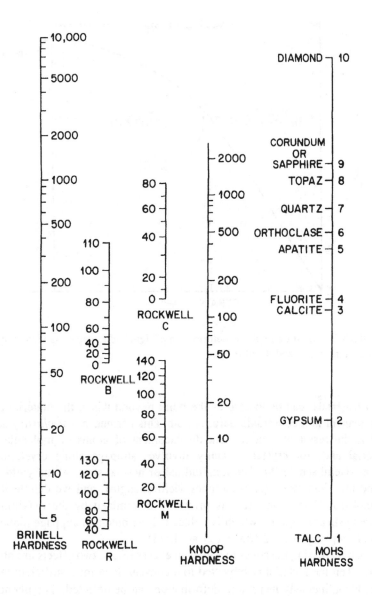

Figure 1.2 An approximate comparison of hardness scales.

In Figure 1.3 we show a typical stress–strain curve extended all the way beyond the linear (elastic) regime to the breaking point. The point B at which the curve first markedly deviates from linearity is called the yield point and, for

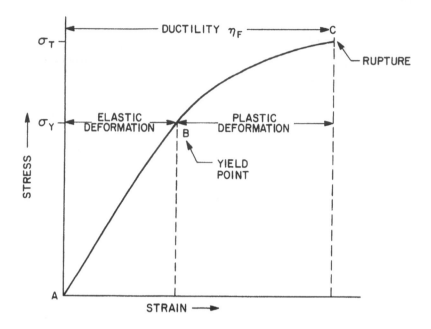

Figure 1.3 Elastic and plastic deformation of rigid bodies depicting yield strength σ_Y, tensile strength σ_T, and ductility η_F.

most materials, can be located as the point beyond which the material acquires a permanent nonrecoverable deformation. This permanent (or *plastic*) deformation is the result of a permanent displacement of atoms or molecules in the material and, for crystals, usually involves slippage along crystallographic planes (Nembach 1975). The longitudinal or normal stress at the yield point is called the yield stress (or sometimes yield strength), and we denote it by the symbol σ_Y. This yield stress is primarily determined by the maximum slip-inducing shear stress σ_S, which is induced along potential slippage planes in the crystal, with $\sigma_S \approx \sigma_Y/2$ (Barrett et al. 1973).

Upon increasing the stress to values in excess of σ_Y, two effects are observed. First, when the material is subjected to a constant load for a sufficient length of time, a continuously increasing deformation can be detected. The phenomenon is known as flow or creep. Second, as a function of increasing load, a breaking point (ductile fracture) is reached. At this point (C in Fig. 1.3) it is possible to define a tensile strength (or cleavage stress) σ_T as the maximum load carried before rupture divided by the original cross-sectional area of the specimen. If the stress required to produce fracture is applied in the plane of the cross section of the material, then the analogous shear stress (or shear strength) σ_S can be directly measured.

However, fracture (or the separation of a body by physical force into two or more parts) can take place by more than one mechanism. Two important limiting cases are those of static load (as in Fig. 1.3) and of dynamic (or impact) load. The former leads most often to ductile (or stress) fracture, which takes place after a region of plastic deformation. The latter usually leads to brittle fracture, taking place without prior plastic deformation along cleavage planes.

Whereas the conditions for the onset of plasticity depend primarily on shear strength $\sigma_S \approx \sigma_Y/2$, the condition for stress fracture is a measure of tensile strength σ_T. Consequently the prediction of whether a material will yield or fracture under static load depends ideally on the ratio $2\ \sigma_S/\sigma_T$. If this ratio is larger than unity, then the material will tend to be brittle; if it is smaller than 1, then the material will tend to yield and undergo stress fracture. Since σ_S is a much more sensitive function of temperature than σ_T, brittleness is also a sensitive function of temperature, usually increasing as temperature falls. Under cyclic stresses a material may fracture for stress magnitudes less than the maximum static value. This phenomenon, known as fatigue, is one of the most common forms of material failure under repeated stressing.

Toughness is the ability of a material to absorb energy during deformation. In a static tensile test, this energy is measured by the area under the stress–strain curve (e.g., of Fig. 1.3) and is the work (per unit volume) required to produce a specific degree of strain. The work required to reach the yield point is called the modulus of resilience W_R, while the maximum amount of energy that can be absorbed before stress fracture is the modulus of toughness W_T. Ductility η_F is the measure of strain required to produce stress fracture. Generally, tough materials have both moderately high yield strength and ductility. The reciprocal of ductility can be used as a qualitative measure of degree of brittleness. Under impact loads, toughness is measured directly as the energy required to fracture a unit volume of material.

Theories of fracture, cracking, cleavage, and so on can be found in the engineering literature; see, for example, the books by Barrett et al. (1973), Jastrzebski (1976), and Atkins (1985).

At high temperatures, of the order of or greater than about half the melting temperature, most solids exhibit a time-dependent plasticity, or creep, when subjected to static stress over long periods of time t. This occurs whether the stress involved is greater or smaller than the yield strength. A typical "creep curve" of deformation versus time t is shown in Fig. 1.4. As a function of t, an instantaneous elastic deformation is followed first by a region of primary creep (in which strain $\eta \sim t^n$ with $n < 1$) and then by a much more extensive region of secondary (or steady-state) creep in which the strain varies linearly with time, $\eta \sim t$.

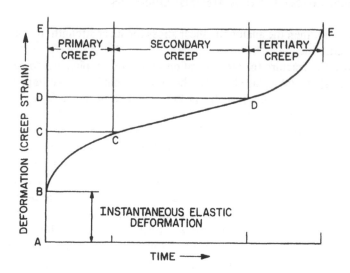

Figure 1.4 Creep (i.e., deformation as a function of time) at constant temperature and stress.

Primary creep is recoverable (that is, reversible) by unloading the specimen. It is called a retarded elastic, or anelastic, effect and can be related to elastic hysteresis and to the damping of elastic vibrations. Perfectly elastic materials, which obey Hooke's law, have no hysteresis, no primary creep region, and no vibrational damping. All real materials, however, show some degree of damping capacity, which may be quantitatively measured by observing the decay of vibrational amplitudes

$$A_n \sim \exp(-n\delta) \tag{74}$$

in which A_n is the amplitude of the nth oscillation and δ is the anelastic logarithmic decrement.

Steady-state creep is highly temperature-dependent and can be related to temperature by the same thermally activated (or Arrhenius) kind of equation as viscosity or diffusion,

$$d\eta/dt \sim \exp(-E_c/kT) \tag{75}$$

where E_c is an activation energy. At constant temperature the creep rate can be approximately related to stress by a power law form

$$d\eta/dt = A\sigma^m \tag{76}$$

in which A and m are material constants. Detailed theories of anelastic and viscoelastic behavior can be found in books by Christensen (1982) and Pipkin (1986) and in the review article by Hadley and Ward (1975).

Finally, beyond the steady-state creep regime a region of tertiary creep exists for which $\eta \sim t^n$ with $n > 1$. This is a region of accelerated creep and is really a process of progressive damage, terminating eventually, when the creep strain reaches the ductility η_F, in fracture.

1.4. INFRARED EMISSION MECHANISMS

The classical theory of electric and magnetic fields in vacuum establishes the existence of electromagnetic waves traveling with a characteristic velocity $c \approx 2.9979 \times 10^8$ m/s. These waves are emitted by electric charges experiencing a nonzero acceleration. The simplest example is that of a charge undergoing simple harmonic motion of angular frequency ω that emits an electromagnetic wave of the same frequency (Ditchburn 1976). For a discussion of infrared optical phenomena, the classical theory of electromagnetic field propagation is generally adequate, and it is not necessary to probe the quantum nature of the pure radiation field as set out, for example, in the book by Heitler (1954).

The motion of electrons and atoms in solids, on the other hand, requires a quantum-mechanical description, although a close analogy often remains between the quantum picture of an electron making transitions between energy levels E_i and $E_j > E_i$ and the classical picture of an electron oscillating with a frequency $\omega_{ij} = (E_j - E_i)/\hbar$ in which $\hbar = 1.0546 \times 10^{-34}$ J·s. For an atom or molecule in quantum level E_j, a transition from E_j to E_i can cause the emission of a photon ω_{ij}. Likewise, in the presence of an electromagnetic field of frequency ω_{ij}, a photon can be absorbed by the system, which is correspondingly excited from level E_i to E_j. The emission process seldom occurs for completely isolated systems, although, in principle, it can. All realizable systems are bathed in radiation and interact with their surroundings.

It is found experimentally that an enclosure that forms part of a body at a uniform temperature T always contains radiation. Moreover, this radiation, if it is in thermal equilibrium with the body, has an energy distribution that depends *only* on the temperature T (i.e., it is independent of the material that forms the enclosure). This distribution, for temperatures at and modestly above room temperature, peaks in the infrared portion of the spectrum. It is referred to as "blackbody" radiation and is the preferred source of incoherent radiation at infrared frequencies.

The coherence of a radiant field is a measure of the correlation in space and time of the field fluctuations. Radiant fields encountered in practice range from the nearly perfect monochromatic (and hence most fully coherent) to the almost "white" and hence almost fully incoherent limit. Experimentally, coherence can be measured by either heterodyne beating or interferometric (fringe) techniques. Theoretically this idea of coherence is best grasped in terms of a model based on the concept of a wave packet. Consider the electric field form

$$E = E_o \exp\left(-t^2/\tau_c^2\right) \exp[i(\omega t - \mathbf{q}\cdot\mathbf{r})] \tag{77}$$

in which t = time, $\mathbf{q}$ = wave vector, and $\mathbf{r}$ = distance. Clearly, as τ_c goes from zero to infinity, the wave train, Eq. (77), changes from a delta function spike to an infinitely long perfectly sinusoidal wave. It follows that τ_c is a measure of coherence. It can be converted to a coherence length l_c by the relationship $l_c = c\tau_c$ involving the wave velocity c.

The two extremes of coherence in infrared sources are represented by thermal (blackbody) sources and laser sources. The former has a coherence time $\tau_c \approx 10^{-14}$ s and therefore a corresponding coherence length of the order of micrometers. Today's narrow-linewidth "single-frequency" lasers, on the other hand, have bandwidths $(2\pi\tau_c)^{-1}$ less than 100 MHz, corresponding to coherence times longer than 10^{-9} s and coherence lengths approaching 1 m.

1.4.1 Thermal (Blackbody) Radiation

Experimentally it is found that the total energy (for all wavelengths) emitted from a blackbody cavity per unit area per second is related to absolute temperature T by

$$E = sT^4 \tag{78}$$

which is known as Stefan's law, with s referred to as Stefan's constant. Additionally it is observed that the distribution of radiation peaks at a frequency ω_{max} that is directly proportional to T. The precise relationship, known as Wien's displacement law, is (Chantry 1984)

$$\hbar\omega_{max} = 1.961 \ T \ \text{cm}^{-1} \tag{79}$$

if T is in kelvins. The detailed spectral shapes for a selection of temperatures are shown in Fig. 1.5.

Theoretically the derivation of the mathematical form of these curves can be obtained by assuming the blackbody to be made up of quantum oscillators each having its own characteristic frequency. Inside the blackbody cavity, at temperature T, there will exist a spectral energy density $U(\omega,T)$, the detailed form of which it is our purpose to calculate.

Consider a quantum oscillator of frequency ω in thermal equilibrium at temperature T. From statistical thermodynamics the mean energy of such an oscillator with energy levels $n\hbar\omega$, $n = 0, 1, 2, 3, \ldots$, is (Landau and Lifshitz 1958)

$$\overline{E} = \frac{\sum\limits_{n=0}^{\infty} n\hbar\omega \, \exp(-n\hbar\omega/kT)}{\sum\limits_{n=0}^{\infty} \exp(-n\hbar\omega/kT)} = \frac{\hbar\omega}{\exp(\hbar\omega/kT) - 1} \tag{80}$$

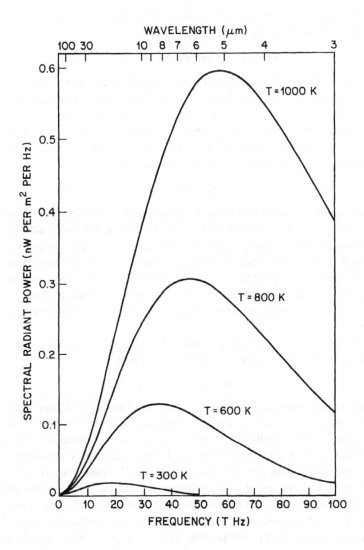

Figure 1.5 Emission spectra of blackbody sources at various temperatures.

Since the number of quantized standing-wave oscillators per unit volume with frequencies between ω and $\omega + d\omega$ is $\omega^2 d\omega / \pi^2 c^3$, it follows immediately that the spectral energy density $U(\omega,T)\, d\omega$ is just this number times the $\overline{E}$ of Eq. (80), that is,

$$U(\omega,T) = (\hbar\omega^3 / \pi^2 c^3) / (e^{\hbar\omega/kT} - 1) \qquad (81)$$

As a function of wavelength $\lambda = 2\pi c/\omega$, this distribution can be reexpressed as $U(\lambda,T) = U(\omega,T)\, d\omega/d\lambda$, or

$$U(\lambda,T) = (16\pi^2 \hbar c/\lambda^5) / (e^{2\pi\hbar c/\lambda kT} - 1) \tag{82}$$

Equations (81) and (82) are alternative forms of Planck's law. It is easily verified that they conform both with Stefan's law of Eq. (78) and Wien's displacement of law of Eq. (79). The distribution is completely independent of the details of the energy-level schemes composing the atomic structure of the blackbody. Since the thermal population N_i of these "atomic" states E_i are derivable from Boltzmann's law of statistical mechanics, namely,

$$N_i \sim \exp(-E_i/kT) \tag{83}$$

and any significant departure from this would imply a violation of the second law of thermodynamics, it follows that a combination of the Boltzmann and Planck laws must tell us something about the manner in which the dynamics of energy-exchange between the radiation field and atomic excitations maintains this balance. Specifically, a population form of Eq. (83) can be maintained in this way only if each excitation process $E_i \rightarrow E_j$ that absorbs a photon of energy $\hbar\omega_{ij} = E_j - E_i$ is exactly balanced by the reverse process, which creates a like-energy photon.

Consider two "atomic" energy states E_i, E_j with equilibrium populations N_i and N_j, respectively. It follows from Eq. (83) that

$$N_j = N_i \exp(-\hbar\omega_{ij}/kT) \tag{84}$$

The number of atoms *stimulated* into excitation from E_i to E_j per second by the radiation field must be proportional to N_i and to $U(\omega,T)$, thereby defining a constant of proportionality B_{ij}. A reverse process, proportional to N_j and $U(\omega, T)$, defines an analogous parameter B_{ji}. If only stimulated transitions were allowed, the above principle of "detailed balance" would require $B_{ij}N_i = B_{ji}N_j$ and, being valid for any $U(\omega,T)$ whatsoever, would not stabilize the Planck distribution of Eq. (81). To induce the latter distribution requires recognition of the existence of an additional *spontaneous* emission probability $A_{ji}N_j$ from the higher energy level that is independent of the radiation field. Including it in the detailed balance equation now leads to

$$B_{ij}N_iU(\omega,T) = B_{ji}N_jU(\omega,T) + A_{ji}N_j \tag{85}$$

which implies, using Eq. (84), that

$$U(\omega,T) = \frac{A_{ji}}{B_{ij}\exp(\hbar\omega_{ij}/kT) - B_{ji}} \tag{86}$$

that is, a radiation distribution equal to that of Eq. (81) if

$$B_{ij} = B_{ji} \qquad (87)$$

and

$$A_{ji} = (\hbar\omega^3 / \pi^2 c^3) B_{ji} \qquad (88)$$

These parameters, known as Einstein coefficients, measure the transitional probabilities for spontaneous and stimulated emission. Quantum mechanically they can be shown to be proportional to the square of the electric dipole moment matrix element between the two atomic eigenfunctions ψ_i and ψ_j that are involved.

If there is only one transition path from E_j to E_i, then $A_{ji} = (\tau_j)^{-1}$, where τ_j measures the lifetime of the state j in the absence of radiation. If the state j can emit at several frequencies, then $(\tau_j)^{-1} = \Sigma_i A_{ji}$. As an example, the value of τ_j for the yellow lines of Na is about 10^{-8} s. For the densities of radiation normally used in infrared experiments, the second term on the right-hand side of Eq. (85) is very small compared to the first.

A blackbody is, by definition, an object that absorbs *all* the radiation incident upon it. If a blackbody is in thermal equilibrium, it follows that the total radiant energy emitted per second per unit area (called the radiance, or radiant emittance R) must exactly equal the flux of radiant energy R_{inc} incident per unit area on the surface. For "non-black" bodies, $R/R_{inc} < 1$, enabling us to define an emission $e < 1$ via the equation (known as Kirchhoff's law)

$$R = eR_{inc} = eR_B \qquad (89)$$

in which R_B is the radiance of a blackbody. Since for all bodies in thermal equilibrium the total power absorbed must be exactly equal to that emitted, we can equally well define an absorptivity a exactly equal to e. In fact, although this equality applies to total radiance, an equivalent equality

$$a(\omega,T) = e(\omega,T) \qquad (90)$$

can be shown (using the second law of thermodynamics) to hold for spectral components $a(\omega,T)$ and $e(\omega,T)$ of absorptivity and emissivity separately.

The total radiance transmitted by a body R_t is equal to the incident radiance R_{inc} minus the absorbed and reflected components. For a slab sample of thickness L, the variation of R_t with L is usually expressed in the form (Lambert's law)

$$R_t(L) = R_t(0) \exp(-\alpha_R L) \qquad (91)$$

which defines the radiance loss coefficient α_R. If reflectivity is small (e.g., a rough surface), it follows that any material can become "black" if L is large enough. In fact, as a material property, "blackness" is often discussed in terms of an "optical thickness" $L_o = 1/\alpha_R$. Materials that are optically thin can therefore be used to provide excellent thermal contact between thermodynamic systems on either side. They are said to be diathermanous materials. At the other extreme, materials that are optically thick can be used to establish adiabatic barriers that prevent heat exchange.

1.4.2. Stimulated Emission and Infrared Lasers

The concepts of the preceding section, used to describe the equilibria inside cavities, can be used just as well to describe the propagation of infrared radiation through an absorbing/emitting medium (Chantry 1984). Consider, for example, propagation in direction z through a medium of two-level atomic constituent units. The probability per second that a photon $\hbar\omega_{ij}$ will be absorbed is, from Eq. (85),

$$P_{ij} = B_{ij}U(\omega,T)N_i \tag{92}$$

while the probability that a photon of this same energy will be emitted is

$$P_{ji} = [A_{ji} + B_{ij}U(\omega,T)]N_j \tag{93}$$

It follows that the increment $dI(\omega)$ of spectral intensity $I(\omega) = cU(\omega,T)$ on passage through a material length increment dz can be written

$$dI(\omega) = [N_j(A_{ji}/4\pi) + (N_j - N_i)B_{ij}\, I(\omega)/c]\, \hbar\omega f(\omega)\, dz \tag{94}$$

in which $f(\omega)$ is a lineshape function centered at ω_{ij}, and A_{ji} has been divided by 4π because the spontaneous contribution is isotropic while the stimulated emission and absorption take place only in the direction of propagation.

Two limiting situations are of interest. The first is the familiar one used for absorption spectroscopy in which the A term in (94) is negligible and N_j and N_i are related by the equilibrium condition of Eq. (84) with $N_j < N_i$. Clearly, in this case, $dI(\omega)$ is negative and absorption results. The second is the case of so-called population inversion $N_j > N_i$. If this can be achieved for some metastable configuration, then both terms in the equation, Eq. (94), for $dI(\omega)$ are positive, and spectral intensity increases with distance traveled z. The A term, representing spontaneous emission, is incoherent, but the B term, representing stimulated emission, is coherent. In the infrared region, once again, the latter is completely dominant, and we have the basic condition for laser (*l*ight *a*mplification by *s*timulated *e*mission of *r*adiation) action. Ignoring the effects of the competing spontaneous process, one can obtain from Eq. (94) an arbitrarily large spectral intensity providing one has a long enough path. In the laboratory this is usually acquired by placing the gain medium inside a resonator (such as a Fabry–Perot cavity) in which the beam can traverse the medium many times.

The essence of laser operation is the production of a population inversion $N_j > N_i$ $(E_j > E_i)$. The secret lies in the fact that the excited states of atoms or of more complex many-body systems do not all decay (spontaneously or otherwise) at the same rate. For example, the magnitude of the electric dipole matrix elements (which primarily control radiative transitions) between two atomic levels depends sensitively on the details of the wave functions of those states. They may, for example, be zero by virtue of some spin or orbital selection rule concerning the relative symmetries of the states involved (forbidden transitions). On the other hand, deexcitations can also take place via nonradiative processes involving, for example, atomic collisions in gases or phonon perturbations in solids.

In most systems, therefore, certain quantum levels are long-lived (or metastable), and the possibility of producing a population inversion exists in principle if the systems can be "zapped" with sufficient energy to substantially depopulate the ground state. In practice, the threshold energies so required are usually prohibitively high (even as pulses), and a little more thought is required for the derivation of efficiently lasing systems. Excellent overviews of laser theory can be found in the books by Ben-Shaul et al. (1981) and Wilson and Hawkes (1983).

One important class of infrared lasers is the gas laser. Consider, for the moment, the situation in a mixture of He and Ne gases through which an electric discharge is flowing. This discharge excites both the He and the Ne to ionized levels, which then both decay via allowed radiative transitions (see Fig. 1.6). There is, however, a fundamental difference in the decay schemes for these two inert gases that is caused by the difference in their electronic configurations as neutral atoms ($1s^2$ for He, $1s^22s^22p^6$ for Ne). It happens that the lowest excited states for He are S states with the same L quantum number ($L = 0$) as the He ground state, while the lowest excited state for Ne is a p-based multiplet with an orbital quantum number different from its s-based ground state. Since for free gaseous atoms L is a good quantum number, allowed electric dipole transitions are subject to the Laporte rule $\Delta L = \pm 1$. It follows that whereas the Ne excitations can quickly decay back to the Ne ground state (leaving the Ne excited states effectively all depopulated), the He first excited states are both highly metastable (with lifetimes > 1 s) and remain significantly populated.

The secret of population inversion in the He–Ne gas laser now depends on the fact that He–Ne collisions can produce an efficient transfer of energy from the metastable He states to the Ne 2s- and 3s-based multiplets with which they are closely energy degenerate (allowing for energy conservation). A strong population inversion then exists between both the Ne 2s- and 3s-based multiplets and the depopulated 2p and 3p multiplets, giving rise to laser lines at 1.15 μm and 3.39 μm, respectively (Fig. 1.6). Other gas lasers (e.g., CO_2–N_2 with technically important laser emissions at 10.4 μm and 9.4 μm) operate on similar principles but acquire population inversion using molecular states that are metastable by reason of selection rules involving vibrational levels in homopolar (e.g., N_2) molecules.

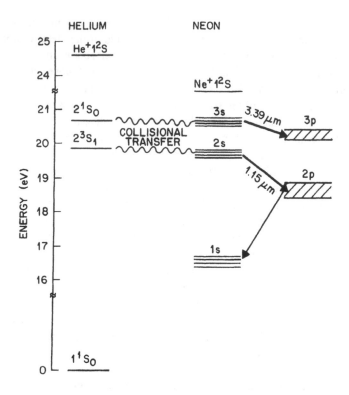

Figure 1.6 The origin of laser action in a He–Ne mixture. The various states of Ne consist of close multiplets, and under strong excitation several fine structure laser lines may be obtained. However, under normal operating conditions only a single member of each set is significant. Note that 1 eV $= 1.6021 \times 10^{-19}$ J.

A second technically important class of infrared laser sources is the solid-state laser containing small concentrations of transition metal or rare earth ions. Prototypical examples are the ruby (Cr^{3+} in Al_2O_3) laser and the Nd^{3+} in YAG ($Y_3Al_5O_{12}$) laser. The general principle here is that the lowest energy excited levels of the transition metal or rare earth ions are sharp because they involve only inner-shell d or f electrons, whereas higher energy bands may be much broader. In addition, the lowest level excitations are relatively long lived (e.g., $\sim 10^{-3}$ s) since, being intra-d-band or f-band in character, they would be forbidden in a spherical environment (i.e., with L a good quantum number). We therefore conceive of a pump energy (Fig. 1.7) that, via the broadband ultraviolet levels, efficiently populates the highly excited states and a decay mechanism by which the levels revert quickly (say $\sim 10^{-7}$ s) to the lowest metastable states. In

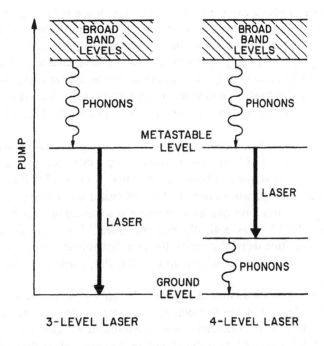

Figure 1.7   Schematic representation of the laser action in solid-state three-level and four-level lasers.

actuality the decay mechanism for these systems is predominantly a phonon-assisted nonradiative process.

In the ruby laser, only the ground level, the metastable level, and broadband level are involved. The laser is consequently termed a three-level laser and relies for population inversion on ground-level depopulation. It therefore has a high threshold energy and is operated under pulsed pump conditions. The metastable level is actually a crystal-field split doublet leading to two laser lines, at 694.4 nm and 693 nm, respectively, when operated at room temperature.

In Nd^{3+} lasers, the need to deplete the ground state is avoided by the presence of a fourth level between the metastable and ground states (Fig. 1.7). In this four-level laser, the population inversion exists between the metastable level and the essentially empty intermediate level, which can decay to the ground state quickly via phonon-assisted processes. Because population inversion can be induced without depleting the ground state, this laser can be operated in the continuous wave (cw) mode and is perhaps the most important solid-state laser for near-infrared purposes. It operates at 1.06 μm.

Other classes of important laser sources include chemical lasers and dye lasers. The former utilize gas-phase reactions, the products of which are often formed selectively in excited states so that one has a built-in pumping mechanism, via exothermic chemical reactions, for providing population inversions. The latter make use of the broad fluorescence spectra possessed by solutions of organic dyes. The large polyatomic molecules have electronic levels that, in solution, are broadened to quasi-continuous bands. Fluorescence from the bottom of the first excited band to states in the ground continuum create an effective four-level laser that is tunable over a fairly extensive range of frequencies (Maeda 1984).

Another increasingly popular method for inducing population inversion is to use an already existing laser to selectively pump a different system and produce what are in effect stimulated fluorescence lines. It is found that a vast range of low-pressure vapors, when confined to a resonator and pumped with (say) intense CO_2 radiation, give out laser emission at a variety of longer wavelengths. They are referred to as optically pumped lasers. However, since molecular collisions can often thermalize away the population inversion, it is usually necessary to operate at very low pressures (<100 Pa), particularly for far-infrared cw operation.

Another important source of coherent infrared radiation is the junction or diode laser, formed at the junction of two semiconductor materials—one p-doped (that is, rich in holes) and one n-doped (electron-rich). When the diode is biased forward, electrons are injected into the p region, where they combine with holes, the energy (which is essentially that of the bandgap) being converted into quanta of radiation. The population inversion mechanism can be qualitatively visualized by looking at Fig. 1.8, which schematically depicts the electron energy as a function of the density of states in (a) an undoped (intrinsic) semiconductor (formally at $T=0$) and (b) a doped semiconductor (under high injection) containing electrons in the conduction band to the electronic Fermi level F_c and holes in the valence band to the hole Fermi level F_v.

Photons with energy greater than the bandgap E_g but smaller than $F_c - F_v$ cannot be absorbed (because the conduction band states are occupied), but these photons can induce downward electron transitions to the valence band. At finite temperatures the electrons and holes are somewhat redistributed, smearing out the sharply defined carrier distributions shown in the figure, but the probability of inducing a downward transition still exceeds that for an upward transition, creating the laser condition. At low current densities, however, the emission is not coherent, and the device acts as an incoherent light source—the light-emitting diode (LED).

The main advantage of the LED or laser diode is the ability to change the operating wavelength by changing the bandgap through choice of semiconductor or degree of alloying. In this manner it has proven possible to obtain emission anywhere from the near ultraviolet to as far out as 30 μm. More details of the

T = 0 K

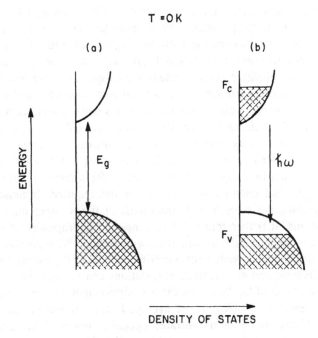

Figure 1.8 Electron energy as a function of the density of states in an intrinsic direct-bandgap semiconductor at $T = 0$ K. (a) Equilibrium and (b) under high injection.

structures and capabilities of laser diodes and light-emitting diodes can be found in Kressel (1981).

In spite of years of development there is still a distinct shortage of strong laser sources in the far infrared beyond a few tens of microns (micrometers in SI). However, recent successes of devices known as free-electron lasers (or FELs) show promise for producing tunable laser beams in this frequency regime. The principle involved concerns the perturbation of electron synchrotron radiation by magnets of alternating polarity, which cause the electron trajectory to "wiggle" and emit radiation (Sessler and Vaughan 1987). Radiation from different "bends" in the electron motion are manipulated to cause them to interfere constructively.

1.5. LINEAR PROPAGATION OF RADIATION IN SOLIDS: MACROSCOPIC THEORY

When plane-wave infrared radiation impinges upon a plane crystal vacuum–solid interface, the electrons and nuclei in the solid are subjected to electric forces that oscillate with the frequency of the radiation. Under the influence of the oscillating electric field component (the magnetic field component has a negligible effect on linear optical properties) these particles act as sources of secondary spherical waves, which combine with the original wave to influence other particles. We know from experimentation that the result is a reflected plane wave with the same phase velocity c as the incident wave together with a transmitted plane wave of phase velocity c/n, where n is defined as the *refractive index* of the solid.

From a microscopic point of view this means that the secondary waves must exactly cancel the incident wave inside the solid and replace it with another plane wave of smaller velocity (since $n > 1$ in a transparent regime). This can be explained theoretically from first principles by using a fully microscopic theory involving vacuum field equations. Fortunately, for a theory of infrared optical properties, the full microscopic approach is not required, because the wavelengths involved are some tens of thousands of times larger than the spacing between atoms. It is therefore valid to consider the properties of macroscopic fields within the material medium (Maxwell fields). Whereas the true microscopic fields fluctuate widely between different points, depending on their proximity to electrons and nuclei, these fluctuations can be smoothed by averaging over each unit cell of the lattice on the assumption that the variation of *average* fields between cells is small. These averaged (macroscopic or Maxwell) fields are then sufficient to define the infrared optical response of the lattice.

In the language of these averaged Maxwell fields, the propagation of electromagnetic waves is governed by the so-called Maxwell equations (see, for example, Hodgson 1970, Bleaney and Bleaney 1976)

$$\nabla \times \mathbf{E} = -\partial \mathbf{B}/\partial t, \qquad \nabla \times \mathbf{H} = \mathbf{J} + (\partial \mathbf{D}/\partial t) \qquad (95)$$

$$\text{cgs:} \qquad \nabla \times \mathbf{E} = (-1/c)(\partial \mathbf{B}/\partial t), \qquad \nabla \times \mathbf{H} = 4\pi \mathbf{J}/c + (1/c)(\partial \mathbf{D}/\partial t) \qquad (95')$$

in which $\mathbf{E}$ is the (macroscopic) electric field, $\mathbf{H}$ the magnetic field, $\mathbf{D}$ the electric displacement, $\mathbf{B}$ the magnetic induction, and $\mathbf{J}$ the electric current density. When combined with the divergence equations for induction and displacment, namely,

$$\nabla \cdot \mathbf{B} = 0, \qquad \nabla \cdot \mathbf{D} = 0 \qquad (96)$$

which are valid in the absence of free charge, this set of equations can be used to probe the electromagnetic propagation in solids. In this context the nature of the particular solid in question is defined by the set of linear material compliances that relate the responses $\mathbf{D}, \mathbf{B},$ and $\mathbf{J}$ to simtuli $\mathbf{E}$ and $\mathbf{H}$ in the manner

$$\mathbf{D} = \epsilon_o \epsilon \cdot \mathbf{E}, \qquad \mathbf{B} = \mu_o \mu \cdot \mathbf{H}, \qquad \mathbf{J} = \sigma \cdot \mathbf{E} \qquad (97)$$

$$\text{cgs:} \qquad \mathbf{D} = \epsilon \cdot \mathbf{E}, \qquad \mathbf{B} = \mu \cdot \mathbf{H}, \qquad \mathbf{J} = \sigma \cdot \mathbf{E} \qquad (97')$$

In these equations, optical frequency electric conductivity σ, permitivity, ϵ, and permeability μ are second-task tensors (the last two dimensionless), while ϵ_o and μ_o are respectively the scalar permittivity and permeability of empty space (and are equal to unity in the cgs system). In describing infrared properties these macroscopic parameters can be related to microscopic concepts (like electron mobilities and atomic or bond polarizabilities) and thereby provide a probe of the condensed state, particularly via their frequency dependence.

In SI units the velocity of light in vacuo is $c = (\epsilon_o\mu_o)^{-1/2}$; the corresponding velocity in an isotropic medium is $(\epsilon\epsilon_o\mu\mu_o)^{-1/2}$ or, equivalently, $c/(\epsilon\mu)^{1/2}$. It follows that the refractive index n is equal to $(\epsilon\mu)^{1/2}$ in general and to $\epsilon^{1/2}$ in nonmagnetic materials. In Gaussian-cgs units the vacuum light propagation velocity c is included explicitly in the Maxwell equations. The refractive index n and the dimensionless permittivity and permeability parameters ϵ and μ are independent of the system of units.

The relationship between displacement $\mathbf{D}$ and field $\mathbf{E}$ inside a medium can be used to define polarization $\mathbf{P}$, the dipole moment per unit volume, in the form

$$\mathbf{D} = \epsilon_o\mathbf{E} + \mathbf{P} \tag{98}$$

$$\text{cgs:} \quad \mathbf{D} = \mathbf{E} + 4\pi\mathbf{P} \tag{98'}$$

from which follows the definition of macroscopic polarizability χ via the equation $\mathbf{P} = \epsilon_o\chi\cdot\mathbf{E}$ (cgs: $\mathbf{P} = 4\pi\chi\cdot\mathbf{E}$) or equivalently

$$\mathbf{D} = \epsilon_o(1 + \chi) \cdot \mathbf{E} \tag{99}$$

$$\text{cgs:} \quad \mathbf{D} = (1 + 4\pi\chi) \cdot \mathbf{E} \tag{99'}$$

Polarizability χ is therefore a second-rank tensor, and $\mathbf{1}$ denotes the unit second-rank tensor with elements $1_{ij} = \delta_{ij}$. Equivalent equations can be defined for magnetic polarization (more commonly known as magnetization):

$$\mathbf{B} = \mu_o\mathbf{H} + \mathbf{M} \tag{100}$$

$$\text{cgs:} \quad \mathbf{B} = \mathbf{H} + 4\pi\mathbf{M} \tag{100'}$$

leading to the concept of magnetic polarizability χ_m via $\mathbf{M} = \mu_o\chi_m\cdot\mathbf{H}$ (cgs: $\mathbf{M} = 4\pi\chi_m\cdot\mathbf{H}$) and the equations

$$\mathbf{B} = \mu_o(1 + \chi_m) \cdot \mathbf{H} \tag{101}$$

$$\text{cgs:} \quad \mathbf{B} = (1 + 4\pi\chi_m) \cdot \mathbf{H} \tag{101'}$$

However, for optical vibrations of infrared frequency, ferromagnetic and paramagnetic moments are unable to follow the rapid oscillations of $\mathbf{H}$ because of their relatively long relaxation times. The remaining diamagnetic moments are so small as to have no appreciable effect on optical behavior, and we therefore set $\chi_m = 0$, $\mu = 1$ for optical frequency magnetic fields.

Using the material equations (97) with $\mu = 1$, Maxwell's equations (95) can be reexpressed as equations in field variables $\mathbf{E}$ and $\mathbf{H}$ alone. Assuming that these fields have the time dependence $\exp(i\omega t)$, they now take the form

$$\nabla \times \mathbf{E} = -i\mu_o\omega\mathbf{H} \tag{102}$$

$$\text{cgs:} \quad \nabla \times \mathbf{E} = -(1/c)i\omega\mathbf{H} \tag{102'}$$

$$\nabla \times \mathbf{H} = i[\epsilon - (i\sigma/\epsilon_o\omega)]\epsilon_o\omega\mathbf{E} = i\epsilon_o\omega\hat{\epsilon} \cdot \mathbf{E} \tag{103}$$

$$\text{cgs:} \quad \nabla \times \mathbf{H} = i[\epsilon - (4\pi i\sigma/\omega)]\omega\mathbf{E}/c = (i\omega/c)\hat{\epsilon} \cdot \mathbf{E} \tag{103'}$$

enabling us to define a dimensionless complex dielectric function $\hat{\epsilon} = \epsilon' - i\epsilon''$ with real and imaginary components

$$\epsilon' = \epsilon, \quad \epsilon'' = \sigma/\epsilon_o\omega \tag{104}$$

$$\text{cgs:} \quad \epsilon' = \epsilon, \quad \epsilon'' = 4\pi\sigma/\omega \tag{104'}$$

For a rigorous solution of these equations subject to the restraints of Eqs. (96), proper note must be taken of boundary conditions. Since this task can rarely be accomplished analytically, it is common to adopt the use of "scalar waves." So universal is this procedure that the fact that it is an approximation is often overlooked. In the scalar wave approach it is assumed that the sizes of objects transmitting and shaping the beam are large on the scale of the infrared wavelength. In this approximation, a solution to the coupled Maxwell equations can be found in the form of a plane wave propagating (say) along the x_3 (or z) axis with $\mathbf{E}$ parallel to x_1 and $\mathbf{H}$ parallel to x_2.

Writing scalar field variables $E(z)$ and $H(z)$ and assuming for the moment an isotropic medium, the Maxwell equations (96), (102), and (103) can be separated in the form

$$\nabla^2 E = -(\omega^2\hat{\epsilon}/c^2)E \tag{105}$$

$$\nabla^2 H = -(\omega^2\hat{\epsilon}/c^2)H \tag{106}$$

with $H = (\epsilon\epsilon_o/\mu_o)^{1/2}E$, or its equivalent in cgs units $H = (\epsilon)^{1/2}E$. Solutions can be formulated as

$$E(z) = E_o \exp[i\omega(t - \hat{n}z/c)] \tag{107}$$

$$H(z) = H_o \exp[i\omega(t - \hat{n}z/c)] \tag{108}$$

where $\hat{n}$ is a complex refractive index related to the complex dielectric constant $\hat{\epsilon}$ by the equation $\hat{n}^2 = \hat{\epsilon}$.

The flow of field energy (or the rate at which energy flows across a unit area normal to the direction of propagation) is given by the Poynting vector, or radiation intensity,

$$\mathbf{I} = \mathbf{E} \times \mathbf{H}^* \tag{109}$$

$$\text{cgs:} \quad \mathbf{I} = (c/4\pi)\mathbf{E} \times \mathbf{H}^* \tag{109'}$$

(where the asterisk denotes a complex conjugate), the observable quantity at infrared frequencies being the time-averaged intensity.

Resolving $\hat{n}$ into real and imaginary components in the manner

$$\hat{n} = n - ik \tag{110}$$

it is customary to refer to n simply as the refractive index and k as the extinction coefficient or absorption index. Using the relationship $\hat{n}^2 = \hat{\epsilon}$ we obtain the equations

$$n^2 - k^2 = \epsilon' = \epsilon \tag{111}$$

and

$$2nk = \epsilon'' = \sigma/\epsilon_o\omega \tag{112}$$

$$\text{cgs:} \quad 2nk = \epsilon'' = 4\pi\sigma/\omega \tag{112'}$$

In these equations the scalar compliances ϵ and σ should, for crystals, be interpreted as the tensor components ϵ_{11} and σ_{11}. However, here we shall maintain an isotropic scalar formalism for simplicity. In the older literature one often finds $\hat{n}$ decomposed as $n(1 - i\kappa)$ in terms of which κ is also known as the extinction coefficient. Note that, since $\hat{n}^2 = \hat{\epsilon}$, the refractive index itself is *not* a tensor.

The terminology of "absorption index" for k arises because the intensity $I = EH^*$ can be expressed via Eqs. (107)–(109) as

$$I \propto \exp(-2\omega kz/c) \tag{113}$$

In other words, the intensity decreases exponentially with distance traveled in a manner controlled by the parameter k. For example, in traveling a distance equal to the wavelength in vacuum, λ, the intensity decreases by a factor $\exp(-4\pi k)$. It is also common to find this intensity decrease (or attenuation) expressed as

$$I \propto \exp(-\alpha z) \tag{114}$$

where α is called the loss coefficient and is a function of λ. This equation should be compared with Eq. (91) defining the radiance loss coefficient, which concerns an integrated loss over all wavelengths. For highly absorbing solids, such as metals, $1/\alpha$ or $4\pi/\alpha$ is often used as a measure of penetration or "skin depth"

Table 1.11 Permittivity and Conductivity Tensor Elements

Triclinic

ϵ_{11}	ϵ_{12}	ϵ_{13}	σ_{11}	σ_{12}	σ_{13}
ϵ_{12}	ϵ_{22}	ϵ_{23}	σ_{12}	σ_{22}	σ_{23}
ϵ_{13}	ϵ_{23}	ϵ_{33}	σ_{13}	σ_{23}	σ_{33}

(Diad $\parallel x_2$) Monoclinic (all classes)

ϵ_{11}	0	ϵ_{13}	σ_{11}	0	σ_{13}
0	ϵ_{22}	0	0	σ_{22}	0
ϵ_{13}	0	ϵ_{33}	σ_{13}	0	σ_{33}

Orthorhombic (all classes)

ϵ_{11}	0	0	σ_{11}	0	0
0	ϵ_{22}	0	0	σ_{22}	0
0	0	ϵ_{33}	0	0	σ_{33}

Tetragonal, Trigonal, and Hexagonal (all classes)

ϵ_{11}	0	0	σ_{11}	0	0
0	ϵ_{11}	0	0	σ_{11}	0
0	0	ϵ_{33}	0	0	σ_{33}

Cubic (all classes) and Isotropic

ϵ_{11}	0	0	σ_{11}	0	0
0	ϵ_{11}	0	0	σ_{11}	0
0	0	ϵ_{11}	0	0	σ_{11}

δ. Evidently, from the above equations, the loss coefficient is related to the absorption index k by

$$\alpha = 2k\omega/c = 4\pi k/\lambda \tag{115}$$

Combining this equation with Eq. (112), we see that the conductivity of a medium is intimately related to its loss (or attenuation) coefficient, namely,

$$\alpha = 4\pi k/\lambda = \sigma/c\epsilon_o n \tag{116}$$

$$\text{cgs:} \quad \alpha = 4\pi k/\lambda = 4\pi\sigma/nc \tag{116'}$$

In a general crystalline environment, the dielectric constant (and electrical conductivity) are symmetric second-rank tensors with at most six independent tensor elements. They can always be diagonalized in a principal axis coordinate frame, after which only three elements remain, for example, ϵ_{ii} ($i = 1, 2, 3$). The principal axes are the conventional crystalline axes for all point groups except the monoclinic and triclinic ones (Table 1.11). Using crystal axis coordinates, the independent tensor elements are shown in Table 1.6 for all symmetry classes. In the principal coordinate system of a transparent crystal, the principal values of refractive index n_i can be defined by the equations $n_i = (\epsilon'_i)^{1/2}$ in terms of the principal components ϵ'_i of dielectric constant assuming $n^2 >> k^2$ in Eq. (111). In accord with this definition, c/n_i are then called the principal velocities of light in the crystal.

In discussing optical properties of solids in terms of electronic structures, we shall, for the most part, consider linearly polarized light waves with their electric vectors parallel to one of the principal axes. The optical phenomena that occur for more general situations form the subject of crystal optics and are discussed in several well-known texts (e.g., Wahlstrom 1979). We see from Table 1.11 that $n_{11} = n_{22} = n_{33}$ for cubic crystals. As far as refractive index is concerned, such a crystal is therefore optically equivalent to an isotropic solid. For hexagonal, tetragonal, and trigonal crystal classes, $n_{11} = n_{22} \neq n_{33}$, and such crystals are said to be uniaxial. For them the quantity $n_{33} - n_{11}$ is often referred to as the uniaxial birefringence, and the x_3 axis is called the optic axis. A plane wave traveling along x_3 has its E vector perpendicular to x_3, and since $n_{11} = n_{22}$ it follows that its phase velocity and absorption index must therefore be independent of its polarization (or E direction). This latter property defines propagation along an *optic axis*.

In triclinic, monoclinic, and orthorhombic crystals, $n_{11} \neq n_{22} \neq n_{33}$. Such systems are said to be *biaxial*, because they possess two different directions (optic axes) for which propagation is independent of polarization. In biaxial crystals these directions do not coincide with any of the principal axes of the dielectric tensor and hence are not determined by crystal symmetry. Indeed, for triclinic and monoclinic systems, even the principal axes of ϵ are not fixed by crystal symmetry alone (see Table 1.11). In fact, these directions may all vary with frequency because they depend on the dielectric tensor elements, which, in general, are functions of frequency. Fortunately, most simple compounds crystallize in high-symmetry crystal lattice structures for which crystal symmetry plays the dominant role.

To this point we have considered only systems that are macroscopically homogeneous or quasi-homogeneous. The equivalent response for completely inhomogeneous materials (such as those consisting of a mixture of two or more kinds of small particles) is a more complex problem (Carr et al. 1985). For radiation wavelengths much smaller than the particle size, the response resembles some average of the separate responses of the individual components. However, at the opposite extreme, for which radiation wavelength is very large on the scale of particle size, the inhomogeneity disappears and the response becomes that of a well-defined effective homogeneous medium. In particular, for aggregates of small metal particles with insulating coatings [such as exists, for example, in gold smoke (Harris and Beasley 1952)], an insulator-to-conductor transition in dynamic response will take place as a function of decreasing radiation wavelength.

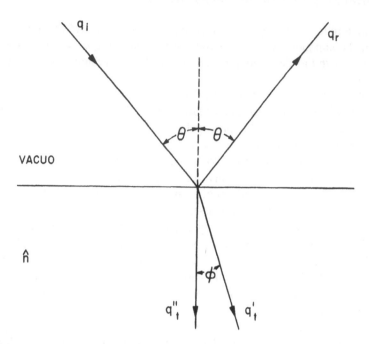

Figure 1.9 Depiction of an electromagnetic wave in vacuo incident upon an absorbing medium at a plane interface. The incident (reflected) wave vectors q_i (q_r) are both real and equal in magnitude. The transmitted wave vector is complex with components q_t' and q_t'', respectively (see text).

1.5.1. Reflection and Refraction

When an external plane wave is incident on the surface of a crystal, part of its intensity is transmitted and part is reflected. Explicit relationships can be derived for the amplitude, direction, and phase of the reflected and refracted waves from an analysis of the boundary conditions that prevail. Full tensor formalism requires knowledge of the direction of the optic axes with respect to the surface normal, and here we shall, for simplicity, revert to a scalar (isotropic medium) representation. If a plane wave in vacuo, with wave vector $\mathbf{q_i}$, is incident at an angle θ to the normal of the planar surface of a medium with complex refractive index $\hat{n} = n - ik$, then an undamped reflected wave $\mathbf{q_r}$ at a reflection angle θ is produced together with a damped refracted (or transmitted) wave $\mathbf{q_t} = \mathbf{q_t'} - i\mathbf{q_t''}$ at angle ϕ. The situation is shown schematically in Fig. 1.9.

The wave-vector boundary conditions determine ϕ, q_t', and q_t'' via the equations

$$q_i \sin \theta = q_t' \sin \phi,$$
$$q_i(\hat{n}^2 - \sin^2 \theta)^{1/2} = q_t' \cos \phi + i q_t'' \tag{117}$$

When $k = 0$ ($\hat{n} = n$; i.e., the medium is transparent), these equations reduce to $q_t' = nq_i$, $q_t'' = 0$, and to the well-known angular relationship known as Snell's law,

$$\sin \theta / \sin \phi = n \qquad (118)$$

By applying similar boundary conditions for **E** and **H**, we find the Fresnel equations for amplitude and phase. They are most concisely written in the form (Chantry 1984, Ditchburn 1976)

$$\hat{r}_{\parallel} = \frac{\hat{n} \cos \theta - \cos \phi}{\hat{n} \cos \theta + \cos \phi}, \qquad \hat{t}_{\parallel} = \frac{2 \cos \theta}{\hat{n} \cos \theta + \cos \phi}$$

$$\hat{r}_{\perp} = \frac{\cos \theta - \hat{n} \cos \phi}{\cos \theta + \hat{n} \cos \phi}, \qquad \hat{t}_{\perp} = \frac{2 \cos \theta}{\cos \theta + \hat{n} \cos \phi} \qquad (119)$$

in which $\hat{r}$ and $\hat{t}$ are the complex amplitude reflection and transmission coefficients E_r/E_i and E_t/E_i for electric field components parallel ($\parallel$) and perpendicular ($\perp$) to the plane of incidence. The division of intensity at the interface ($I_i = I_r + I_t$) is also easily derived as

$$I_r/I_i = (1/2)(|\hat{r}_{\parallel}|^2 + |\hat{r}_{\perp}|^2)$$

$$I_t/I_i = (n/2)(\cos \phi / \cos \theta)(|\hat{t}_{\parallel}|^2 + |\hat{t}_{\perp}|^2) \qquad (120)$$

where these two equations, using (119), are seen to sum to unity, as they must.

When the medium is close to transparent, Snell's and Fresnel's equations are often combined in the alternative forms

$$r_{\parallel} = \frac{\tan(\theta - \phi)}{\tan(\theta + \phi)}, \qquad t_{\parallel} = \frac{2 \sin \phi \cos \theta}{\sin(\theta + \phi) \cos(\theta - \phi)}$$

$$r_{\perp} = -\frac{\sin(\theta - \phi)}{\sin(\theta + \phi)}, \qquad t_{\perp} = \frac{2 \sin \phi \cos \theta}{\sin(\theta + \phi)} \qquad (121)$$

From these we see that when $\theta + \phi$ is a right angle, then $r_{\parallel} = 0$. This means that if an unpolarized beam of plane-wave radiation is incident at this special angle θ_B of incidence (known as the Brewster angle), then the reflected wave is 100% polarized. Using Snell's law, we easily find that $\theta_B = \tan^{-1} n$.

It follows that studies of reflected and transmitted radiation can be used to study both the real and imaginary components of refractive index or of dielectric constants, including their frequency dependencies. Observations of transmitted light are, of course, normally made in air, after the radiation has passed through both surfaces of a parallel plate (or lamellar) material sample. For highly absorbing media, such plates are necessarily thin; that is, with a thickness that does not greatly exceed the skin depth.

1.5.2.   Kramers–Kronig Relations

It can be shown that the real (ϵ') and imaginary (ϵ'') parts of the complex dielectric constant $\hat{\epsilon}$ are not completely independent quantities over their entire range of frequency response. The physical basis of this connection is the principle of causality—the fact that the response of a body at any point to an applied field $\mathbf{E}$ at time $t = 0$ can depend only on the fields at that point for $t \leq 0$.

Let $D(t)$ denote a component of electrical displacement $\mathbf{D}$ parallel to one of the principal axes of dielectric response as a function of time t. $E(t)$ then denotes the component of electric field $\mathbf{E}$ in the same direction. The general linear relation between these quantities is

$$D(t) = \int_0^\infty F(t')E(t - t') \, dt' \tag{122}$$

The physical significance of the response function $F(t)$ is made clear if we consider the response of the system to a single delta function pulse of electric field at $t = 0$. By direct use of Eq. (122) the response is given by $D(t) = F(t)$, from which it follows that $F(t)$ is precisely the electrical displacement due to this delta function stimulus. By causality it must be zero for $t < 0$, and physically it must also return to zero at long enough times $t \to \infty$. In between, however, it will have structure that depends on the dielectric response function $\hat{\epsilon}(\omega)$ of the system.

The exact relationship may be probed (Hogdson 1970) by considering the response to an oscillating field $E(t) = E_0 \exp(+i\omega t)$ for which Eq. (122) becomes

$$D(t) = E_0 \exp(i\omega t) \int_0^\infty F(t') \exp(-i\omega t') \, dt' \tag{123}$$

The relevant component of dielectric response $\hat{\epsilon}(\omega) = D(\omega)/E(\omega)$ therefore follows as

$$\hat{\epsilon}(\omega) = \int_0^\infty F(t) \exp(-i\omega t) \, dt \tag{124}$$

with real and imaginary components

$$\epsilon'(\omega) = \int_0^\infty F(t) \cos(\omega t) \, dt \tag{125}$$

$$\epsilon''(\omega) = \int_0^\infty F(t) \sin(\omega t) \, dt \tag{126}$$

Inverting these equations now gives an explicit relationship for the response function in the form

$$F(t) = \frac{1}{\pi} \int_0^\infty \left[\epsilon'(\omega) \cos(\omega t) + \epsilon''(\omega) \sin(\omega t) \right] d\omega \qquad (127)$$

The causality condition $F(t) = 0$ when $t < 0$ is therefore satisfied if for $t > 0$

$$\int_0^\infty \epsilon'(\omega) \cos(\omega t) \, d\omega = \int_0^\infty \epsilon''(\omega) \sin(\omega t) \, d\omega \qquad (128)$$

leading via Eq. (127) to the forms

$$\frac{\pi}{2} F(t) = \int_0^\infty \epsilon'(\omega) \cos(\omega t) \, d\omega = \int_0^\infty \epsilon''(\omega) \sin(\omega t) \, d\omega \qquad (129)$$

Using (129) in (125) and (126) now enables us to derive the Kramers–Kronig relationships between $\epsilon'(\omega)$ and $\epsilon''(\omega)$:

$$\epsilon'(\omega) = 1 + \frac{2}{\pi} \int_0^\infty ds \, \epsilon''(s) \left(\frac{s}{s^2 - \omega^2} \right) \qquad (130)$$

$$\epsilon''(\omega) = \frac{2}{\pi} \int_0^\infty ds \, \epsilon'(s) \left(\frac{\omega}{\omega^2 - s^2} \right) \qquad (131)$$

The significance of the 1 in the first of these equations is that it represents the value of $\epsilon'(\omega)$ as $\omega \to \infty$. In each integral, the principal part is implied.

Using Eqs. (111) and (112), the Kramers–Kronig relations can also be cast in terms of refractive index n and absorption index k in the form

$$n^2(\omega) - k^2(\omega) = 1 + \frac{2}{\pi} \int_0^\infty \frac{2sn(s)k(s)}{s^2 - \omega^2} \, ds \qquad (132)$$

$$2n(\omega)k(\omega) = \frac{2}{\pi} \int_0^\infty \frac{[n^2(s) - k^2(s)]\omega}{\omega^2 - s^2} \, ds \qquad (133)$$

If we know (say) a measured absorption coefficient as a function of frequency for all frequencies, then we can use these relations to evaluate the complete frequency dependence of refractive index without the need for further measurement. In Fig. 1.10 we show $n(\omega)$ and $k(\omega)$ for CdS as computed, using the Kramers–Kronig relations, from an experimental reflectance spectrum at normal

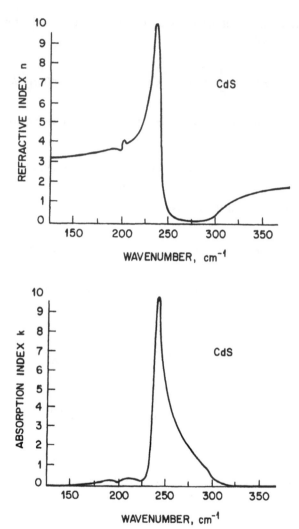

Figure 1.10 Real and imaginary parts of the Reststrahl response in CdS. [After Balkanski (1971).]

incidence [i.e., $r_{\parallel} = -r_{\perp} = (\hat{n} - 1)/(\hat{n} + 1)$ from the Fresnel equations (119) with $\theta = \phi = 0$].

Insulating and semiconductor crystals have, in general, two broad ranges of absorption as a function of frequency. The lower frequency range is associated with lattice vibrational (or Reststrahl) absorption bands and occurs in the far-infrared. The response shown in Fig. 1.10 is a typical example. The higher frequency absorption band usually occurs at or above optical frequencies and is associated with electronic transitions between the valence and conduction bands. Between these two regions there is usually an extended frequency range for which $\epsilon''(\omega)$ is small. In this range of quasi-transparency the optical properties are dominated by the real part of the dielectric response function. If $\epsilon''(\omega)$ can be represented as a sum of peaks at frequencies $s = \omega_i$, and the optical "window" frequency ω is far from these peaks, then the Kramers–Kronig relation (130) can be approximately recast as

$$\epsilon'(\omega) - 1 = \sum_i \frac{S_i \omega_i^2}{\omega_i^2 - \omega^2} \tag{134}$$

in which S_i is a dimensionless oscillator strength. This form is known as the *Sellmeier equation*, and it is also often seen in its vacuum-wavelength ($\lambda = 2\pi c/\omega$) equivalent form

$$\epsilon'(\lambda) - 1 = \sum_i \frac{S_i \lambda^2}{\lambda^2 - \lambda_i^2} \tag{135}$$

The number of terms i required for a satisfactory description of $\epsilon'(\omega)$ in the window regime depends on the material; but three terms are usually sufficient (two representing electronic oscillators and one representing lattice vibrations).

Another important concept derives from the high-frequency limit of Eq. (130). When ω is much greater than *all* the electronic resonances (including those involving the inner shell electrons), then we find the form

$$\epsilon'(\omega) = 1 - \omega_p^2/\omega^2 \tag{136}$$

in which

$$\omega_p^2 = \frac{2}{\pi} \int_0^\infty s\epsilon''(s)\, ds \tag{137}$$

is called the plasma frequency. In this limit the electronic behavior corresponds to that of a free electron gas for which a better-known form is (Ziman 1964)

$$\omega_p^2 = Ne^2/m\epsilon_0 \tag{138}$$

$$\text{cgs:} \qquad \omega_p^2 = 4\pi Ne^2/m \tag{138'}$$

where e and m are the charge and mass of an electron and N is the number of electrons per cubic centimeter.

1.6. CHEMICAL THERMODYNAMICS

In Section 1.2.3 we defined the eight thermodynamic potentials sufficient to determine a nonmagnetic macroscopic system with conjugate state variables

(T,S), (σ,η), and $(\mathbf{E},\mathbf{D})$. In discussions of chemical stability and phase changes, the most common experimental configuration is one for which applied electric fields are absent $(\mathbf{E} = \mathbf{0})$ and pressure $p = -\sigma_{ii}$ $(i = 1, 2, 3)$ is held constant. Expressed in terms of pressure p and volume V (instead of stress and strain) for an assumed isotropic material, we see from Eqs. (37) and (38) that under these conditions it is either the enthalpy

$$H = U + pV \qquad (139)$$

or Gibbs free energy

$$G = U + pV - TS \qquad (140)$$

which remains constant in the form

$$dH = T\,dS + V\,dp = 0 \qquad (141)$$

or

$$dG = -S\,dT + V\,dp = 0 \qquad (142)$$

depending on whether adiabatic $(dS = 0)$ or isothermal $(dT = 0)$ restrictions are maintained. It is also conventional in chemical thermodynamics to express energies per unit mass (usually per mole) rather than per unit volume, which has been more convenient for definitions of dielectric and elastic compliances in the earlier sections.

Thus, for example, in the case of a phase change at constant T and p, the Gibbs energy remains constant while the enthalpy changes. The change of enthalpy $dH = T\,dS = dQ$ then records the heat transfer that takes place during the phase change. Since sublimation, fusion, and vaporization all take place isothermally and isobarically (and can, at least in ideal circumstances, be conceived of as occurring reversibly), enthalpy change records heat transfer for all of these cases. Consequently enthalpy is often referred to as the *heat function*.

Between any two temperatures T_1 and T_2 we can more generally express the heat transfer at constant pressure as

$$H_2 - H_1 = \int_{T_1}^{T_2} c_p\,dT + \sum \Delta H_{\text{tr}} \qquad (143)$$

in which $c_p = (T\,dS/dT)_p$ is the specific heat (or heat capacity per unit mass) at constant pressure and ΔH_{tr} is the heat transfer taking place at each of the phase transitions (if any) that occur within this temperature range $T_1 \rightarrow T_2$. Also, since $(dS/dT)_p = c_p/T$, it is possible to evaluate the change in entropy $S_2 - S_1$ between two temperatures in the form

$$S_2 - S_1 = \int_{T_1}^{T_2} \frac{c_p}{T}\,dT + \sum \Delta S_{\text{tr}} \qquad (144)$$

with an obvious notation in analogy with Eq. (143). The most frequently tabulated values of $H_2 - H_1$ and $S_2 - S_1$ are for $T_2 = 25°C$ (i.e., room temperature) and $T_1 = 0°C$ with the constant pressure fixed at 1 atm (101,235 Pa).

Another frequent tabulation is for ΔHf_T and ΔGf_T, the energies (enthalpy and Gibbs, respectively) of formation at the temperature T (usually with $T = 0$ or T = room) [see, for example, Knight (1970)]. These quantities represent the change in enthalpy (or Gibbs free energy) when 1 gram formula weight of the solid in question is formed adiabatically (or isothermally) at the relevant temperature from the elements, each in its appropriate standard reference state (the condensed phase under a pressure of 1 atm). From these heats of formation it is then possible, by combining Eqs. (139) and (140) to the form $G = H - TS$, to determine an associated *thermal entropy* S_T^0 via the equation

$$Gf_T = Hf_T - TS_T^0 \qquad (145)$$

1.6.1. Vapor Pressure

If a liquid is introduced into an evacuated volume, it will tend to evaporate. If the volume is large enough, then the evaporation will progress to completion, leaving an unsaturated vapor. On a pressure–volume plot (as in Fig. 1.11) this state is represented by the point A. If the vapor is then compressed isothermally, the pressure will rise until a point is reached at which condensation begins (point B in Fig. 1.11). At this point the liquid and vapor are in equilibrium, and the associated pressure is called the *vapor pressure*. Condensation takes place as a volume change at constant temperature and pressure (line BC in the figure), and the equilibrium between liquid and vapor persists throughout. Finally, when condensation is complete (point C in the figure), further compression again increases the pressure in a manner (line CD in the figure) that monitors the PV behavior of the liquid state of this same temperature. The entire line ABCD is called an *isotherm*.

As the temperature is raised, the BC condensation part of the isotherm becomes shorter and shorter until, at a temperature $T = T_{crit}$, it vanishes and there is no longer any distinction between liquid and vapor. All points of type B (Fig. 1.11), at which the vapor is saturated, are said to lie on the liquid saturation curve. In the same manner, all points of type C, at which the liquid is saturated, are said to lie on the liquid saturation curve. These two curves meet at the *critical point*, which is a singular point on the "critical isotherm" defined by $T = T_{crit}$, $p = p_{crit}$, and $V = V_{crit}$.

At temperatures much lower than those shown on Fig. 1.11 the horizontal BC part of the isotherm now defines a phase change between the solid phase and the vapor. This portion of the isotherm then represents a sublimation transition in which the solid and the vapor are in equilibrium throughout. There must obviously, therefore, be one such BC line that exactly marks the boundary between the solid–vapor and liquid–vapor regimes. This particular line is associated with

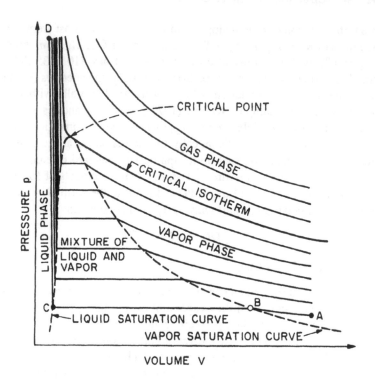

Figure 1.11   Isotherms of a pure substance.

a *triple point*, which is more easily sketched on a pressure–temperature diagram (Fig. 1.12). It represents a region in which vapor, liquid, and solid are all in equilibrium with each other.

Theoretical calculations of vapor pressures are easy to envisage in principle, but exceedingly difficult to carry out in practice with accuracy. To proceed, it is first necessary to redefine the thermodynamic potentials to include the possibility of diffusion from one phase to another. This is because we are now concerned with systems that are in diffusive as well as thermal contact and can consequently exchange matter as well as energy. If we consider a system in equilibrium under certain constraints that contains j different phases (with n_j moles per constituent phase j), then an increment dn_j to the jth phase increases the relevant thermodynamic potential by a proportional amount $\mu_j \, dn_j$ that defines a porportionality constant μ_j called the *chemical potential* (McClelland 1973).

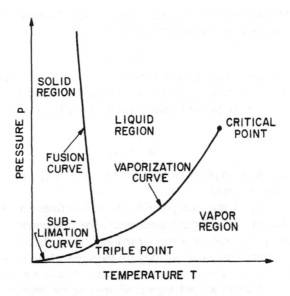

Figure 1.12 p–T diagram for a pure substance.

For example, if we perform a reversible process in which T, p, and n_j change by infinitesimal amounts dT, dp, and dn_j, respectively, then the resulting change in the Gibbs free energy is

$$dG = \frac{\partial G}{\partial T}\, dT + \frac{\partial G}{\partial p}\, dp + \frac{\partial G}{\partial n_j}\, dn_j \qquad (146)$$

Using Eq. (142) and the above definition of chemical potential, it follows that

$$\frac{\partial G}{\partial T} = -S, \qquad \frac{\partial G}{\partial p} = V, \qquad \frac{\partial G}{\partial n_j} = \mu_j \qquad (147)$$

so that Eq. (146) can be recast as

$$dG = -S\, dT + V\, dp + \mu_j\, dn_j \qquad (148)$$

which now denotes the expanded form of (142) in the presence of diffusion. The analogous extension of the enthalpy equation (141) is

$$dH = T\, dS + V\, dp + \mu_j\, dn_j \qquad (149)$$

It follows from Eq. (148) that, during a phase change at constant temperature and pressure, thermodynamic equilibrium exists if

$$\mu_j \, dn_j = 0 \tag{150}$$

where j is summed over the phases present. For example, in a phase transition between vapor (v) and solid (s), where we assume that the total system is closed, that is, $dn_v + dn_s = 0$, the equilibrium condition of Eq. (150) can be satisfied only if

$$\mu_v = \mu_s \tag{151}$$

If $\mu_s < \mu_v$, then the solid phase alone is stable. Conversely, if $\mu_s > \mu_v$, then the vapor phase alone is stable.

Now, at least in principle, it is possible to calculate from a microscopic model of the solid (by use of statistical mechanics) the free energy per constituent molecular unit of the solid. In a similar manner, using the statistical theory of gases we can calculate the same quantity for the vapor phase of the same formal chemical composition. The quantum-mechanical details are complicated (involving vibrational and rotational degrees of freedom and possibly, for the solid, additional complexities involving electronic band excitations) and obviously cannot be given here. However, the essential point is that these free energies are [e.g., via Eqs. (148) and (149)] directly related to μ_s and μ_v. In addition, the calculation for the vapor explicitly involves the pressure p. It follows that by equating these theoretical forms for μ_v and μ_s the resulting equation can be solved for p, which now by definition is the vapor pressure of the solid in question (Moelwyn-Hughes 1967). Similar calculations can be performed for liquid vapor pressure but are even more difficult in the light of the complexity of the liquid state.

1.6.2. Phase Changes

The most familiar phase changes, such as melting, vaporization, and sublimation, involve a heat transfer known as the latent heat of transformation. If the two phases in question are designated as 1 and 2, then this latent heat L can be related to the change of entropy S_2-S_1 at the transition temperature T via the equation $L = T(S_2$-$S_1)$. The existence of a latent heat and a corresponding entropy change implies, via the relationship $S = -(\partial G/\partial T)_p$, that such transitions involve discontinuities in the first derivatives of the free-energy functions. As a result they are known as first-order phase changes.

Writing the Gibbs function G as a function of p and T, we can, by use of the phase change equality $G_1 = G_2$, consider two neighboring isotherms in the "horizontal" transition region of Fig. 1.11 in the form

$$G_1(T,p) = G_2(T,p)$$
$$G_1(T + dT, p + dp) = G_2(T + dT, p + dp) \tag{152}$$

By subtraction and use of Eq. (142) it follows that

$$dG_1 = -S_1\, dT + V_1\, dp = dG_2 = -S_2\, dT + V_2\, dp \qquad (153)$$

from which we deduce the so-called vapor-pressure or Clapeyron equation (Zemansky 1951)

$$dp/dT = (S_2 - S_1)/(V_2 - V_1) = L/T(V_2 - V_1) \qquad (154)$$

In this equation it is essential to understand that dp/dT is not simply taken from the equation of state of either of the separate phases 1 and 2. It refers to the very special change of p and T that maintains the coexistence of the two phases.

If V_2 is a gas phase, then $V_2 \gg V_1$, and the equation of state of the vapor will be some perturbed form of the ideal gas relationship $pV_2 = RT$, where $R = Nk$ is the gas constant [8.314 J/(mole·deg)]. It follows that in this case Eq. (154) can be approximately recast as

$$dp/dT = Lp/RT^2 \qquad (155)$$

Integrating this equation between p_1, T_1 and p_2, T_2 gives

$$\ell n(p_2/p_1) = -(L/R)(T_2^{-1} - T_1^{-1}) \qquad (156)$$

which is called the Clausius–Clapeyron equation.

This equation is equally applicable for sublimation or vaporization (i.e., phase 1 equal to a solid or to a liquid, respectively) with an appropriately defined latent heat L. The implied curves, $\ln p = (-L/RT) +$ constant, for solid and liquid vapor pressures appear as straight lines on a $\ln p$ versus $1/T$ plot (Fig. 1.13) that meet at the triple point ($T = T_{tp}$, $p = p_{tp}$) at which the vapor pressure of both liquid and solid phases are identical. Equation (154), though not (155) and (156), is also valid for fusion. In this context dp/dT does not refer to vapor pressure, because no vapor is involved, but represents the reciprocal for the rate of change of melting point with pressure.

On approach to the critical point in Fig. 1.11, the latent heat $L \rightarrow 0$. In this limit the first-order derivatives of the Gibbs free energy are no longer discontinuous. Thus, for example, $S_1 = S_2$, where $S = -(\partial G/\partial T)_p$, and $V_1 = V_2$, where $V = (\partial G/\partial p)_T$. At the critical point itself it is the quantities involving the second derivatives of G that suffer discontinuities, such as the specific heat $c_p = T(\partial S/\partial T)_p = -T(\partial^2 G/\partial T^2)_p$ and thermal expansion $\beta = V^{-1}(\partial V/\partial T)_p = V^{-1}[\partial^2 G/(\partial p\, \partial T)]$. Such a phase change, with zero latent heat, is therefore called a second-order transition. Near such a transition we can write

$$T\, dS_1(T,p) = T\, dS_2(T,p) \qquad (157)$$

or equivalently

$$T\left(\frac{\partial S_1}{\partial T}\right)_p dT + T\left(\frac{\partial S_1}{\partial p}\right)_T dp = T\left(\frac{\partial S_2}{\partial T}\right)_p dT + T\left(\frac{\partial S_2}{\partial p}\right)_T dp \qquad (158)$$

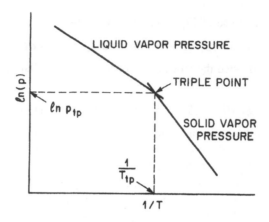

Figure 1.13 Graphical determination of the triple point from vapor pressure plots for solid and liquid phases.

Since $(\partial S/\partial p)_T = -(\partial V/\partial T)_p = \partial^2 G/(\partial T\, \partial p)$, we can use the above definitions of specific heat c_p and thermal expansion β to recast Eq. (158) in the form

$$(c_p)_1\, dT - TV\beta_1\, dp = (c_p)_2\, dT - TV\beta_2\, dp \tag{159}$$

or equivalently

$$\frac{dp}{dT} = \frac{(c_p)_2 - (c_p)_1}{TV(\beta_2 - \beta_1)} \tag{160}$$

In an exactly analogous manner, starting from $dV_1 = dV_2$ one can derive the alternative form

$$dp/dT = (\beta_2 - \beta_1)/(K_2 - K_1) \tag{161}$$

in which $K = V^{-1}(\partial V/\partial p)_T$ is the isothermal compressibility. Equations (160) and (161) together are known as Ehrenfest's equations.

Both first- and second-order phase transitions also occur in many guises within the solid state itself. Among the many examples we may cite ferromagnets, antiferromagnets, ferroelastics, ferroelectrics, antiferroelectrics, superconductivity, Jahn–Teller transitions, and order–disorder transitions in alloys. Microscopic theories of second-order phase transitions, which calculate the precise mathematical forms of the singularities of the various second-order Gibbs function derivatives on approach to a critical point, are now very well developed in the physics literature. Although different second-order transitions have many qualitative characteristics in common, there is no universal theory of critical instabilities, the different numerical exponents involved in the various singularities depending on the grosser characteristics (such as dimensionality and range) of the forces giving rise to the transition (Ma 1976).

One final comment should be made concerning the "phase transition" from supercooled liquid to glass. This "glass transition" at $T \approx T_g$ is not a phase change in the conventional thermodynamic sense but represents a slowing down of the fluid relaxational motion to a point at which the relaxation time becomes far greater than the time of observation (Brawer 1985). The glass phase is therefore, in a sense, a "snapshot" (frozen in time) of the fluid phase. The glass "transition" would not take place at all if the thermodynamic processes were carried out quasi-statistically. Both the supercoded liquid and glass represent phases that are not absolutely stable (they are said to be metastable). Cooling rates necessary to produce the glass phase vary enormously from material to material—from rates in excess of 10^6 K/s for many amorphous metals to as low as a few degrees per hour for the best glass formers. The glass transition itself is not sharp, like a first- or second-order transition would be, and no thermodynamic functions change discontinuously. They require instead an interval of a few degrees in the region of the glass transition to transform from their fluid to their glass forms.

1.6.3. Solubility and Durability

The general statistical mechanical and thermodynamic descriptions of systems with two or more components (i.e., solutions) are of widespread interest in physical chemistry. The simplest examples can be given for two-component systems, for which the component present in the greater amount is called the solvent and the component present in the smaller amount is called the solute. The solubility of a solid substance in a liquid can then be defined as the mass of dissolved substance contained as solute in a given volume of solution that is in equilibrium with an excess of the undissolved substance. In other words, it is the concentration of solute that can exist in equilibrium with the undissolved solute in a particular solvent at a particular temperature. This solubility can be measured as a (dimensionless) mole fraction or as the mass of solute per unit volume of solution.

Thermodynamically, solubility is most readily probed in the limit of perfectly random mixing. If there are n_1 moles of solute and n_2 moles of solvent, then it can be simply established from statistical mechanics (McClelland 1973) that the resulting entropy of perfect mixing is

$$\Delta S_{\text{mix}} = -R \sum_i n_i \, \ell n \, (x_i) \tag{162}$$

where $x_i = n_i/(n_1 + n_2)$ with $i = 1, 2$ is a mole fraction, and we note that ΔS_{mix} is a positive quantity because $x_i < 1$. Since, from Eq. (142), $\Delta S_{\text{mix}} =$

$-(\partial \Delta G/\partial T)_p$, it follows that the change in Gibbs free energy ΔG, to the extent that it is dominated by the entropy term (i.e., neglecting changes in internal energy U and volume V), is given by

$$\Delta G = RT \sum_i n_i \ell n \ (x_i) \tag{163}$$

Using Eq. (148) it now follows that the chemical potential for each of the two components can be expressed in the form

$$\mu_i = \mu_i^0 + RT \ \ell n \ (x_i) \tag{164}$$

in which μ_i^0 refers to the pure component i at the temperature and pressure of the solution.

If we measure solubility s as a mole fraction, then s is equal to that mole fraction x_1 of solute that is in equilibrium with the undissolved solute. From the general condition for equilibrium given by Eq. (150) involving the equality of chemical potentials, it follows that s is defined thermodynamically by the equation (Moelwyn-Hughes 1967)

$$(\mu_1)_{\text{solid}} = (\mu_1)_{\text{liq}} + RT \ \ell n \ s \tag{165}$$

where $(\mu_1)_{\text{liq}} = \mu_1^0$ is the chemical potential for the pure liquid form of component 1.

A free-energy calculation on model solid and liquid forms can now be used to deduce the relevant chemical potentials from fundamental microscopical quantities relating to the structure of solids and liquids, after which, by use of Eq. (165), the solubility s can be obtained in terms of the same quantities. Accurate estimates, on the other hand, are extremely difficult to obtain. Not only are microscopical calculations for chemical potentials extremely complicated, but Eq. (165) is itself extremely idealistic. Real solutions may differ in many ways from our simplistic assumptions. Not only can ΔU_{mix} and ΔV_{mix} contribute to the free energy of mixing, but the mixing itself may be far from ideal. Even in the absence of chemical reactions in the solution, interactions between dissimilar atomic or molecular constituents can cause a marked preferential clustering on a local scale.

Among compounds that dissociate completely in ionic form in solution, many dissolve to only a minute extent. In this situation the equilibrium between the ions in solution and the undissolved solid is often treated in terms of the law of mass action, in which the rates of forward and reverse reactions are equal. For example, the behavior of a slightly soluble salt $A_m B_n$ can be represented by the equilibrium

$$A_m B_n(\text{solid}) \rightleftharpoons m A^{n+}(\text{liq}) + n B^{m-}(\text{liq}) \tag{166}$$

which represents the dissociation of the ionic solid into A^{n+} cations and B^{m-} anions in solution. In such a situation the equilibrium concentrations are determined by the *solubility product* K_{sp} according to (Knight 1970)

$$K_{sp} = x_A^m x_B^n \qquad (167)$$

in which x_A and x_B signify the equilibrium mole fractions of cations and anions, respectively. If the solvent contains no ion in common with the solute, then the solubility s follows from the relationships

$$s = x_A/m = x_B/n \qquad (168)$$

Equating the Gibbs free energy difference per mole ΔG^o between dissolved and undissolved pure solvent to the entropy of mixing energy increment $T \Delta S_{mix}$ at equilibrium, we find

$$\begin{aligned} \Delta G^o &= RT(m \ln x_A + n \ln x_B) \\ &= RT \ln x_A^m x_B^n = RT \ln K_{sp} \end{aligned} \qquad (169)$$

from which it follows that the solubility product can be expressed directly in terms of this (negative) Gibbs energy difference ΔG^o in the form

$$K_{sp} = \exp(\Delta G^o/RT) \qquad (170)$$

In the procedure whereby a solid dissolves in a solvent, the concept of solubility is concerned only with the end (that is, equilibrium) composition of the saturated solution. Of greater importance as a material property is the rate at which the solid dissolves, particularly the initial rate when the solvent is pure. The latter can be measured either as a mass loss per unit surface area per unit time or as a decrease in sample thickness per unit time. The two measures are obviously simply related, and either can be used as a measure of *chemical durability*. From a fundamental theoretical point of view, chemical durabilities are even more difficult to calculate than solubilities, because they involve nonequilibrium thermodynamic concepts, and few computations from first principles are yet to be found in the literature. Experimentally, on the other hand, measurements are easy to perform as functions of temperature and pressure, for a wide variety of solvents.

REFERENCES

Atkins, A.G. (1985). *Elastic and Plastic Fracture*, Halsted Press, New York.
Balkanski, M. (1971). *Light Scattering in Solids*, Flammarion, Paris.
Barrett, C.R., Nix, W.D., and Tetelman, A.S. (1973). *The Principles of Engineering Materials*, Prentice-Hall, Englewood Cliffs, N.J.

Ben-Shaul, A., Haas, Y., Kompa, K.L., and Levine, R.D. (1981). *Lasers and Chemical Change*, Springer-Verlag, Heidelberg.

Bleaney, B.I., and Bleaney, B. (1976). *Electricity and Magnetism*, Oxford Univ. Press, Oxford.

Boardman, A.D., O'Connor, D.E., and Young, P.A. (1973). *Symmetry and Its Applications in Science*, Wiley, New York.

Bradfield, G. (1970). *Ultrasonics 8:*112.

Brawer, S. (1985). *Relaxation in Viscous Liquids and Glasses*, Am. Ceram. Soc., Columbus, Ohio.

Carr, G.L., Perkowitz, S., and Tanner, D.B. (1985). In: *Infrared and Millimeter Waves* (K.J. Button, Ed.), Academic, New York, Vol. 13, Pt. IV, p. 171.

Chantry, G.W. (1984). *Long-Wave Optics*, Academic, New York, Vol. 1.

Christensen, R.M. (1982). *Theory of Viscoelasticity: An Introduction*, Academic, New York.

Ditchburn, R.W. (1976). *Light*, Wiley, New York.

Hadley, D.W., and Ward, I.M. (1975). *Rep. Progr. Phys. 38:*1143.

Harris, L., and Beasley, J.K. (1952). *J. Opt. Soc. Am. 42:*134.

Heitler, W. (1954). *The Quantum Theory of Radiation*, Clarendon Press, Oxford.

Hodgson, J.N. (1970). *Optical Absorption and Dispersion in Solids*, Chapman and Hall, London.

Jastrzebski, Z.D. (1976). *The Nature and Properties of Engineering Materials*, Wiley, New York.

Kittel, C. (1976). *Introduction to Solid State Physics*, Wiley, New York.

Knight, A.R. (1970). *Introductory Physical Chemistry*, Prentice-Hall, Englewood Cliffs, N.J.

Kressel, H. (1981). In: *Fundamentals of Optical Fiber Communications* (M.K. Barnoski, Ed.), Academic, New York, p. 187.

Landau, L.D., and Lifshitz, E.M. (1958). *Statistical Physics*, Pergamon, London.

Lines, M.E., and Glass, A.M. (1977). *Principles and Applications of Ferroelectrics and Related Materials*, Clarendon Press, Oxford.

Ma, S.-K. (1976). *Modern Theory of Critical Phenomena*, Benjamin, Reading, Mass.

Maeda, M. (1984). *Laser Dyes*, Academic, New York.

Margenau, H., and Murphy, G.M. (1956). *The Mathematics of Physics and Chemistry*, Van Nostrand, Princeton, N.J.

McClelland, B.J. (1973). *Statistical Thermodynamics*, Chapman and Hall, London.

Moelwyn-Hughes, E.A. (1967). *A Short Course in Physical Chemistry*, Elsevier, New York.

Narasimhamurty, T.S. (1981). *Photoelastic and Electro-Optic Properties of Crystals*, Plenum, New York.

Nemback, E. (1975). In: *Treatise on Solid State Chemistry* (N.B. Hannay, Ed.), Plenum, New York.

Nye, J.F. (1960). *Physical Properties of Crystals*, Oxford University Press, Oxford.

Pipken, A.C. (1986). *Lectures on Viscoelastic Theory*, Springer-Verlag, Berlin.

Pollard, H.F. (1977). *Sound Waves in Solids*, Pion, London.

Sessler, A.M., and Vaughan, D. (1987). *Am. Sci. 75:*34.

Wahlstrom, E.E. (1979). *Optical Crystallography*, Wiley, New York.
Wilson, J., and Hawkes, J.F.B. (1983). *Optoelectronics: An Introduction*, Prentice-Hall, London.
Zemansky, M.W. (1951). *Heat and Thermodynamics*, McGraw-Hill, New York.
Ziman, J.M. (1964). *Principles of the Theory of Solids*, Cambridge Univ. Press, Cambridge.

2

Interaction of Light with Matter: Theoretical Overview

Malcolm E. Lines, *AT&T Bell Laboratories, Murray Hill, New Jersey*

2.1. MICROSCOPIC THEORY OF DISPERSION AND ABSORPTION IN INSULATORS

2.1.1. Polaritons and the Reststrahl

In the macroscopic approach to radiation propagation developed in Section 1.5, the origins of the mechanisms that produce attenuation were not detailed. To do this requires a microscopic discussion of the relevant elementary excitations with which the radiation interacts. In insulators and semiconductors these are of two distinct kinds, involving lattice vibrational modes and electronic excitations, respectively. Here we consider first the interactions with lattice vibrations (or phonons) that are responsible for the strong Reststrahl absorption bands in the far-infrared. That subset of optic phonons that involves a transverse fluctuating electric dipole (polar phonons) can couple directly to the $\mathbf{E}$ component of electromagnetic radiation. The result is a mixed phonon–photon mode termed a *polariton* with a frequency that depends both on the wave vector $\mathbf{q}$ and on polarization (Mills and Burstein 1974).

Phonon modes in a solid can be separated into optical and acoustic branches with dispersion relations defined throughout the first Brillouin zone of the relevant reciprocal lattice. The acoustic mode has a frequency that goes to zero as the wave vector $\mathbf{q} \to 0$ (with a slope that defines acoustic velocity), whereas the optical modes all approach finite frequency values in this same limit. The first important point to recognize in discussing interactions with electromagnetic radiation is that the characteristic wave-vector magnitude $q = \omega/c$ for infrared radiation is only some 10^{-5} times that of a typical Brillouin zone dimension. It follows that, in the present context, our description of pure optic phonons can be taken to be completely independent of q.

Accordingly, we can adequately represent the motion of an optical mode lattice vibration by the relevant motion of ions in a representative unit cell of the lattice. Let us consider the jth polar mode of the mth lattice cell of a crystal lattice and assume it to be of damped harmonic form (Lines and Glass 1977)

$$\ddot{\xi}_{j,m} + \omega_j^2 \, \xi_{j,m} + \gamma_j \, \dot{\xi}_{j,m} = Z_j E_m \tag{1}$$

where $\xi_{j,m}$ is a normal-mode displacement coordinate, $\omega_j^2 \, \xi_{j,m}$ is a restoring force, $\gamma_j \, \dot{\xi}_{j,m}$ is a damping force, Z_j is an effective charge, and E_m is the Maxwell field at the site of the jth cell. For mathematical simplicity we adopt a scalar formalism that is adequate as long as we recognize that each set of parameters can take up to three different values corresponding to the three principal axes of the response tensor.

In examining the details of mixed phonon–photon (or polariton) response in the Reststrahl frequency regime, it is usually sufficient to lump all the (primarily ultraviolet) electronic oscillators together into a single effective high-frequency oscillator, which, for the moment, may also be formally included in any summation over j. With this formalism, the instantaneous value of electronic polarization (dipole moment per unit volume) is

$$P_m = \sum_j \frac{z_j \, \xi_{j,m}}{v} \tag{2}$$

in which v is the unit-cell volume and z_j are also effective charges [which may differ from Z_j owing to local field effects (Born and Huang 1954)]. Looking for plane-wave solutions,

$$P_m, E_m, \, \xi_{j,m} \sim \exp[i(\omega t - qm)] \tag{3}$$

with respective amplitudes P, E, ξ_j, we can solve Eq. (1) for amplitude ξ_j and insert it into Eq. (2) to get the relationship

$$P = \sum_j \left\{ \frac{z_j Z_j / v}{\omega_j^2 - \omega^2 + i\omega\gamma_j} \right\} E \tag{4}$$

between polarization and field amplitudes P and E. Using the definition

$$\epsilon_o E + P = \hat{\epsilon}\epsilon_o E \tag{5}$$

$$\text{cgs:} \qquad E + 4\,\pi P = \hat{\epsilon}E \tag{5'}$$

of dielectric response $\hat{\epsilon}$ now leads directly to

$$\hat{\epsilon} = 1 + \sum_j \frac{S_j \omega_j^2}{\omega_j^2 - \omega^2 + i\omega\gamma_j} \tag{6}$$

in which the parameters

$$S_j = z_j Z_j / \omega_j^2 v \epsilon_o \qquad (7)$$

$$\text{cgs:} \qquad S_j = 4\pi z_j Z_j / \omega_j^2 v \qquad (7')$$

are introduced as convenient dimensionless measures of mode strength. Note that the real part of Eq. (6) is essentially the Sellmeier equation (134) of Chapter 1, but now with a microscopic formalism for oscillator strength.

For an interpretation of the infrared (Reststrahl) absorption bands it is acceptable to assume $\omega \ll \omega_\infty$, where $j = \infty$ formally labels the effective electronic oscillator. We can accordingly replace the high-frequency electronic contribution to $\hat{\epsilon}$ by its real limiting value S_∞. This, in turn, is conventionally included with the vacuum term to define a high-frequency or "electronic" dielectric constant $\epsilon_\infty = 1 + S_\infty$. High frequency in this context means high with respect to Reststrahl frequencies but low in comparison to exciton or other electronic frequencies. The dielectric function in the Reststrahl range now becomes

$$\hat{\epsilon} = \epsilon_\infty + \sum_j{}' \frac{S_j \omega_j^2}{\omega_j^2 - \omega^2 + i\omega\gamma_j} \qquad (8)$$

in which $\sum'$ now runs over only polar phonon modes.

Solution of the Maxwell equations of Section 1.5 (Chapter 1) are now of two types depending on whether $\hat{\epsilon} = 0$ or $\nabla \cdot \mathbf{E} = 0$ in the $\nabla \cdot \mathbf{D} = 0$ Eq. (96) of Chapter 1. The former are q-independent modes with $\mathbf{E} \parallel \mathbf{q}$ (longitudinal or LO modes), while the latter have $\mathbf{E} \perp \mathbf{q}$ and lead to wave equations (105) and (106), of Chapter 1, which describe a transverse wave motion with dispersion

$$q^2 c^2 = -\omega^2 \hat{\epsilon} \qquad (9)$$

with $\hat{\epsilon}$ given by Eq. (8), above. These modes are mixed phonon–photon modes, and their quanta are referred to as polaritons.

The general character of polariton frequency dispersion is most easily envisaged for the case of negligible damping and is shown schematically in Fig. 2.1 for the case of three undamped phonons. The curves show regions of photonlike behavior (i.e., constant slopes) at very high and very low frequencies, with slopes determined by the high- and low-frequency refractive indices $n(\infty)$ and $n(0)$, respectively, joined continuously to horizontal phononlike curves at intermediate frequencies. The latter, marked ω_{Tj} in Fig. 2.1, are the transverse optic (TO) phonons with frequencies equal to the ω_j of Eq. (1) in the case of negligible damping. The frequencies marked ω_{Lj} in the figure are those of the pure longitudinal optic (LO) phonons whose dispersionless branches are shown dashed in the plot.

The electromagnetic waves are seen to couple only to the TO phonons. It follows that the LO modes are not normally observed in a direct fashion by infrared spectroscopy. We note also that there are gaps (where the vibrations are

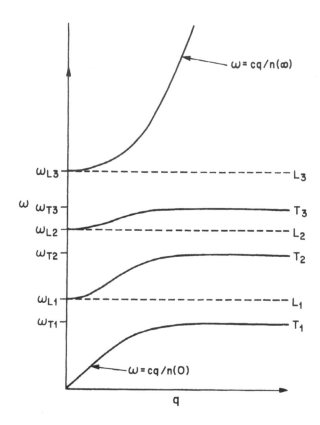

Figure 2.1 A schematic of longitudinal (L_i) and transverse (T_i) polariton dispersion for the case of three undamped modes $i = 1,2,3$; ω_{Li} labels longitudinal mode frequency and ω_{Ti} transverse mode frequency in the elastic limit.

completely damped) between pairs ω_{Tj} and ω_{Lj} of TO and LO frequencies. Near the TO frequencies the energy of the coupled modes (or polaritons) is carried primarily by the elastic part of the field, whereas at very high frequencies the ions cannot take part in the motion to any appreciable extent (owing to their large inertia) and the solutions are essentially radiative. The corresponding curves for the dielectric constant $\hat{\epsilon}(\omega)$ are shown in Fig. 2.2, and an actual experimental spectrum in Fig. 2.3.

Figure 2.3 shows the transverse solutions for the three modes that fit the dielectric spectrum of $SiTiO_3$ in the Reststrahl regime. They are plotted for real frequencies, and the figure clearly shows the photonlike behavior at high and low frequencies. Now, however, there are no completely nonpropagating gaps, and

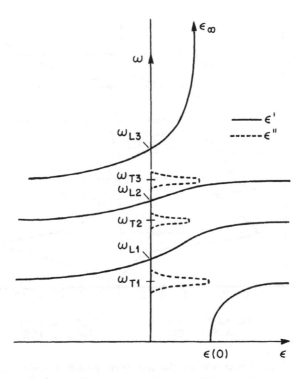

Figure 2.2 The real and imaginary parts of the dielectric function $\epsilon(\omega) = \epsilon' - i\epsilon''$ for the three-mode model of Fig. 2.1. Some damping has now been included to prevent ϵ'' from shrinking into a series of δ functions.

transverse vibrational modes exist at all frequencies although there are still regions where these modes are highly damped.

The polariton frequencies are defined as the poles (or divergencies) of the dielectric function. Thus, from Eq. (8), they take the form

$$\omega(\text{polariton}) = \left(\omega_j - \frac{1}{4}\gamma_j^2 \right) + \frac{1}{2}i\gamma_j \qquad (10)$$

For a quantitative interpretation of measured infrared spectra, the damping parameter γ_j will generally be frequency-dependent. The physical origin of this damping is the coupling of the phonon branch in question to the other phonon branches. It is, for example, not difficult to conceive of a situation in which there is a high density of these coupled phonons at one or more particular frequencies. In such cases γ_j will tend to peak at these special frequencies to reflect the enhanced leakage of energy from the mode in question (j) near these resonances.

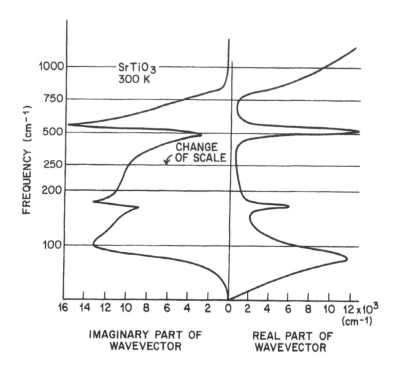

Figure 2.3 Dispersion curves for the transverse polariton modes in SrTiO₃. The curves (plotted for real frequency) are to be compared with the idealistic model of Figs. 2.1 and 2.2. [After Barker (1967).]

Monatomic lattices, such as that of pure diamond, possess no polar optic phonon branches and therefore do not absorb in the infrared in lowest order. It is this feature (together with its strength and durability) that makes diamond such a valuable infrared window material.

2.1.2. Multiphonon Absorption

Besides the main Reststrahl bands, ionic crystals have other weaker absorption bands in the infrared that can be explained by processes in which a photon is absorbed by the lattice with the creation of two or more phonons. Such effects are produced both by the anharmonicity of the basic phonon modes and by any nonlinearity in the relationship between P and E (which defines higher order susceptibilities). The relative importance of the two mechanisms in specific instances is still a much debated question.

In general, two-phonon absorption can be centered at frequencies that are either the sum or the difference of phonon frequencies that correspond to the classical combination frequencies of normal modes mixed by nonlinearities (Mitra 1985). The sum frequency absorption corresponds to the creation of two phonons, whereas the difference frequency absorption corresponds to the creation of one phonon and the annihilation of another. Even in homopolar crystals (like silicon), which do not exhibit first-order Reststrahl bands, these combination bands can arise via electrical nonlinearities (Johnson 1959; Lax and Burstein 1955) although their integrated intensity is extremely small compared to ionic Reststrahl bands. More commonly they are observed as wings, or structured tails, on either side of the Reststrahl bands.

Most detailed study has been carried out for the sum resonances contributing to the high-frequency "multiphonon edge," which tails off into the optic window from the top of the Reststrahl band. Although pronounced structure is usually observed on this "edge," reflecting density of states peaks in the characteristics of the phonon band dispersion, there is a persistent progressive exponential decrease in intensity as one progresses out from the primary first-order Reststrahl bands through the regimes dominated by two-, three-, four-, . . . phonon processes. An example is shown for ZnSe in Fig. 2.4.

Not surprisingly, the detailed theory of such high-order nonlinear processes is rather complicated, and derivations from first principles have met with limited success. Nevertheless, a qualitative understanding of the underlying exponential form of the edge is not difficult to grasp from the most general arguments. For example, we recognize that the absorption in any multiphonon spectral regime will be dominated by the lowest order processes (that is, the minimum number of phonons) consistent with energy conservation. This immediately implies that in the n-phonon summation regime $\omega \approx n\omega_0$ (where ω_0 is a typical optic phonon frequency) the loss coefficient α will be dominated by a term $\alpha_n \sim g^n$, where g is a measure of the dominant nonlinearity or anharmonicity present. This immediately leads us to anticipate a factor $\exp(-b\omega/\omega_0)$ in the multiphonon tail, where b is a frequency-independent material parameter.

Turning to temperature dependence, one also expects from quite general arguments that $\alpha_{n+1}/\alpha_n \sim n(\omega_0) + 1$, where $n(\omega_0)$ is the Bose–Einstein thermal occupancy function

$$n(\omega_o) = [exp \ (\hbar\omega_o/kT \ ) \ - \ 1]^{-1} \tag{11}$$

implying the presence of an additional factor $[n(\omega_0) + 1]^{\omega/\omega_o}$. A more detailed analysis (Bendow 1977, 1978; Lipson et al. 1976) leads to the form

$$\alpha(\omega,T) = A_o \frac{[n(\omega_o) + 1]^{\omega/\omega_o}}{n(\omega) + 1} \exp\left[-\frac{b\omega}{\omega_o}\right] \tag{12}$$

with A_o a frequency-independent material parameter.

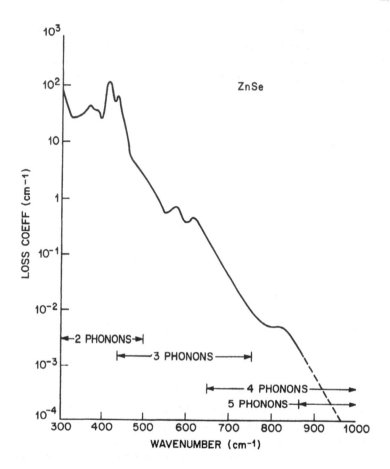

Figure 2.4   The infrared multiphonon absorption edge of ZnSe in the two-phonon to five-phonon regime. [After Miles 1977).]

The frequency dependence of the exponential prefactor is quite modest compared with that of the exponential itself, and for room-temperature spectra (particularly in glasses for which lack of translational invariance diminishes the tail structural features), the simple approximation

$$\alpha(\text{multiphonon}) = A \exp(-a/\lambda) \qquad (13)$$

often suffices, where A and a are constant for any particular sample and $a = (2\pi c/\omega_o)b$ correlates well with estimates of ω_o obtained from a simple electrostatic model of polar vibrations (Lines 1986).

2.1.3. Impurities and Induced Polarization

The introduction of defects into a crystal lattice has two primary effects on the normal modes of lattice vibration. One is a perturbation of their frequencies (or energies), and the other is a perturbation of their mode strengths (Bell 1972, Mermin 1979). The latter effect is particularly dramatic in crystals that have no first-order Reststrahl bands (that is, are optically inactive to first order) in their pure state. Impurities or defects destroy lattice symmetry and modify the electronic structure in their vicinity in a manner that creates a polar character if none was present before. Under electromagnetic stimulation this then induces an electronic moment that couples directly to the E-vector of the radiation. The result is an induced first-order Reststrahl absorption. If the induced moment does not seriously perturb the normal-mode lattice vibrational energies of the pure crystal, then the outcome is a sequence of optical absorption maxima close to the fundamental optical vibrational frequencies of the pure crystal—or, more accurately, at each peak in the density of optical modes as a function of frequency. The impurity or defect therefore induces a "defect optical activity" in each of the previously inactive modes and renders them visible to infrared spectroscopy via a first-order process.

Generally, however, the introduction of defects also perturbs the normal-mode frequencies of the bulk lattice modes (Elliott et al. 1974). Again, two qualitatively different manifestations can be distinguished. The first concerns those bulk modes that lie well within the vibrational bands of the pure system and are consequently easily propagated. These are usually only very slightly perturbed, although in some cases a resonance can appear that is characterized by an increase in the defect vibrational amplitude. The second, and much more dramatic, effect concerns those bulk lattice modes near the edge of the phonon bands of the pure crystal. In the presence of a defect or impurity these modes may be very markedly shifted in frequency, being pulled away from the band entirely. They represent "localized modes" that do not propagate at all but remain pinned to the defect site with an amplitude that decreases rapidly as a function of distance from the site. Such localized modes are most likely to appear in the presence of impurity atoms that are very light compared to the mass of the host atoms. In this case they pull modes off the "top" (that is, the high-frequency end) of the bulk vibrational bands and into the multiphonon wing of the optic window. Being optically active, they produce absorption peaks at the localized mode frequencies and hence give rise to features that are not related in any obvious fashion to the bulk phonon modes of the pure crystal (Mitra 1985).

As the concentration of defects is increased, the various infrared response features broaden and increase in complexity. Eventually the physical picture goes over to that of a mixed crystal or alloy if translational periodicity is maintained in the atomic site positions or to a multicomponent glass if it is not. In

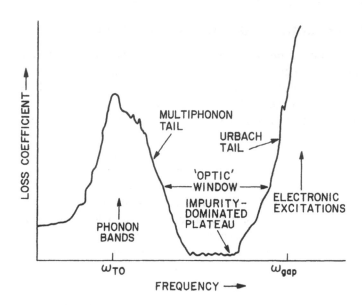

Figure 2.5 Schematic attenuation spectrum for a typical heteropolar crystal in the infrared/visible range of frequency response.

either case the full details of the Reststrahl bands become immensely complex. In some simple binary alloys (GaP_xAs_{1-x}, for example), the major features of the resulting Reststrahl bands do approximate the sum of the component Reststrahl spectra (weighted by composition). In others (such as $Zn_xCd_{1-x}S$), a basically single-mode spectrum is maintained with a frequency that varies continuously between the end members.

In all cases, however, as the radiation frequency increases from the Reststrahl regime, one finally encounters the multiphonon edge leading to an optic window for which absorption is small. Within this window, which persists until the onset of bulk electronic excitations (Fig. 2.5), the major sources of optical attenuation for a pure crystal would not result from absorption mechanisms at all, but from scattering processes (see Section 2.3). In real crystals, however, window regime attenuation is normally impurity-dominated, this time by the very low frequency (with respect to the bandgap) localized electronic excitations that can take place within unfilled d or f shells of transition metal or rare earth contaminants, or by the very high frequency (with respect to the Reststrahl) vibrational excitations of molecular contaminants containing hydrogen.

Well inside the impurity-dominated optic window regime, the refractive index [via the Sellmeier (135) equation of Chapter 1] passes through a point of inflection $d^2n/d\lambda^2 = 0$ as it progresses from a phonon-dominated dielectric response toward an electron-dominated one. This "zero material dispersion" point $\lambda = \lambda_0$ is of great importance in the field of fiber optics (Nassau and Wemple 1982), because it is the wavelength for which light pulses spread the least as they propagate.

2.1.4. Electronic Interband Absorption

If a large number of similar atoms are brought together to form a crystal, each single atomic energy level is broadened by the interatomic interactions into a band of electronic levels that can be thought of as a continuum of energy states. For inner-shell electrons the bandwidths are narrow. On the other hand, the outermost electrons are subjected to large interaction energies of the same order of magnitude (several electronvolts) as the excitation energies of the outer electrons in the individual isolated atoms. It is a description of these broad outermost-electron bands that is necessary in considering the interaction of electromagnetic radiation with *electrons* in solids. We shall first concentrate on nonmetals, for which the individual bands are separated in energy and are formally completely filled or empty.

In Fig. 2.6 we depict schematically the development of electronic band structure as a crystal like CaS is assembled starting from a gas of neutral argon atoms. In Ar gas the energy levels are atomic, filled up to the $3s$ and $3p$ levels. The lowest empty level is $4s$, and above this (labeled "vacuum" in the figure) is the ionization energy necessary for a current to flow. As we bring the Ar atoms together to form a cubic lattice with a nearest-neighbor spacing equal to that of CaS, each sharp electronic level broadens into a band. The energy now required to generate a current is that between the top of the filled $3p$ "valence" band and the bottom of the empty $4s$ "conduction" band. This energy is called the "bandgap" for the solid and is a threshold energy above which electronic interband absorption can occur. Possible excitations to "bound electron levels" or "excitons" below the bandgap are neglected for the moment (but see Section 2.1.6). Finally we transfer protons between adjacent nuclei to transform the solid Ar lattice to CaS. This causes a redistribution of electrons that splits each Ar band into two (see Fig. 2.6). The final bandgap E_g is then the one between the top of the $S(3p)$ valence band and the bottom of the Ca($4s$) conduction band.

In a more accurate picture the band wave functions become mixed to some extent in the sense that, for example, the formally $S(3p)$ band is no longer exactly made up of solely $S(3p)$-based wave functions. Indeed, for strongly covalent materials this mixing may be so strong that a more realistic description can be cast in terms of linear combinations of atomic states (i.e., hybridized orbitals such as the tetrahedrally directed sp^3-hybridized orbitals in Si or AlP), with the bands labeled accordingly (Harrison 1980). Nevertheless, the essential picture of a formally filled valence band and an empty conduction band is maintained.

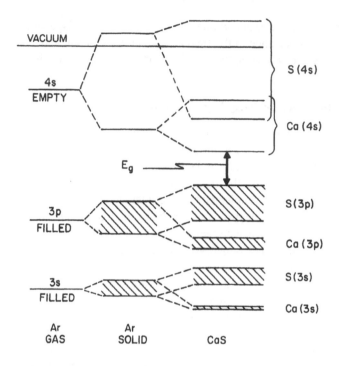

Figure 2.6 Schematic of the manner in which the electronic bands of CaS result by compaction and nuclear charge transfer from the sharp levels of Ar gas (see text). Filled bands are hatched.

The wave functions and the energies of valence or conduction electrons in crystals can be calculated by several approximate methods, all in agreement as to the qualitative forms. Each electron is usually assumed to move in the potential field of the nuclei plus the averaged distribution of other electrons. This "one-particle" approximation leads to a Schrödinger equation for a single-particle wave function ψ. The potential function $V(r)$ in this equation has the periodicity of the lattice, and the wave functions consequently possess this same property. They can be expressed in the so-called Bloch form,

$$\psi_{\mathbf{k}}(\mathbf{r}) = e^{i\mathbf{k}\cdot\mathbf{r}}u(\mathbf{k},\mathbf{r}) \tag{14}$$

in which $u(\mathbf{k},\mathbf{r})$ has the full periodicity of the lattice. The corresponding energy levels $E(\mathbf{k})$, in which $\mathbf{k}$ is a wave vector, have discontinuities on certain planes in $\mathbf{k}$-space that are determined by the symmetry of the lattice and define the boundaries of Brillouin zones. The first Brillouin zone is that zone which contains all reciprocal space closer to the origin $\mathbf{k} = 0$ than to any other reciprocal lattice vector. It is often convenient to consider only this first Brillouin zone, because Eq. (14) can be reexpressed solely within it in the form

$$\psi_{s,\mathbf{k}}(\mathbf{r}) = e^{i\mathbf{k}\cdot\mathbf{r}}U_s(\mathbf{k},\mathbf{r}) \tag{15}$$

with s being a band index, the corresponding energy $E_s(\mathbf{k})$ having a separate branch for each band. The band index s, as well as the special high-symmetry locations within the first Brillouin zone, are usually labeled in a fashion that designates the group theoretical (or symmetry) properties of the mode and lattice (Jones 1960). Some calculated energy bands are shown in Fig. 2.7 for Ge.

The calculation of the one-electron optical properties arising from the interaction of radiation with these energy bands can now be carried out by treating the electromagnetic field as a time-dependent perturbation that induces excitations from filled valence bands to empty conduction band levels. The first requisite is a knowledge of the manner in which **E** and **H** fields interact with electrons. The force **F** on an electron of charge e due to an electromagnetic field is given by (Bleaney and Bleaney 1976)

$$\mathbf{F} = e(\mathbf{E} + \mathbf{v}\times\mathbf{B}) \tag{16}$$

$$\text{cgs:} \qquad \mathbf{F} = e[\mathbf{E} + (\mathbf{v}/c) \times \mathbf{B}] \tag{16'}$$

where **v** is the electron velocity. This force **F**, in the equation of electronic motion, must be equated with an effective rate of change of momentum due to the field.

The formalism is given its most succinct form by introducing the electromagnetic vector potential **A**, which is defined by

$$\mathbf{B} = \mu_0\mu\cdot\mathbf{H} = \nabla \times\mathbf{A} \tag{17}$$

$$\text{cgs:} \qquad \mathbf{B} = \mu\cdot\mathbf{H} = \nabla \times\mathbf{A} \tag{17'}$$

Using Maxwell's equation [(95) of Chapter 1], this implies

$$\mathbf{E} = -\partial\mathbf{A}/\partial t - \nabla\phi \tag{18}$$

$$\text{cgs:} \qquad \mathbf{E} = -(1/c)\,\partial\mathbf{A}/\partial t - \nabla\phi \tag{18'}$$

where ϕ is called the scalar potential. Since mathematically $\nabla \times \nabla \phi = 0$ for any scalar function ϕ, it follows that **A** and ϕ are not uniquely related to **E** and **H**. The different possible choices are called gauges, and the invariance of **E** and **H** under these transformations is called gauge invariance. Using a gauge in which ϕ is set equal to zero, and considering the plane wave $\mathbf{E} = \mathbf{E}_o$ $\exp(i\omega t - i\mathbf{q}\cdot\mathbf{r})$, it follows that

$$\mathbf{A} = (i/\omega)\mathbf{E}_o\,\exp(i\omega t - i\mathbf{q}\cdot\mathbf{r}) \tag{19}$$

$$\text{cgs:} \qquad \mathbf{A} = (ic/\omega)\mathbf{E}_o\,\exp(i\omega t - i\mathbf{q}\cdot\mathbf{r}) \tag{19'}$$

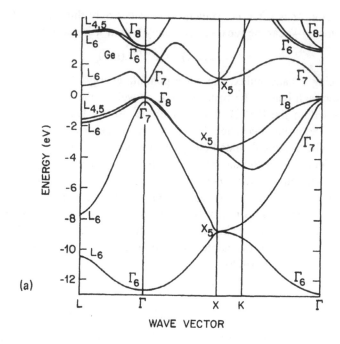

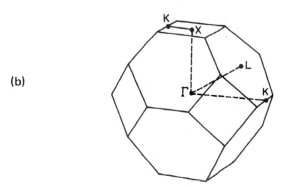

Figure 2.7 (a) Calculated band structure of Ge as a function of wave vector along the lines LΓ, ΓX, XK, and KΓ of the first Brillouin zone (b). [From Chelikowsky and Cohen (1976).]

Since **v** in Eq. (16) is directly related to current density **J** by **J** $= e\mathbf{v}$, with e equal to charge, the Maxwell equations can now be further incorporated to express **F** as a function of **A** alone. In this way it is found (Heitler 1954) that the influence of the electromagnetic force $\mathbf{F}$ on the motion of an electron is simply to replace its momentum operator $\mathbf{p}$ by $\mathbf{p} - e\mathbf{A}$ (cgs: $\mathbf{p} - e\mathbf{A}/c$).

The lowest order interaction term in the electronic Hamiltonian therefore follows from the embellished kinetic energy $(\mathbf{p} - e\mathbf{A})^2/2m$ form as

$$\mathsf{H}_{int} = -(e/m)\mathbf{A} \cdot \mathbf{p} = (e/m)i\hbar \, \mathbf{A} \cdot \nabla \qquad (20)$$

$$\text{cgs:} \qquad \mathsf{H}_{int} = -(e/mc)\mathbf{A} \cdot \mathbf{p} = (e/mc)i\hbar \, \mathbf{A} \cdot \nabla \qquad (20')$$

where m is electron mass and we have used the quantum equivalence $\mathbf{p} = -i\hbar\nabla$. Treating H_{int} as a time-dependent perturbation now leads, via the golden rule (Harrison 1980), to a transition rate W_{if} between an initial state ψ_i of energy E_i and a final state ψ_f of energy E_f of the form

$$W_{if} = (2\pi/\hbar) \langle \psi_f \mid \mathsf{H}_{int} \mid \psi_i \rangle^2 \delta(E_f - E_i - \hbar\omega) \qquad (21)$$

or equivalently, using Eqs. (19) and (20),

$$W_{if} = \left(\frac{2\pi e^2 \hbar E_o^2}{m^2 \omega^2} \right) \mid \int \psi_f^* e^{\, i\mathbf{q} \cdot \mathbf{r}} \mathbf{n} \cdot \nabla \psi_i d\tau \mid^2 \delta(E_f - E_i - \hbar\omega) \qquad (22)$$

where $d\tau$ is an element of volume and $\mathbf{E}_o = \mathbf{n} E_o$ with $\mathbf{n}$ a unit vector. Substituting the Bloch forms (15) with $\mathbf{k} = \mathbf{k}_f$ and $\mathbf{k} = \mathbf{k}_i$, respectively, for states ψ_f and ψ_i, we see immediately that the matrix element in Eq. (22) is zero unless

$$\mathbf{q} - \mathbf{k}_f + \mathbf{k}_i = 0 \qquad (23)$$

Since, in the optical frequency regime, the wave vector $\mathbf{q}$ extends only over a tiny fraction of the first Brillouin zone near the origin, this selection rule (23) corresponds to essentially "vertical" transitions in Fig. 2.7 between filled and empty bands s and s' in the reduced zone. Using the symbol $\mathbf{M}$ to represent matrix elements in the form (Hodgson 1970)

$$\mathbf{M}(\mathbf{k}) = \int u_s^*(\mathbf{k},\mathbf{r}) \nabla \, u_{s'}(\mathbf{k},\mathbf{r}) \, d\tau \qquad (24)$$

with $q = 0$, the transition rate (22) now becomes

$$W_{if} = \left(\frac{2\pi e^2 \hbar E_o^2}{m^2 \omega^2} \right) \mid \mathbf{n} \cdot \mathbf{M}(\mathbf{k}) \mid^2 \delta(E_f - E_i - \hbar\omega) \qquad (25)$$

The association with dielectric response is now made by relating the rate of absorption of energy $\hbar\omega \, W_{if}(\hbar\omega)$ with conductivity σ via the loss coefficient α of Eqs. (114) and (116) of Chapter 1, namely,

$$\hbar\omega W_{if}(\hbar\omega) = \sigma_{if}(\omega) \langle E^2 \rangle \qquad (26)$$

leading to a contribution to optical conductivity

$$\sigma_{if}(\omega) = \left(\frac{\pi e^2 \hbar^2}{m^2 \omega}\right) \mid \mathbf{n} \cdot \mathbf{M}(\mathbf{k}) \mid^2 \delta(E_f - E_i - \hbar\omega) \qquad (27)$$

associated with direct transitions of an electron between final and initial states of wave vector $\mathbf{k}$. The total response is now obtained by summing over the wave-vector states in the first Brillouin zone. Since the density of Bloch states in $\mathbf{k}$-space is $(1/2\pi)^3$ and each state contains two electrons, the final formula for the optical conductivity is

$$\sigma(\omega) = \left(\frac{e^2 \hbar^2}{4\pi^2 m^2 \omega}\right) \int d^3\mathbf{k} \mid \mathbf{n} \cdot \mathbf{M}(\mathbf{k}) \mid^2 \delta(E_{s'}(\mathbf{k}) - E_s(\mathbf{k}) - \hbar\omega) \quad (28)$$

The imaginary component of dielectric response $\epsilon''(\omega)$ follows from Eq. (112) of Chapter 1, and the corresponding real part $\epsilon'(\omega)$ can then be determined by use of the Kramers–Kronig relation (130) of Chapter 1.

Theoretically, direct computation of the matrix elements $\mathbf{M}(\mathbf{k})$ of Eq. (24) is difficult, and they are frequently replaced by an average value in attempts to interpret measured spectra. In addition, use is made of the property of the delta function in replacing the Brillouin zone volume integral by a surface integral in the form

$$\sigma(\omega) = \left(\frac{e^2 \hbar^2}{4\pi^2 m^2 \omega}\right) \langle [\mathbf{n} \cdot \mathbf{M}(\mathbf{k})]^2 \rangle \int \frac{dS}{\nabla_k [E_{s'}(\mathbf{k}) - E_s(\mathbf{k})]} \qquad (29)$$

where the integral is over the surface defined by

$$E_{s'}(\mathbf{k}) - E_s(\mathbf{k}) - \hbar\omega = 0 \qquad (30)$$

Often it is the integrand denominator in Eq. (29) that produces the major features of the electronic absorption spectra. In particular, the $\mathbf{k}$ values at which the denominator vanishes produce singularities called *critical points*, and several classifications of critical point can be made according to the manner in which the denominator approaches zero as the critical point is approached. Critical points are expected to produce recognizable features in the electronic absorption spectra of all crystals. In particular, unless the relevant matrix element $\mathbf{M}(\mathbf{k})$ is zero, all insulators and semiconductors are expected to exhibit a critical point at the lowest allowed direct (i.e., $\mathbf{k}_{s'} = \mathbf{k}_s$) transition, which is called the fundamental absorption edge.

In Fig. 2.8 we compare the calculated and measured electron density of states for Ge (compare Fig. 2.7) together with calculated and measured reflectivity spectra that sample the dielectric response in the interband excitation regime. The critical point features, particularly the one at the fundamental absorption

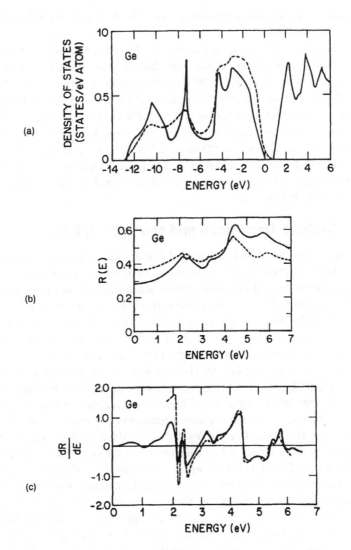

Figure 2.8 Comparison of the calculated (full curves) and measured (dashed curves) results for (a) the density of states, (b) the reflectivity $R(E)$, and (c) the reflection derivative dR/dE, of Ge. [After Chelikowski and Cohen (1976).]

edge $\hbar\omega = E_g$, are best seen in the derivative reflectivity spectrum, which is also shown in Fig. 2.8 and exhibits clear critical point features at 0.7 (band edge), 2.1, 2.3, 3.2, 4.4, 5.5, and 5.9 eV, all understandable, at least semiquantitatively, from the band structure. Measurements of piezoreflectance and electroreflectance (i.e, modulation spectroscopy) can also be used to help identify critical features (Aspnes 1980).

The theory to this point has assumed the presence of completely filled valence bands and empty conduction bands. For metals, semimetals, or narrow-bandgap semiconductors at room temperature, this will not be a valid assumption. For these cases the theory is not difficult to modify to allow, via the Fermi–Dirac distribution function (Ziman 1964), for the effects of thermal population. Metals in particular will exhibit pronounced low-energy features associated with *intra*band transitions. At higher energies, however, interband features again predominate and may be particularly pronounced for metals like Cu, Ag, and Au, which possess narrow filled *d*-electron bands a few electron-volts below the Fermi level.

2.1.5. Indirect Transitions and the Urbach Edge

The discussion to this point has assumed that the electronic excitations involved in the interband spectral response are sharp. In actuality they are broadened both by electron–electron and electron–phonon scattering. In many respects the effects of this relaxation can be incorporated within the theory in a phenomeno-logical fashion by introducing energy widths. The basic theory of direct transitions then remains valid as long as these widths are small on the scale of the bandwidths; if not, the wave-vector selection rule ceases to be valid.

However, transitions caused by electron–phonon interactions are not limited to near-vertical (**k**-conserved) direct excitations in the reduced Brillouin zone, because phonons (unlike infrared radiation) have a range of **k** comparable with that of the electron. An electron–phonon transition can be associated with either the emission or absorption of a phonon for which the total (electron plus phonon) wave vector must be conserved. It follows that the electronic **k** vector need not be conserved in these processes, which are termed "indirect," because they require two excitation processes (electron and phonon). For this reason they occur in second-order perturbation theory and are consequently weak. They would therefore be of little consequence except for one important fact: Many materials possess what is termed an indirect bandgap. By this we mean that the energy maximum of the valence band and the energy minimum of the conduction band lie at different points in the reciprocal space of the Brillouin zone. Germanium (see Fig. 2.7) is just such a material, the energy at the conduction band L-point being about 0.15 eV lower than that at the zone center Γ-point, which corresponds to the valence band maximum. It follows that the indirect bandgap E_{ind} for these materials is less than the direct bandgap E_{dir}, which corresponds to the fundamental absorption edge. As a consequence, phonon-assisted indirect transitions can produce an absorption tail extending into the optic window at energies below the fundamental edge.

The theoretical formula for phonon-induced tail absorption must sum over all branches of the phonon spectrum and include occupation factors for both phonon absorption and emission. Consequently it is extremely complex. In practice, however, many solids are observed to have absorption tails of simple exponential (or Urbach) from

$$\alpha(E) = C \exp[(E - E_1)/E_o] \tag{31}$$

in which C, E_1, and E_o are wavelength-independent material parameters. For crystals these parameters are usually strongly temperature-dependent in a manner that is approximately represented by rewriting Eq. (31) in the form

$$\alpha(\omega) = C' \exp[c'(\omega - \omega_o)/kT] \tag{32}$$

with C', c', and ω_o now only weakly temperature-dependent at most. The temperature dependence in Eq. (32) is generally in accord with the expectations of a thermally excited disorder, like phonons, but it does not appear that phonons alone are responsible for all observed tails, and many theories exist (and much controversy remains) concerning the detailed origin of the Urbach form (Ihm and Phillips 1983, Sa-Yakanit and Glyde 1987).

The situation in glasses is interestingly different (Zallen 1983). Although an edge of form (31) is commonly observed, it does not conform with the temperature dependence of Eq. (32) but is much less sensitive to temperature. The explanation resides in the structural disorder present in the glass, which adds to, and usually dominates, thermally induced disorder. Glass disorder removes all translational invariance, and with it all **k** selection rules, allowing all pairs of electron states $\mathbf{k}_v - \mathbf{k}_c$ (v = valence band, c = conduction band) to contribute to absorption to some degree. Not only does this temperature-stabilize the Urbach edge, but also it tends to smooth out the sharp critical-point features of the electronic spectrum including the direct threshold at ω_o in Eq. (32). In fact, for a glass, the absorptive part of amorphous response can be recast approximately in terms only of the density-of-states functions n_v (E) and n_c (E) in the form

$$\alpha(\omega) \sim \int n_v(E) \, n_c(E + \hbar\omega) \, dE \tag{33}$$

In particular, if the band edges are taken to be approximately parabolic, that is, $n_v(E) \sim (E_v - E)^{1/2}$ and $n_c(E) \sim (E - E_c)^{1/2}$, then Eq. (33) near the threshold $\hbar\omega_o = E_c - E_v$ integrates to

$$\alpha(\omega) \sim (\omega - \omega_o)^2 \tag{34}$$

Such an absorptive regime, called the Tauc edge (Tauc 1970), is often seen in glasses on the high-frequency side of the Urbach edge. An example is shown in Fig. 2.9 for the amorphous chalcogenide As_2S_3.

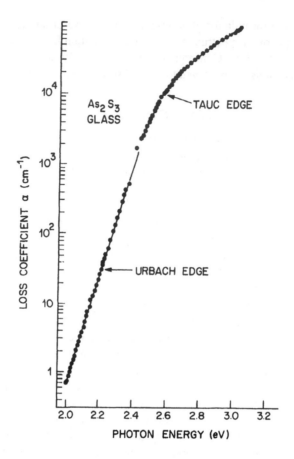

Figure 2.9 The Urbach and Tauc electron absorption edges for As_2S_3 glass. [After Tauc (1974).]

2.1.6. Excitonic Effects

In electronic absorption processes, the excited electron and the vacated ''hole'' are created at the same site, so their electrostatic interaction should not be neglected as it was in our earlier development. The consequences of including it are most pronounced near the band edge, because it is found that for this situation, bound electron–hole excitations, called excitons, can then exist below the conduction band minimum (Egri 1985).

If the wave function of the exciton is extremely localized (e.g., extends only over one unit cell), then the exciton resembles an atomic excitation perturbed only by weak interactions with neighboring atoms (which, however, can induce "hopping"). These so-called Frenkel excitons exist in organic complexes and in some very-wide-bandgap halides. The more usual situation (Wannier excitons) involves exciton wave functions that extend over many lattice cells. For this case the electrostatic electron–hole interaction is screened by crystal polarization and is reduced in magnitude by the dielectric constant ϵ; that is, $V(r) = -K/r$, with

$$K = e^2/4\pi\epsilon\epsilon_o \qquad (35)$$

$$\text{cgs:} \quad K = e^2/\epsilon \qquad (35')$$

Let us consider Wannier excitons formed from electrons and holes near the extrema of the band energies $E_v(\mathbf{k})$ and $E_c(\mathbf{k}')$. For the simplest case of a direct exciton ($\mathbf{k} = \mathbf{k}'$) near the Brillouin zone center $\mathbf{k} = 0$, the energy bands to terms quadratic in k take the form

$$E_c(k) = E_o + \hbar^2 k^2/2m_e^* \qquad (36)$$

$$E_v(k) = -\hbar^2 k^2/2m_h^*$$

The electron and hole consequently behave respectively as particles with charges $-|e|$ and $|e|$ and masses (assumed isotropic) m_e^* and m_h^*. In the presence of the potential $V(r)$, a hydrogen-like Hamiltonian results with electron mass replaced by a reduced mass $\mu = m_e^* m_h^*/(m_e^* + m_h^*)$. This Hamiltonian has eigenvalues

$$E_n(k) = E_o - K^2\mu/2\hbar^2 n^2 + \hbar^2 k^2/2(m_e^* + m_h^*) \qquad (37)$$

in which $n = 1,2,3, \ldots$ is a principal quantum number.

Above the threshold energy E_o there is a continuum of states corresponding to the band model modified by the electron–hole interaction (Elliott 1957). Below E_o are the exciton levels n with the binding energy of the lowest ($n = 1$) level, namely, $K^2\mu/2\hbar^2$, being typically in the range 0.001–0.1 eV. The optical conductivity associated with direct transitions between a ground state Ψ_0 and an exciton state Ψ can be derived by methods analogous to those leading to Eq. (28) and is

$$\sigma(\omega) = \left(\frac{e^2\hbar^2}{4\pi^2 m^2 \omega}\right) \Big| \int \Psi^* \mathbf{n}\cdot\nabla\Psi_0 d\tau \Big|^2 \delta(E - E_o - \hbar\omega) \qquad (38)$$

Formally it consists of discrete lines merging into a continuum as depicted in Fig. 2.10. In reality the exciton lines are broadened (for example, by exciton–phonon coupling), and only a few exciton peaks are discernible. Below the lowest exciton peak the absorption drops off rapidly to join the Urbach edge. Although the major features of excitonic effects are observed below the bandgap, the electron–hole interaction is present for all electron–hole pairs produced by ex-

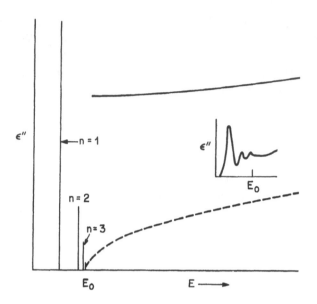

Figure 2.10 A schematic absorption edge spectrum (dashed line) and how it is modified by the electron–hole interaction (solid line). Only the $n = 1$, 2, and 3 excitons are shown (as unbroadened delta functions with heights proportional to their oscillator strengths). An actual observed spectrum would more closely resemble the one shown in the inset.

citing any electron to an excited state and must perturb the entire valence to conduction band spectrum to some extent.

2.2 FREE CARRIER OPTICAL PROPERTIES

The interband transitions discussed in Section 2.1.4 presuppose the existence of an energy gap between a filled valence band and an empty conduction band. If one or more of these bands are partly filled by electrons or holes, then *intra*band transitions can be induced within these bands and give rise to radiation absorption phenomena down to zero frequency. The transitions involved are necessarily indirect, because any change of energy within a single band must be associated with a change of crystal momentum.

Although a complete quantum theory is necessary for a theoretical interpretation that includes the full complexity of transitions between degenerate bands, a semiclassical treatment gives essentially the correct results for the case of a single band response. Consider an electron (or hole) of effective mass m^* and carrier velocity v in the presence of radiation with electric field $E = E_o e^{i\omega t}$. It has, in a simplified scalar formalism, an equation of motion

$$m^*(dv/dt) + m^*(v/\tau) = eE \tag{39}$$

where τ is a relaxation time such that, in the absence of the field, velocity fluctuations decay exponentially as $v(t) = v(0)e^{-t/\tau}$. Writing $v = v_o e^{i\omega t}$, we find the solution for the complex velocity amplitude v_o:

$$v_o = eE_o \tau / m^*(1 + i\omega\tau) \tag{40}$$

leading to an induced current density $J = Nev_o$ (where there are N conducting electrons or carriers per unit volume) and consequently to a complex conductivity $\hat{\sigma}(\omega) = J/E_o$ of the form

$$\hat{\sigma}(\omega) = \sigma'(0)/(1 + i\omega\tau) \tag{41}$$

in which $\sigma'(0)$ is the dc conductivity

$$\sigma'(0) = Ne^2\tau/m^* \tag{42}$$

Equation (41) can be resolved into its real and imaginary components by writing $\hat{\sigma} = \sigma' + i\sigma''$, where

$$\sigma'(\omega) = \sigma'(0)/(1 + \omega^2\tau^2) \tag{43}$$

$$\sigma''(\omega) = -\omega\tau\sigma'(0)/(1 + \omega^2\tau^2) \tag{44}$$

These equations are referred to as the Drude conductivity relations.

If, as a generalization of Eq. (104) of Chapter 1 to complex conductivity, we write

$$\hat{\epsilon} = \epsilon' - i\hat{\sigma}/\epsilon_o\omega \tag{45}$$

$$\text{cgs:} \qquad \hat{\epsilon} = \epsilon' - 4\pi i\hat{\sigma}/\omega \tag{45'}$$

then σ'' can be associated with a free carrier real component of dielectric constant,

$$\epsilon' = n^2 - k^2 = \sigma''/\epsilon_o\omega = -\tau\sigma'(0)/\epsilon_o(1 + \omega^2\tau^2) \tag{46}$$

$$\text{cgs:} \qquad \epsilon' = n^2 - k^2 = 4\pi\sigma''/\omega = -4\pi\tau\sigma'(0)/(1 + \omega^2\tau^2) \tag{46'}$$

and, via Eq. (116) Chapter 1, σ' with an induced free carrier loss coefficient

$$\alpha = \sigma'/c\epsilon_o n = \sigma'(0)/c\epsilon_o n(1 + \omega^2\tau^2) \tag{47}$$

$$\text{cgs:} \qquad \alpha = 4\pi\sigma'/nc = 4\pi\sigma'(0)/cn(1 + \omega^2\tau^2) \tag{47'}$$

Associated with this loss coefficient is a skin depth $\delta = \alpha^{-1}$, which decreases rapidly as dc conductivity increases.

2.2.1. Infrared Absorption in Metals

The dielectric constant in metals can be written as the sum of an intraband component $\hat{\epsilon}_f$ (with the subscript f denoting free electron) and an interband component $\hat{\epsilon}_o$ (Hodgson 1970):

$$\hat{\epsilon}(\omega) = \hat{\epsilon}_f(\omega) + \hat{\epsilon}_b(\omega) \qquad (48)$$

The imaginary (absorptive) part of the interband component is zero until the radiation energy $\hbar\omega$ approaches the bandgap—a frequency normally in the visible or near-infrared but which can be in the mid-infrared for semimetals. Below the bandgap E_g the real part of $\hat{\epsilon}_b$ tends toward a frequency-independent value.

In the infrared, optical absorption and dispersion for most metals is dominated by the carrier component of $\hat{\epsilon}_f(\omega)$. Thus, utilizing the Drude formalism of Eqs. (41), (42), and (45), we can approximate

$$\hat{\epsilon}(\omega) \approx - (iNe^2\tau/\epsilon_o m^*\omega) / (1 + i\omega\tau) \qquad (49)$$

$$\text{cgs:} \qquad \hat{\epsilon}(\omega) \approx - (4\pi iNe^2\tau/m^*\omega)/(1 + i\omega\tau) \qquad (49')$$

which can be rewritten as

$$[\hat{\epsilon}(\omega)]^{-1} = (i\omega - \omega^2\tau)/\omega_p^2\tau \qquad (50)$$

where ω_p is the plasma frequency

$$\omega_p^2 = Ne^2/m^*\epsilon_o \qquad (51)$$

$$\text{cgs:} \qquad \omega_p^2 = 4\pi Ne^2/m^* \qquad (51')$$

Most experiments probing $\hat{\epsilon}(\omega)$, or its inverse, for metals take the form of a reflection measurement at the metal surface. The expectation from Eq. (50), that $[\hat{\epsilon}(\omega)]^{-1}$ should have a real component $\sim \omega^2$ and an imaginary component $\sim \omega$, is generally confirmed, although difficulties are encountered in making direct comparisons. The most important of these is the anomalous skin effect. It is encountered whenever the skin depth α^{-1} is small compared to the electronic mean free path $v\tau$. Since the skin depth measures essentially the maximum distance over which the applied field can be considered approximately constant inside the metal, the condition $\alpha^{-1} << v\tau$ implies that $J(r)$ is no longer proportional to $E(r)$ but that the response of electrons is nonlocal (that is, the electron samples an entire range of fields between scattering events). Anomalous skin effect corrections are essential in carrying out any quantitative comparisons with local response theories such as that leading to Eq. (50)

2.2.2. Carrier Absorption in Semiconductors

In semiconductors the assumption of a filled valence band and an empty conduction band adopted in Section 2.1.4 never holds precisely. Either via the mechanism of thermal activation in intrinsic semiconductors or via dopants in extrinsic semiconductors, carriers always exist in the form of electrons, holes, or both (Bassani et al. 1974). Their effect is to contribute an intraband component to dielectric response, but this effect is much smaller than its counterpart in metals because of the smaller carrier concentrations. This leads to greater skin depths and to a general absence of anomalous skin-depth complications in semiconductors. Also, compared to metals, the relaxation times τ are much longer, making the inequality $\omega\tau \gg 1$ valid at infrared frequencies in most semiconductors (Pidgeon 1980).

The dielectric constant $\hat{\epsilon}(\omega)$ can again be expressed, as in Eq. (48), as the sum of interband and intraband components (with additional lattice phonon contributions if necessary). Normally, below the bandgap, the absorptive response is dominated by $\hat{\epsilon}_f$ but is small compared to the total response (i.e., absorption index k is small compared with refractive index n), so we may write [see Eqs. (111) and (112) of Chapter 1]

$$n \approx [\epsilon'_f(\omega) + \epsilon'_b(\omega)]^{1/2} \qquad (52)$$

$$2nk \approx \epsilon''_f \qquad (53)$$

Let us consider first the absorptive part. Using the Drude formulas of Eqs. (42)–(45) we have

$$2nk \approx \epsilon''_f = \sigma'/\epsilon_o\omega \approx Ne^2/\epsilon_o\tau m^*\omega^3 \qquad (54)$$

$$\text{cgs:} \quad 2nk \approx \epsilon''_f = 4\pi\sigma'/\omega \approx 4\pi Ne^2/\tau m^*\omega^3 \qquad (54')$$

In general N is temperature-dependent, and more precise theoretical formulas for ϵ''_f are complicated. Nevertheless the ω^{-3} frequency dependence is common to many of them. Experimentally a power law frequency dependence ω^{-s} is observed, with s in the range 2.5–4, the value of the exponent seemingly depending on the character of the scattering mechanisms that control the relaxation time τ.

We can also probe the real part of the semiconductor response of Eq. (52) by use of the Drude equations. Making use of Eqs. (42) and (46) we can reexpress Eq. (52) in the form

$$\epsilon' \approx n^2 = \epsilon_\infty (1 - \omega_p^2/\omega^2) \qquad (55)$$

where ω_p is the generalized plasma frequency

$$\omega_p^2 = Ne^2/m^*\epsilon_o\epsilon_\infty \qquad (56)$$

$$\text{cgs:} \quad \omega_p^2 = 4\pi Ne^2/m^*\epsilon_\infty \qquad (56')$$

and we have assumed $n \gg k$ and have included all the non-carrier contributions in a "high-frequency" term ϵ_∞. It is assumed that this "high frequency" is still well below the onset of interband excitations and can therefore be construed simply as a frequency-independent quantity.

From Eq. (55) it can be seen that the dielectric constant is positive at high frequencies and negative at low frequencies, changing sign when $\omega = \omega_p$. Negative values of ϵ' imply an absence of propagating waves inside the medium (i.e., total reflection at the surface). At frequency values $\omega > \omega_p$, propagating waves do exist with a refractive index $n(\omega) = [\epsilon'(\omega)]^{1/2}$. For a range of frequencies immediately above ω_p, however, the value of n is very small, and external waves can still be totally reflected for angles of incidence greater than a critical angle $\theta_c = \sin^{-1} n(\omega)$.

For normal incidence, the theoretical reflectivity falls abruptly from 1 to 0 at ω_p (the plasma edge) before recovering to more normal semiconductor values as ω rises toward near-infrared values (Fig. 2.11). Experimental observations follow this qualitative picture, although, because of the finite relaxation times of the plasma electrons, the peak value of reflectivity is never exactly unity nor the minimum exactly zero. Consequently, the experimental plasma edge, though

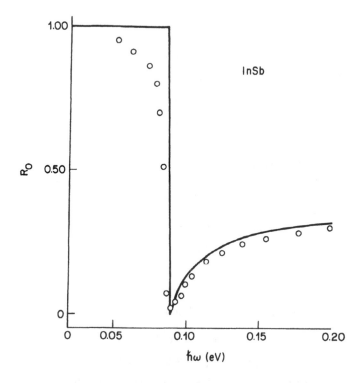

Figure 2.11 The plasma edge of n-type InSb (showing experimental points and a theoretical curve) where R_o is reflectance at normal incidence. [After Spitzer and Fan (1957).]

often still dramatic, is not quite so sharp as simple theory would indicate. The plasma edge also exists in metals but, because of the much larger carrier concentration N for that case, is usually found in the ultraviolet so that the infrared optical properties of metals are normally well within the quasi-totally reflective part of the spectrum.

2.2.3. Free Carrier Magneto-optical Properties

The equation of motion, Eq. (39), when modified by the addition of an applied magnetic dc field H, becomes [see Eq. (16)]

$$m^*(dv/dt) + m^*\gamma v = e(E + v \times B) \tag{57}$$

$$\text{cgs:} \quad m^*(dv/dt) + m^*\gamma v = e(E + c^{-1}v \times B) \tag{57'}$$

in which $\gamma = 1/\tau$ and $B = \mu_o H$. If the magnetic field lies along the x_3 axis, then this vector equation can be resolved along the three orthogonal Cartesian axes x_i, $i = 1,2,3$, in the form

$$m^*(\dot{v}_1 + \gamma v_1) = e(E_1 + Bv_2) \tag{58a}$$

$$m^*(\dot{v}_2 + \gamma v_2) = e(E_2 - Bv_1) \tag{58b}$$

$$m^*(\dot{v}_3 + \gamma v_3) = eE_3 \tag{58c}$$

and has periodic solutions $E = E_o e^{i\omega t}$, $v = v_o e^{i\omega t}$, with $Nev_o = \sigma \cdot E_o$ defining a second-rank tensor conductivity, with components

$$\hat{\sigma}_{11} = \hat{\sigma}_{22} = \left(\frac{Ne^2}{m^*}\right) \frac{\gamma + i\omega}{(\gamma + i\omega)^2 + \omega_c^2} \tag{59a}$$

$$\hat{\sigma}_{12} = -\hat{\sigma}_{21} = \left(\frac{Ne^2}{m^*}\right) \frac{\omega_c}{(\gamma + i\omega)^2 + \omega_c^2} \tag{59b}$$

$$\hat{\sigma}_{33} = \left(\frac{Ne^2}{m^*}\right) \frac{1}{\gamma + i\omega} \tag{59c}$$

where ω_c is called the cyclotron frequency and is given by

$$\omega_c = eB/m^* = e\mu_o H/m^* \tag{60}$$

$$\text{cgs:} \quad \omega_c = eH/m^*c \tag{60'}$$

We note that the conductivity along the axis of the applied magnetic field H (namely, $\hat{\sigma}_{33}$) is unchanged from its form in the absence of a field. The "transverse" conductivity components, on the other hand, are perturbed in an interesting way by the field (Palik and Furdyna 1970).

The effect of this transverse perturbation is most easily seen by considering a plane-polarized beam to be the sum of two circularly polarized components

$E_0^{\pm} = (E_0)_1 \pm i(E_0)_2$. For these circular components one can define dielectric responses (Pidgeon 1980)

$$\hat{\epsilon}^{\pm} = (n^{\pm} - ik^{\pm})^2 = (\epsilon')^{\pm} - \hat{\sigma}^{\pm}/\epsilon_0\omega \qquad (61)$$

with corresponding refractive indices $n^{\pm}$ and absorption indices $k^{\pm}$. Combining Eqs. (59) and (61), we now find that these two counterrotating modes have different n and k values that can be derived from

$$(n^{\pm} - ik^{\pm})^2 = \epsilon_\infty \left[1 - \frac{\omega_p^2}{\omega(\omega \mp \omega_c - i\gamma)} \right] \qquad (62)$$

where ϵ_∞ once again includes all the non-carrier contributions.

In the most common situation, where $n \gg k$ and $\omega \gg \gamma$, we easily extract the real and imaginary parts in the form

$$(n^{\pm})^2 = \epsilon_\infty \left[1 - \frac{\omega_p^2}{\omega(\omega \mp \omega_c)} \right] \qquad (63)$$

$$2n^{\pm}k^{\pm} = \epsilon_\infty \omega_p^2\gamma/\omega(\omega \mp \omega_c) \qquad (64)$$

It is immediately apparent from these equations that pronounced dielectric anomalies occur when $\omega = \omega_c$. They are referred to as "cyclotron resonance effects."

Even away from cyclotron resonance, however, the differing refractive and absorption indices for the two counterrotating polarizations have important consequences. The differing refractive indices imply that the E^+ and E^- radiation components travel with different velocities through the medium. This results in a rotation of the plane of polarization for plane-polarized light, an effect known as Faraday rotation. Light propagating parallel to a dc magnetic field is therefore subjected to a rotation of its polarization. The sense of the rotation is defined to be positive if it is in the direction of positive current in the coil that produces the magnetic field. This rotation is measured quantitatively by the rotation angle per unit length

$$\theta = (\omega/2c)(n^- - n^+) \qquad (65)$$

Using Eq. (63), this rotation can be written in terms of the plasma and cyclotron frequencies in the form

$$\theta = \epsilon_\infty \omega_p^2\omega_c/2nc \; \omega^2 \qquad (66)$$

where n is the mean of n^+ and n^- and we have assumed $\omega^2 \gg \omega_c^2$. Using Eqs. (56) and (60) for ω_p^2 and ω_c respectively, enables us finally to conclude

$$\theta = Ne^3\mu_0H/2ncm^{*2}\omega^2\epsilon_0 \qquad (67)$$

$$\text{cgs:} \qquad \theta = 2\pi Ne^3H/nc^2m^{*2}\omega^2 \qquad (67')$$

Thus the rotation is directly proportional to the induction $B = \mu_o H$, and consequently it is customary to define it as $\theta = VB$ with the constant of proportionally V being known as the Verdet constant.

Not only do counterpolarized components of radiation have differing velocities when traveling parallel to a static magnetic field, they also, via Eq. (64), are unequally absorbed or attenuated. This effect induces an elliptical polarization of an initially plane-polarized light beam, the magnitude of which can be expressed by

$$\Delta = (\omega/2c)(k^+ - k^-) \tag{68}$$

Using Eqs. (63) and (64), we find that Δ bears a very simple relationship to the rotation angle θ in the form $\theta/\Delta = \omega\tau/2$.

For plane-polarized light propagating perpendicular to $\mathbf{H}$, it is easy to establish that the relevant refractive index $(n_\perp, n_\parallel)$ depends on the direction of polarization $(\mathbf{E}\perp\mathbf{H}, \mathbf{E}\parallel\mathbf{H})$. The difference $n_\perp - n_\parallel$ is a quadratic function of $\mathbf{H}$ and is responsible for the so-called Voigt (sometimes spelled Voight) effect, which is a degree of ellipticity $\omega(n_\perp - n_\parallel)/c$ caused by this magnetically induced birefringence.

For materials possessing a spontaneous bulk magnetic moment $\mathbf{M}$, it is the *internal* magnetic field that produces a Faraday rotation (even in the absence of an applied $\mathbf{H}$). The Faraday rotation $\theta = F$ for such ferromagnets- or ferrimagnets is expressed directly as a rotation angle per unit distance of travel for light propagating parallel to the magnetic axis. Because the internal field is proportional to the magnetic moment magnitude M, it is possible to relate the rotation angle θ linearly to M. The proportionality factor $K = \theta/M$ is called *Kundt's constant*. Since magneto-optical devices require an efficient Faraday mechanism coupled with low attenuation, it is common practice to use, as a figure of merit for magneto-optical materials, the ratio of F or V to the loss coefficient $\alpha = 4\pi k/\lambda$ of Eq. (115) of Chapter 1. Finally, it is interesting to note (Ferré and Gehring 1984) that the Voigt effect, being proportional to H^2, also exists spontaneously in an antiferromagnet.

2.2.4. Electron–Phonon Coupling

The theory of the present section (2.2) has thus far assumed the existence of essentially "free" carriers, subject only to "collision events" that determine their relaxation times. Although it is not appropriate to go beyond such a model in any detail in an account of the present kind, it is well to point out that a number of electron–phonon coupled effects do embellish the simple picture.

The electron–phonon interaction can manifest itself through both collective and single-"particle" effects (Harper et al. 1973). The normal modes of oscillation of a completely free electron gas are the plasma modes ω_p, which are longitudinal excitations. Assuming that the positive ions form a uniform background of compensating charge within a rigid framework, the repulsive coulombic forces between electrons induce an essentially macroscopic depolarizing field that couples the plasma modes to longitudinal optic (LO) phonon modes ω_{LO}. The resulting mixed electron–phonon modes are therefore of longitudinal form and consequently cannot couple directly to electromagnetic radiation. However, on application of a magnetic field **H**, the LO-phonon depolarization field perpendicular to **H** can couple the plasma motion with the cyclotron excitations in a manner that can interact with radiation propagating in the Voigt configuration, that is, with wave vector $\mathbf{q} \perp \mathbf{H}$. If the concentration of carriers is adjusted to make $\omega_p < \omega_{LO}$, then this coupled plasma–cyclotron mode can be tuned to resonance with the LO mode (i.e., $\omega_p^2 + \omega_c^2 = \omega_{LO}^2$) as a function of applied magnetic field via Eq. (60); see Fig. 2.12. Far from their crossover point, the two curves can be thought of as dominantly of electronic or phonon origin. Near the resonant crossover the curves repel by an amount proportional to their coupling strength, and both modes adopt a strongly mixed electron–phonon nature. They can be observed by their effects upon the dispersion and attenuation of infrared radiation, or by Raman scattering.

These excitations are of very long wavelength and hence involve long-range quasi-macroscopic polarization effects. Electron–phonon coupling effects can

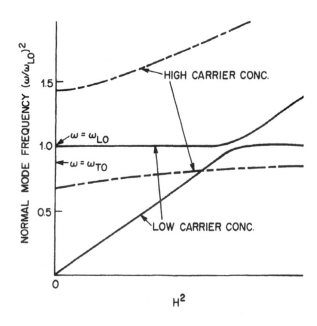

Figure 2.12   Coupled mode absorption frequencies as a function of the square of the magnetic field H for examples of low and high carrier concentrations.

also manifest themselves in a perturbation of single-electron motion (Klinger 1968). In this context they produce an increase in electron mass by tending to drag ion cores along with the electronic motion. The result is a so-called dressed electron or *polaron*. This coupling of a single mobile electron to its surrounding strain field is largest in ionic crystals, for which the coulombic forces themselves are most pronounced. Two types of polarons are commonly distinguished. The first, called the *large polaron,* is for a dressed electron that can propagate as a quasi-free carrier in a well-defined electronic band. The second, or *small,* polaron is one that can move only between trapped sites by means of thermally activated hopping (at higher temperatures) or by tunneling mechanisms (at low temperatures).

Polaron effects can be studied both by magnetoabsorption and by cyclotron resonance, since they lead to an effective mass $m*$ [in equations like Eq. (60), for example], which is magnetic field dependent. The general phenomenon of the absorption of angular momentum (i.e, circularly polarized light) by matter, leading to a nonthermal population of magnetic energy levels, is called *optical pumping.* The procedure has proven very fruitful in obtaining general information on band structures and on carrier relaxation mechanisms as well as in a study of polaron effects.

2.3. LIGHT SCATTERING

The problem of finding the electrostatic potential distribution for a dielectric sphere of volume V and dielectric constant ϵ_2 placed in an infinite background medium of permittivity ϵ_1 in the presence of a uniform electric field **E** is familiar from any standard text on electrostatics [e.g., Bleaney and Bleaney (1976)]. Solving Laplace's equation for this geometry reveals that the sphere is equivalent to a point dipole,

$$\mathbf{p} = \frac{3\ \epsilon_0 \epsilon_1 (\epsilon_2 - \epsilon_1)\ VE}{\epsilon_2 + 2\epsilon_1} \tag{69}$$

at the sphere center, as far as its effect on the host medium ϵ_1 is concerned. It follows that optical radiation incident on a sphere of this kind (with radius small compared to the wavelength of the light, λ) will induce in this sphere an oscillating dipole of this magnitude. Such an oscillating dipole is the source of scattered radiation in all directions with field amplitude (in SI units)

$$E' = \frac{|\ddot{p}|\ cos\ \theta}{4\pi\epsilon_0 c^2 r} \tag{70}$$

where r is distance and θ is the scattering angle with respect to the direction of the incident radiation (Fig. 2.13). This in turn gives rise to a scattered intensity

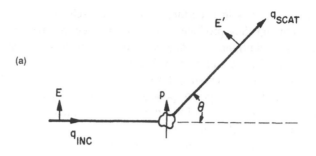

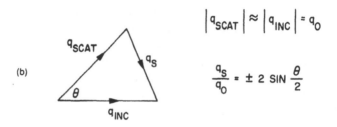

Figure 2.13 (a) Schematic of induced dipole scattering (E', q_{scat}) from an element of volume dielectrically perturbed by a polarized incident ray with electric field **E** and wave vector q_{inc}. (b) The vector diagram of the Bragg momentum conservation law that defines the "scattering vector" q_s.

(or power per unit area) that can be calculated as the mean Poynting flux $I = (\epsilon_o/\mu_o)^{1/2}E'^2$. In the direction θ, therefore, combining this equation with Eqs. (70), we find an intensity

$$I(\theta) = \left(\frac{\epsilon_o}{\mu_o}\right)^{1/2} \frac{\ddot{p}^2 \cos^2 \theta}{16\pi^2 \epsilon_o^2 c^4 r^2} \tag{71}$$

Defining an incident intensity $I_o = (\epsilon_o/\mu_o)^{1/2}E^2$ and using Eq. (69) together with $\ddot{E} = -\omega^2 E = -(2\pi c/\lambda)^2 E$ transforms Eq. (71) into

$$I(\theta) = I_o \frac{9\pi^2 V^2 \epsilon_1^2 (\epsilon_2 - \epsilon_1)^2 \cos^2 \theta}{r^2 \lambda^4 (\epsilon_2 + 2\epsilon_1)^2} \tag{72}$$

which is now independent of the (cgs or SI) unit system. If we consider unpolarized light (with components both in and perpendicular to the plane of Fig. 2.13, then Eq. (72) obviously generalizes to (Fabelinskii 1968)

$$I(\theta) = I_o \frac{9\pi^2 V^2 \epsilon_1^2}{2r^2 \lambda^4} \left(\frac{\epsilon_2 - \epsilon_1}{\epsilon_2 + 2\epsilon_1}\right)^2 (1 + \cos^2 \theta) \tag{73}$$

An important feature of this relationship, which will occur throughout this section, is the λ^{-4} wavelength dependence known as *Rayleigh's law*.

In solids and liquids the variations $\epsilon_2 - \epsilon_1 = \Delta\epsilon$, caused by any static or dynamic fluctuations in the medium, are small in amplitude compared to ϵ_1 and are uniformly spaced throughout the sample volume L^3. For this case Eq. (73) can now be expressed as

$$I(\theta) = \frac{I_o L^3 \pi^2}{2 r^2 \lambda^4} (1 + \cos^2 \theta) \left\langle \int \Delta\epsilon(r) \, \Delta\epsilon(0) \, d^3r \right\rangle \qquad (74)$$

in which $\Delta\epsilon(r)$ is the dielectric fluctuation at the point r, the integral d^3r runs over the spatial extent of the fluctuation, and $\langle \cdot \cdot \cdot \rangle$ indicates an ensemble (or thermal) average.

Evidently, if the medium is perfectly homogeneous, $\Delta\epsilon(r) = 0$ for all r, and no scattering results. Microscopically this implies a full coherence between light rays scattered by different molecules, which interfere in such a way as to cancel the resultant in every direction except that ($\theta = 0$) of the incident beam. However, inhomogeneities of any kind, whether random (e.g., impurity atoms) or periodic (e.g., phonons) induce an angularly dependent scattering of the above form. Summing Eq. (74) over all angles θ, the total power P_s, scattered in traversing a cube L^3 of material is (Chu 1974)

$$P_s = \frac{8\pi^3 L^3 I_o}{3\lambda^4} \left\langle \int \Delta\epsilon(r) \, \Delta\epsilon(0) \, d^3r \right\rangle \qquad (75)$$

When expressed in terms of incident power $P(0) = I_o L^2$, this can be related to the loss coefficient α of Eq. (114) of Chapter 1 in the form

$$P_s = P(0) - P(L) = P(0)(1 - e^{-\alpha L}) \qquad (76)$$

to give, for small losses $\alpha L \ll 1$,

$$\alpha = \alpha_s = \frac{8\pi^3}{3\lambda^4} \left\langle \int \Delta\epsilon(r) \, \Delta\epsilon(0) \, d^3r \right\rangle \qquad (77)$$

This loss from scattered radiation will, in general, add to the attenuation from absorption phenomena of the types discussed in Section 2.1.

2.3.1. Rayleigh and Mie Scattering

For any static spatial fluctuations in ϵ, or equivalently in refractive index n, conservation of energy requires the scattered light to be equal in frequency to the incident light (elastic scattering). If the scattering centers are sufficiently remote from each other that they can be considered independent sources we have two distinct situations referred to, respectively, as the Rayleigh and Mie scattering regimes. The former, represented by Eq. (77), concerns situations for which the spatial extent of the scattering fluctuations is small compared to λ. The latter covers scattering for which the spatial extent of the scattering centers approaches, or even exceeds, the radiation wavelength.

For Rayleigh scattering the formalism is particularly simple in that it is independent of the detailed shape of the scattering center. In particular, we can rewrite the ensemble average in Eq. (77) in the manner

$$\left\langle \int \Delta\epsilon(r) \, \Delta\epsilon(0) \, d^3r \right\rangle = \langle (\Delta\epsilon)^2 \rangle \, v_\epsilon \qquad (78)$$

or in words, as the product of a mean square fluctuation magnitude and an associated correlation volume v_ϵ of unspecified shape. Mie scattering is a much more difficult problem, because it does depend on the shape of the scattering center. The simplest solutions are obtained when spherical scatterers (say, of radius a) are assumed and were first set out by Mie (1908). Although the details of the calculations are beyond the scope of this section [see Bayvel and Jones (1981) for a more recent discussion], two general features can be described. First, the angular dependence $\cos^2 \theta$ of Eq. (72) becomes increasingly skewed toward forward scattering ($\theta = 0$) as a/λ increases, and the pattern develops a ripple substructure. Second, the integrated loss α increases first in the Rayleigh fashion as $(a/\lambda)^4$ but then less rapidly, going through a maximum near $\lambda = a$ and finally settling down for $a/\lambda \gg 1$ to a basically λ-independent value but again with a ripple substructure that "echoes" the sphere size a.

In the context of scattering in solids and liquids, the $a/\lambda \ll 1$ limit is usually valid, although exceptions may occur near critical points (critical opalescence) or near phase separations, where the fluctuation "correlation length" ℓ_c, defined by $\Delta\epsilon(r) \sim \Delta\epsilon(0)e^{-r/\ell_c}$, tends to diverge. The Rayleigh criterion certainly holds for single atomic or single molecular impurities in a crystal. More important, the condition also holds for diffusing density fluctuations in liquids and their frozen-in counterparts in glasses. Thus, whereas the elastic scattering in good-quality single crystals is usually relatively small compared to the scattering from propagating phonon sources (of the kind discussed in Sections 2.3.3 and 2.3.4), it is nearly always completely dominant in glasses.

In glasses, this Rayleigh scattering "fills in" the bottom of the optic window (see Fig. 2.14 and compare Fig. 2.5) in a manner that makes the ultimate minimum attenuation within the window almost completely of Rayleigh scattering origin (Nassau and Lines 1986). In ideally clean (i.e., impurity- and defect-free) and compositionally homogeneous glasses, the dominant sources of this Rayleigh scattering are density fluctuations $\Delta\rho$ that were in thermal equilibrium at the density-freezing "fixation" temperature T_F of the melt. Using equilibrium

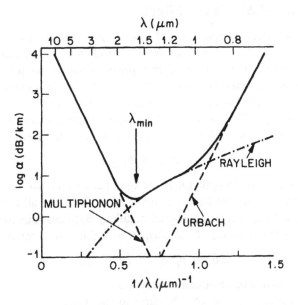

Figure 2.14 Schematic of the theoretical attenuation for a glass as a function of reciprocal wavelength on a log-linear plot, emphasizing the separate component contributions from the Urbach edge, Rayleigh scattering, and the multiphonon tail. The scale is qualitatively appropriate for an oxide glass.

thermodynamics at $T = T_F$, their magnitude can be approximately related to compressibility K in the manner (Schroeder 1977; Lines 1984)

$$\langle(\Delta\epsilon)^2\rangle_{v_\epsilon} = \left(\frac{\partial\epsilon}{\partial\rho}\right)^2 \langle(\Delta\rho)^2\rangle_{v_\rho} \approx \rho^2\left(\frac{\partial\epsilon}{\partial\rho}\right)^2 kT_F K_T(T_F) \qquad (79)$$

where $K_T(T_F)$ is the static isothermal compressibility at fixation. Using Eqs. (77)–(79), it follows that the Rayleigh loss α_ρ from frozen density fluctuations is

$$\alpha_\rho = \left(\frac{8\pi^3}{3\lambda^4}\right) \rho^2 \left(\frac{\partial\epsilon}{\partial\rho}\right)^2 kT_F K_T(T_F) \qquad (80)$$

Putting λ equal to the value λ_{min} at which attenuation is lowest (Fig. 2.14), this equation then approximately establishes the minimum value of attenuation α_{min} possible for the glass in question.

2.3.2. The Photoelastic Effect

To obtain numerical estimates from Eq. (80), it is necessary to know the sensitivity of the optical dielectric response ϵ to changes in density ρ. This is a particular example of the more general phenomenon of photoelasticity, or the dependence of the dielectric tensor ϵ_{ij} (which is scalar for a glass) on strain $\eta_{k\ell}$. The photoelastic compliance, which measures this effect, is traditionally cast (Narasimhamurty 1981) in terms of the impermeability tensor $\mathbf{B} = \epsilon^{-1}$ by expressing small changes $\Delta\mathbf{B}$ resulting from strain in the form

$$\Delta B_{ij} = P_{ijk\ell}\, \eta_{k\ell} \tag{81}$$

where $P_{ijk\ell}$ are the strain-optic (or elasto-optic) constants. Since both B_{ij} and $\eta_{k\ell}$ are *symmetric* tensors, the 81 components of $P_{ijk\ell}$ are not all independent. Expressing B_{ij} and $\eta_{k\ell}$ in their contracted notations B_i [analogous to σ_i of Eq. (29) of Chapter 1] and η_j of Eq. 30 of Chapter 1 enables us to rewrite Eq. (81) as

$$\Delta B_i = p_{ij}\, \eta_j \qquad i,j, = 1, \ldots ,6 \tag{82}$$

with at most 36 independent components p_{ij}.

One can equally well relate ΔB_{ij} (ΔB_i) to stress $\sigma_{k\ell}$ (σ_j):

$$\Delta B_{ij} = q_{ijk\ell}\sigma_{k\ell} \tag{83}$$

$$\Delta B_i = q_{ij}\sigma_j, \qquad i,j = 1, \ldots ,6 \tag{84}$$

to define stress-optic (or piezo-optic) constants $q_{ijk\ell}$ (q_{ij}). The q and p compliances are related via the elastic compliance s and stiffness c of Section 1.3.1 in the form

$$q_{ijk\ell} = P_{ijmn}S_{mnk\ell}, \qquad P_{ijmn} = q_{ijk\ell}c_{k\ell mn} \tag{85}$$

They are nonzero for all 32 crystal point groups and also for the isotropic solid. The matrix forms of p_{ij} and q_{ij} for all crystal classes are shown in Table 2.1.

The strain components that correspond to a density fluctuation in an isotropic material are

$$\eta_{11} = \eta_{22} = \eta_{33} = \frac{1}{3}(\Delta V/V) = -\frac{1}{3}(\Delta\rho/\rho) \tag{86}$$

where V is the volume. It follows that

$$\Delta B = -(\Delta\epsilon/\epsilon^2) = -\frac{1}{3}(p_{11} + 2p_{12})(\Delta\rho/\rho) = -p(\Delta\rho/\rho) \tag{87}$$

where

$$p = \frac{1}{3}(p_{11} + 2p_{12}) \tag{88}$$

Table 2.1 Elasto-optic and Piezo-optic Matrices

Triclinic, classes C_1, C_i $(1, \bar{1})$ (36 constants)

p_{11}	p_{12}	p_{13}	p_{14}	p_{15}	p_{16}	q_{11}	q_{12}	q_{13}	q_{14}	q_{15}	q_{16}
p_{21}	p_{22}	p_{23}	p_{24}	p_{25}	p_{26}	q_{21}	q_{22}	q_{23}	q_{24}	q_{25}	q_{26}
p_{31}	p_{32}	p_{33}	p_{34}	p_{35}	p_{36}	q_{31}	q_{32}	q_{33}	q_{34}	q_{35}	q_{36}
p_{41}	p_{42}	p_{43}	p_{44}	p_{45}	p_{46}	q_{41}	q_{42}	q_{43}	q_{44}	q_{45}	q_{46}
p_{51}	p_{52}	p_{53}	p_{54}	p_{55}	p_{56}	q_{51}	q_{52}	q_{53}	q_{54}	q_{55}	q_{56}
p_{61}	p_{62}	p_{63}	p_{64}	p_{65}	p_{66}	q_{61}	q_{62}	q_{63}	q_{64}	q_{65}	q_{66}

Monoclinic, classes C_s, C_2, C_{2h} $(m, 2, 2/m)$ (20 constants); diad $\parallel x_3$

p_{11}	p_{12}	p_{13}	0	0	p_{16}	q_{11}	q_{12}	q_{13}	0	0	q_{16}
p_{21}	p_{22}	p_{23}	0	0	p_{26}	q_{21}	q_{22}	q_{23}	0	0	q_{26}
p_{31}	p_{32}	p_{33}	0	0	p_{36}	q_{31}	q_{32}	q_{33}	0	0	q_{36}
0	0	0	p_{44}	p_{45}	0	0	0	0	q_{44}	q_{45}	0
0	0	0	p_{54}	p_{55}	0	0	0	0	q_{54}	q_{55}	0
p_{61}	p_{62}	p_{63}	0	0	p_{66}	q_{61}	q_{62}	q_{63}	0	0	q_{66}

Orthorhombic, classes D_2, C_{2v}, D_{2h} $(222, mm2, mmm)$ (12 constants)

p_{11}	p_{12}	p_{13}	0	0	0	q_{11}	q_{12}	q_{13}	0	0	0
p_{21}	p_{22}	p_{23}	0	0	0	q_{21}	p_{22}	p_{23}	0	0	0
p_{31}	p_{32}	p_{33}	0	0	0	q_{31}	q_{32}	q_{33}	0	0	0
0	0	0	p_{44}	0	0	0	0	0	q_{44}	0	0
0	0	0	0	p_{55}	0	0	0	0	0	q_{55}	0
0	0	0	0	0	p_{66}	0	0	0	0	0	q_{66}

Trigonal, classes C_3, S_6 $(3, \bar{3})$ (12 constants)

p_{11}	p_{12}	p_{13}	p_{14}	$-p_{25}$	p_{62}	q_{11}	q_{12}	q_{13}	q_{14}	$-q_{25}$	$2q_{62}$
p_{12}	p_{11}	p_{13}	$-p_{14}$	p_{25}	$-p_{62}$	q_{12}	q_{11}	q_{13}	$-q_{14}$	q_{25}	$-2q_{62}$
p_{31}	p_{31}	p_{33}	0	0	0	q_{31}	q_{31}	p_{33}	0	0	0
p_{41}	$-p_{41}$	0	p_{44}	p_{45}	p_{52}	q_{41}	$-q_{41}$	0	q_{44}	q_{45}	$2q_{52}$
$-p_{52}$	p_{52}	0	$-p_{45}$	p_{44}	p_{41}	$-q_{52}$	q_{52}	0	$-q_{45}$	q_{44}	$2q_{41}$
$-p_{62}$	p_{62}	0	p_{25}	p_{14}	$\tfrac{1}{2}(p_{11}-p_{12})$	$-q_{62}$	q_{62}	0	q_{25}	q_{14}	$(q_{11}-q_{12})$

Trigonal, classes C_{3v}, D_3, D_{3d} $(3m, 32, \bar{3}m)$ (8 constants)

p_{11}	p_{12}	p_{13}	p_{14}	0	0	q_{11}	q_{12}	q_{13}	q_{14}	0	0
p_{12}	p_{11}	p_{13}	$-p_{14}$	0	0	q_{12}	q_{11}	q_{13}	$-q_{14}$	0	0
p_{31}	p_{31}	p_{33}	0	0	0	q_{31}	q_{31}	q_{33}	0	0	0
p_{41}	$-p_{41}$	0	p_{44}	0	0	q_{41}	$-q_{41}$	0	q_{44}	0	0
0	0	0	0	p_{44}	p_{41}	0	0	0	0	q_{44}	$2q_{41}$
0	0	0	0	p_{14}	$\tfrac{1}{2}(p_{11}-p_{12})$	0	0	0	0	q_{14}	$(q_{11}-q_{12})$

(continued)

Table 2.1   Elasto-optic and Piezo-optic Matrices (Continued)

Tetragonal, classes C_4, S_4, C_{4h} (4, $\bar{4}$, 4/m) (10 constants)

p_{11}	p_{12}	p_{13}	0	0	p_{16}	q_{11}	q_{12}	q_{13}	0	0	q_{16}
p_{12}	p_{11}	p_{13}	0	0	$-p_{16}$	q_{12}	q_{11}	q_{13}	0	0	$-q_{16}$
p_{31}	p_{31}	p_{33}	0	0	0	q_{31}	q_{31}	q_{33}	0	0	0
0	0	0	p_{44}	p_{45}	0	0	0	0	q_{44}	q_{45}	0
0	0	0	$-p_{45}$	p_{44}	0	0	0	0	$-q_{45}$	q_{44}	0
p_{61}	$-p_{61}$	0	0	0	p_{66}	q_{61}	$-q_{61}$	0	0	0	q_{66}

Tetragonal, classes D_{2d}, C_{4v}, D_4, D_{4h} ($\bar{4}2m$, 4mm, 422, 4/mmm) (7 constants)

p_{11}	p_{12}	p_{13}	0	0	0	q_{11}	q_{12}	q_{13}	0	0	0
p_{12}	p_{11}	p_{13}	0	0	0	q_{12}	q_{11}	q_{13}	0	0	0
p_{31}	p_{31}	p_{33}	0	0	0	q_{31}	q_{31}	q_{33}	0	0	0
0	0	0	p_{44}	0	0	0	0	0	q_{44}	0	0
0	0	0	0	p_{44}	0	0	0	0	0	q_{44}	0
0	0	0	0	0	p_{66}	0	0	0	0	0	q_{66}

Hexagonal, classes C_{3h}, C_6, C_{6h} ($\bar{6}$, 6, 6/m) (8 constants)

p_{11}	p_{12}	p_{13}	0	0	$-p_{61}$	q_{11}	q_{12}	q_{13}	0	0	$-2q_{61}$
p_{12}	p_{11}	p_{13}	0	0	p_{61}	q_{12}	q_{11}	q_{13}	0	0	$2q_{61}$
p_{31}	p_{31}	p_{33}	0	0	0	q_{31}	q_{31}	q_{33}	0	0	0
0	0	0	p_{44}	p_{45}	0	0	0	0	q_{44}	q_{45}	0
0	0	0	$-p_{45}$	p_{44}	0	0	0	0	$-q_{45}$	q_{44}	0
p_{61}	$-p_{61}$	0	0	0	$\frac{1}{2}(p_{11}-p_{12})$	q_{61}	$-q_{61}$	0	0	0	$(q_{11}-q_{12})$

Hexagonal, classes D_{3h}, C_{6v}, D_6, D_{6h} ($\bar{6}m2$, 6mm, 622, 6/mmm) (6 constants)

p_{11}	p_{12}	p_{13}	0	0	0	q_{11}	q_{12}	q_{13}	0	0	0
p_{12}	p_{11}	p_{13}	0	0	0	q_{12}	q_{11}	q_{13}	0	0	0
p_{31}	p_{31}	p_{33}	0	0	0	q_{31}	q_{31}	q_{33}	0	0	0
0	0	0	p_{44}	0	0	0	0	0	q_{44}	0	0
0	0	0	0	p_{44}	0	0	0	0	0	q_{44}	0
0	0	0	0	0	$\frac{1}{2}(p_{11}-p_{12})$	0	0	0	0	0	$(q_{11}-q_{12})$

Cubic, classes T, T_h (23, m3) (4 constants)

p_{11}	p_{12}	p_{13}	0	0	0	q_{11}	q_{12}	q_{13}	0	0	0
p_{13}	p_{11}	p_{12}	0	0	0	q_{13}	q_{11}	q_{12}	0	0	0
p_{12}	p_{13}	p_{11}	0	0	0	q_{12}	q_{13}	q_{11}	0	0	0
0	0	0	p_{44}	0	0	0	0	0	q_{44}	0	0
0	0	0	0	p_{44}	0	0	0	0	0	q_{44}	0
0	0	0	0	0	p_{44}	0	0	0	0	0	q_{44}

(*continued*)

Table 2.1 Elasto-optic and Piezo-optic Matrices (Continued)

Cubic, classes T_d, O, O_h ($\overline{4}3m$, 432, $m3m$) (3 constants)

$$
\begin{bmatrix}
p_{11} & p_{12} & p_{12} & 0 & 0 & 0 \\
p_{12} & p_{11} & p_{12} & 0 & 0 & 0 \\
p_{12} & p_{12} & p_{11} & 0 & 0 & 0 \\
0 & 0 & 0 & p_{44} & 0 & 0 \\
0 & 0 & 0 & 0 & p_{44} & 0 \\
0 & 0 & 0 & 0 & 0 & p_{44}
\end{bmatrix}
\qquad
\begin{bmatrix}
q_{11} & q_{12} & q_{12} & 0 & 0 & 0 \\
q_{12} & q_{11} & q_{12} & 0 & 0 & 0 \\
q_{12} & q_{12} & q_{11} & 0 & 0 & 0 \\
0 & 0 & 0 & q_{44} & 0 & 0 \\
0 & 0 & 0 & 0 & q_{44} & 0 \\
0 & 0 & 0 & 0 & 0 & q_{44}
\end{bmatrix}
$$

Isotropic solids (2 constants)

$$
\begin{bmatrix}
p_{11} & p_{12} & p_{12} & 0 & 0 & 0 \\
p_{12} & p_{11} & p_{12} & 0 & 0 & 0 \\
p_{12} & p_{12} & p_{11} & 0 & 0 & 0 \\
0 & 0 & 0 & \tfrac{1}{2}(p_{11}-p_{12}) & 0 & 0 \\
0 & 0 & 0 & 0 & \tfrac{1}{2}(p_{11}-p_{12}) & 0 \\
0 & 0 & 0 & 0 & 0 & \tfrac{1}{2}(p_{11}-p_{12})
\end{bmatrix}
\qquad
\begin{bmatrix}
q_{11} & q_{12} & q_{12} & 0 & 0 & 0 \\
q_{12} & q_{11} & q_{12} & 0 & 0 & 0 \\
q_{12} & q_{12} & q_{11} & 0 & 0 & 0 \\
0 & 0 & 0 & (q_{11}-q_{12}) & 0 & 0 \\
0 & 0 & 0 & 0 & (q_{11}-q_{12}) & 0 \\
0 & 0 & 0 & 0 & 0 & (q_{11}-q_{12})
\end{bmatrix}
$$

We can therefore write

$$\rho \frac{\partial \epsilon}{\partial \rho} = p\epsilon^2 = pn^4 \tag{89}$$

from which, by substitution into Eq. (80), we obtain a final form for density fluctuation Rayleigh loss in a glass:

$$\alpha_\rho \approx (8\pi^3/3\lambda^4)n^8 p^2 kT_F K_T(T_F) \tag{90}$$

We note, in particular, an extreme sensitivity to refractive index. For glasses, p is typically in the range $0.15-0.3$; for crystals the range is wider.

In crystals the p_{ij} compliances can be of either sign, and there is much confusion in the literature, where often only the magnitude is recorded. There are, however, some general empirical trends for normal strain components p_{ij}, $i,j = 1,2,3$, that can help to resolve the difficulty in many cases. The p_{ij} appear to be almost always positive for crystals with low cation and anion coordination (particularly planar and chainlike compounds). For high-coordination, three-dimensionally connected structures, the p_{ij} are positive for halides but decrease in value as anion valency Z_A increases to become near zero for many chalcogenides ($Z_A = 2$) and almost universally negative for pnictides ($Z_A = 3$) and elemental semiconductors (Weinstein et al. 1981; Lines 1986).

2.3.3. Spontaneous and Stimulated Brillouin Scattering

Whereas dielectric fluctuations that result from static or diffusive sources induce an elastic (or quasi-elastic) scattering of light, those from propagating modulations of refractive index (such as phonons) give rise generally to inelastic components. Such propagating contributions can be broadly separated into two categories, one involving acoustic vibrational modes (Brillouin scattering) and the other optical modes of vibration (Raman scattering).

In acoustical vibrations, all atoms within a single unit cell (say, of volume τ_0) vibrate in unison, so that one can proceed without regard for the details of intracell structure. Summing over unit cells i and j, we can recast the ensemble average $F = \langle \int \Delta\epsilon(r) \, \Delta\epsilon(0) \, d^3r \rangle$ of Eq. (74) as

$$F = N^{-1} \sum_i \sum_j \langle \Delta\epsilon_i \, \Delta\epsilon_j \rangle_{\tau_0} = N^{-1} \sum_i \sum_j \frac{\partial\epsilon_i}{\partial\eta_i} \frac{\partial\epsilon_j}{\partial\eta_j} \langle \Delta\eta_i \Delta\eta_j \rangle_{\tau_0} \quad (91)$$

in which Σ_i and Σ_j run over N cells, η_j is the normal strain at the jth cell site, and we shall use a scalar formalism for simplicity. Utilizing the definition Eq. (82), $B = \epsilon^{-1} = p\eta$, of the elasto-optic constant p in the form

$$-(\partial\epsilon/\partial\eta_j) = \epsilon_j^2 p_j = n^4 p \quad (92)$$

reduces Eq. (91) to

$$F = N^{-1} \sum_i \sum_j n^8 p^2 \langle \Delta\eta_i \Delta\eta_j \rangle \, \tau_0 \quad (93)$$

Introducing acoustic phonons $\Delta Q(q) = N^{-1/2} \sum_j e^{-iq\ell_0 j} \Delta Q_j$, where unit cell volume $\tau_0 = \ell_0^3$, the jth component of normal strain follows as

$$\Delta\eta_j = \ell_0^{-1}(\Delta Q_{j+1} - \Delta Q_j) = N^{-1/2} \sum_q iq e^{iq\ell_0 j} \Delta Q(q) \quad (94)$$

and its correlation as

$$\langle \Delta\eta_i \, \Delta\eta_j \rangle = N^{-1} \sum_q \sum_{q'} (-qq') e^{iq\ell_0(i+j)} \langle \Delta Q(q) \, \Delta Q(q') \rangle \quad (95)$$

Inserting Eq. (95) into Eq. (93) leads now to

$$F = N^{-1} \sum_j \sum_q q^2 n^8 p^2 \, e^{iq\ell_0 j} \langle | \Delta Q(q) |^2 \rangle_{\tau_0} \quad (96)$$

The spatial dependence of the cell summation j in this equation, and therefore also in $I(\theta)$ of Eq. (74), is sinusoidal with a wavelength $2\pi/q$ for each q. Each acoustic phonon q can consequently contribute a nonzero amount to scattering only if q is equal to the difference $q_s = q_{inc} - q_{scat}$ of incident and scattered light

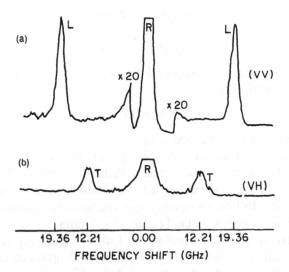

Figure 2.15 Spectrum of Rayleigh–Brillouin scattered light from fused silica as observed in (a) the parallel VV and (b) cross-polarized VH orientational configurations. R, L, and T mark respectively the Rayleigh line and the longitudinal and transverse Brillouin lines. [After Schroeder (1977).]

vector (Fig. 2.13). This is the *Bragg momentum conservation* rule. With energy conservation restricting $|q_{scat}| \approx |q_{inc}| = q_o$ (say), since light velocity c far exceeds sound velocity v, we obtain directly from the geometry of Fig. 2.13b,

$$q_s/q_o = \pm 2 \sin(\theta/2) \tag{97}$$

which, putting $q_s = \omega_s/v$ and $q_o = \omega_o/c$, give frequency shifts ω_s of the Brillouin lines from the Rayleigh line ω_o of

$$\omega_s = \pm 2n\omega_o(v/c) \sin(\theta/2) \tag{98}$$

Two inelastic Brillouin lines at $\omega_o + \omega_s$ and $\omega_o - \omega_s$, referred to respectively as the anti-Stokes (phonon annihilation) and Stokes (phonon creation) lines, therefore bracket the Rayleigh line for each angle θ. In actuality there are two such pairs (corresponding to longitudinal and transverse acoustic phonons) for each scattering angle θ. They can be studied separately by using polarized light in the parallel (VV) and cross-polarized (VH) orientational configurations (Dil 1982). A typical Brillouin spectrum is shown in Fig. 2.15.

Substituting F [i.e., $\langle \int \Delta\epsilon(r) \, \Delta\epsilon(0) \, d^3r \rangle$] of Eq. (96) into Eq. (77) now provides us with an estimate for Brillouin power loss,

$$\alpha(\text{Bril}) = (8\pi^3/3\lambda^4)q_s^2 n^8 p^2 \langle |Q(q)|^2 \rangle \tau_o \tag{99}$$

Using the high-temperature harmonic oscillator result

$$\left\langle \mid Q(q) \mid \right\rangle^2 = kT/\rho\tau_0\omega_s^2 \tag{100}$$

and relating v to adiabatic compressibility K_S via $v^2 = 1/\rho K_S$ (Pollard 1977) reduces (99) to its final form

$$\alpha(\text{Bril}) = (8\pi^3/3\lambda^4)n^8p^2kTK_S\ (T) \tag{101}$$

where T is the temperature of the sample.

This equation for Brillouin loss, which is valid for all except the very lowest temperatures, is of closely similar form to that of Eq. (90) for Rayleigh density loss α_p. This similarity had led to much discussion concerning the ratio of total Rayleigh to total Brillouin scattering loss (particularly in viscous fluids and glasses), a quantity known as the Landau–Placek ratio.

In full detail the intensity equations for Brillouin scattering in crystals naturally depend on the complete tensor form of the elasto-optic constants. The most frequently used experimental arrangement studies 90° scattering [see, for example, Schroeder (1980)], and proper attention must be given to the polarization characteristics of both the incident and collected beams. In general, the positions of the Brillouin lines carry information about acoustic velocities and adiabatic compressibilities; the intensities of the lines, information about the photoelastic constants. The line widths, which are typically several hundred megahertz, measure the acoustic phonon lifetimes.

Although the Stokes (S) and anti-Stokes (AS) lines are essentially equally intense at room temperature, the equality is not exact, because the S line involves phonon excitation (which can take place even at $T = 0$) whereas the AS line involves phonon annihilation and can be nonzero only if the phonon spectrum is thermally excited (i.e., its intensity goes to zero as $T \rightarrow 0$). In detail, the intensity ratio is controlled by the Bose–Einstein thermal occupation number $n(\omega_s) = [\exp(\hbar\omega_s/kT) - 1]^{-1}$ in the form

$$I(S)/I(AS) = [1 + n(\omega_s)]/n(\omega_s) \tag{102}$$

The contribution $\sim n(\omega_s)$ is referred to as spontaneous Brillouin scattering; being thermally activated, it is incoherent and is completely dominant at room temperature. However, the unity term in the Stokes intensity numerator represents a coherent scattering contribution. It can become very important in nonlinear processes at high incident intensities I_0, because it provides a stimulated gain mechanism $P(L) = P(0)\exp(GI_0L)$, which increases exponentially with I_0 and can outweigh the incoherent loss $P(L) = P(0)e^{-\alpha L}$ if the stimulated Brillouin gain coefficient G times I_0 is greater than the total integrated loss α from all other sources (Stolen 1979). Note, however, that since $\omega_s = 0$ for $\theta = 0$ [Eq. (98)], such gain can be produced in a fiber only in a backward configuration $\theta = \pi$, that is, with the pump beam propagating in a direction opposite to that of the carrier beam.

From an applications point of view, the photoelastic coupling between acoustic modes and radiation can be used to generate acoustically established phase gratings. These lead to a class of devices such as optical switches and modulators. The criteria for the selection of good acousto-optical materials have been discussed by Pinnow (1970) and, to a significant degree, parallel those that provide for a large Brillouin scattering intensity. The detailed form of the acousto-optic figure of merit, however, depends somewhat on the precise device operation envisaged, although it tends to have the general form $n^a p^2/\rho v^b$, in which $a \approx 6$–7, $b \approx 1$–3, and v is the acoustic velocity.

2.3.4. Raman Scattering

Raman scattering, which is the scattering of light by optical-mode lattice vibrations, is theoretically described in a manner that parallels the development for Brillouin scattering in Section 2.3.3, except that the correlation $\langle \Delta\epsilon(r)\, \Delta\epsilon(0)\, d^3 r \rangle$ of Eq. (74) must now be expanded in terms of optical mode variables instead of strain. We can again derive for crystals a Bragg momentum conservation law $q_s = q_{inc} - q_{scat}$, for example, but in other respects the details are very different (Anderson 1971).

First, for the small wave vectors (on the scale of the Brillouin zone) that are relevant for first-order light scattering, the optical mode frequencies do not go to zero, like their acoustical counterparts, but approach finite nonzero polariton values ω_i with a negligible wave-vector dependence. The Bragg condition therefore translates to a frequency dependence of scattered radiation ω_s with respect to incident radiation ω_o of the form

$$\omega_s = \omega_o \pm \omega_i \tag{103}$$

which is angularly independent, in contrast with the Brillouin finding of Eq. (98). Second, many different "intracell" optical vibrational modes i are Raman-active, but to differing degrees depending on the relative efficiency with which a particular local distortion can perturb the components of the dielectric response tensor ϵ_{jk}. As a result, a whole host of Stokes and anti-Stokes Raman lines can, for more complex crystal structures, bracket the Rayleigh–Brillouin triplet of Fig. 2.15. Third, not all optical phonon modes can perturb ϵ_{jk} in higher symmetry crystal structures. Consequently, Raman-active vibrations (defined as modes that can induce a first-order or single-phonon scattering) are determined by selection rules. In detail, a particular phonon is active only if it transforms under the symmetry operations of the point group of the lattice in the same way as the components of ϵ_{jk} do. The full selection rules are given in Lines and Glass (1977). Fourth, since the Raman frequency shifts are normally measured in hundreds of cm^{-1} (as opposed to hundredths of a cm^{-1} for Brillouin shifts) the intensity difference of the Stokes and anti-Stokes lines, as expressed by Eq. (102), can be large at room temperature for the higher frequency lines.

The selection rules for infrared absorption and those for Raman scattering are quite different. The distinction is particularly dramatic for centrosymmetric structures, where all modes that change sign under inversion symmetry can be shown to be Raman-inactive. These include all the infrared-active polar modes, so that for such crystals infrared and Raman spectroscopy are complementary techniques for the study of optic phonons. In some simple high-symmetry structures (like the alkali halides), all the optic phonons are Raman-inactive and no first-order Raman spectrum exists. For these, second-order Raman scattering, involving the excitation of two phonons with closely equal and opposite wave vectors $\mathbf{q}$, is seen. Such a spectrum is allowed for all values $\mathbf{q}$ in the first Brillouin zone and is consequently much more diffuse than a first-order spectrum and reflects to a significant degree the joint density of two-phonon states.

A nonlinear stimulated Raman scattering mechanism can also be defined that is exactly analogous to that for Brillouin spectra (Stolen 1979). One essential difference exists, however, concerning the fact that the Raman frequency shift is essentially independent of scattering angle. It follows that both forward and backward gain configurations (with pump beam propagating respectively with or against the carrier beam) can be defined.

On passage into a glassy, or amorphous, state, the disappearance of periodicity relaxes all the selection rules derivable from structural invariance under translation and rotation (Shuker and Gammon 1970). As a result, *all* phonon modes (both optical and acoustic) of all wave vectors become first-order Raman-active to some degree. Many major features of a crystalline Raman spectrum remain recognizable in broadened forms, with a line shape that largely reflects the density of states of the optical spectral branch in question, but now a completely new feature appears. It is the "acoustic Raman" spectrum (Lines 1987), which extends in frequency from zero to the zone-edge frequency, which, for the typical acoustic branch, is 200–400 cm^{-1}. At the lower extreme of this range the true Brillouin peak is still present, because it remains well defined even in a homogeneous medium. In Fig. 2.16 we schematically represent the differences in a room-temperature light-scattering spectrum between the crystalline and amorphous phases of nominally the same material (which possesses one Raman-active optical vibrational branch in its phonon spectrum). The peak in the acoustic Raman component of the amorphous spectrum occurs typically in the range 30–80 cm^{-1} and results from competition between the thermal population of acoustic phonons (which increases without bound as $q \rightarrow 0$) and a density of states (which decreases to zero as $q \rightarrow 0$). The acoustic Raman spectrum is consequently very temperature-dependent.

Finally we should mention that the discussion of Raman scattering so far presented has been limited in two respects. First, some optic phonons that are

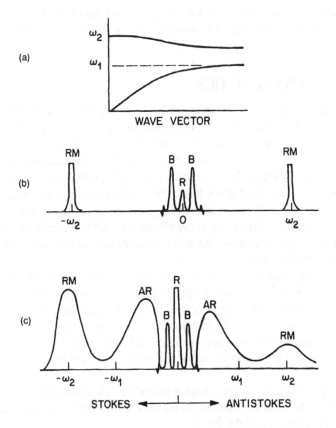

Figure 2.16 Room-temperature Rayleigh (R), Brillouin (B), and Raman (RM) spectra of (b) the crystalline phase and (c) the amorphous phase of a material with the lattice vibrational dispersion shown in (a). The amorphous spectrum (c) contains "acoustic Raman" peaks (AR) that are produced by the Raman activity induced in the acoustical branch of the vibrational spectrum by structural disorder. Note also the increase in intensity of the Rayleigh peak in spectrum (c) due to this same disorder.

Raman-active for light of one particular plane of polarization may be inactive for light polarized in the perpendicular plane. Consequently, Raman spectra should be analyzed for their dependence on polarization for both incident and scattered light in order to extract as much information as possible about the vibrational modes involved. Second, we have so far assumed that the incident light frequency ω_0 is far removed from that at which electronic interband excitations can be induced (i.e., is well within the optic window). When incident light energy $\hbar\omega_0$ does approach the bandgap energy E_g, then not only can absorption begin to take place, but also the incipient electronic excitations can themselves modulate ϵ_{jk} and influence the Raman scattering intensity. There is a tendency for the

Raman amplitude to diverge as $(\hbar\omega_{\mathrm{o}} - E_{\mathrm{g}})^{-1}$, and the phenomenon is known as *resonant Raman scattering*. An excellent review of the field has been given by Martin and Falikov (1975).

2.4. NONLINEAR OPTICS

In previous sections (except for the brief mention of stimulated Brillouin and stimulated Raman scattering), it was assumed that the intensity of incident infrared radiation is small enough that nonlinear terms in the relations between polarization **P** and radiation field **E** can be ignored. For radiation produced by incoherent light sources, this is generally true. However, with the advent of the laser, the experimental situation has been transformed, because small nonlinear response by a single atom can lead to large effects when many atoms act coherently. With the use of lasers, the study of nonlinear effects concerning the mixing of two or more optical frequencies or the generation of optical harmonics became possible for the first time.

On the other hand, not all properties classified under the general heading "nonlinear optical properties" require laser sources, because not all require the sole use of high-frequency fields. Some, like the Pockels and Kerr electro-optical effects, probe the response of optical radiation to low-frequency (or even static) electric field perturbations. Phenomena of this kind, which are also included in this section, were known and studied for many years before laser light sources became available.

A fundamental macroscopic description of nonlinear processes can be set out by expanding the electric polarization vector **P** as a power series in field **E**, strain $\boldsymbol{\eta}$, and temperature T, in the form

$$\mathbf{P} = \mathbf{P}(\text{linear}) + \mathbf{P}(\text{nonlinear}) \tag{104}$$

with

$$P_i \text{ (linear)} = \epsilon_0 \chi_{ij} E_j + e_{ijk}\, \eta_{jk} + \Pi_i T \tag{105}$$

and

$$P_i \text{ (nonlinear)} =$$
$$\epsilon_0 \chi_{ijk} E_j E_k + \phi_{ijk\ell}\, \eta_{jk} E_\ell + \xi_{ij} T E_j + \epsilon_0 \chi_{ijk\ell} E_j E_k E_\ell + \cdots \tag{106}$$

where the repeated index summation convention is again assumed. Of the linear terms, only the one involving the *pyroelectric* response $\Pi_i T$ has not yet been discussed in detail. It is of major importance only for spontaneously polar materials at low frequencies [particularly in ferroelectric materials near their Curie temperatures (Lines and Glass 1977)] but plays no significant role in the context of infrared optics except for application in the detection of radiation (i.e., measuring charge flow resulting from small radiation-induced changes of temperature).

We are here concerned with the nonlinear terms of Eq. (106) for the case where at least one of the $\mathbf{E}$-components is an optical frequency electric field. The higher-order compliances χ_{ijk}, ϕ_{ijkl}, ξ_{ij}, and so on are then referred to, respectively, as second-order nonlinear susceptibility, photoelastic compliance, and thermo-optic coefficient. The photoelastic compliance is responsible for the interaction between acoustical and optical waves (e.g., Brillouin scattering) and has been discussed already in Sections 2.3.2 and 2.3.3. The thermo-optic coefficient gives rise to the thermal self-focusing of infrared radiation but will not be discussed in detail here. The primary concern of this section will therefore involve the restricted expansion of $\mathbf{P}$ in terms of $\mathbf{E}$ alone:

$$P_i = \epsilon_o[\chi_{ij}E_j + \chi_{ijk}E_jE_k + \chi_{ijk\ell}E_jE_kE_\ell + \ldots] \tag{107}$$

When inserted into the wave equation (105) of Chapter 1, it gives

$$\nabla^2\mathbf{E} = -(\omega^2\hat{\epsilon}/c^2)\,\mathbf{E} = \mu_o\ddot{\mathbf{D}} = (1/c^2)\,(\ddot{\mathbf{E}} + \ddot{\mathbf{P}}/\epsilon_o) \tag{108}$$

cgs: $\quad \nabla^2\mathbf{E} = -(\omega^2\hat{\epsilon}/c^2)\,\mathbf{E} = (1/c^2\ddot{\mathbf{D}}) = (1/c^2)(\ddot{\mathbf{E}} + 4\pi\ddot{\mathbf{P}}) \tag{108'}$

from which it is seen that the nonlinear terms in Eq. (107) act, via $\ddot{\mathbf{P}}$, as sources of the harmonics of any Fourier component introduced into the medium. For example, a quadratic source term arising from

$$P_i \sim (E_j \cos \omega_1 t)^2 = (E_j^2/2)(1 + \cos 2\omega_1 t) \tag{109}$$

generates second harmonics with $\omega = 2\omega_1$ and a dc component $\omega = 0$. In similar fashion, the third-order term $\sim(E_j \cos \omega_1 t)^3$ generates third harmonics with $\omega = 3\omega_1$ and so on. The nonlinear terms also mix two or more frequencies, which are simultaneously introduced into the medium, for example, via

$$P_i \sim (E_j \cos \omega_1 t + E_k \cos \omega_2 t)^2 \tag{110}$$

which has Fourier components at $\omega_1 + \omega_2$, $\omega_1 - \omega_2$, $2\omega_1$, $2\omega_2$, and $\omega = 0$. The higher the order of the effect, the more intense the radiation required to observe it. In practice, the upper limits of intensity are set by the heat produced by the radiation inducing it, which eventually damages the medium.

Quantitatively we define Fourier complex field-amplitude components $E_j(\omega)$ according to (Hodgson 1970)

$$E_j(t) = E_j \cos(\omega t - \phi) = (E_j/2)[e^{i(\phi - \omega t)} + e^{-i(\phi - \omega t)}]$$
$$= (1/2)[E_j(\omega)^{-i\omega t} + E_j^*(\omega)e^{i\omega t}] \tag{111}$$

where $E_j^*(\omega) = E_j(-\omega)$, and corresponding polarization amplitudes from the analogous equation

$$P_i(t) = (1/2)[P_i(\omega)e^{-i\omega t} + P_i^*(\omega)e^{i\omega t}] \tag{112}$$

Wavelength dependence is neglected, because infrared wavelengths are large compared with the dimensions of polarizable entities in the medium. With these definitions of stimulus and response, we can now write Eq. (107) in detail as (Chemla and Jerphagnon 1980, Chemla 1980)

$$P_i(\omega) = \epsilon_0 \{\chi_{ij}(\omega)E_j(\omega) + g_1(\omega)\chi_{ijk}(-\omega; \omega_1, \omega_2)E_j(\omega_1)E_k(\omega_2)\}$$
$$+ g_2(\omega)\chi_{ijk\ell}(-\omega; \omega_1', \omega_2', \omega_3')E_j(\omega_1')E_k(\omega_2')E_\ell(\omega_3') + \cdots \quad (113)$$

where a summation over all distinct sets of frequencies ω_1, ω_1', etc. is implied. In this equation, the field frequencies, having both positive and negative values, are such that $\omega = \omega_1 + \omega_2 = \omega_1' + \omega_2' + \omega_3'$, the negative sign before the semicolon indicating that the wave vectors of the interacting fields must sum to zero. The g factors take account of degeneracies that arise in the number of distinguishable permutations of the frequencies and are given in Table 2.2. In nonsubscripted notation, the higher order susceptibilities are often expressed as $\chi^{(2)} \equiv \chi_{ijk}$, $\chi^{(3)} \equiv \chi_{ijk\ell}$, etc. Some of the more important nonlinear processes that can be defined using the expansion of Eq. (113) are listed in Table 2.2.

2.4.1. Second-Order Nonlinear Optical Response

In addition to universal frequency permutation symmetries of the form (Armstrong et al. 1962)

$$\chi_{ijk}(-\omega_3; \omega_1, \omega_2) = \chi_{jki}(\omega_1; -\omega_3, \omega_2) = \chi_{kji}(\omega_2; \omega_1, -\omega_3)$$
$$= \chi_{ijk}(-\omega_3; \omega_2, \omega_1) \quad (114)$$

and

$$\chi_{ijk}(-\omega_3; \omega_1, \omega_2) = \chi_{ijk}^*(\omega_3; -\omega_1, \omega_2) \quad (115)$$

the second-order susceptibility must also conform to the spatial symmetry restrictions dictated by the relevant point group. In particular, χ_{ijk} necessarily vanishes for all centrosymmetric systems.

Table 2.2 Nonlinear Optical Processes

Susceptibility	Process	g factor
$\chi_{ijk}(-2\omega; \omega, \omega)$	Second-harmonic generation	$g_1 = 1/2$
$\chi_{ijk}(0; \omega, -\omega)$	Optical rectification	$g_1 = 1/2$
$\chi_{ijk}(-\omega; 0, \omega)$	Linear electro-optic (or Pockels) effect	$g_1 = 2$
$\chi_{ijk}(-\omega; \omega_1, \omega_2)$	Parametric (sum or difference) mixing	$g_1 = 1$
$\chi_{ijk\ell}(-3\omega; \omega, \omega, \omega)$	Third-harmonic generation	$g_2 = 1/4$
$\chi_{ijk\ell}(-\omega; \omega, 0, 0)$	Quadratic electro-optic (or Kerr) effect	$g_2 = 3/4$
$\chi_{ijk\ell}(-2\omega; \omega, \omega, 0)$	Field-induced second harmonic generation	$g_2 = 3/4$
$\chi_{ijk\ell}(-\omega; \omega_1, \omega_1, \omega_2)$	Three-wave mixing	$g_2 = 3/4$
$\chi_{ijk\ell}^\dagger(-\omega; \omega_1, -\omega_1, \omega_2)$	Stimulated Raman (or Brillouin) scattering	$g_2 = 3/2$

†Imaginary component.

A further type of symmetry holds when some of the fields involved in the interaction are indistinguishable [e.g., simple harmonic generation $\chi_{ijk}(-2\omega; \omega, \omega)$]. In this case it can be shown that χ_{ijk} is invariant under the permutation of the last two of the Cartesian indices j and k. It is consequently possible to introduce for it a contracted notation in which the indices jk of χ_{ijk} are reduced in the form

$$11 \rightarrow 1, \qquad 22 \rightarrow 2, \qquad 33 \rightarrow 3, \qquad 23 = 32 \rightarrow 4,$$
$$31 = 13 \rightarrow 5, \qquad 12 = 21 \rightarrow 6 \tag{116}$$

equivalent to the Voigt convention for stress of Eq. (29) of Chapter 1. There is consequently a maximum of 18 independent second harmonic susceptibility components $\chi_{ij}^{(2)}$ ($i = 1,2,3; j = 1, \ldots ,6$) for any process that depends on this nonlinearity. The matrix formulation for $\chi_{ij}^{(2)}$ is identical to that of Table 1.10 (Chapter 1) for the piezoelectric stress compliance d_{ij}. The sign convention is also just that given for d_{ij} in Section 1.3.3 (Chapter 1).

This equivalence has led, in the nonlinear optics literature, to a rather confusing adoption of the piezoelectric symbol d_{ijk} for χ_{ijk} (and $d_{ij}^{(2)}$ for $\chi_{ij}^{(2)}$). To be precise (and even more confusing), the exact correspondence involves an additional factor 1/2 according to

$$d_{ijk} = (1/2) \chi_{ijk} \tag{117}$$

$$d_{ij}^{(2)} = (1/2) \chi_{ij}^{(2)} \tag{118}$$

with the 1/2 introduced to eliminate the $g_1 = 1/2$ factor in equations for harmonic generation and optical rectification (Table 2.2) in which context the notation was first developed. With this notation, the equation for second harmonic generation, for example, becomes

$$P_i(2\omega) = \epsilon_o d_{ijk} E_j(\omega) E_k(\omega); \qquad i,j,k = 1,2,3 \tag{119}$$

or, in contracted form,

$$P_i(2\omega) = \epsilon_o d_{ij}^{(2)} E_j^{(2)}(\omega); \qquad i = 1,2,3; j = 1, \ldots 6 \tag{120}$$

in which $E_j^{(2)}, j = 1, \ldots ,6$, denote the quadratic field components E_1^2, E_2^2, E_3^2, $2E_2 E_3, 2E_3 E_1, 2E_1 E_2$, respectively.

Two additional observations have been made concerning second-order susceptibility. The first, due to Kleinman (1962), is that if all three frequencies ω_i in Eq. (114) are in a transparent (optic window) regime, for which frequency dispersion effects are negligible, then χ_{ijk} (or equivalently d_{ijk}) becomes symmetrical with respect to the interchange of any of its subscripts. In contracted d-notation, this implies

$$d_{12} = d_{26}, \qquad d_{13} = d_{35}, \qquad d_{14} = d_{25} = d_{36}$$
$$d_{15} = d_{31}, \qquad d_{16} = d_{21}, \qquad d_{23} = d_{34}, \qquad d_{24} = d_{32} \tag{121}$$

Experimentally this invariance to Cartesian tensor coordinate permutation has been found to be well obeyed in the relevant frequency regime. The second, due to Miller (1964), is the realization that a recasting of Eq. (119) in the form

$$E_i(2\omega) = \epsilon_o^{-1}\, \delta_{ijk}P_j(\omega)P_k(\omega) \tag{122}$$

with 'Miller's delta' δ_{ijk} derivable from d_{ijk} via the relationship

$$d_{ijk} = \epsilon_o\chi_{i\ell}(2\omega)\chi_{jm}(\omega)\chi_{kn}(\omega)\delta_{\ell mn} \tag{123}$$

is in general a more useful parameterization. This is because not only is δ fairly constant from material to material (while four order of magnitude are necessary to scale the variation of **d** among nonlinear systems), but it is also virtually frequency-independent even near a strong absorption band edge (Wynne 1971).

A theoretical understanding of the latter property is provided by a simple scalar anharmonic oscillator model for electronic motion (Kurtz and Robinson 1967) in a field E. For example, the Hamiltonian

$$\mathbb{H} = p^2/2m + (1/2)m\omega_e^2\,\xi^2 + D\xi^3 - Ee\xi \tag{124}$$

with p and ξ electronic momentum and displacement variables, m and e effective mass and charge parameters, ω_e the harmonic response frequency, and D a measure of lowest order anharmonicity, yields the equation of motion

$$\ddot{\xi} + \omega_e^2\xi + (3D/m)\xi^2 = eE/m \tag{125}$$

for which the field dependence of response is readily evaluated. Assuming a frequency-dependent field $E(\omega) \sim e^{i\omega t}$, we derive solutions for linear susceptibility

$$\chi(\omega) = Ne^2/\epsilon_o m\,(\omega_e^2 - \omega^2) \tag{126}$$

and for lowest order nonlinear susceptibility

$$\chi^{(2)}(-2\omega;\omega,\omega) = \frac{N(3De^3/\epsilon_o m^3)}{(4\omega^2 - \omega_e^2)(\omega^2 - \omega_e^2)^2} \tag{127}$$

where there are N oscillators per unit volume. Combining these two equations, we can derive the ratio

$$\frac{\chi^{(2)}(-2\omega;\omega,\omega)}{\epsilon_o^2\chi(2\omega)\chi(\omega)\chi(\omega)} = \frac{2\delta}{\epsilon_o} = \frac{-3D}{N^2e^3} \tag{128}$$

which directly implies a frequency-independent δ within the model. It would therefore appear that, within the context of harmonic generation and optical rectification, Miller's δ is a more fundamental compliance to pursue than the second-order susceptibility itself.

Any spatial or temporal modulation of d_{ijk} or δ_{ijk} will scatter second harmonic light in a manner analogous to that of Sections 2.3.3 and 2.3.4, for which a modulation of linear susceptibility (or dielectric constant) induces Raman and Rayleigh scattering of fundamental light. Such inelastic and elastic scattering of second harmonic light is referred to as hyper-Raman and hyper-Rayleigh scattering and possesses selection rules that differ from those of ordinary Raman and Rayleigh scattering.

2.4.2. Third-Order Nonlinear Optical Response

The third-order nonlinear susceptibility $\chi^{(3)}$ in the optical frequency range is most commonly probed by a measure of third harmonic generation

$$P_i(3\omega) = \epsilon_o C_{ijk\ell}(-3\omega; \omega, \omega, \omega)E_j(\omega)\, E_k(\omega)E_\ell(\omega) \qquad (129)$$

in which $C_{ijk\ell} = (1/4)\, \chi_{ijk\ell}$ is defined to remove the $g_2 = 1/4$ degeneracy factor in Eq. (113). This is because, although the full tensor $C_{ijk\ell}(-\omega; \omega_1, \omega_2, \omega_3)$ has $3^4 = 81$ elements, the third harmonic generation tensor (with $\omega_1 = \omega_2 = \omega_3$) is necessarily invariant under the permutation of the jkl indices. It can therefore be written in a reduced notation:

$$111 \rightarrow 1, \qquad 222 \rightarrow 2, \qquad 333 \rightarrow 3, \qquad (112) \rightarrow 4,$$
$$(223) \rightarrow 5, \qquad (331) \rightarrow 6, \qquad (113) \rightarrow 7, \qquad (221) \rightarrow 8, \qquad (130)$$
$$(332) \rightarrow 9, \qquad (123) \rightarrow 0$$

where (112), for example, implies all the permutations 112, 121, 211. The reduced matrix $C_{ij}^{(3)}$ has 30 elements at most ($i = 1,2,3; j = 0, \ldots, 9$) and far fewer for the more symmetrical crystal structures. Its form for the various crystal classes is shown in Table 2.3. The full tables of elements for the 81-component $C_{ijk\ell}$ tensor in its most general form have been given by Butcher (1965) and Singh (1986). In reduced notation, Eq. (129) becomes

$$P_i(3\omega) = C_{ij}^{(3)}E_j^{(3)}(\omega) \qquad (131)$$

where $E_j^{(3)}$, with $j = 1,2,3, \ldots, 9,0$, denote the cubic field components E_1^3, E_2^3, E_3^3, $3E_1^2 E_2$, $3E_2^2 E_3$, $3E_3^2 E_1$, $3E_1^2 E_3$, $3E_2^2 E_1$, $3E_3^2 E_2$, and $6E_1E_2E_3$, respectively.

Unlike the second-order susceptibility, which vanishes for centrosymmetric crystal structures, the third-order (and higher-order odd) susceptibilities exist for all crystal classes, including an isotropic medium. Consequently, third-order nonlinear processes can occur in solids, liquids, free atoms, and molecules. Also, since they remain invariant under inversion, they require no polar axis sign convention for their complete definition.

Among the many other combinations of high-frequency nonlinear response that can be generated by $C_{ijk\ell}$, one other is of particular interest, because it

Table 2.3 Third-Order Harmonic Compliance Matrices

Triclinic (all classes)

$$
\begin{bmatrix}
C_{11} & C_{12} & C_{13} & C_{14} & C_{15} & C_{16} & C_{17} & C_{18} & C_{19} & C_{10} \\
C_{21} & C_{22} & C_{23} & C_{24} & C_{25} & C_{26} & C_{27} & C_{28} & C_{29} & C_{20} \\
C_{31} & C_{32} & C_{33} & C_{34} & C_{35} & C_{36} & C_{37} & C_{38} & C_{39} & C_{30}
\end{bmatrix}
$$

Monoclinic (all classes)

$$
\begin{bmatrix}
C_{11} & C_{12} & 0 & C_{14} & 0 & C_{16} & 0 & C_{18} & C_{19} & 0 \\
C_{21} & C_{22} & 0 & C_{24} & 0 & C_{26} & 0 & C_{28} & C_{29} & 0 \\
0 & 0 & C_{33} & 0 & C_{35} & 0 & C_{37} & 0 & 0 & C_{30}
\end{bmatrix}
$$

Orthorhombic (all classes)

$$
\begin{bmatrix}
C_{11} & 0 & 0 & 0 & 0 & C_{16} & 0 & C_{18} & 0 & 0 \\
0 & C_{22} & 0 & C_{24} & 0 & 0 & 0 & 0 & C_{29} & 0 \\
0 & 0 & C_{33} & 0 & C_{35} & 0 & C_{37} & 0 & 0 & 0
\end{bmatrix}
$$

Tetragonal
Classes 4 (C_4), $\bar{4}(S_4)$, 4/m (C_{4h})

$$
\begin{bmatrix}
C_{11} & C_{12} & 0 & C_{14} & 0 & C_{16} & 0 & C_{18} & C_{19} & 0 \\
-C_{12} & C_{11} & 0 & C_{18} & 0 & -C_{19} & 0 & -C_{14} & C_{16} & 0 \\
0 & 0 & C_{33} & 0 & C_{35} & 0 & C_{35} & 0 & 0 & 0
\end{bmatrix}
$$

Classes 422 (D_4), 4mm (C_{4v}), $\bar{4}$2m (D_{2d}), 4/mmm (D_{4h})

$$
\begin{bmatrix}
C_{11} & 0 & 0 & 0 & 0 & C_{16} & 0 & C_{18} & 0 & 0 \\
0 & C_{11} & 0 & C_{18} & 0 & 0 & 0 & 0 & C_{16} & 0 \\
0 & 0 & C_{33} & 0 & C_{35} & 0 & C_{35} & 0 & 0 & 0
\end{bmatrix}
$$

Trigonal
Classes 3(C_3), $\bar{3}(S_6)$

$$
\begin{bmatrix}
C_{11} & C_{12} & 0 & C_{12} & C_{15} & C_{16} & -C_{15} & C_{11} & C_{19} & 2 \times C_{25} \\
-C_{12} & C_{11} & 0 & C_{11} & C_{25} & -C_{19} & -C_{25} & -C_{12} & C_{16} & 2 \times C_{15} \\
C_{31} & C_{32} & C_{33} & -3C_{32} & C_{35} & 0 & C_{35} & -3C_{31} & 0 & 0
\end{bmatrix}
$$

Classes 32 (D_3), 3m (C_{3v}), $\bar{3}$m (D_{3d})

$$
\begin{bmatrix}
C_{11} & 0 & 0 & 0 & C_{15} & C_{16} & -C_{15} & C_{11} & 0 & 0 \\
0 & C_{11} & 0 & C_{11} & 0 & 0 & 0 & 0 & C_{16} & 2C_{15} \\
C_{31} & 0 & C_{33} & 0 & C_{35} & 0 & C_{35} & -3C_{31} & 0 & 0
\end{bmatrix}
$$

(*continued*)

Table 2.3 Third-Order Harmonic Compliance Matrices (Continued)

Hexagonal
Classes $6(C_6)$, $\bar{6}(C_{3h})$, $6/m$ (C_{6h})

$$
\begin{bmatrix}
C_{11} & C_{12} & 0 & C_{12} & 0 & C_{16} & 0 & C_{11} & C_{19} & 0 \\
-C_{12} & C_{11} & 0 & C_{11} & 0 & -C_{19} & 0 & -C_{12} & C_{16} & 0 \\
0 & 0 & C_{33} & 0 & C_{35} & 0 & C_{35} & 0 & 0 & 0
\end{bmatrix}
$$

Classes 622 (D_6), 6mm (C_{6v}), $\bar{6}m2(D_{3h})$, 6/mmm (D_{6h})

$$
\begin{bmatrix}
C_{11} & 0 & 0 & 0 & 0 & C_{16} & 0 & C_{11} & 0 & 0 \\
0 & C_{11} & 0 & C_{11} & 0 & 0 & 0 & 0 & C_{16} & 0 \\
0 & 0 & C_{33} & 0 & C_{35} & 0 & C_{35} & 0 & 0 & 0
\end{bmatrix}
$$

Cubic
Classes 23(T), $m3(T_h)$

$$
\begin{bmatrix}
C_{11} & 0 & 0 & 0 & 0 & C_{16} & 0 & C_{18} & 0 & 0 \\
0 & C_{11} & 0 & C_{16} & 0 & 0 & 0 & 0 & C_{18} & 0 \\
0 & 0 & C_{11} & 0 & C_{16} & 0 & C_{18} & 0 & 0 & 0
\end{bmatrix}
$$

Classes 432(O), $\bar{4}3m$ (T_d), m3m (O_h)

$$
\begin{bmatrix}
C_{11} & 0 & 0 & 0 & 0 & C_{16} & 0 & C_{16} & 0 & 0 \\
0 & C_{11} & 0 & C_{16} & 0 & 0 & 0 & 0 & C_{16} & 0 \\
0 & 0 & C_{11} & 0 & C_{16} & 0 & C_{16} & 0 & 0 & 0
\end{bmatrix}
$$

Isotropic

$$
\begin{bmatrix}
C_{11} & 0 & 0 & 0 & 0 & C_{11} & 0 & C_{11} & 0 & 0 \\
0 & C_{11} & 0 & C_{11} & 0 & 0 & 0 & 0 & C_{11} & 0 \\
0 & 0 & C_{11} & 0 & C_{11} & 0 & C_{11} & 0 & 0 & 0
\end{bmatrix}
$$

controls the field dependence of refractive index. If we consider the generation of a nonlinear $P(\omega)$ at the same frequency as the driving field $E(\omega)$, then it can be added directly to linear response in the form

$$
P_i(\omega) = \epsilon_o[\chi_{ij}(\omega)E_j(\omega) + 3C_{ijk\ell} \, (-\omega; \omega, -\omega, \omega)
$$
$$
E_j(\omega)E_k^*(\omega) \, E_\ell(\omega)] \tag{132}
$$

Consider, for example, a linearly polarized beam with $i=j=k=\ell=1$. Via Eq. (132) it gives rise to a response $P_1(\omega) = \epsilon_o^2 n^2 E_1(\omega)$, where

$$
n^2 = n_o^2 + 6C_{1111} \, | \, E_1(\omega) \, |^2 \tag{133}
$$

which approaches its field-independent value n_o as $E_1 \rightarrow 0$. Writing $n = n_o + n_2 \mid E_1(\omega) \mid ^2$, it follows, to the first order in C, that the lowest order nonlinear refractive index is given by

$$n_2 = (3)(C_{1111}/n_o) \tag{134}$$

Equivalent results for other field configurations can be found in Singh (1986) and references therein. The increase in index n with intensity leads to self-focusing effects observed in laser beams.

2.4.3. The Pockels and Kerr Electro-optical Effects

The first- and second-order (Pockels and Kerr) electro-optical effects, describing the perturbation of optical response by low-frequency electric fields, have been studied since the last century and consequently have developed with their own nomenclature, which differs from (although, of course, can be directly related to) that of the high-frequency formalism of Eq. (106). The low-frequency electro-optic coefficients are usually defined in terms of pertubations (by electric field) of the impermeability tensor $\mathbf{B} = \mathbf{\epsilon}^{-1}$. The origin of this lies in the fact that the optical properties of crystals can be expressed in terms of a refractive index ellipsoid (or optical indicatrix) that, in a Cartesian reference frame x_i, has the equation

$$B_{ij}^0 x_i x_j = 1 \tag{135}$$

in which $B_{ij}^0 = 1/n_{ij}^2$ define the components n_{ij} of refractive index. In a principal coordinate system, Eq. (135) reduces to $x_i^2/n_i^2 = 1$, where n_i are the principal components of refractive index.

In the presence of a low-frequency field $\mathbf{E}$, the indicatrix B_{ij} is perturbed from its zero-field form B_{ij}^0 in a manner that enables us to define a linear electro-optic tensor r_{ijk} by the relation

$$B_{ij} - B_{ij}^0 = \Delta B_{ij} = r_{ijk}E_k \tag{136}$$

Using a contracted notation,

$$B_1 = B_{11}, \qquad B_2 = B_{22}, \qquad B_3 = B_{33}, \qquad B_4 = B_{23},$$
$$B_5 = B_{31}, \qquad B_6 = B_{12} \tag{137}$$

Eq. (136) can be expressed as

$$\Delta B_i = r_{ij}E_j, \qquad i = 1, \ldots ,6; j = 1,2,3 \tag{138}$$

with a maximum number of 18 independent r_{ij} matrix elements.

In measuring the Pockels effect it is important to recognize that unless the crystal is effectively clamped (e.g., by using frequencies that are higher than the mechanical resonances of the sample), Eq. (138) requires modification because

E_j will induce a strain via the inverse piezoelectric effect, which in turn will produce a change of refractive index through the phenomenon of photoelasticity. The change in B_i resulting from both the Pockels electro-optical effect and the photoelastically induced contributions can be written as either

$$B_i = r_{ij}' E_j + q_{ik} \sigma_k \qquad (139)$$

or

$$B_i = r_{ij} E_j + p_{ik} \eta_k \qquad (140)$$

where q_{ik} (p_{ik}) are the stress (strain) photoelastic constants defined in Section 2.3.2, r_{ij} is the primary (or true) linear electro-optical effect, and the free crystal equivalent r_{ij}' (zero stress) is related to this true value by

$$r_{ij}' = r_{ij} + p_{ik} d_{jk} \qquad (141)$$

in which d_{jk} are piezoelectric constants. The symmetry restrictions on the electro-optic matrices r_{ij} are closely related to those for piezoelectricity and are set out for the 20 "active" crystal classes in Table 2.4.

Pockels effects are also definable in terms of induced polarization **P** via the equation

$$\Delta B_{ij} = f_{ijk} P_k \qquad (142)$$

The tensor f_{ijk} is simply related to r_{ijk} by use of the relationship $P_k = \epsilon_0(\epsilon_{kk}-1)E_k$ [cgs: $4\pi P_k = (\epsilon-1)E_k$] to give

$$f_{ijk} = r_{ijk}/\epsilon_0(\epsilon_{kk}-1) \qquad (143)$$

$$\text{cgs:} \qquad f_{ijk} = 4\pi r_{ijk}/(\epsilon_{kk}-1) \qquad (143')$$

The compliances f vary less widely between different materials than do their r counterparts.

Higher order low-frequency electro-optical responses can readily be defined, the most frequently quoted of which is the quadratic electro-optic (or Kerr) tensor $K_{ijk\ell}$,

$$\Delta B_{ij} = K_{ijk\ell} E_k E_l \qquad (144)$$

or, in contracted notation,

$$\Delta B_i = K_{ij} E_j^{(2)}, \qquad i = 1, \dots, 6; j = 1, \dots, 6 \qquad (145)$$

in which $E_j^{(2)}$ are the quadratic field components as defined earlier in association with Eq. (120). The Kerr tensor is nonzero for all symmetry groups and, as with the Pockels effect, can be defined for both clamped and free crystals. The form of its contracted matrix configuration as a function of crystal class is the same as that for the elasto-optical compliance p_{ij} shown in Table 2.1. In general, the Kerr

Table 2.4   Nonzero Linear Electro-optic Matrices

Triclinic

Class 1 (C_1)

$$
\begin{matrix}
r_{11} & r_{12} & r_{13} \\
r_{21} & r_{22} & r_{23} \\
r_{31} & r_{32} & r_{33} \\
r_{41} & r_{42} & r_{43} \\
r_{51} & r_{52} & r_{53} \\
r_{61} & r_{62} & r_{63}
\end{matrix}
$$

Monoclinic

Class 2 (C_2) (diad $\parallel x_3$)

$$
\begin{matrix}
0 & 0 & r_{13} \\
0 & 0 & r_{23} \\
0 & 0 & r_{33} \\
r_{41} & r_{42} & 0 \\
r_{51} & r_{52} & 0 \\
0 & 0 & r_{63}
\end{matrix}
$$

Class m (C_s) m $\perp x_3$

$$
\begin{matrix}
r_{11} & r_{12} & 0 \\
r_{21} & r_{22} & 0 \\
r_{31} & r_{32} & 0 \\
0 & 0 & r_{43} \\
0 & 0 & r_{53} \\
r_{61} & r_{62} & 0
\end{matrix}
$$

Orthorhombic

Class 222 (D_2)

$$
\begin{matrix}
0 & 0 & 0 \\
0 & 0 & 0 \\
0 & 0 & 0 \\
r_{41} & 0 & 0 \\
0 & r_{52} & 0 \\
0 & 0 & r_{63}
\end{matrix}
$$

Class mm2 (C_{2v})

$$
\begin{matrix}
0 & 0 & r_{13} \\
0 & 0 & r_{23} \\
0 & 0 & r_{33} \\
0 & r_{42} & 0 \\
r_{51} & 0 & 0 \\
0 & 0 & 0
\end{matrix}
$$

Tetragonal

Class $\bar{4}$ (S_4)

$$
\begin{matrix}
0 & 0 & r_{13} \\
0 & 0 & -r_{13} \\
0 & 0 & 0 \\
r_{41} & -r_{51} & 0 \\
r_{51} & r_{41} & 0 \\
0 & 0 & r_{63}
\end{matrix}
$$

Class 4 (C_4)

$$
\begin{matrix}
0 & 0 & r_{13} \\
0 & 0 & r_{13} \\
0 & 0 & r_{33} \\
r_{41} & r_{51} & 0 \\
r_{51} & -r_{41} & 0 \\
0 & 0 & 0
\end{matrix}
$$

Class 4mm (C_{4v})

$$
\begin{matrix}
0 & 0 & r_{13} \\
0 & 0 & r_{13} \\
0 & 0 & r_{33} \\
0 & r_{51} & 0 \\
r_{51} & 0 & 0 \\
0 & 0 & 0
\end{matrix}
$$

Tetragonal

Class $\bar{4}2m$ (D_{2d}) (Diad $\parallel x_1$)

$$
\begin{matrix}
0 & 0 & 0 \\
0 & 0 & 0 \\
0 & 0 & 0 \\
r_{41} & 0 & 0 \\
0 & r_{41} & 0 \\
0 & 0 & r_{63}
\end{matrix}
$$

Class 422 (D_4)

$$
\begin{matrix}
0 & 0 & 0 \\
0 & 0 & 0 \\
0 & 0 & 0 \\
r_{41} & 0 & 0 \\
0 & -r_{41} & 0 \\
0 & 0 & 0
\end{matrix}
$$

(*continued*)

Table 2.4 Nonzero Linear Electro-optic matrices (Continued)

Trigonal

Class 3 (C_3)
$$\begin{pmatrix} r_{11} & -r_{22} & r_{13} \\ -r_{11} & r_{22} & r_{13} \\ 0 & 0 & r_{33} \\ r_{41} & r_{51} & 0 \\ r_{51} & -r_{41} & 0 \\ -r_{22} & -r_{11} & 0 \end{pmatrix}$$

Class 32 (D_3)
$$\begin{pmatrix} r_{11} & 0 & 0 \\ -r_{11} & 0 & 0 \\ 0 & 0 & 0 \\ r_{41} & 0 & 0 \\ 0 & -r_{41} & 0 \\ 0 & -r_{11} & 0 \end{pmatrix}$$

Class 3m (C_{3v})
$m \perp x_1$
$$\begin{pmatrix} 0 & -r_{22} & r_{13} \\ 0 & r_{22} & r_{13} \\ 0 & 0 & r_{33} \\ 0 & r_{51} & 0 \\ r_{51} & 0 & 0 \\ -r_{22} & 0 & 0 \end{pmatrix}$$

Hexagonal

Class 6 (C_6)
$$\begin{pmatrix} 0 & 0 & r_{13} \\ 0 & 0 & r_{13} \\ 0 & 0 & r_{33} \\ r_{41} & r_{51} & 0 \\ r_{51} & -r_{41} & 0 \\ 0 & 0 & 0 \end{pmatrix}$$

Class 6mm (C_{6v})
$$\begin{pmatrix} 0 & 0 & r_{13} \\ 0 & 0 & r_{13} \\ 0 & 0 & r_{33} \\ 0 & r_{51} & 0 \\ r_{51} & 0 & 0 \\ 0 & 0 & 0 \end{pmatrix}$$

Class 622 (D_6)
$$\begin{pmatrix} 0 & 0 & 0 \\ 0 & 0 & 0 \\ 0 & 0 & 0 \\ r_{41} & 0 & 0 \\ 0 & -r_{41} & 0 \\ 0 & 0 & 0 \end{pmatrix}$$

Hexagonal

Class $\bar{6}m2$ (D_{3h})
$m \perp x_2$
$$\begin{pmatrix} r_{11} & 0 & 0 \\ -r_{11} & 0 & 0 \\ 0 & 0 & 0 \\ 0 & 0 & 0 \\ 0 & 0 & 0 \\ 0 & -r_{11} & 0 \end{pmatrix}$$

Class $\bar{6}$ (C_{3h})
$$\begin{pmatrix} r_{11} & -r_{22} & 0 \\ -r_{11} & r_{22} & 0 \\ 0 & 0 & 0 \\ 0 & 0 & 0 \\ 0 & 0 & 0 \\ -r_{22} & -r_{11} & 0 \end{pmatrix}$$

Cubic

Classes 23 (T) and $\bar{4}3m$ (T_d)
$$\begin{pmatrix} 0 & 0 & 0 \\ 0 & 0 & 0 \\ 0 & 0 & 0 \\ r_{41} & 0 & 0 \\ 0 & r_{41} & 0 \\ 0 & 0 & r_{41} \end{pmatrix}$$

effect bears the same relation to the Pockels effect as electrostriction does to inverse piezoelectricity. It is used in centrosymmetric crystals and in liquids for the purpose of light modulation (Kerr cells).

A number of fundamental theories dealing with the microscopical origins of these various low- and high-frequency nonlinear effects can be found in the literature (Chemla 1980). They are, however, beyond the scope of this handbook.

2.4.4. Polymeric Materials

Developments in the field of nonlinear optics have led to important applications in information processing, telecommunications, and integrated optics (Glass 1984). The search for materials with large electro-optical nonlinearities for these purposes centered until recently on polar semiconductors and insulators. Lately it has become apparent that organic and polymeric materials, with large delocalized π-electron systems, possess nonlinear responses much larger than those of their inorganic counterparts (Williams 1983). Of particular importance is the fact that the origin of polymeric nonlinearities is in the polarizability of π-electron clouds and does not rely (as it usually does in inorganic crystals) on the degree of rearrangement of nuclear coordinates. Thus the potential usefulness of polymers for high-frequency applications contrasts with the bandwidth limitations of most inorganic compounds. In addition, the ability to tailor charge asymmetry and to provide optical transparency in spectral regions compatible with useful light sources represents enormous potential for designing materials with specific properties and for particular applications.

An indication of the general superiority of organic materials over their inorganic counterparts in achieving large nonlinearities is shown in Fig. 2.17, where Miller's δ (as a measure of $\chi^{(2)}$) and electro-optic coefficient f of Eq. (143) (as a measure of $\chi^{(3)}$) are compared for a selection of materials of both types. The larger response of organic materials, combined with their design flexibility, seems likely to push polymeric materials to the forefront in future device applications in the nonlinear optics field.

REFERENCES

Anderson, A. (1971). *Raman Effect*, Marcel Dekker, New York, Vol. 1.

Armstrong, J. A., Bloembergen, N., Ducuing, J., and Pershan, P. S. (1962). *Phys. Rev.* *127*:1918.

Aspnes, D. E. (1980). In: *Handbook on Semiconductors* (M. Balkanski, Ed.), North-Holland, Amsterdam, Vol. 2, p. 109.

Barker, A. S. (1967). In: *Ferroelectricity* (E. F. Weller, Ed.), Elsevier, New York.

Bassani, F., Iadonisi, G., and Preziosi, B. (1974). *Rep. Progr. Phys. 37*:1099.

Bayvel, A. P., and Jones, A. R. (1981). *Electromagnetic Scattering and Its Applications*, Applied Science Publishers, London.

Bell, R. J. (1972). *Rep. Progr. Phys. 35*:1315.

Bendow, B. (1977). *Ann. Rev. Mat. Sci. 7*:23.

Bendow, B. (1978). In: *Solid State Physics* (H. Ehrenreich, F. Seitz, and D. Turnbull, Eds.), Academic, New York, Vol. 33, p. 249.

Bleaney, B. I., and Bleaney, B. (1976). *Electricity and Magnetism*, Oxford Univ. Press, Oxford.

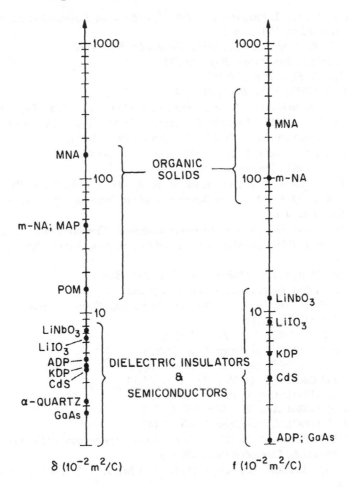

Figure 2.17 Comparative quantities for selected tensor components of second harmonic generation (left) and linear electro-optical effect (right). [After Garito and Singer (1982).]

Born, M., and Huang, K. (1954). *Dynamical Theory of Crystal Lattices*, Oxford Univ. Press, Oxford.

Butcher, P. N. (1965). Nonlinear Optical Phenomena, Bull. 200, Engineering Expt. Station, Ohio State Univ., Columbus.

Chelikowsky, J. R., and Cohen, M. L. (1976). *Phys. Rev. B 14*:556.

Chemla, D. S. (1980). *Rep. Progr. Phys. 43*:1191.

Chemla, D. S., and Jerphagnon, J. (1980). *Handbook on Semiconductors*, North-Holland, Amsterdam, Vol. 2, p. 545.

Chu, B. (1974). *Laser Light Scattering*, Academic, New York.

Dil, J. G. (1982). *Rep. Progr. Phys. 45:*285.

Egri, I. (1985). *Phys. Rep. 119:*363.

Elliott, R. J. (1957). *Phys. Rev. 108:*1384.

Elliott, R. J., Krumhansl, J. A., and Leath, P. L. (1974). *Rev. Mod. Phys. 46:*465.

Fabelinskii, I. L. (1968). *Molecular Scattering of Light*, Plenum, New York.

Ferré, J., and Gehring, G. A. (1984). *Rep. Progr. Phys. 47:*513.

Garito, A. F., and Singer, K. D. (1982). *Laser Focus 80:*59.

Glass, A. M. (1984). *Science 226:*657.

Harper, P. G., Hodby, J. W., and Stradling, R. A. (1973). *Rep. Progr. Phys. 36:*1.

Harrison, W. A. (1980). *Electronic Structure and the Properties of Solids*, W. H. Freeman, San Francisco.

Heitler, W. (1954). *The Quantum Theory of Radiation*, Clarendon Press, Oxford.

Hodgson, J. N. (1970). *Optical Absorption and Dispersion in Solids*, Chapman and Hall, London.

Ihm, J., and Phillips, J. C. (1983). *Phys. Rev. B27:*7803.

Johnson, F. (1959). *Proc. Roy. Soc. Lond. A73:*265.

Jones, H. (1960). *The Theory of Brillouin Zones and Electronic States in Crystals*, North-Holland, Amsterdam.

Kleinman, D. A. (1962). *Phys. Rev. 128:*1761.

Klinger, M. I. (1968). *Rep. Progr. Phys. 31:*225.

Kurtz, S. K., and Robinson, F. N. H. (1967). *Appl. Phys. Lett. 10:*62.

Lax, M., and Burstein, E. (1955). *Phys. Rev. 97:*39.

Lines, M. E. (1984). *J. Appl. Phys. 55:*4052.

Lines, M. E. (1986). *Ann. Rev. Mat. Sci. 16:*113.

Lines, M. E. (1987). *J. Non-Cryst. Solids 89:*143.

Lines, M. E., and Glass, A. M. (1977). *Principles and Applications of Ferroelectrics and Related Materials*, Clarendon Press, Oxford.

Lipson, H. G., Bendow, B., Massa, N. E., and Mitra, S. S. (1976). *Phys. Rev. B 13:*2614.

Martin, R. M., and Falikov, L. M. (1975). In: *Topics in Applied Physics* (M. Cardona, Ed.), Springer-Verlag, Berlin, Vol. 8.

Mermin, N. D. (1979). *Rev. Mod. Phys. 51:*591.

Mie, G. (1908). *Ann. Phys. 25:*377.

Miles, P. A. (1977). *Appl. Opt. 16:*2891.

Miller, R. C. (1964). *Appl. Phys. Lett. 5:*17.

Mills, D. L., and Burstein, E. (1974). *Rep. Progr. Phys. 37:*817.

Mitra, S. (1985). In: *Handbook of Optical Constants of Solids* (E. D. Palik, Ed.), Academic, New York.

Narasimhamurty, T. S. (1981). *Photoelastic and Electro-Optic Properties of Crystals*, Plenum, New York.

Nassau, K., and Lines, M. E. (1986). *Opt. Eng. 25:*602.

Nassau, K., and Wemple, S.H. (1982). *Electron Lett. 18:*450.

Palik, E. D., and Furdyna, J. K. (1970). *Rep. Progr. Phys. 33:*1193.

Pidgeon, C. R. (1980). In: *Handbook on Semiconductors* (M. Balkanski, Ed.), North-Holland, Amsterdam, Vol. 2, p. 223.

Pinnow, D. A. (1970). *IEEE J. Quantum Electron. QE-6:*223.

Pollard, H. F. (1977). *Sound Waves in Solids,* Pion, London.

Sa-Yakanit, V., and Glyde, A. R. (1987). *Comments Cond. Mat. Phys. 13:*35.

Schroeder, J. (1977). In: *Treatise on Materials Science and Technology* (M. Tomozawa and R. H. Doremus, Eds.), Academic, New York, Vol. 12.

Schroeder, J. (1980). *J. Non-Cryst. Solids 40:*549.

Shuker, R., and Gammon, R. W. (1970). *Phys. Rev. Lett. 25:*222.

Singh, S. (1986). In: *Handbook of Laser Science and Technology,* Vol. III, *Optical Materials* (M. J. Weber, Ed.), CRC Press, Boca Raton, Fl. Pt. 1.

Spitzer, W. G., and Fan, H. Y. (1957). *Phys. Rev. 106:*882.

Stolen, R. H. (1979). In: *Optical Fiber Telecommunications* (S. E. Miller and A. G. Chynoweth, Eds.), Academic, New York.

Tauc, J. (1970). *Mat. Res. Bull. 5:*721.

Tauc, J. (1974). *Amorphous and Liquid Semiconductors,* Plenum, London.

Weinstein, B. A., Zallen, R., Slade, M. A., and deLozanne, A. (1981). *Phys. Rev. B 24:*4652.

Williams, D. J. (1983). *Nonlinear Optical Properties of Organic and Polymeric Materials,* American Chemical Society, Washington, D.C.

Wynne, J. (1971). *Phys. Rev. Lett. 27:*17.

Zallen, R. (1983). *The Physics of Amorphous Solids,* Wiley, New York.

Ziman, J. M. (1964). *Principles of the Theory of Solids,* Cambridge Univ. Press, Cambridge.

3

A Survey of Experimental Methods: Physical Property Determination

Malcolm E. Lines *AT&T Bell Laboratories, Murray Hill, New Jersey*

3.1. ELASTIC COMPLIANCES

The most frequently used techniques for measuring elastic stiffness and compliance moduli are (roughly in order of decreasing accuracy) ultrasonic wave transmission, resonance vibrations in rods, light scattering from ultrasonic waves, static deformation, Brillouin scattering, diffuse X-ray scattering, and neutron scattering (Pollard 1977, Hearmon 1961, Wooster 1962, Musgrave 1970, Vacher and Boyer 1972, Schreibner et al. 1973, Elbaum 1981). Various thermodynamic restraints (Section 1.2.3) should be recognized in accurate work. However, such differences are often small and are seldom designated in tables except for unusual circumstances. Brillouin scattering has the advantage that it can be used when only small samples (of perhaps a few millimeters) are available. Neutron scattering is included because it alone samples elastic responses at time intervals smaller than the mean phonon lifetime (i.e., in the collision-free or "zero-sound" regime). Complete and accurate measurements over the frequency range that bridges zero sound to the more common collision-dominated "first sound" are available for rather few materials. Variations with temperature and pressure in the normal or first-sound domain are more frequently available.

Nonlinear elastic moduli can be measured as deviations from Hooke's law or, equivalently, in terms of the stress or strain dependence of acoustic velocities as measured by ultrasonic or hypersonic (e.g., Brillouin) methods (Hearmon 1979). Beyond the elastic and plastic deformation regimes a whole science of fracture toughness (Atkins and Mai 1986) has grown up that explores the eventual failure of materials by crack propagation. Notched bars under various loading arrangements can be studied by such techniques as cinematography, electron microscopy, electric potential measurements, mechanical displacement gauge measurement, and transducer vibrational detection (Brown and Srawley 1964).

3.2. PIEZOELECTRIC COMPLIANCES

Methods of measurement fall primarily into four categories as follows:

1. Use of the direct effect $P_i = d_{ijk}\sigma_{jk}$ of Eq. (57) of Chapter 1 by measuring the charge generated by an applied mechanical stress at frequencies below the elastic resonant frequencies of the sample (Haussühl et al. 1977).
2. Use of the converse effect $\eta_{jk} = d_{ijk}E_i$ of Eq. (58) of Chapter 1 by measuring the strain resulting from application of a low-frequency voltage to the specimen (Yamaguchi and Hamano 1981).
3. The effect of piezoelectric interaction on the dielectric impedance of an electroded crystal as a function of frequency, particularly at or near one or more of the piezoelastically active vibrational modes (Berlincourt et al. 1963, Dankov et al. 1979).
4. Analysis of the difference of sound velocities at constant field E and constant displacement D (Smith and Welsh 1971).

3.3. MEASUREMENT OF RADIANT FLUX

It is difficult to construct a true absolute radiometer because it has to compare an unknown radiant power with a known standard under strictly equivalent conditions. Two approaches have emerged. The first relies on calorimetry, or the measurement of temperature rise due to the absorption of radiant power (Blaney 1980). The second relies on the use of secondary radiometers that have previously been calibrated against a known source of radiant power (such as a blackbody). For application at room temperature the most frequently used secondary radiometers are pyroelectric detectors (Lines and Glass 1977, Hadni 1983) and Golay cells (Golay 1947). The former utilizes the charge flow from a spontaneously polar material upon change of temperature. The Golay cell is a pneumatic device that records the change in gas pressure in an enclosed chamber on receipt of radiant energy.

3.4. LINEAR OPTICAL RESPONSE

The reflection (and transmission) of plane-wave electromagnetic radiation at a plane interface between air and an absorbing material is described by the complex coefficients $r_{\parallel}, r_{\perp}$ (and $t_{\parallel}, t_{\perp}$) of Eq. (119) of Chapter 1. Since the amplitude of plane-wave radiation is also most generally defined by two complex E components, a full description of such a reflection or transmission event involves eight parameters. Optical spectroscopic techniques that are capable of a complete measurement of this kind (polarimetry, Hauge 1979) are still not widely available. The most frequently used methods for sampling linear optical response (i.e., refractive index n and extinction coefficient k) concentrate either upon the size of the polarization ellipse defined by the complex E fields or upon its shape. The former techniques (Potter 1985) concern themselves only with intensity relationships like those of Eq. (120) of Chapter 1 and are referred to as *reflection* or *transmission spectroscopy*. The complementary technique, which focuses on the shape of the polarization ellipse (but not its size), is called *ellipsometry* (Aspnes 1985). Although ellipsometry is in principle a more powerful technique than reflectometry and refractometry (since it can derive both n and k directly as a function of λ in a single measurement), it has to date been less widely used than the older and more traditional intensity probes, particularly at longer wavelengths where Fourier transform scanning interferometry (Chantry 1984) remains a popular method. In the far-infrared, where the low spectral intensity of thermal sources leads to large statistical errors in Fourier spectrometry, an accent on the development of cw far-infrared lasers is now significantly affecting the field (Tachi 1985) even though they cover the spectrum only incompletely.

3.5. OPTICAL LOSS COEFFICIENT α

Total attenuation, as measured by transmission intensity, is not directly associated with the extinction coefficient k by Eq. (115) of Chapter 1, because it includes losses from other mechanisms (such as scattering and nonlinear processes) as well. In particular, within the optic window the optical loss coefficient is completely dominated by effects not associated with linear response. Since optical attenuation in the window regime varies over a range of many decades, more than one technique is required for its measurement. The common procedure of placing a 1-mm- to 1-cm-thick sample in one arm of a standard spectrograph is adequate for loss coefficients $\alpha \approx 10^{-2}-10^{2}$ cm^{-1}. At higher attenuation values, thinner samples are necessary, and etalon effects can become significant. For appreciably smaller α values, the simplest technique is to measure (directly or indirectly) the heat generated by the passage of a laser beam (laser calorimetry). Unfortunately, many of the data in the literature are distorted by a failure to account adequately for reflection and surface absorption factors, and care must be exercised in the use of such data for estimating the true bulk attenuation (Hordvik 1977, Klein 1981, Temple 1985). In the case of optical fiber materials of extremely high transparency, transmission measurements can sometimes be made directly over fiber lengths of 100 m or more. At high transparency and high light intensities, nonlinear effects (e.g., two-photon processes) can become important and lead to a power-dependent α. Observations of such effects are again normally made by transmission (Bechtel and Smith 1976) or laser calorimetric methods (Bass et al. 1978).

3.6. INFRARED MAGNETO-OPTICAL SPECTROSCOPY

Magneto-optical spectroscopy employs a combination of a laser (of design relevant for the wavelength regime under study) with a high-field magnet. Fields up to about 20–30 T are available as stationary fields, and pulsed fields can be produced nondestructively to about 50 T (Miura et al. 1981). By the use of destructive methods [such as explosively driven magnetic flux compression techniques (Fowler et al. 1976)], fields in excess of 300 T, or 3 megagauss, have been obtained. Using solid-state lasers in the near-infrared and molecular gas lasers in the far-infrared, the most commonly used techniques for the study of magneto-optic effects are as follows:

1. Magnetoreflection (Kido 1983)
2. Magnetotransmission; Alfven wave propagation (Hiruma et al. 1983)
3. Cyclotron resonance (Kido et al. 1981)
4. Faraday rotation in spontaneously magnetic materials (Dillon 1958; Yang et al. 1980)

Recent work in the megagauss field range has been reviewed by Miura (1984).

3.7. PHOTOELASTIC COMPLIANCES p_{ijkl} AND q_{ijkl}

The compliances q_{ijkl} and p_{ijkl}, which relate impermeability B_{ij} to stress σ_{kl} and strain η_{kl}, respectively, are most commonly measured by piezobirefringence (Ramachandran and Ramaseshan 1961), acoustical diffraction (Dixon 1971), or Brillouin scattering (Vacher and Boyer 1972). Piezobirefringence measures the refractive index changes resulting from homogeneous static (or quasi-static) deformations. In piezoelectric crystals, care must be exercised to separate the intrinsic piezo-optical effect $\Delta B_{ij} = q_{ijkl}\sigma_{kl}$ from that induced indirectly via coupled piezoelectric ($P_i = d_{ijk}\sigma_{jk}$) and Pockels ($\Delta B_{ij} = f_{ijk}P_k$) contributions.

In both the acousto-optical and Brillouin scattering methods, light is inelastically deflected by interaction with acoustic phonons. In the former the acoustic wave is coherently generated by a transducer. In the latter the phonons involved are incoherently generated by thermal excitations, and analysis requires the use of thermodynamic scattering theory (Born and Huang 1954). In recent years elasto-optical measurements in opaque wavelength regimes have been performed by surface Brillouin scattering (Sandercock 1972) and by stress-induced Raman scattering (Chandrasekhar et al. 1978). In such an opaque regime, measurements of piezoabsorption (i.e., the imaginary part of the elasto-optic tensor) are also possible (Adachi and Hamaguchi 1980).

3.8. NONLINEAR OPTICAL COMPLIANCES d_{ijkl} AND C_{ijkl}

Observation of optical nonlinearities involving wholly high frequency fields (e.g., second and third harmonic generation, multiwave mixing) are carried out primarily by use of the following methods:

1. Maker fringes (Maker et al. 1962, Jerphagnon and Kurtz 1970); used most frequently for materials that are transparent in the visible
2. Wedge techniques (Wynne and Bloembergen 1969); used most frequencly for materials transmitting in the infrared
3. Phase-matching techniques (Bjorkholm and Siegman 1967, Midwinter and Warner 1965); limited to compliance elements that are phase-matchable
4. Raman scattering (Johnston and Kaminow 1969, Yang 1976); for crystals with simultaneously Raman-active and infrared-active modes
5. Optical difference mixing (Maker and Terhune 1965, Levenson and Bloembergen 1974)
6. Change of refractive index (Maker et al. 1964, Milam and Weber 1976); for third-order nonlinearities only
7. Third harmonic generation (Meredith 1981); for third-order nonlinearities only
8. Interference methods (Okada et al. 1977); for relative sign determination of the nonlinear optic coefficients

Note that crystals that are inversion images of each other have opposite signs of d_{ijk}. The absolute sign of these second harmonic coefficients therefore depends on a definition of the $+$ or $-$ direction of the polar axis. The standard convention conforms with that used for piezoelectricity (Section 1.3.3, Chapter 1) with an electrode at the $+$ end of the polar axis developing a negative charge upon compression along that axis. Third harmonic compliances C_{ijkl} are invariant under inversion and are therefore independent of the sign of the polar axis.

3.9. ELECTRO-OPTICAL COMPLIANCES
$r_{ijk}, f_{ijk}, K_{ijkl}$

The two most frequently used techniques for determining electro-optical compliances are by measuring the phase changes (Kaminow 1974) or intensities (Nelson and Turner 1968) of the frequency-shifted components of light when the sample is subjected to perturbation by a low-frequency electric field of wavelength larger than the sample dimensions. If the frequency of the perturbing field is greater than those of the mechanical resonances of the sample, then constant-strain electro-optical components are measured. If the frequency is below these resonances, then constant-stress components result. Although the quadratic (Kerr) compliances K_{ijkl} are nonzero for all materials, they are usually measured only in systems for which the linear components are absent by symmetry.

REFERENCES

Adachi, S., and Hamaguchi, C. (1980). *Phys Rev. B 21*:1701.
Aspnes, D.E. (1985). In: *Handbook of Optical Constants of Solids* (E.D. Palik, Ed.), Academic, New York. p. 89.

Atkins, A.G., and Mai, J.W. (1986). *J. Mat. Sci. 21*:1093.

Bass, M., Van Stryland, E.W., and Stewart, A.F. (1978). In: *Laser Induced Damage in Optical Material* (A.J. Glass and A.H. Guenther, Eds.), U.S. Government Printing Office, Washington, D.C., p. 19.

Bechtel, J.H., and Smith, W.L. (1976). *Phys. Rev.B 13*:3515.

Berlincourt, J., Jaffe, H., and Shiozawa, L.R. (1963). *Phys. Rev. 129*:1009.

Bjorkholm, J.E., and Siegman, A.E. (1967). *Phys. Rev. 154*:851.

Blaney, T.G. (1980). *Proc. SPIE 234*:22.

Born, M., and Huang, K. (1954). *Dynamical Theory of Crystal Lattices*, Oxford Univ. Press, Oxford.

Brown, W.F., Jr. and Srawley, J.E. (1964). Fracture Toughess Testing and its Applications, A.S.T.M. Tech. Publ. 381, Philadelphia, p. 133.

Chandrasekhar, M., Grimsditch, M.H., and Cardona, M. (1978). *J. Opt. Soc. Am. 68*:523.

Chantry, G.W. (1984). *Long-Wave Optics*, Academic, New York, Vol. 1.

Dankov, L.A., Pado, G.S., Kobyakov, I.B., and Berdnik, V.V. (1979). *Fiz. Tverd. Tela. 21*:2570 (transl. *Sov. Phys. Solid State 21*:1481).

Dillon, J.F., Jr. (1958). *J. Appl. Phys. 29*:539.

Dixon, R.W. (1971). In: *The Physics of Opto-electronic Materials* (W.A. Albers, Ed.), Plenum, New York, p. 131.

Elbaum, C. (1981). *J. Phys.* (Paris), *42 Suppl.*:C5–855.

Fowler, C.M., Caird, R.S., Garn, W.B., and Erickson, D.J. (1976). *I.E.E.E. Trans. Mag. 12*:1018.

Golay, M. (1947). *Rev. Sci. Instrum. 18*:347.

Hadni, A. (1983). In: *Infrared and Millimeter Waves* (K.J. Button, Ed.), Academic, New York, Vol. 8, Pt. I, p. 173.

Hauge, P.S. (1979). *Surface Sci. 96*:108.

Haussühl, S., Eckstein, J., Recker, K., and Wallrafen, F. (1977). *J. Cryst. Growth 40*:200.

Hearmon, R.F.S (1961). *An Introduction to Applied Anisotropic Elasticity*, Oxford Univ. Press, Oxford.

Hearmon, R.F.S. (1979). In: *Landolt-Bornstein* (K.H. Hellwege, Ed.), Springer-Verlag, Berlin, New Series, Vols. III/11 and III/18.

Hiruma, K., Kido, G., and Miura, N. (1983). *J. Phys. Soc. Jpn. 52*:2550.

Hordvik, A. (1977). *Appl. Opt. 16*:2827.

Jerphagnon, J., and Kurtz, S.K. (1970). *J. Appl. Phys. 41*:1667.

Johnston, W.D., and Kaminow, I.P. (1969). *Phys. Rev. 188*:1209.

Kaminow, I.P. (1974). *Introduction to Electro-Optic Devices*, Academic, New York.

Kido, G. (1983). In: *High Field Magnetism* (M. Date, Ed.), North-Holland, Amsterdam, p. 339.

Kido, G., Miura, N., Akihiro, M., Katayama, H., and Chikazumi, S. (1981). In: *Physics in High Magnetic Fields* (S. Chikazumi and N. Miura, Eds.), Springer-Verlag, Berlin, p. 72.

Klein, P.A. (1981). *Opt. Eng. 20*:790.

Levenson, D., and Bloembergen, N. (1974). *Phys. Rev. B 10*:4447.

Lines, M.E., and Glass, A.M. (1977). *Principles and Applications of Ferroelectrics and Related Materials*, Clarendon Press, Oxford.

Maker, P.D., and Terhune, R.W. (1965). *Phys. Rev. 137*:A801.

Maker, P.D., Terhune, R.W., Nisenoff, M., and Savage, C.M. (1962). *Phys. Rev. Lett. 8*:21.

Maker, P.D., Terhune, R.W., and Savage, C.M. (1964). *Phys. Rev. Lett. 12*:507.

Meredith, G.R. (1981). *Phys. Rev. B 24*:5522.

Midwinter, J.E., and Werner, J. (1965). *J. Appl. Phys. 16*:1667.

Milam, D., and Weber, M.J. (1976). *J. Appl. Phys. 47*:2497.

Miura, N. (1984). In: *Infrared and Millimeter Waves* (K.J. Button, Ed.), Academic, New York, Vol. 12, Pt. II, p. 73.

Miura, N., Kido, G., Miyajima, H., Nakao, K., and Chikazumi, S. (1981). In: *Physics in High Magnetic Fields* (S. Chikazumi and N. Miura, Eds.), Springer-Verlag, Berlin, p. 12.

Musgrave, M.J.P. (1970). *Crystal Acoustics*, Holden-Day, San Francisco.

Nelson, D.F., and Turner, E.H. (1968). *J. Appl. Phys. 39*:3337.

Okada, M., Takizawa, K., and Ieiri, S. (1977). *J. Appl. Phys. 48*:205.

Pollard, H.F. (1977). *Sound Waves in Solids*, Pion, London.

Potter, R.F. (1985). In: *Handbook of Optical Constants of Solids* (E.D. Palik, Ed.), Academic, New York, p. 11.

Ramachandran, G.N., and Ramaseshan, S. (1961). In: *Handbuch der Physik* (S. Flugge, Ed.), Springer-Verlag, Berlin, Vol. XXV/1.

Sandercock, J.R. (1972), *Phys. Rev. Lett. 28*:237.

Schreibner, E., Anderson, O.L., and Soga, M. (1973). *Elastic Constants and Their Measurement*, McGraw-Hill, New York.

Smith, R.T., and Welsh, F.S. (1971). *J. Appl. Phys. 42*:2219.

Tachi, M. (1985). In: *Infrared and Millimeter Waves* (K.J. Button, Ed.), Academic, New York, Vol. 13, Pt. IV, p. 265.

Temple, P.A. (1985). In: *Handbook on Optical Constants of Solids* (E.D. Palik, Ed.), Academic New York, p. 135.

Vacher, R., and Boyer, L. (1972). *Phys. Rev. B 6*:639.

Wooster, W.A. (1962). *Diffuse X-Ray Reflections from Crystals*, Claredon Press, Oxford.

Wynne, J.J., and Bloembergen, N. (1969). *Phys. Rev. 188*:1211.

Yamaguchi, T., and Hamano, K. (1981). J. Phys. Soc. Jpn. 50:3956.

Yang, F.M., Miura, N., Kido, G., and Chikazumi, S. (1980). *J. Phys. Soc. Jpn. 48*:71.

Yang, T.T. (1976). *Appl. Phys. 11*:167.

4

Physical Properties of Infrared Optical Materials

James Steve Browder   *Jacksonville University, Jacksonville, Florida*

Stanley S. Ballard   *University of Florida, Gainesville, Florida*

Paul Klocek   *Texas Instruments, Inc., Dallas, Texas*

4.1. OPTICAL PROPERTIES: DEFINITIONS AND UNITS

4.1.1. Wavelength Units

The SI unit for wavelength that is appropriate for the infrared region is the micrometer (μm), formerly called micron; 1 μm $= 10^{-6}$ m. For the visible and ultraviolet regions, a more convenient unit may be the nanometer (nm) formerly called the millimicron; 1 nm $= 10^{-9}$ m $= 10^{-3}$ μm. The oldest wavelength unit is probably the angstrom (Å); 1 Å $= 10^{-10}$ m $= 10^{-1}$ nm $= 10^{-4}$ μm.

A less common unit is the wavenumber or reciprocal centimeter (cm^{-1}). It has the advantage in some situations that it is proportional to the energy of the photon, as is the electronvolt (eV). Figure 4.1 provides a ready comparison of nm, cm^{-1}, and eV.

4.1.2. Transmission Range

When light is incident on a dielectric boundary, part of the energy is reflected, part is absorbed, and part is transmitted. The optical absorption of a material is a function of its chemical and physical structure and varies with wavelength and thickness.

There are losses due to reflection at the two surfaces and also losses due to absorption within the material. External transmittance is a measure of the sum of these two types of losses; it is the ratio of the intensity of the light that has passed through a material and is observable on the other side to the intensity of the

141

incident light. Transmittance is a pure number less than unity and may be expressed as a percentage.

The ratio of the intensity of the light arriving at the second interface to that leaving the first interface is the internal transmittance. For a particular sample, the internal transmittance, which relates only to absorption losses unless there is bulk scattering, is larger than the external transmittance. The internal transmittance can be calculated in the following manner:

Internal transmittance = external transmittance + 2× reflection loss per surface

Long- and short-wavelength transmission cutoffs are often given for samples of known thickness. The wavelength cutoff usually corresponds to an external transmittance of 10%. These "cutoffs" may be sensitive functions of temperature and impurity content, particularly for semiconductor materials.

4.1.3. Reflection

A useful datum may be the total reflection loss for the two surfaces of a plate or window, ordinarily for normal incidence.

The reflectance, when light strikes an interface between two materials, is defined as the intensity of the reflected beam divided by the intensity of the beam that is incident at a given angle on the interface. Thus values of reflectance will lie in the range 0–1.0, or they may be given in percentages.

4.1.4. Absorption

Sometimes absorption curves are given instead of transmission curves. The (internal) absorption is expressed in terms of an absorption coefficient α (also called absorption constant or loss coefficient). It is defined by the relationship

$$\alpha = -\left(\frac{1}{t}\right)\ell n\left(\frac{I}{I_o}\right)$$

where I/I_o represents the internal transmittance and t is the thickness. Thus, the units of absorption coefficient are reciprocal lengths.

4.1.5. Refractive Index, *n*

The refractive index of a substance is the ratio of the speed of light in a vacuum to the speed of light in that substance. Refractive index is a function of both wavelength and the temperature of the sample but is not strongly affected by the sample purity.

Both tables and graphs of refractive index versus wavelength at a given temperature (usually room temperature) are useful, as are data on the temperature

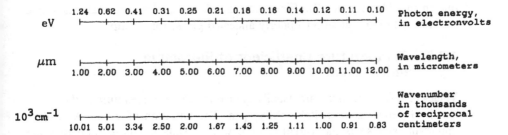

Figure 4.1 Comparison (to scale) of wavelength and energy units.

coefficient of refractive index, *dn/dt*. Figure 5.3 of Chapter 5 shows several typical "dispersion curves," as they are sometimes called.

Many crystals possess the property of double refraction or birefringence, by which a single incident beam is split into two refracted beams. The natural crystals calcite (or Iceland spar, $CaCO_3$) and quartz (SiO_2) are doubly refracting, as are such synthetic crystals as sodium nitrate ($NaNO_3$) and ammonium dihydrogen phosphate ($NH_4H_2PO_4$). When a beam of unpolarized light is incident normally upon the end face of a calcite crystal, the ray is split. Ray O passes through the crystal without deviation—it is called the *ordinary ray* because it obeys the law of refraction. Ray E is deviated despite its perpendicular incidence and hence is called the *extraordinary ray*. The emergent beams are parallel to the incident beam and to each other; if the crystal is rotated around the O beam, the E beam rotates with the motion. Each beam is linearly polarized, and the two polarization axes are mutually perpendicular.

There is a direction in the crystal along which no splitting of the refracted beam occurs; this is called the *optic axis*.

The direction taken by the rays within the crystal depends upon the angle of incidence and the orientation of the optic axis. The speed of the ordinary ray does not depend upon its direction, and the refractive index, n_O, of calcite for this ray has the constant value of 1.658, as measured for the sodium D line. The speed of the extraordinary ray varies with its direction in the crystal. If this ray happens to advance along the optic axis, its speed is the same as that of the ordinary ray, and the two rays coincide. If it travels in any other direction, its speed is greater, with a maximum value in a direction at right angles to the optic axis. The refractive index of calcite for the extraordinary ray, as measured for the D line, varies from 1.658 along the optic axis to 1.486 at right angles to the axis. The refractive index for the extraordinary ray is customarily defined as the value that is farthest from the refractive index for the ordinary ray, and this is the value quoted in tables for n_E. Birefringence may also occur in a normally isotropic material with internal strain—poorly annealed glass, for example.

4.1.6. Temperature Coefficient of Refractive Index, *dn/dt*

Values of refractive index are usually quoted for room temperature at the wavelength of yellow sodium light, 0.589 μm. The variation of n with λ is often important, and so is the variation of n with t. Data on these properties are reported in this handbook. For some materials, $(1/n)\,(dn/dt)$ is reported, rather than dn/dt.

4.1.7. Dispersion

The dispersion of a substance is defined as the derivative of the refractive index with respect to wavelength, dn/dλ. Figure 5.4 of chapter 5 shows the dispersion of several materials for purposes of comparison only; the information needed for any calculations of dispersion should be taken from the refractive index data tables.

Dispersion is the slope of the n versus λ curve at the wavelength of interest. Thus it is usually a negative number in regions where the material is tranparent and away from regins of large absorption. Absolute values of dn/dλ are plotted in Fig. 5.4.

4.1.8. Zero-Dispersion Wavelength

Zero-dispersion wavelength is the wavelength for which the second derivative of refractive index with respect to wavelength, $d^2n/d\lambda^2$, is zero. At this wavelength the phase velocity of a plane wave traveling in the material will remain unchanged. If $d^2n/d\lambda^2$ is nonzero, the phase velocity changes nonlinearly, which causes light pulses to spread in the material.

4.2. THERMAL PROPERTIES: DEFINITIONS AND UNITS

4.2.1. Melting Temperature

Melting temperature is that temperature at which the phase of a material changes from solid to liquid. At atmospheric pressure in an open system, some optical materials melt; some, such as cadmium sulfide, sublime; others, such as gallium arsenide, dissociate; and still others, such as the glasses, soften.

Temperatures are now given in kelvins (K), previously called degrees Kelvin (°K). The earlier custom was to use degrees Celsius (°C; also called degrees

Centigrade at that time), where K = °C + 273. The size of the degree is the same on the Celsius and Kelvin temperature scales.

4.2.2. Specific Heat

When heat is applied to a material, its temperature increases, but at different rates for different materials. The specific heat capacity of a substance is defined as the amount of heat necessary to raise the temperature of a unit mass of the substance by 1°; the units are cal/(g·C°). In the cgs system, the value for water is exactly 1. In energy units, joules per gram per kelvin, or J/(g·K), the specific heat capacity of water is 4.184.

The specific heat of a substance is defined as the ratio of its specific heat capacity to that of water. Thus, specific heat is dimensionless regardless of the system of units employed. Since the value varies somewhat with temperature, it is customary to list the temperature or temperature range for which the specific heat was measured.

4.2.3. Thermal Conductivity

When heat flows at a uniform rate through a regularly shaped solid in a direction perpendicular to the faces of the solid, the quantity of heat that passes through a section of the face in unit time is proportional to the area of the section and the temperature gradient, that is, the difference in temperature per unit thickness in the direction of heat flow. The constant of proportionality is called *thermal conductivity*. Thermal conductivity is a function of the temperature and the transport properties of the material. Historically, it has usually been given in units of cal/(s·cm·°C). The SI units are watts per meter per kelvin: W/(m·K). Dividing these values by 418.4 converts them to the earlier cgs units. For some crystalline materials, the thermal conductivity varies with the direction of heat flow through the crystal. In such cases it is usual to give two values, one (*p*) for heat flow parallel to the *c* axis of the crystal, and the other (*s*) for heat flow perpendicular to the *c* axis.

4.2.4. Thermal Expansion

When a material is heated, its dimensions change. Usually the dimensions increase with increasing temperature. The coefficient of linear thermal expansion, α, is a measure of the proportional change in length of a given sample for a temperature change of 1°; it is defined as the change in length divided by the original length and the change in temperature:

$$\alpha = \frac{1}{L}\frac{\Delta L}{\Delta T}$$

The units of α are hence reciprocal temperature, K^{-1}, since the kelvin and the Celsius degree are equal in magnitude. Unless a change of phase occurs, α at ambient temperatures does not vary strongly with temperature. Its value may vary with direction in a crystal, in which case values (p) parallel to and (s) perpendicular to the c axis should be quoted. The coefficient of volume expansion, β, is equal to 3α for isotropic materials.

4.3. OTHER PHYSICAL PROPERTIES: DEFINITIONS AND UNITS

4.3.1. Molecular Weight

Units are not ordinarily specified for atomic or molecular weights. The quoted numbers are multiplies of the atomic mass unit (amu), which is 1/12 of the mass of a neutral atom of isotope 12 of carbon.

4.3.2. Specific Gravity

Specific gravity is a measure of the density of a material. It is the ratio of the density of a substance to the density of water, both at a temperature of 4°C. The values given are usually for room temperature, unless a temperature is specified. Specific gravity is a dimensionless quantity. The density of a material in the metric system (g/cm^3) is numerically the same as its specific gravity. The density in the English system (lb/ft^3) is numerically 62.4 times the specific gravity.

4.3.3. Hardness

Several different methods have been used to measure and specify this well-known physical property of materials. Basically, hardness is a measure of the amount of force required to plastically deform the material. Hardness values are now most commonly given as Knoop numbers: A pyramidal diamond point is pressed into the sample with a known force, the indentation made is then measured, and the Knoop value is calculated. The point is so designed that the surface being tested is not work-hardened along the direction from which the hardness value is obtained. The Knoop number varies slightly with the indenter load and with temperature. Knoop numbers vary from values as small as 4 for a soft material such as potassium bromide to values as high as 2000 for a hard material such as sapphire and 7000 for diamond.

Moh values of hardness are arrived at by determining which materials can scratch other materials. Moh value is determined relative to a value of 1 for talc and 10 for diamond.

Vickers values of hardness are determined by pressing a pyramidal diamond indenter into the material. The value is then determined by dividing the indenter load (in kilograms) by the pyramidal area of the indenter (in square millimeters).

Rockwell hardness and Brinell hardness values are not quoted in this report. Rockwell hardness is related to a specific measuring instrument; Brinell hardness values are analogous to Vickers values. The Brinell indenter is a section of a sphere.

4.3.4. Solubility

When a solid sample is placed in a solvent, the molecular or crystalline structure may be broken down so that the solute (the sample) is dissolved by the solvent. Solubility is a measure of the amount of solute that is "absorbed" by a given amount of solvent. Water is usually taken as the solvent. A material is said to be insoluble when its solubility is less than 10^{-3} g per 100 g of water.

Due to the specific requirements of individual applications and the lack of standardized testing, further data on solubility or chemical durability and weathering are not included in this book. The reader is encouraged to use the references in the book to obtain data on durability in acids and alkali and on weathering, which is corrosion, particularly of the surface of the material, by the atmosphere—fog, humidity, sulfur dioxide, etc.

4.3.5. Dielectric Constant

From the classical point of view, when two plane-parallel plates are separated by a distance that is small compared to the dimensions of the plates and the volume between the plates is evacuated, the application of a potential difference will set up a homogeneous electric field between the plates. The capacitance of this system is the ratio of the total surface charge to the potential difference between the plates. When the volume between the plates is filled with an insulating material, the charge on the plates remains the same, but the potential difference between the plates is lowered and the capacitance of the system is correspondingly increased. The dielectric constant of the insulating material is the ratio of the potential difference when the volume between the plates is evacuated to the potential difference when the volume is filled with the insulating material. The values given in Table 5.11 are the relative dielectric constants of the materials, that is, the ratios of the dielectric constants of the material to that of a vacuum. Dielectric constant is dimensionless; in certain crystals it varies with direction. If measurements are made with other than direct current, the frequency should be noted.

4.3.6. Debye Temperature

The Debye (characteristic) temperature, Θ_D, represents the temperature above which a given crystal behaves classically; that is, it is the temperature above which thermal vibrations are more important than quantum effects. It is not usually considered to be an intrinsic property of the material. It is rather an interpolated, temperature-dependent parameter whose data are used to construct a curve for the specific heat versus temperature, in the best theoretical model available (the Debye theory accurately gives the heat capacity due to lattice vibrations for temperatures less than $\Theta_D/100$). The Debye temperature, which is usually determined by low-temperature calorimetric measurements, is given in kelvins and depends explicitly on the mass and implicitly on the characteristic atomic separation of the material.

4.3.7. Bandgap

As atoms bond to form a solid, their discrete electronic energy levels spread into bands. Because of the Pauli principle, the lower energy bands are occupied and the higher are empty, at least near absolute zero. The highest energy occupied band is termed the "valence band," because its states are usually those most involved in bonding. In an insulator or semiconductor, the valence band is separated in energy from the lowest empty band. That lowest empty band is called the "conduction band," because an electron in an unfilled band is free to move with the addition of arbitrarily small amounts of energy.

The interval separating the valence and conduction bands in an insulator or semiconductor depends on the electron wave vector. The smallest value of that interval is the bandgap. Customary units for energy bands and bandgaps are electron volts, rydbergs (1 Ry = 13.6058 eV), and (less commonly), hartrees (1 Ha = 27.2116 eV).

The conduction band is not necessarily separated from the valence band; when the two overlap in energy, a metal results. Hence, in a metal the valence and conduction bands are both partially occupied, even at absolute zero. An insulator in the same circumstances has a completely vacant conduction band.

The main difference between insulators and semiconductors is simply the size of the bandgap. The semiconductor bandgap is much the narrower, typically 1 eV or less. Semiconductor technology exploits that small bandgap by the introduction of impurities (doping) that produce a sequence of impurity states in the gap. These allow controlled conduction by making empty electron states available at small enough energy cost that thermal effects are usually enough to promote the electrons.

The bandgap of a material has a pronounced effect upon its absorption, photoconductivity, and photoluminescence properties. Large-bandgap insulators, for example, absorb only in the ultraviolet, because the incoming photon must provide enough energy to promote an electron across the bandgap. Thus, dielectric materials tend to be transparent in the visible and infrared.

4.3.8. Elastic Constants

Hooke's law states that for small deformations the stress acting on a solid is proportional to the strain existing within it. The ratio of stress to strain is called the *elastic modulus*: the three common elastic moduli are Young's, sometimes called the stretch modulus; bulk modulus, which is the reciprocal of compressibility; and shear or rigidity modulus. The names indicate the nature of the deformation that causes strain in the body.

Young's modulus is concerned with that elastic deformation of a body in which an applied force results in a change in length of the body; Young's modulus is defined as the ratio of stress to strain, where the stress is the force per unit area perpendicular to which the force is applied, and the strain is the resulting fractional change in length, which may be either positive or negative, depending on whether stretching or compression is involved. Since strain is a dimensionless number, the units of elastic modulus are those of stress: force per unit area, or pressure.

Bulk modulus is concerned with that elastic deformation of a body in which an applied force results in a change in volume of the body. Bulk modulus is the ratio of the pressure applied to a body to the volume strain, which is the resulting fractional decrease in volume of the body. Thus, bulk modulus applies to fluids as well as to solids.

The modulus of rigidity, or *shear modulus*, is concerned with that elastic deformation of a body in which an applied force results in a change in the shape of the body. The modulus of rigidity is the ratio of the shearing stress to the shear strain, where the shearing stress is the force per unit area parallel to which the force is applied, and the shear strain is the angle of shear in radians. Torsional displacement can be regarded as "rotational shear."

The SI units for pressure, and hence for elastic moduli, are pascals (Pa); 1 Pa = 1 N/m^2. These relate to cgs units by the equivalence 1 Pa = 10 $dynes/cm^2$ and to British units by 1 pound per square inch (psi) = 6895 Pa. Because of the values involved, elastic constants may be given in gigapascals (1 GPa = 10^9 Pa) or terapascals (1 TPa = 10^{12} Pa). These numbers will seem unfamiliar to engineers who have used psi for many years.

This discussion of elastic properties is applicable only when the elastic deformation of the crystal is small, that is, less than 1% in research studies and less than 2% for practical applications. When a crystal is deformed to such an extent that Hooke's law is no longer obeyed, the crystal is said to be plastically deformed. The *elastic limit* is used to indicate the stress above which plastic deformation occurs and below which Hooke's law is obeyed. For metals, the stress–strain curve shows an abrupt change at the elastic limit; for most optical materials the stress–strain curve gradually changes from a straight line to a curve with a decreasing slope. In the apparent absence of a unique departure from Hooke's law, an *apparent elastic limit* must be defined. This limit is taken as that stress on the stress–strain curve where the slope is half the slope at the origin of the stress–strain curve.

A *modulus of rupture* is quoted for some materials. This value is strongly dependent upon the history of the particular sample and should be used only for approximate calculations. The presence of small cracks or barely macroscopic cleavages can significantly change the apparent rupture strength.

When a material elastically elongates under tensile stress, the dimensions perpendicular to the stress direction become shorter by an amount proportional to the fractional change in length. The ratio of the fractional change in width to the fractional change in length is called *Poisson's ratio*; it is a dimensionless constant of the material.

Hooke's law involves the stress acting on a solid and the strain existing within it. The tensor components of stress are linear functions of the components of strain; the constants of proportionality between the components of stress and the components of strain are called the *elastic stiffness constants*, or the elastic coefficients. These coefficients are designated by the quantities c_{hk}, where h and k have values from 1 to 6. For cubic crystals there are three independent c_{hk}'s, namely, c_{11}, c_{12}, and c_{44}. For tetragonal crystals, such as titanium dioxide, there are five independent c_{hk}'s, namely, c_{11}, c_{12}, c_{13}, c_{33}, and c_{44}.

The components of strain may alternatively be considered as linear functions of the components of stress; then, the constants of proportionality between the respective components are called *elastic compliance constants* and are denoted by s_{hk}, where h and k have values from 1 to 6. For cubic crystals, there are three independent s_{hk}'s, namely s_{11}, s_{12}, s_{44}. As in the case of c_{hk}, there are five independent s_{hk}'s for tetragonal crystals and six for hexagonal crystals.

For a cubic crystal, the equations that relate c_{hk} with s_{hk} are

$$s_{11} = \frac{c_{11} + c_{12}}{(c_{11} + 2c_{12})(c_{11} - c_{12})}$$

$$s_{12} = \frac{-c_{12}}{(c_{11} + 2c_{12})(c_{11} - c_{12})}$$

$$s_{44} = 1/c_{44}$$

The three engineering moduli are related to the c_{kk} by the following equations:

$$\text{Young's modulus} = \frac{(c_{11} + 2c_{12})\,(c_{11} - c_{12})}{c_{11} + c_{12}}$$

$$\text{Modulus of rigidity} = c_{44}$$

$$\text{Bulk modulus} = \frac{c_{11} + 2c_{12}}{3}$$

In general, crystals are anisotropic with respect to their elastic properties; that is, the values of these moduli differ with direction in the crystal. A measure of the anisotropy of a cubic crystal is given by the *anisotropy factor*, A; it is defined as $A = 2c_{44}/(c_{11} - c_{12})$. This factor equals unity for elastically isotropic materials, and for elastically anisotropic crystals it has values either greater than or less than unity.

4.4. UNIQUE PHYSICAL PROPERTIES OF GLASSES

Glass materials are treated separately from crystalline because of some fundamental differences in the nature of their state. Although these differences are generally not obvious to the eye or in application, they are essential to understanding the glass material, its physical properties, and how it can be used.

The term *glass* requires some definition. It usually refers to a solid formed by quenching a molten liquid. In contrast, the term *amorphous* usually refers to a solid lacking any long-range atomic order, that is, one that is disordered or nonperiodic. All glasses are amorphous, but the reverse is not correct. This is due to the specific nature of glass, as will be explained momentarily. The term *vitreous* is usually considered synonymous with glass.

It is generally true that any crystalline solid can be made amorphous if cooled sufficiently rapidly from above its melting point. An example of this is amorphous metals, which are usually cooled on the order of 10^6 K/sec. These are not glasses, however, because they do not possess a glass transition (T_g) temperature. Glasses will show a continuous decrease in volume as they are cooled from high temperatures through a supercooled liquid region until their T_g is reached where the rate of change in the volume changes (usually decreases) but remains continuous. T_g is the temperature at the intersection of the glass solid and liquid phases; it is dependent on cooling rate and increases with increasing cooling rate. Crystalline materials, on the other hand, will have the same continuous decrease in volume as they are cooled from high temperatures until their melting point is reached; at this point a discontinuous change in volume occurs as the material crystallizes. Once crystallized, the material will again show a continuous change in volume, which usually decreases. These changes in volume for both crystalline and glass materials are expressed in terms of their volume thermal expansions. The lack of a discontinuous change in the volume of a glass material is due to its viscoelastic property. As a glass liquid cools, its viscosity continuously increases. The viscosity eventually becomes so great that the mass transport required for crystallization has time constants on the order of years, and thus crystallization is frustrated.

The viscoelastic property of a glass is an essential difference between its solid state and that of a crystal. Unlike crystals, glasses do not melt from a solid to a liquid but become less and less viscous. Glasses can, however, devitrify (crystallize) if held above their T_g for sufficient time. The time and temperature required for devitrification and the composition of the resulting phases depend on the composition of the glass. The parameters to be used for considering an application with a glass are therefore different from a crystal's melting point. The three commonly used parameters for a glass are strain point, annealing point, and softening point. A specific viscosity is assigned to each of these three parameters: $10^{14.6}$, $10^{13.4}$, and $10^{7.6}$ poise, respectively. The *strain point* is the temperature at which the glass will begin to flow rather than fracture. The *annealing point* is the temperature at which residual stress or strain in the glass will substantially relieve itself in a few minutes. The *softening point* is the temperature at which the glass could be worked or molded. For use as an optical component where dimensional stability is essential, a glass should not be considered for an application involving a temperature above its T_g. For this reason it is the T_g that is listed in the tables of Chapter 7.

Another essential difference between crystals and glasses is the lack of long-range order in the structure of a glass. Crystals have periodic arrangements of atoms that can be reduced to a unit cell that when translated reproduces the crystal. The types of structures crystals assume greatly influence their physical properties, owing to the orientational differences in the structures. Glasses, on the other hand, are isotropic. Their physical properties are the same in all directions within the glass. Thus the physical properties listed in the tables of Chapter 7 can be used regardless of the orientation of the glass material. The meaning of the physical properties as discussed in prior sections apply to both glass and crystalline materials. The exception in the case of the melting point has been discussed above. While the meanings of the physical properties are the same, the random structure of the glasses modifies their values with respect to a crystalline analog. While the bond type, bond angle, and bond strength in a glass are usually similar to those of a crystal analog, they do not have discrete values like the crystal. They are broadened by the lack of periodicity in the glass structure. This results in a broadening of the density of states, resulting in smoother features in the properties of the material such as bandgap and electronic absorption edge.

While glasses will be considered isotropic, a caution must be noted. Due to the cooling rate dependence of T_g, nonuniform cooling of a glass will cause variations in T_g and therefore in the density and the corresponding index of refraction. This index variation will result in birefringence. Glass birefringence is similar to the orientationally dependent index of refraction discussed in Chapter 1 and Section 4.1.5 of this chapter. The glass birefringence is due to density differences that result in residual stresses and strains. Such residual strains can lead to a weakening of these already brittle materials. On the other hand, control of these residual stresses, particularly at the surface of a glass object, can be used to strengthen it. Thus, residual stresses can render a glass anisotropic, resulting in index variations that cause light scattering and loss of imaging capability, elastic modulus variations, strength variations, hardness variations, and, generally speaking, variations in all physical properties.

4.4.1. Optical Properties of Glasses

The optical properties of glasses are described in section 4.1, along with those of crystalline materials.

4.4.2. Thermal Properties of Glasses

Glass transition temperature (T_g)—the temperature at which there is a change in the rate of change of the volume of a glass on cooling or heating. It is the temperature below which the glass is a solid and above which the glass is a liquid. The value of T_g is dependent on the cooling or heating rate. T_g is usually determined by differential thermal or calorimetric techniques.

Softening point—the temperature at which the viscosity of the glass is approximately $10^{7.6}$ poise. At or above this temperature, the glass can be worked. It is determined as the temperature at which a fiber of the glass viscously elongates under its own weight by 1 mm/min as described in ASTM Designation C-338.

Annealing point—the temperature at which the viscosity of the glass is approximately $10^{13.4}$ poise. At the annealing point, stresses in the glass will be relieved in minutes. It is determined as the temperature at which fibers viscously elongate or beams viscously bend at stresses and rates as described in ASTM Designations C-336 and C-598.

Strain Point—the temperature at which the viscosity of the glass is approximately $10^{14.6}$ poise. At the strain point, stresses in the glass will be substantially relieved in hours. The strain point is determined by extrapolation of the annealing point data to elongation rates 31.6 times smaller.

Coefficient of Thermal Expansion—see Section 4.2.4. For glasses, it is determined using ASTM Designation E-228.

For other thermal properties—see Section 4.2.

4.4.3. Other Physical Properties of Glasses

Solubility—see Section 4.3.4.

Dielectric Constant—see Section 4.3.5. For glasses, it is determined using ASTM Designation D-150.

Young's Modulus—see Section 4.3.8. For glasses, it is determined using ASTM Designation C-623.

For other physical properties, see the appropriate sections of this chapter.

5

Physical Property Comparisons of Infrared Optical Materials

James Steve Browder *Jacksonville University, Jacksonville, Florida*

Stanley S. Ballard *University of Florida, Gainesville, Florida*

Paul Klocek *Texas Instruments, Inc., Dallas, Texas*

5.1 INTRODUCTION

This chapter includes data on a large number of optical materials. The data were assembled by searching available sources such as existing handbooks and other compilations and also the scientific and technical literature into 1990. Data are collated and presented in formats intended to be readily useful to the reader, who is presumed to be an engineer, designer, or other applied scientist. Thus, lengthy discussions, even of the possible accuracy of the data, are largely avoided. It is hoped that the reader/user will be able to extract the needed information rapidly from the tables and graphs.

SI or metric units are used where feasible. However, common usage still favors cgs or even English units in some cases. Happily, in many cases there is no problem with competing units. The reader is referred to Tables 1.1–1.5 of Chapter 1.

Lists of materials can be strictly alphabetical (for quickest reference) or grouped into "families": oxides, halides, chalcogenides, for example. They could also be grouped according to the amount of anisotropy for various optical and other physical properties, for example, as glasses, cubic crystals, other single crystals, poly- or microcrystalline compacts. Different users facing differing selection problems would prefer one or another of the possible groupings. Obviously, it is impractical to use them all in this handbook. In Tables 5.1 and 5.2 we have chosen to list the materials alphabetically in two categories, crystals and glasses.

Our data sources have supplied specific data for the large number of physical properties listed in Table 5.3. Data for all these properties are not available for each of the materials of Tables 5.1 and 5.2, as will be noted when examining data sheets and accompanying figures and tables in Chapters 6 and 7. These data sheets are arranged in the same alphabetical order for the materials as given in Tables 5.1 and 5.2.

Table 5.1 Crystalline Infrared Optical Materials: Names and Chemical Formulas

ALON, $Al_{23}O_{27}N_5$ (Raytran)
Aluminum nitride, AlN
Aluminum oxide (ruby), Al_2O_3
Aluminum oxide (sapphire), Al_2O_3
Ammonium dihydrogen phosphate (ADP), $NH_4H_2PO_4$
Barium fluoride, BaF_2
Barium titanate, $BaTiO_3$
Boron nitride, BN
Boron phosphide, BP
Cadmium fluoride, CdF_2
Cadmium phosphide, Cd_3P_2
Cadmium selenide, CdSe
Cadmium sulfide, CdS
Cadmium telluride, CdTe
Calcium carbonate, $CaCO_3$ (calcite)
Calcium fluoride, CaF_2
Calcium lanthanum sulfide, $CaLa_2S_4$
Cesium bromide, CsBr
Cesium iodide, CsI
Cuprous chloride, CuCl
Diamond II A, C
Gallium antimonide, GaSb
Gallium arsenide, GaAs
Gallium phosphide, GaP
Gallium selenide, GaSe
Germanium, Ge
Indium antimonide, InSb
Indium arsenide, InAs
Indium phosphide, InP
Lanthanum fluoride, LaF_3
Lead fluoride, PbF_2
Lead selenide, PbSe
Lead sulfide, PbS
Lead telluride, PbTe
Lithium fluoride, LiF
Lithium niobate, $LiNbO_3$
Magnesium aluminate (spinel) $MgAl_2O_4$
Magnesium aluminate (Raytran spinel), $MgAl_2O_4$

(continued)

Table 5.1 Crystalline Infrared Optical Materials: Names and Chemical Formulas (Continued)

Magnesium fluoride, MgF_2
Magnesium fluoride (Irtran 1), MgF_2
Magnesium oxide, MgO
Magnesium oxide (magnesia) MgO
Potassium aluminosilicate (mica), $KAl_3Si_3O_{10}(OH)_2$
Potassium bromide, KBr
Potassium chloride, KCl
Potassium dihydrogen phosphate (KDP), KH_2PO_4
Potassium iodide, KI
Rubidium bromide, RbBr
Rubidium chloride, RbCl
Rubidium iodide, RbI
Selenium, Se
Silicon, Si
Silicon carbide, SiC
Silicon dioxide (crystal quart), SiO_2
Silicon nitride, Si_3N_4
Silver bromide, AgBr
Silver chloride, AgCl
Sodium chloride, NaCl
Sodium fluoride, NaF
Strontium barium niobate, $SrBaNb_3O_6$
Strontium fluoride, SrF_2
Strontium titanate, $SrTiO_3$
Tellurium, Te
Thallium bromide, TlBr
Thallium bromoiodide (KRS-5), $Tl(Br,I)$
Thallium chloride, TlCl
Titanium dioxide, TiO_2
Yttrium oxide, Y_2O_3
Yttrium oxide (Raytran yttria), Y_2O_3
Yttrium oxide, lanthanum-doped, Y_2O_3-La
Zinc chloride, $ZnCl_2$
Zinc selenide, ZnSe
Zinc selenide (CVD Raytran), ZnSe
Zinc sulfide, ZnS
Zinc sulfide (CVD Raytran, multispectral grade), ZnS
Zinc sulfide (CVD Raytran, standard grade), ZnS
Zinc sulfide (Irtran 2), ZnS
Zirconium dioxide, ZrO_2
Zirconium dioxide, yttrium oxide doped, ZrO_2-Y_2O_3

CVD = chemical vapor deposition; Irtran = trade name, Eastman Kodak Company; Polytran = trade name, Harshaw Company; Raytran = trade name, Raytheon Company.

Table 5.2   Glass Infrared Optical Materials: Names and Chemical Formulas

Arsenic trisulfide, As_2S_3
Beryllium fluoride, BeF_2
BZYbT (fluoride)
Schott IRG 11 (calcium aluminate)
Corning 9753 (calcium aluminosilicate)
Schott IRG N6 (calcium aluminosilicate)
Ohara HTF-1 (fluoride)
Schott IRG 9 (fluorophosphate)
SiO_2 (fused silica)
Corning 7940 (fused silica)
Corning 7957 (fused silica)
Corning EO (GaBiPb oxide)
Corning 9754 (germanate)
Schott IRG 2 (germanate)
TI 1173 ($Ge_{28}\ Sb_{12}\ Se_{60}$)
AMTIR-1 ($Ge_{33}\ As_{12}\ Se_{55}$)
$Ge_{25}\ Se_{75}$
Schott IRG 100 (chalcogenide)
HBL (fluoride)
Schott IRG 3 (lanthan dense flint)
Schott IRG 7 (lead silicate)
Corning 7971 (titanium silicate)
ZBL (fluoride)
ZBLA (fluoride)

Table 5.3   Physical Properties

Optical properties
  Transmission
  Reflection
  Absorption
  Refractive index
  Temperature coefficient of refractive index
  Dispersion
  Zero dispersion wavelength
Thermal properties
  Melting or softening temperature
  Specific heat
  Thermal conductivity
  Thermal expansion
Other physical properties
  Molecular weight

(continued)

Table 5.3   Physical Properties (Continued)

Specific gravity
Hardness
Solubility
Dielectric constant
Debye temperature
Bandgap
Elastic constants
   Young's modulus
   Bulk modulus
   Shear modulus
   Elastic limit
   Rupture modulus
   Poisson's ratio
   Elastic stiffness
   Elastic compliance

5.2.   COMPARISONS OF PROPERTIES

Tables 5.1 and 5.2 list a large number of optical materials with transmission at least partially in the infrared. How does the designer/engineer, faced with building a specific device, choose among them? Surely the first criterion would be high transmission in the appropriate region of the broad infrared spectrum (see Figures 5.1 and 5.2, which indicate the transmission of crystalline and glass materials, respectively). This may narrow the choice considerably. Other optical properties (see Table 5.3, physical properties) should also be considered, depending on the end use: window, lens, prism, filter substrate, etc. Figures 5.3 and 5.4 may be useful here.

The choice of possible materials will also depend on the type of device and its working environment: outer space or battlefield on earth, for instance. Table 5.3 also lists thermal properties and a group called "other physical properties" for lack of a more specific designation. Individual properties are treated in tables listing values for several materials to facilitate comparisons by the engineer and the choice of possible candidate materials; see Tables 5.4–5.30. The materials are listed in Tables 5.4–5.30 in order of the increasing value of the individual property.

After completion of the selection/decision process just sketched, it may well be that there is no single material that fully satisfies the design criteria. The engineer must then resort to compromises. The comparison tables should assist

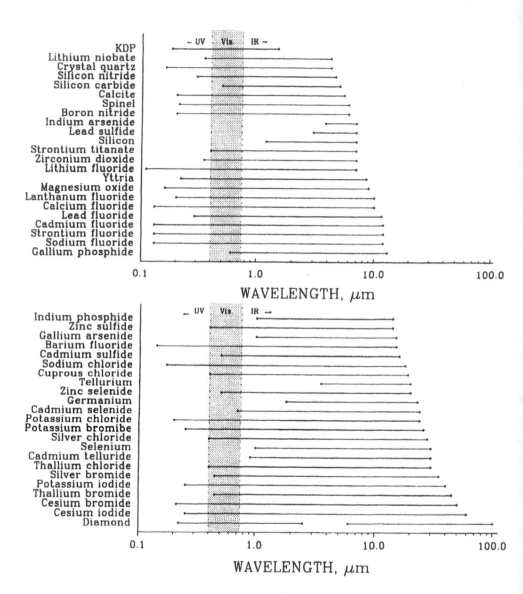

Figure 5.1 Transmission regions of crystalline infrared optical materials. The long- and short-wavelength cutoffs indicate external transmittance of 10% for a 1 mm thick sample. Transmittances for other thicknesses may be given in the data sheets of Chapter 6.

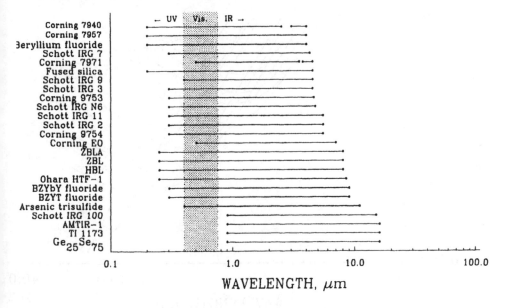

Figure 5.2   Transmission regions of glass infrared optical materials. Data plotted from Table 5.17.

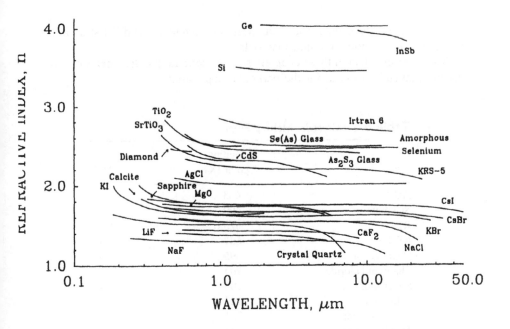

Figure 5.3   Refractive index versus wavelength for several infrared optical materials.

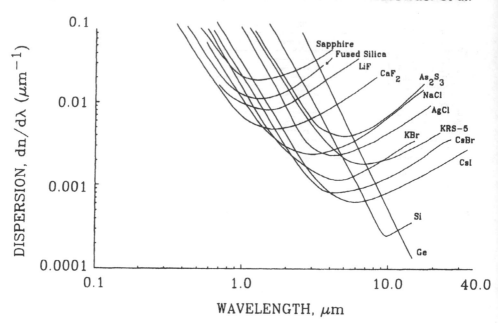

Figure 5.4 Dispersion versus wavelength for several infrared optical materials.

in making these decisions. The data sheets of Chapters 6 and 7 list additional properties of the more common materials.

The extensive data in these chapters should facilitate the designer's choice of the most suitable material for the particular application.

Table 5.4 Crystals: Melting Point

Material	Melting point (K)	Ref.
ADP	463	1
Selenium	490	2
KDP	525.6	3
KRS-5	687	1
Cuprous chloride	695	4
Thallium chloride	703	5
Silver bromide	705	4
Tellurium	723	5
Silver chloride	731	4
Thallium bromide	733	5

(continued)

Table 5.4 Crystals: Melting Point (Continued)

Material	Melting point (K)	Ref.
Indium antimonide	796	6
Cesium iodide	894	4
Cesium bromide	909	4
Rubidium iodide	920	3
Rubidium bromide	966	3
Rubidium chloride	991	3
Gallium antimonide	993	7
Potassium iodide	996	3
Potassium bromide	1000	3
Potassium chloride	1050	3
Sodium chloride	1070	3
Lead fluoride	1100	3
Lithium fluoride	1140	3
Strontium fluoride	1190	4
Lead telluride	1190	3
Germanium	1210	4
Indium arsenide	1215	4
Gallium selenide	1233	—
Sodium fluoride	1270	3
Cadmium telluride	1320	4
Indium phosphide	1330	4
Lead selenide	1340	3
Cadmium fluoride	1370	4
Lead sulfide	1390	3
Mica	1473–1573	8
Gallium arsenide	1511	9
Irtran 1[a]	1528	1
Magnesium fluoride	1528	3
Lithium niobate	1530	3
Barium fluoride	1550	4
Calcite	1610	4
Cadmium selenide	1623	4
Calcium fluoride	1630	4
Cadmium sulfide	1670	10
Silicon	1690	9
Quartz	1740	3
Gallium phosphide	1740	9
Lanthanum fluoride	1760	3
Zinc selenide	1790	5
Barium titanate	1870	4

(continued)

Table 5.4 Crystals: Melting Point (Continued)

Material	Melting point (K)	Ref.
Titanium dioxide	2090	5
Zinc sulfide	2100	5
Irtran 2[a]	2103	11
Sapphire	2300	4
Silicon nitride	>2300	3
Ruby	2313	12
Spinel	2320	3
Strontium titanate	2350	5
Raytran[b] spinel	2408	13
Raytran ALON[b]	2443	13
Yttria	2650	5
Raytran[b] yttria	2723	13
GTE lanthanum-doped yttria	2737	14
Boron phosphide	3000	15
Silicon carbide	~3000	3
Yttria-doped zirconia	~3000	5
Magnesia	3053	16
Aluminum nitride	3070	17
Magnesium oxide	3070	3
Boron nitride	3500	4
Diamond	3770	4

[a]Trade name, Eastman Kodak Co.
[b]Trade name, Raytheon Co.

Table 5.5 Crystals: Specific Heat

Material	Specific heat [cal/(g K)]	Ref.
Zinc selenide	0.0090 (24 K)	5
Gallium antimonide	0.019	4
Lead telluride	0.036 (240 K)	18
Lead selenide	0.042 (240 K)	19
Thallium bromide	0.045 (293 K)	5
Cesium iodide	0.048 (293 K)	4
Tellurium	0.048 (573 K)	5
Indium antimonide	0.0496 (273 K)	20

(continued)

Table 5.5   Crystals: Specific Heat (Continued)

Material	Specific heat [cal/(g K)]	Ref.
Lead sulfide	0.050 (273 K)	3
Thallium chloride	0.052 (273 K)	5
Cadmium telluride	0.056 (300 K)	21
Rubidium iodide	0.058 (283 K)	3
Indium arsenide	0.060 (273 K)	4
Cesium bromide	0.063 (293 K)	4
Silver bromide	0.070 (273 K)	4
Lead fluoride	0.072 (282 K)	3
Indium phosphide	0.073 (273 K)	4
Germanium	0.074 (273–373 K)	4
Rubidium bromide	0.074 (283 K)	3
Potassium iodide	0.075 (270 K)	3
Barium titanate	0.077 (175 K)	4
Selenium	0.077 (293.5 K)	3
Gallium arsenide	0.078 (300 K)	22
CVD zinc selenide	0.081 (296 K)	23
Calcium lanthanum sulfide	0.082 (300 K)	24
Silver chloride	0.085 (273 K)	4
Cadmium sulfide	0.0882 (273 K)	4
Barium fluoride	0.096 (274 K)	25
Rubidium chloride	0.10 (283 K)	3
Yttria-doped zirconia	0.10 (273 K)	5
Potassium bromide	0.104 (273 K)	3
GTE lanthanum-doped yttria	0.1078 (325 K)	26
Raytran yttria	0.11 (300 K)	27
Yttria	0.11 (300 K)	5
CVD zinc sulfide (standard grade)	0.112 (296 K)	23
Irtran 1	0.12 (298 K)	28
Diamond	0.12 (298 K)	4
Zinc sulfide	0.12 (273 K)	5
Lanthanum fluoride	0.121	3
Zinc chloride	0.123 (271 K)	29
CVD zinc sulfide (multispectral grade)	0.124 (298 K)	30
Strontium fluoride	0.13 (274 K)	31
Strontium titanate	0.13 (300 K)	5
Silicon carbide	0.14 (273 K)	3
Lithium niobate	0.15	3
Aluminum nitride	0.16 (276 K)	32
Potassium chloride	0.162 (273 K)	3
Quartz	0.17 (273 K)	3

(continued)

Table 5.5   Crystals: Specific Heat (Continued)

Material	Specific heat [cal/(g K)]	Ref.
Titanium dioxide	0.17 (298 K)	5
Sapphire	0.18 (298 K)	4
Silicon	0.18 (298 K)	3
Raytran ALON	0.185 (300 K)	33
Gallium phosphide	0.20 (400 K)	4
Sodium chloride	0.20 (273 K)	3
Calcite	0.203 (273 K)	4
Calcium fluoride	0.204 (273 K)	4
Mica	0.208 (293–373 K)	8
KDP	0.21 (306 K)	3
Magnesium oxide	0.21 (273 K)	3
Silicon nitride	0.21 (533 K)	34
Irtran 2	0.23	28
Magnesia	0.24	16
Magnesium fluoride	0.24 (298 K)	3
Raytran spinel	0.248	13
Spinel	0.26 (441 K)	35
Sodium fluoride	0.26 (273 K)	3
Boron nitride	0.35 (473 K)	4
Lithium fluoride	0.37 (283 K)	5

Table 5.6   Crystals: Thermal Conductivity

Material	Thermal conductivity (W/(mK))	Ref.
Mica	0.25–0.59	8
KRS-5	0.544	36
Thallium bromide	0.59	5
Lead sulfide	0.67	3
Thallium chloride	0.75	5
Cesium bromide	0.94	4
Cesium iodide	1.1	4
Silver chloride	1.12	4
Silver bromide	1.21	4
ADP	1.3	5
Selenium (trigonal)	1.3	5
KDP	1.3	5
Barium titanate	1.34	4

(*continued*)

Table 5.6   Crystals: Thermal Conductivity (Continued)

Material	Thermal conductivity (W/(mK))	Ref.
Strontium fluoride	1.42	37
Calcium lanthanum sulfide	1.7	24
Potassium iodide	2.1	3
Lead telluride	2.4	3
Lead selenide	4.2	3
Calcite	4.64	4
Potassium bromide	4.8	3
Lanthanum fluoride	5.1	3
Lithium niobate	5.6	3
GTE lanthanum-doped yttria	6.02	26
Quartz	6.2	3
Cadmium telluride	6.3	4
Tellurium	6.3	5
Sodium chloride	6.5	3
Potassium chloride	6.7	3
Rubidium chloride	7.6	3
Titanium dioxide	8.8	5
Cadmium selenide	9	4
Sodium fluoride	9.2	3
Rubidium iodide	9.9	3
Calcium fluoride	10	4
Yttria-doped zirconia	10.5	5
Strontium titanate	11.2	5
Lithium fluoride	11.3	3
Barium fluoride	11.7	4
Rubidium bromide	12.2	3
Raytran ALON	12.6	33
Spinel	13.8	3
Raytran yttria	14	33
Irtran 1	14.6	28
Raytran spinel	14.6	33
Irtran 2	15.4	28
Cadmium sulfide	16	4
CVD zinc sulfide (standard grade)	17	23
CVD zinc selenide	18	23
Zinc selenide	19	5
Magnesium fluoride	21	3
CVD zinc sulfide (multispectral grade)	27	30
Yttria	27	5
Zinc sulfide	27	5

Table 5.6   Crystals: Thermal Conductivity (Continued)

Material	Thermal conductivity (W/(mK))	Ref.
Sapphire	33	4
Silicon nitride	33	3
Indium antimonide	35.58	3
Magnesia	40.6	16
Gallium antimonide	44	4
Ruby	46	12
Indium arsenide	50	4
Gallium arsenide	55	22
Germanium	59	4
Magnesium oxide	59	3
Indium phosphide	70	4
Gallium phosphide	110	38
Silicon	163	3
Aluminum nitride	320	17
Boron phosphide	350	15
Silicon carbide	490	3
Boron nitride	1300	4
Diamond	2600	39

Table 5.7   Crystals: Thermal Expansion

Material	Thermal expansion ($\times 10^{-6}$ K^{-1})	Ref.
Diamond	0.8 (293 K)	39
Boron nitride	1.2 (300 K)	4
Silicon nitride	2.1 (293–400K) (p)	3
	1.1 (273–400K) (s)	3
Aluminum nitride	2.7 (300 K)	17
Silicon carbide	2.8 (300 K)	3
Boron phosphide	3.0 (300 K)	4
Silicon	3.1 (373–473 K)	3
Lithium niobate	4.1 (300 K) (p)	3
Indium phosphide	4.5	4
Indium antimonide	4.9 (270 K)	3
Cadmium sulfide	5 (323–773 K) (s)	40
	3.5 (323–773 K) (p)	40
Gallium phosphide	5.3 (300 K)	9
Indium arsenide	5.3 (273 K)	4

(*continued*)

Table 5.7 Crystals: Thermal Expansion (Continued)

Material	Thermal expansion ($\times 10^{-6}$ K^{-1})	Ref.
Sapphire	5.6 (293 K) (p)	4
	5.0 (293 K) (s)	4
Gallium arsenide	5.7 (300 K)	9
Raytran ALON	5.8 (303–473 K)	13
Raytran yttria	5.8 (293–473 K)	13
Ruby	5.8 (293–323 K)	12
Cadmium telluride	5.9 (300 K)	10
Spinel	5.9 (313 K)	3
Germanium	6.0 (300 K)	9
Yttria	6.4 (300 K)	5
CVD zinc sulfide (multispectral grade)	6.5 (208–473 K)	30
Raytran spinel	6.5 (303–473 K)	13
CVD zinc sulfide (standard grade)	6.6 (273 K)	41
Zinc sulfide	6.7 (300 K)	5
Gallium antimonide	6.9	4
Irtran 2	6.9 (300 K)	42
CVD zinc selenide	7.1 (273 K)	41
Cadmium selenide	7.3 (320 K)	43
Zinc selenide	7.5 (300 K)	5
Lead selenide	7.7 (303 K)	3
Magnesia	8 (373 K)	16
GTE lanthanum-doped yttria	8.12 (298–1273 K)	44
Yttria-doped zirconia	8.8 (293 K)	5
Lead telluride	9 (303 K)	3
Titanium dioxide	9.2 (313 K) (p)	5
	7.1 (313 K (s)	5
Strontium titanate	9.4	5
Cuprous chloride	10 (313–413 K)	4
Irtran 1	10.4 (300 K)	41
Magnesium oxide	11.0 (300 K)	3
Quartz	13.4 (273–353 K) (s)	3
	8.0 (273–353 K) (p)	3
Magnesium fluoride	14 (310 K) (p)	3
	8.9 (310 K) (s)	3
Calcium lanthanum sulfide	14.7 (293–673 K)	24
Lanthanum fluoride	17 (288 K) (s)	3
Lead sulfide	18	3
Barium fluoride	18.4 (273–573 K)	45
Strontium fluoride	18.4 (293 K)	4
Calcium fluoride	18.9 (300 K)	46

(*continued*)

Table 5.7 Crystals: Thermal Expansion (Continued)

Material	Thermal expansion ($\times$ 10^{-6} K^{-1})		Ref.
Barium titanate	19	(283–343 K)	4
Calcite	25	(273 K) (*p*)	4
	−5.8	(273 K) (*s*)	4
Cadmium fluoride	27	(293–333 K)	4
Mica	27		47
Tellurium	27.5	(303 K) (*s*)	5
	−1.52	(303 K) (*p*)	5
Lead fluoride	29	(283 K)	3
Silver chloride	30	(293–333 K)	4
Sodium fluoride	32	(298 K)	3
Silver bromide	35	(298–333 K)	4
Rubidium chloride	36	(293 K)	3
Lithium fluoride	37	(273–373 K)	3
Potassium chloride	37	(273 K)	3
Rubidium bromide	37.7	(300 K)	48
Potassium bromide	39	(273 K)	3
Rubidium iodide	39	(293 K)	3
ADP	39.3	(297–407 K)	49
Sodium chloride	40	(273 K)	3
Potassium iodide	41	(273 K)	3
KDP	42	(300 K) (*p*)	3
	27	(300 K) (*s*)	3
Cesium bromide	47	(273 K)	4
Selenium	49	(273–294 K)	3
Cesium iodide	50	(298–323 K)	4
Thallium bromide	51	(293–333 K)	5
Thallium chloride	53	(292–333 K)	5
KRS-5	58	(293–373 K)	5

Table 5.8 Crystals: Specific Gravity

Material	Specific gravity [g/cm^3]	Ref.
ADP	1.803	65
Potassium chloride	1.984	50
Sodium chloride	2.165	51
Boron nitride	2.25	52
Silicon	2.33	53

Table 5.8 Crystals: Specific Gravity (Continued)

Material	Specific gravity [g/cm³]	Ref.
KDP	2.338	54
Sodium fluoride	2.558	55
Lithium fluoride	2.635	56
Quartz	2.648	88
Calcite	2.71	57
Potassium bromide	2.75	58
Mica	2.78	59
Rubidium chloride	2.80	60
Zinc chloride	2.91	61
Boron phosphide	2.97	57
Potassium iodide	3.13	62
Calcium fluoride	3.18	63
Irtran 1	3.18	64
Magnesium fluoride	3.18	65
Silicon carbide	3.217	53
Silicon nitride	3.24	88
Aluminum nitride	3.26	66
Rubidium bromide	3.35	67
Diamond	3.51	68
Rubidium iodide	3.55	69
Spinel	3.55	70
Raytran spinel	3.57	13
Magnesium oxide	3.58	71
Magnesia	3.585	16
Raytran ALON	3.69	13
Ruby	3.98	57
Sapphire	3.98	57
CVD zinc sulfide (standard grade)	4.08	23
CVD zinc sulfide (multispectral grade)	4.09	72
Irtran 2	4.09	64
Gallium phosphide	4.13	59
Cuprous chloride	4.14	73
Strontium fluoride	4.24	88
Titanium dioxide	4.25	74
Cesium bromide	4.44	59
Cesium iodide	4.51	75
Calcium lanthanum sulfide	4.61	13
Lithium niobate	4.644	65

(*continued*)

Table 5.8 Crystals: Specific Gravity (Continued)

Material	Specific gravity [g/cm^3]	Ref.
Selenium	4.81	88
Cadmium sulfide	4.82	59
Barium fluoride	4.89	76
Yttria	5.01	77
Gallium selenide	5.03	—
Raytran yttria	5.03	13
GTE lanthanum-doped yttria	5.13	78
CVD zinc selenide	5.27	23
Gallium arsenide	5.316	59
Germanium	5.35	79
Zinc selenide	5.42	80
Silver chloride	5.56	81
Cadmium phosphide	5.60	82
Gallium antimonide	5.619	59
Indium antimonide	5.78	83
Cadmium selenide	5.81	84
Barium titanate	5.90	57
Lanthanum fluoride	5.94	59
Cadmium telluride	6.20	59
Tellurium	6.24	88
Zirconium dioxide	6.27	85
Silver bromide	6.473	81
Cadmium fluoride	6.64	57
Thallium chloride	7.004	86
KRS-5	7.371	88
Lead sulfide	7.5	87
Thallium bromide	7.557	88
Lead selenide	8.10	56
Lead telluride	8.164	56
Lead fluoride	8.24	89

Table 5.9 Crystals: Hardness

Material	Hardness (kg/mm^2)	Load (g)	Direction	Type	Ref.
Potassium bromide	7	200	<100>	Knoop	65
Silver bromide	7			Knoop	90
Potassium chloride	7.2	200	<110>	Knoop	91
Silver chloride	9.5	200		Knoop	57

(continued)

Table 5.9 Crystals: Hardness (Continued)

Material	Hardness (kg/mm^2)	Load (g)	Direction	Type	Ref.
Thallium bromide	11.9	500	<110>,<100>	Knoop	91
Thallium chloride	12.8	500	<110>,<100>	Knoop	91
Sodium chloride	15.2	200	<110>	Knoop	91
Tellurium	18			Knoop	88
Cesium bromide	19.5	200		Knoop	92
KRS-5	40.2	200		Knoop	91
Cadmium selenide	44.9			Knoop	59
Cadmium telluride	56			Knoop	59
Sodium fluoride	60			Knoop	65
Barium fluoride	82	500		Knoop	57
Lithium fluoride	102–113	600		Knoop	91
CVD zinc selenide	105	100		Knoop	93
Cadmium sulfide	122	25	parallel	Knoop	59
Strontium fluoride	130		<111>$_b$	Knoop	88
Zinc selenide	137			Knoop	94
CVD zinc sulfide (multispectral grade)	160	100		Knoop	72
Calcium fluoride	160–178		<110>,<100>	Knoop	95
Zinc sulfide (beta)	178			Knoop	94
Lead fluoride	200		<111>$_b$	Knoop	65
Indium antimonide	225			Knoop	59
CVD zinc sulfide (standard grade)	230	100		Knoop	93
Indium arsenide	330			Knoop	59
Irtran 2	355			Knoop	96
Magnesium fluoride	415			Knoop	65
Indium phosphide	430			Knoop	59
Gallium antimonide	469			Knoop	92
Calcium lanthanum sulfide	570			Knoop	13
Irtran 1	575			Knoop	96
Barium titanate	580			Vickers	97
Strontium titanate	595			Knoop	88
Raytran yttria	650			Knoop	13
Magnesium oxide	692	600	⊥ to cleavage planes	Knoop	91
Gallium arsenide	721			Knoop	59
GTE lanthanum-doped yttria	730			Knoop	44

(*continued*)

Table 5.9 Crystals: Hardness (Continued)

Material	Hardness (kg/mm^2)	Load (g)	Direction	Type	Ref.
Quartz	741	500		Knoop	91
Germanium	800			Knoop	59
Gallium phosphide	845			Knoop	9
Yttria	875			Knoop	88
Titanium dioxide	879	500		Knoop	88
Magnesia	910–990	500	<100>,<110>	Vickers	16
Spinel	1140	1000		Knoop	65
Silicon	1150			Knoop	88
Aluminum nitride	1200			Knoop	17
Sapphire	1370	1000		Knoop	57
Raytran spinel	1645			Knoop	13
Raytran ALON	1970			Knoop	13
Ruby	2000			Knoop	12
Silicon carbide	2130–2755			Knoop	88
Silicon nitride	3400			Knoop	88
Boron phosphide	3600			Knoop	57
Diamond	5700–10,400			Knoop	57
Calcite	3			Moh number	97
Cadmium fluoride	~3			Moh number	95
Lanthanum fluoride	4.5			Moh number	65
Cuprous chloride	2–2.5			Moh number	98
Rubidium iodide	1.0			Moh number	88
Selenium (trigonal)	2.6			Moh number	88
Lithium niobate	5			Moh number	99
Yttria-doped zirconia	7.5–8.5			Moh number	94

Table 5.10 Crystals: Solubility

Material	Solubility (g/(100g H$_2$O))	Ref.
Barium titanate	Insoluble	57
Boron nitride	Insoluble	57
Cadmium selenide	Insoluble	59
Cadmium sulfide	Insoluble	100
Cadmium telluride	Insoluble	59
Diamond	Insoluble	57
Gallium antimonide	Insoluble	100

(continued)

Table 5.10   Crystals: Solubility (Continued)

Material	Solubility (g/(100g H$_2$O))	Ref.
Gallium arsenide	Insoluble	100
Gallium phosphide	Insoluble	102
Germanium	Insoluble	100
Indium antimonide	Insoluble	59
Indium arsenide	Insoluble	59
Irtran 1	Insoluble	42
Irtran 2	Insoluble	42
Lanthanum fluoride	Insoluble	65
Lead selenide	Insoluble	65
Lead sulfide	Insoluble	65
Lead telluride	Insoluble	65
Lithium niobate	Insoluble	65
Magnesium fluoride	Insoluble	65
Magnesium oxide	Insoluble	65
Mica	Insoluble	102
Quartz	Insoluble	102
Raytran spinel	Insoluble	13
Raytran yttria	Insoluble	13
Sapphire	Insoluble	100
Selenium (trigonal)	Insoluble	88
Silicon	Insoluble	88
Silicon carbide	Insoluble	88
Silver bromide	Insoluble	57
Silver chloride	Insoluble	57
Spinel	Insoluble	65
Strontium titanate	Insoluble	101
Tellurium	Insoluble	88
Yttria	Insoluble	88
Yttria-doped zirconia	Insoluble	94
Zinc sulfide	Insoluble	94
Titanium dioxide	0.001	88
Zinc selenide	0.001 (298 K)	94
Calcite	0.0014 (298 K)	57
Calcium fluoride	0.0017 (299 K)	100
Cuprous chloride	0.0061	59
Strontium fluoride	0.011 (273 K)	101
KRS-5	0.05 (293 K)	100
Thallium bromide	0.05 (293 K)	101
Lead fluoride	0.064 (293 K)	65
Barium fluoride	0.17	100
Lithium fluoride	0.27 (291 K)	100

(continued)

Table 5.10 Crystals: Solubility (Continued)

Material	Solubility (g/(100g H$_2$O))		Ref.
Thallium chloride	0.32	(293 K)	101
Sodium fluoride	4.22	(291 K)	101
Cadmium fluoride	4.4	(293 K)	57
ADP	22.7	(273 K)	100
KDP	33	(298 K)	65
Potassium chloride	34.7	(293 K)	100
Sodium chloride	35.7	(273 K)	101
Cesium iodide	44		59
Potassium bromide	53.48	(273 K)	100
Rubidium chloride	77	(273 K)	65
Rubidium bromide	98	(278 K)	65
Cesium bromide	124.3	(298 K)	100
Potassium iodide	127.5	(273 K)	100
Rubidium iodide	152	(290 K)	88

Note: Insoluble = solubility < 0.001 (g/(100g H$_2$O)).

Table 5.11 Crystals: Dielectric Constant

Material	Dielectric constant	Frequency (Hz)	Temp.	Ref.
Quartz	4.34(s),4.27(p)	3 × 10^7	290–295 K	97
Potassium chloride	4.64	1 × 10^6	302 K	104
Magnesium fluoride	4.87(p),5.45(s)	95 × 10^3–42 × 10^6		105
Potassium bromide	4.9	1 × 10^2–1 × 10^{10}	298 K	103
Rubidium chloride	4.9			106
Potassium iodide	4.94	2 × 10^6		107
Irtran 1	5.1	8.5 × 10^9–12.0 × 10^9		28
Mica	5.4	1 × 10^2–1.5 × 10^{10}	299 K	107
Cesium iodide	5.65	1 × 10^6	298 K	104
Diamond	5.68	1.6 × 10^6		108
Sodium chloride	5.9	1 × 10^2	298 K	103
Selenium	6.0	1 × 10^2–1 × 10^{10}	298 K	107
Sodium fluoride	6	2 × 10^6	292 K	107
Cesium bromide	6.51	2 × 10^6	298 K	104
Calcium fluoride	6.76	1 × 10^5		104
Barium fluoride	7.33	2 × 10^6		103
Gallium selenide	7.443	High		109

(*continued*)

Table 5.11 Crystals: Dielectric Constant (Continued)

Material	Dielectric constant	Frequency (Hz)	Temp.	Ref.
Strontium fluoride	7.69	2×10^6		103
Zinc sulfide	7.75	3×10^5	298 K	110
Cadmium sulfide	7.77(s),9.43(p)	1×10^4	298 K	111
Calcite	8.5(s),8.0(p)	1×10^4	290–294 K	97
Irtran 2	8.0	8.5×10^9–12.0×10^9		28
Spinel	8.0–9.0			107
CVD zinc sulfide (multispectral grade)	8.393	Low		112
Sapphire	10.55(p),8.6(s)	1×10^2–1×10^8	298 K	103
Lithium fluoride	9.0	1×10^2–1×10^{10}	298 K	103
Gallium phosphide	9.036	Optical	300 K	113
Aluminum nitride	9.1	Zero		17
Indium phosphide	9.52	Optical		114
Magnesium oxide	9.65	1×10^2–1×10^8	298 K	103
Cuprous chloride	10.0	3×10^6	293 K	115
Gallium arsenide	10.88	High		22
Boron phosphide	11.0	Zero		15
Cadmium telluride	11.0	1–1×10^5		104
Indium arsenide	11.8	Optical		116
Yttria-doped zirconia	12.0	1×10^6		117
GTE lanthanum-doped yttria	12.2	1×10^5		118
Silver chloride	12.3	1×10^6	293 K	103
Lead fluoride	13	1×10^6		103
Silicon	13	9.37×10^9		107
Yttria	13		293 K	119
Silver bromide	13.1	1×10^6	300 K	120
Gallium antimonide	14.44	Optical		—
Indium antimonide	15.7	Optical	300 K	121
Germanium	16.6	9.37×10^9		107
Lead sulfide	17.9	1×10^6	298 K	107
KDP	21(p)	1×10^2–1×10^8		122
Lead selenide	21	Optical		19
Lead telluride	28.5	Optical		18
Lithium niobate	29			123
Thallium bromide	30.3	1×10^3–1×10^7	298 K	107

(continued)

Table 5.11   Crystals: Dielectric Constant (Continued)

Material	Dielectric constant	Frequency (Hz)	Temp.	Ref.
Thallium chloride	31.9	2×10^6		107
KRS-5	32.9–32.5	$1 \times 10^2 – 1 \times 10^7$	298 K	104
Titanium dioxide	200–160	$1 \times 10^4 – 1 \times 10^7$	298 K	103
Strontium titanate	306	$1 \times 10^2 – 1 \times 10^5$	298 K	103
Barium titanate	1240–1100	$1 \times 10^2 – 1 \times 10^8$	298 K	104

Table 5.12   Crystals: Debye Temperature

Material	Debye temperature (K)	Ref.
Selenium	90	124
Cesium iodide	93.6	125
Rubidium iodide	103	124
Lead telluride	128	126
Rubidium bromide	131	124
Potassium iodide	132	124
Silver bromide	144	124
Lead selenide	144	126
Cesium bromide	148.8	127
Tellurium	153	124
Rubidium chloride	165	124
Potassium bromide	174	124
Silver chloride	183	124
Lead sulfide	194	126
Indium antimonide	200	126
Lead fluoride	219	128
Potassium chloride	235	124
Zinc selenide	246	129
Barium fluoride	282	130
Zinc sulfide	315	124
Sodium chloride	321	124
Gallium arsenide	360	22
Germanium	370	126
Strontium fluoride	380	131
Indium phosphide	420	132
Gallium phosphide	446	38
Quartz	470	124
Sodium fluoride	492	124
Calcium fluoride	510	126

(continued)

Table 5.12   Crystals: Debye Temperature (Continued)

Material	Debye temperature (K)	Ref.
Boron nitride	600	133
Silicon	640	124
Lithium fluoride	732	126
Titanium dioxide	760	124
Magnesium oxide	946	126
Aluminum nitride	950	17
Boron phosphide	985	15
Diamond	1860	133

Table 5.13   Crystals: Young's Modulus

Material	Young's modulus (GPa)	Ref.
Silver chloride	0.14	134
Cesium iodide	5.3	134
Cesium bromide	15.85	134
KRS-5	15.85	134
Potassium bromide	26.87	134
Thallium bromide	29.49	134
Potassium chloride	29.63	134
Potassium iodide	31.49	134
Thallium chloride	31.69	134
Silver bromide	31.97	135
Barium titanate	33.76	134
Cadmium telluride	36.52	136
Sodium chloride	39.96	134
Indium antimonide	42.79	134
Barium fluoride	53.05	134
Gallium antimonide	63.32	134
Lithium fluoride	64.77	134
Zinc sulfide	66.14	137
Zinc selenide	69.6	10
CVD zinc selenide	70.3	138
CVD zinc sulfide (standard grade)	74.5	138
Calcium fluoride	75.79	134
Gallium arsenide	82.68	139
CVD zinc sulfide (multispectral grade)	87.6	138
Calcite	88.19 (s), 72.35 (p)	134
Calcium lanthanum sulfide	95	24

(continued)

Table 5.13 Crystals: Young's Modulus (Continued)

Material	Young's modulus (GPa)	Ref.
Irtran 2	96.46	42
Quartz	97.2 (*p*), 76.5 (*s*)	134
Strontium fluoride	99.91	140
Gallium phosphide	102.6	9
Germanium	102.66	134
Irtran 1	114.37	42
Silicon	130.91	134
Raytran yttria	164	13
GTE lanthanum-doped yttria	166.5	141
Raytran spinel	193	13
Magnesium oxide	248.73	134
Raytran ALON	317	13
Sapphire	344.5	134
Silicon carbide	386	142
Diamond	1050	39

Table 5.14 Crystals: Bulk Modulus

Material	Bulk modulus (GPa)	Ref.
Potassium bromide	15.02	134
Potassium chloride	17.36	134
KRS-5	19.77	134
Thallium bromide	22.46	134
Thallium chloride	23.56	134
Sodium chloride	24.32	134
Strontium fluoride	24.65	145
Cadmium telluride	25	143
Quartz	36.4	97
Zinc selenide	40	144
Indium antimonide	43.27	134
Silver bromide	44.03	146
Silver chloride	44.03	134
Gallium antimonide	56.43	134
Indium arsenide	57.9	116
Sodium fluoride	62.00	95
Lithium fluoride	62.06	134
Gallium arsenide	75.5	22
Germanium	77.86	134
Zinc sulfide	80.41 (273 K)	144
Calcium fluoride	83.03	134

(continued)

Table 5.14 Crystals: Bulk Modulus (Continued)

Material	Bulk modulus (GPa)	Ref.
Silicon	101.97	134
Calcite	129.53	134
GTE lanthanum-doped yttria	144.56	141
Magnesium oxide	154.34	134
Barium titanate	161.92	134
Boron phosphide	200	15
Sapphire	207	134
Aluminum nitride	220	17
Silicon carbide	273 (293 K)	142
Diamond	440–590	39

Table 5.15 Crystals: Shear Modulus

Material	Shear modulus (GPa)	Ref.
Potassium bromide	5.08	134
Potassium iodide	6.20	134
Potassium chloride	6.24	134
Silver chloride	7.10	134
Thallium bromide	7.58	134
Thallium chloride	7.58	134
Sodium chloride	8.97	134
Quartz	31.14	134
Calcium fluoride	33.76	134
Lithium fluoride	55.12	134
Germanium	67.04	134
Silicon	79.92	134
Barium titanate	126.09	134
Sapphire	148.14	134
Magnesium oxide	154.34	134

Table 5.16 Crystals: Poisson's Ratio

Material	Poisson's Ratio	Ref.
Diamond	0.10–0.29	*
Magnesium oxide	0.18	*
ALON	0.24	*
Raytran ALON	0.24	13
Irtran 1	0.25–0.36	28
Irtran 2	0.25–0.36	28

(*continued*)

Table 5.16 Crystals: Poisson's Ratio (Continued)

Material	Poisson's Ratio	Ref.
Strontium fluoride	0.25	*
Raytran spinel	0.26	13
Spinel	0.26	*
Lead selenide	0.27	*
Sapphire	0.27	*
Germanium	0.278	*
Calcium fluoride	0.28	*
CVD zinc selenide	0.28	138
Silicon	0.28	*
Zinc selenide	0.28	*
CVD zinc sulfide (standard grade)	0.29	138
Raytran yttria	0.29	13
Zinc sulfide	0.29	*
GTE lanthanum-doped yttria	0.308	141
Gallium arsenide	0.31	*
Gallium phosphide	0.31	*
CVD zinc sulfide (multispectral grade)	0.318	138

*Calculated (see page 150).

Table 5.17 Glasses: Transmission Range

Material	Transmission range (μm)
Beryllium fluoride (BeF$_2$)	0.2–4
Corning 7940 (fused silica)	0.2–2.5; 3.0–4.0
Corning 7957 (fused silica)	0.2–4.0
Schott IRG 7 (lead silicate)	0.3–4.25
SiO$_2$ (fused silica)	0.2–4.5
Schott IRG 3 (lanthan dense flint)	0.3–4.5
Corning 7971 (titanium silicate)	0.5–3.5; 3.75–4.5
Schott IRG 9 (fluorophosphate)	0.4–4.5
Corning 9753 (calcium aluminosilicate)	0.3–4.6
Schott IRG N6 (calcium aluminosilicate)	0.3–4.75
Corning 9754 (germanate)	0.3–5.5
Schott IRG 2 (germanate)	0.3–5.5
Schott IRG 11 (calcium aluminate)	0.3–5.5
Corning EO (GaBiPb oxide)	0.5–7
HBL (fluoride)	0.25–8.0
ZBL (fluoride)	0.25–8.0

(*continued*)

Table 5.17 Glasses: Transmission Range (Continued)

Material	Transmission range (μm)
ZBLA (fluoride)	0.25–8.0
Ohara HTF-1 (fluoride)	0.25–8.5
BZYbT (fluoride)	0.3–9
Arsenic trisulfide (As_2S_3)	0.6–11
Schott IRG 100 (chalcogenide)	0.9–15
$Ge_{25}Se_{75}$	0.9–16
AMTIR-1 ($Ge_{33}As_{12}Se_{55}$)	0.9–16
TI-1173 ($Ge_{28}Sb_{12}Se_{60}$)	0.9–16

Table 5.18 Glasses: Index of Refraction

Material	Index of refraction	Ref.
Beryllium fluoride (BeF_2)	1.27 at 0.5893 μm	147
SiO_2 (fused silica)	1.4590 at 0.5780 μm	148
	1.4099 at 3.37 μm	148
Corning 7940 (fused silica)	1.46233 at 0.5 μm	149
	1.38903 at 4.0 μm	149
Corning 7957 (fused silica)	1.4633 at 0.486 μm	150
	1.4566 at 0.656 μm	150
Corning 7971 (titanium silicate)	1.4892 at 0.486 μm	151
	1.4801 at 0.656 μm	151
Schott IRG 9 (fluorophosphate)	1.4902 at 0.4861 μm	152
	1.4583 at 3.303 μm	152
HBL (fluoride)	1.514 at 0.5893 μm	153
ZBLA (fluoride)	1.516 at 0.5893 μm	154
Ohara HTF-1 (fluoride)	1.525 at 0.48 μm	155
	1.49 at 3.3 μm	155
ZBL (fluoride)	1.5233 at 0.5893 μm	154
BZYbT (fluoride)	1.535 at 0.6 μm	156
Schott IRG 7 (lead silicate)	1.5735 at 0.4861 μm	157
	1.5164 at 3.303 μm	157
Schott IRG N6 (calcium alumninosilicate)	1.5892 at 0.5876 μm	158
	1.5451 at 3.303 μm	158
Corning 9753 (calcium aluminosilicate)	1.597 at 0.75 μm	159
	1.565 at 3.0 μm	159
Corning 9754 (germanate)	1.6601 at 0.5893 μm	160
	1.617 at 3.5 μm	160
Schott IRG 11 (calcium aluminate)	1.6809 at 0.5876 μm	161
	1.6156 at 4.258 μm	161

(*continued*)

Table 5.18   Glasses: Index of Refraction (Continued)

Material	Index of refraction	Ref.
Schott IRG 3 (lanthan dense flint)	1.8633 at 3.303 μm	162
	1.7764 at 0.4861 μm	162
Schott IRG 2 (germanate)	1.9129 at 0.4861 μm	163
	1.8253 at 3.303 μm	163
Corning EO (GaBiPb oxide)	2.34 at 0.5 μm	164
	2.23 at 4.0 μm	164
$Ge_{25}Se_{75}$	2.37 at 3 μm	165
	2.355 at 10 μm	165
Arsenic trisulfide (As_2S_3)	2.4777 at 1.0 μm	166
	2.3816 at 10.0 μm	166
AMTIR-1 ($Ge_{33}As_{12}Se_{55}$)	2.5187 at 3.0 μm	167
	2.4976 at 10.0 μm	167
Schott IRG 100 (chalcogenide)	2.6262 at 3.0 μm	168
	2.6004 at 10.0 μm	168
TI-1173 ($Ge_{28}Sb_{12}Se_{60}$)	2.626 at 3 μm	165
	2.600 at 10 μm	165

Table 5.19   Glasses: Glass Transition Temperature

Material	T_g (K)	Ref.
Arsenic trisulfide (As_2S_3)	470	169
$Ge_{25} Se_{75}$	513	165
Beryllium fluoride (BeF_2)	523	147
Corning EO (GaBiPb oxide)	573	164
TI-1173 ($Ge_{28} Sb_{12} Se_{60}$)	573	165
Schott IRG 100 (chalcogenide)	624*	168
ZBL (fluoride)	579	154
ZBLA (fluoride)	583	154
HBL (fluoride)	585	154
BZYbT (fluoride)	617	170
AMTIR-1 ($Ge_{33} As_{12} Se_{55}$)	635	167
Ohara HTF-1 (fluoride)	658	155
Corning 9754 (germanate)	970	160
Corning 9753 (calcium aluminosilicate)	1073	159
Corning 7971 (titanium silicate)	1163	151
Corning 7940 (fused silica)	1263	149
Corning 7957 (fused silica)	1266	150
SiO_2 (fused silica)	1273	171

*Softening temperature.

Table 5.20 Glasses: Specific Heat

Material	Specific heat [cal/(gK)]	Ref.
TI-1173 ($Ge_{28} Sb_{12} Se_{60}$)	0.066	172
AMTIR-1 ($Ge_{33} As_{12} Se_{55}$)	0.07	167
HBL (fluoride)	0.097 (250–550 K)	170
BZYbT (fluoride)	0.106 (298 K)	170
Schott IRG 2 (germanate)	0.108	163
Schott IRG 3 (lanthan dense flint)	0.115 (293–373 K)	162
Arsenic trisulfide As_2S_3	0.118	173
ZBL (fluoride)	0.125 (250–530 K)	170
Corning 9754 (germanate)	0.13	160
ZBLA (fluoride)	0.145 (482–540 K)	170
Schott IRG 7 (lead silicate)	0.151 (293–373 K)	157
Schott IRG 9 (fluorophosphate)	0.166 (293–373 K)	152
Corning 7940 (fused silica)	0.177	149
Schott IRG 11 (calcium alumninate)	0.179 (293–373 K)	161
Corning 7957 (fused silica)	0.18	150
Corning 7971 (titanium silicate)	0.183	151
Schott IRG N6 (calcium aluminosilicate)	0.193 (293–373 K)	158
Corning 9753 (calcium aluminosilicate)	0.19 (323 K)	159

Table 5.21 Glasses: Thermal Conductivity

Material	Thermal conductivity (W/(mK))	Ref.
Arsenic trisulfide (As_2S_3)	0.1674 (313 K)	174
AMTIR-1 ($Ge_{33}As_{12}Se_{55}$)	0.25	167
Schott IRG 100 (chalcogenide)	0.3 (293–573 K)	168
TI-1173 ($Ge_{28}Sb_{12}Se_{60}$)	0.301	172
Schott IRG 7 (lead silicate)	0.73	157
Schott IRG 3 (lanthan dense flint)	0.87	162
Schott IRG 9 (fluorophosphate)	0.88	152
Schott IRG 2 (germanate)	0.91	163
Corning 9754 (germanate)	1.0	160
Schott IRG 11 (calcium aluminate)	1.13	161
Corning 7957 (fused silica)	1.255	150
Corning 7971 (titanium silicate)	1.31	151
Schott IRG N6 (calcium aluminosilicate)	1.36	158
Corning 7940 (fused silica)	1.38	149
SiO_2 (fused silica)	1.6 (373 K)	175
Corning 9753 (calcium aluminosilicate)	2.3 (323 K)	159

Table 5.22 Glasses: Thermal Expansion

Material	Thermal expansion ($\times 10^{-6}$ K^{-1})	Ref.
Corning 7971 (titanium silicate)	0.0 (278–308 K)	151
SiO_2 (fused silica)	0.42 (298 K)	175
	0.75 (423 K)	175
Corning 7957 (fused silica)	0.39 (173–473 K)	150
	0.52 (273–473 K)	150
Corning 7940 (fused silica)	0.52 (278–308 K)	149
	0.57 (273–473 K)	149
	0.48 (173–473 K)	149
Corning 9753 (calcium aluminosilicate)	5.95 (298–573 K)	159
	7.2 (298–973 K)	159
Schott IRG 9 (fluorophosphate)	6.1 (293–573 K)	152
Corning 9754 (germanate)	6.2 (298–573 K)	160
Schott IRG N6 (calcium aluminosilicate)	6.3 (293–573 K)	158
Beryllium fluoride (BeF_2)	6.8	147
Schott IRG 3 (lanthan dense flint)	8.1 (293–573 K)	162
Schott IRG 11 (calcium aluminate)	8.2 (298–573 K)	161
Schott IRG 2 (germanate)	8.8 (293–573 K)	163
Schott IRG 7 (lead silicate)	9.6 (293–573 K)	157
Corning EO (GaBiPb oxide)	11.1	164
AMTIR-1 ($Ge_{33}As_{12}Se_{55}$)	12.0	167
TI-1173 ($Ge_{28}Sb_{12}Se_{60}$)	15.0	172
Schott IRG 100 (chalcogenide)	15.0 (293–573 K)	168
Ohara HTF-1 (fluoride)	16.1 (293–373 K)	155
HBL (fluoride)	17.7	154
ZBLA (fluoride)	18.7	154
ZBL (fluoride)	18.8	154
$Ge_{25}Se_{75}$	22.0	169
Arsenic trisulfide (As_2S_3)	24.6 (306 K)	176

Table 5.23 Glasses: Specific Gravity

Material	Specific gravity [g/cm^3]	Ref.
Beryllium fluoride (BeF_2)	1.982	147
SiO_2 (fused silica)	2.202	177
Corning 7940 (fused silica)	2.202	149
Corning 7957 (fused silica)	2.203	150
Corning 7971 (titanium silicate)	2.205	151

(continued)

Table 5.23 Glasses: Specific Gravity (Continued)

Material	Specific gravity [g/cm^3]	Ref.
Corning 9753 (calcium aluminosilicate)	2.798	159
Schott IRG N6 (calcium aluminosilicate)	2.81	158
Schott IRG 7 (lead silicate)	3.06	157
Schott IRG 11 (calcium aluminate)	3.12	161
Arsenic trisulfide (As$_2$S$_3$)	3.43	178
Corning 9754 (germanate)	3.581	160
Schott IRG 9 (fluorophosphate)	3.63	152
Ohara HTF-1 (fluoride)	3.88	155
AMTIR-1 (Ge$_{33}$As$_{12}$Se$_{55}$)	4.40	167
Schott IRG 3 (lanthan dense flint)	4.47	162
ZBLA (fluoride)	4.61	154
TI-1173 (Ge$_{28}$Sb$_{12}$Se$_{60}$)	4.67	172
Schott IRG 100 (chalcogenide)	4.67	168
ZBL (fluoride)	4.79	154
Schott IRG 2 (germanate)	5.00	163
HBL (fluoride)	5.78	154
BZYbT (fluoride)	6.43	170
Corning EO (GaBiPb oxide)	8.19	164

Table 5.24 Glasses: Hardness

Material	Hardness (kg/mm^2)	Load (g)	Type	Ref.
Arsenic trisulfide (As$_2$S$_3$)	109	100	Knoop	92
Ge$_{25}$Se$_{75}$	146	—	Vickers	165
Schott IRG 100 (chalcogenide)	150	200	Knoop	168
TI-1173 (Ge$_{28}$Sb$_{12}$Se$_{60}$)	166	—	Vickers	165
AMTIR-1 (Ge$_{33}$As$_{12}$Se$_{55}$)	170	—	Knoop	167
HBL (fluoride)	228	—	Vickers	179
ZBLA (fluoride)	267	—	Vickers	179
BZYbT (fluoride)	276	—	Vickers	179
ZBL (fluoride)	288	—	Vickers	179
Ohara HTF-1 (fluoride)	311	—	Knoop	155
Schott IRG 9 (fluorophosphate)	346	200	Knoop	152
Schott IRG 7 (lead silicate)	379	200	Knoop	157
Corning 7971 (titanium silicate)	459	200	Knoop	151
SiO$_2$ (fused silica)	461	200	Knoop	180
Schott IRG 2 (germanate)	481	200	Knoop	163
Corning 7940 (fused silica)	500	200	Knoop	149

(continued)

Table 5.24   Glasses: Hardness (Continued)

Material	Hardness (kg/mm^2)	Load (g)	Type	Ref.
Corning 7957 (fused silica)	500	200	Knoop	150
Schott IRG 3 (lanthan dense flint)	541	200	Knoop	162
Corning 9754 (germanate)	560	100	Knoop	160
Corning 9753 (calcium aluminosilicate)	601	500	Knoop	159
Schott IRG 11 (calcium aluminate)	608	200	Knoop	161
Schott IRG N6 (calcium aluminosilicate)	623	200	Knoop	158

Table 5.25   Glasses: Solubility

Material	Solubility (g/(100g H$_2$O))	Ref.
Arsenic trisulfide (As$_2$S$_3$)	Insoluble	100
TI-1173 (Ge$_{28}$Sb$_{12}$Se$_{60}$)	Insoluble	172
AMTIR-1 (Ge$_{33}$As$_{12}$Se$_{55}$)	Insoluble	167
Corning 7971 (titanium silicate)	Insoluble	151
Schott IRG 3 (lanthan dense flint)	0.012 (293 K)	162
Schott IRG 2 (germanate)	0.012 (293 K)	163
Schott IRG 11 (calcium aluminate)	0.14 (293 K)	161
Schott IRG 7 (lead silicate)	0.171 (293 K)	157
Schott IRG N6 (calcium aluminosilicate)	0.213 (293 K)	158
Schott IRG 9 (fluorophosphate)	0.38 (293 K)	152
Ohara HTF-1 (fluoride)	0.16% weight loss in H$_2$O	155
Beryllium fluoride (BeF$_2$)	Soluble	147

Table 5.26   Glasses: Dielectric Constant

Material	Dielectric constant	Frequency (Hz)	Temperature	Ref.
Corning 7971 (titanium silicate)	3.99	100	298 K	151
	4.00	100	473 K	151
SiO$_2$ (fused silica)	3.78	10^2–10^{10}	300 K	107
Corning 7957 (fused silica)	3.91	10^3		150
Corning 7940 (fused silica)	4.00	10^3	298 K	149
	4.10	10^3	673 K	149
Schott IRG 7 (lead silicate)	6.7	10^6	298 K	157

(continued)

Table 5.26 Glasses: Dielectric Constant (Continued)

Material	Dielectric constant	Frequency (Hz)	Temperature	Ref.
Arsenic trisulfide (As_2S_3)	8.1	10^3-10^6		181
Corning 9753 (calcium aluminosilicate)	8.87	10^6	298 K	159
	9.51	10^6	773 K	159
Schott IRG N6 (calcium aluminosilicate)	9.2	10^6	298 K	158
Beryllium fluoride (BeF_2)	9.7	10^6	298 K	147
Corning 9754 (germanate)	10.08	10^3	673 K	160
	10.35	10^3	823 K	160
Schott IRG 9 (fluorophosphate)	10.4	10^6	298 K	152
Schott IRG 11 (calcium aluminate)	11.5	10^6	298 K	161
ZBL (fluoride)	11.8	(high)		182
Schott IRG 3 (lanthan dense flint)	14.8	10^6	298 K	162
Schott IRG 2 (germanate)	15.6	10^6	198 K	163
Corning EO (GaBiPb oxide)	30			164

Table 5.27 Glasses: Young's Modulus

Material	Young's modulus (GPa)	Ref.
Arsenic trisulfide (As_2S_3)	15.85	183
Schott IRG 100 (chalcogenide)	21.0	168
$Ge_{25} Se_{75}$	21.36	165
AMTIR-1 ($Ge_{33}As_{12}Se_{55}$)	22.07	167
TI 1173 ($Ge_{28} Sb_{12} Se_{60}$)	22.1	165
Schott IRG 7 (germanate)	59.7	157
ZBLA (fluoride)	60.2	184
Ohara HTF-1 (fluoride)	64.2	155
Corning 7971 (titanium silicate)	67.52	151
BZYbT (fluoride)	70.0	179
Corning 7957 (fused silica)	72.4	150
Corning 7940 (fused silica)	73.0	149
SiO_2 (fused silica)	73.03	185
Schott IRG 9 (fluorophosphate)	77.0	152
Corning 9754 (germanate)	84.1	160
Schott IRG 2 (germanate)	95.9	163
Corning 9753 (calcium aluminosilicate)	98.58	159
Schott IRG 3 (lanthan dense flint)	99.9	162
Schott IRG N6 (calcium aluminosilicate)	103.2	158
Schott IRG 11 (calcium aluminate)	107.5	161

Table 5.28 Glasses: Bulk Modulus

Material	Bulk modulus (GPa)	Ref.
$Ge_{25}Se_{75}$	13.69	*
TI-1173 ($Ge_{28}Sb_{12}Se_{60}$)	14.16	*
Schott IRG 100 (chalcogenide)	14.64	*
AMTIR-1 ($Ge_{33}As_{12}Se_{55}$)	15.99	*
Corning 7971 (titanium silicate)	34.45	151
Schott IRG 7 (lead silicate)	35.04	*
Corning 7957 (fused silica)	35.49	*
Corning 7940 (fused silica)	36.9	149
SiO_2 (fused silica)	37 (293 K)	185
ZBLA (fluoride)	40.8	184
Ohara HTF-1 (fluoride)	48.64	*
Schott IRG 9 (fluorophosphate)	60.55	*
Corning 9754 (germanate)	66.75	*
Schott IRG 2 (germanate)	73.32	*
Corning 9753 (calcium aluminosilicate)	74.68	*
Schott IRG N6 (calcium aluminosilicate)	76.79	*
Schott IRG 3 (lanthan dense flint)	78.17	*
Schott IRG 11 (calcium aluminate)	82.95	*

$$*\text{Calculated from: Bulk modulus} = \frac{\text{Young's modulus}}{3\left[1 - (2 \times \text{Poisson's ratio})\right]}$$

Table 5.29 Glasses: Shear Modulus

Material	Shear modulus (GPa)	Ref.
Arsenic trisulfide (As_2S_3)	6.48	183
Schott IRG 100 (chalcogenide)	8.33	*
$Ge_{25}Se_{75}$	8.61	*
TI-1173 ($Ge_{28}Sb_{12}Se_{60}$)	8.91	*
AMTIR-1 ($Ge_{33}As_{12}Se_{55}$)	8.97	167
ZBLA (fluoride)	24.0	184
Schott IRG 7 (lead silicate)	24.55	*
Ohara HTF-1 (fluoride)	25.08	*
Corning 7971 (titanium silicate)	28.94	151
Schott IRG 9 (fluorophosphate)	29.89	*
Corning 7940 (fused silica)	31.0	149
Corning 7957 (fused silica)	31.0	150
Corning 9754 (germanate)	35.41	160
SiO_2 (fused silica)	36.38	185

(*continued*)

Physical Property Comparison of IR Optical Materials

Table 5.29 Glasses: Shear Modulus (Continued)

Material	Shear modulus (GPa)	Ref.
Schott IRG 2 (germanate)	37.40	*
Corning 9753 (calcium aluminosilicate)	38.58	159
Schott IRG 3 (lanthan dense flint)	38.81	*
Schott IRG N6 (calcium aluminosilicate)	40.44	*
Schott IRG 11 (calcium aluminate)	41.86	*

$$*\text{Calculated from: Shear modulus} = \frac{\text{Young's modulus}}{2(1 + \text{Poisson's ratio})}$$

Table 5.30 Glasses: Poisson's Ratio

Material	Poisson's ratio	Ref.
Corning 7957 (fused silica)	0.16	150
SiO$_2$ (fused silica)	0.17	185
Corning 7940 (fused silica)	0.17	149
Corning 7971 (titanium silicate)	0.17	151
Schott IRG 7 (lead silicate)	0.216	157
TI-1173 (Ge$_{28}$ Sb$_{12}$ Se$_{60}$)	0.24	165
Ge$_{25}$ Se$_{75}$	0.24	165
ZBLA (fluoride)	0.250	184
Schott IRG 100 (chalcogenide)	0.261	168
AMTIR-1 (Ge$_{33}$ As$_{12}$ Se$_{55}$)	0.27	167
Schott IRG N6 (calcium aluminosilicate)	0.276	158
Corning 9753 (calcium aluminosilicate)	0.28	159
Ohara HTF-1 (fluoride)	0.28	155
Schott IRG 2 (germanate)	0.282	163
Schott IRG 11 (calcium aluminate)	0.284	161
Schott IRG 3 (lanthan dense flint)	0.287	162
Schott IRG 9 (fluorophosphate)	0.288	152
Corning 9754 (germanate)	0.290	160

Note: References for this chapter are located in the Appendix, p. 593.

6

Physical Properties of Crystalline Infrared Optical Materials

James Steve Browder   *Jacksonville University, Jacksonville, Florida*

Stanley S. Ballard   *University of Florida, Gainesville, Florida*

Paul Klocek   *Texas Instruments, Inc., Dallas, Texas*

After an optical material has been chosen for a particular application, the engineer may wish to know more about several of its physical properties. All available data are given in Chapters 6 and 7.

The materials are arranged in alphabetical order, as in Tables 5.1 and 5.2 of Chapter 5. The sources of the data on the individual data sheets in Chapter 6 and 7 are identified in the comparison tables of Chapter 5. The sources for the tables and figures in Chapters 6 and 7 are given in the tables and figure legends. For some of the materials, a great deal of data are given, including transmission curves, refractive index tables, and so on. For others, few data were found in our literature search.

Several properties are defined or discussed in the earlier chapters, as are the units usually used with each of them. For anisotropic crystals the values may be quoted for two directions: parallel to the c axis (*"p"*) and perpendicular to the c or optic axis (*"s"*). It can be assumed that data are given for the samples at room temperature (say 20°C or 293 K), unless it is indicated otherwise.

ALON, Al$_{23}$O$_{27}$N$_5$ (Raytran)

Specific Gravity
   3.69 [g/cm^3]
Crystal Class
   Cubic
Transmission
   (See Fig. 6.1.)
Absorption Coefficient
   (See Fig. 6.2.)
Refractive Index
   1.793 at 0.589 μm
   1.66 at 4 μm
Melting Temperature
   2443 K
Thermal Conductivity
   12.6 W/(m·K) at 300 K
Thermal Expansion
   5.8 × 10^{-6} K^{-1} at 303–473 K
Specific Heat
   0.185 [cal/(g·K)] at 300 K
Hardness
   1970 kg/mm^2 (Knoop number)

Elastic Moduli
   Young's modulus
   317 GPa
   Poisson's ratio
   0.24
Notes: Raytran ALON is an extremely durable crystalline material that is fabricated by using a Raytheon-developed powder-processing technique. The material combines mechanical and optical properties similar to those of sapphire with the advantages of an isotropic cubic crystal structure. It has an approximate composition of Al$_{23}$O$_{27}$N$_5$. Windows up to 0.5 in. thick and 5 in. in diameter or 10 in. long by 5 in. wide can be made. Hemispherical domes up to 5 in. in diameter are available. Rods and tubes can also be made.

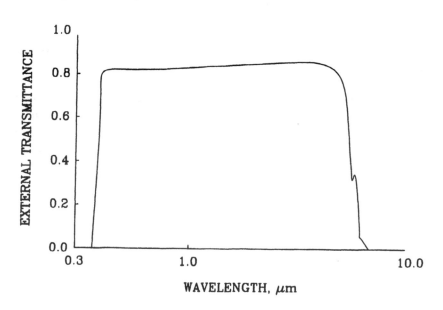

Figure 6.1 The external transmittance of ALON, thickness 2.36 mm. [Adapted from PBN-84-744, Raytheon Company Research Division, Lexington, Mass. August 1987.]

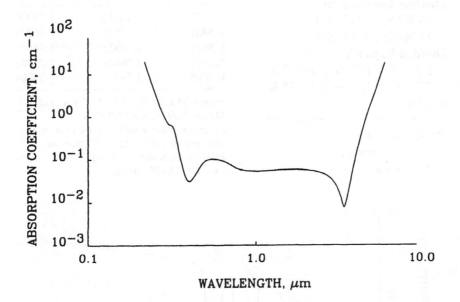

Figure 6.2   Absorption coefficient of ALON. [Adapted from information given by Raytheon Company Research Division, Lexington, Mass., August 1987.]

Aluminum Oxide (Ruby), Al₂O₃

Specific Gravity
  3.98 [g/cm^3]
Transmission
  (See Fig. 6.3.)
Refractive Index
  (See Table 6.1.)
Melting Temperature
  2313 K
Thermal Conductivity
  46 W/(m·K) at 273 K
  13 W/(m·K) at 673 K
Thermal Expansion
  5.8×10^{-6} K^{-1} at 293–323 K
  7.7×19^{-6} K^{-1} at 293–773 K
Hardness
  2000 kg/mm^2 (Knoop number)
Elastic moduli
  Rupture modulus
  0.448 GPa

Notes: Ruby, one of the earliest laser rod materials, is essentially sapphire with a 0.05% by weight chromium impurity. It is mechanically rugged and is not hygroscopic. [*See also* Aluminum oxide (sapphire).]

Table 6.1 Refractive Index of Ruby at 22°C

λ (μm)	n_O	n_E
0.4358	1.78115	1.77276
0.5461	1.77071	1.76258
0.5876	1.76822	1.76010
0.6678	1.76445	1.75641
0.7065	1.76302	1.75501

Source: M.J. Dodge, I.H. Malitson, and A.I. Mahan, *Appl. Opt. 8*:1703 (1969).
Several refractive index values at high temperatures are given by T.W. Houston, L.F. Johnson, P. Kisliuk, and D.J. Walsh, *J. Opt. Soc. Am. 53*:1286 (1963).

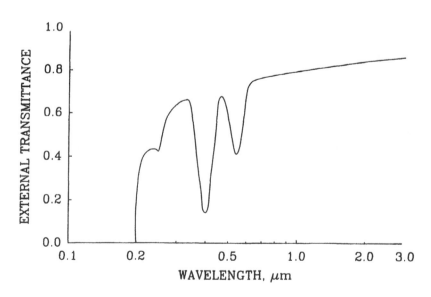

Figure 6.3 The external transmittance of ruby, thickness 6.10 mm. [Adapted from D.E. McCarthy, *Appl. Opt. 6*: 1896 (1967).]

Aluminum Oxide (Sapphire), Al_2O_3

Molecular Weight
101.94

Specific Gravity
3.98 [g/cm^3]

Crystal Class
Trigonal

Transmission
Long-wavelength limit 6.5 μm
Short-wavelength limit 0.17 μm
(See Figs. 6.4 and 6.5.)

Absorption Coefficient
(See Fig. 6.6.)

Reflection Loss
12% for two surfaces for the ordinary ray at 4 μm

Reflectivity
(See Fig. 6.7.)

Refractive Index
(See Table 6.2.)

Temperature Coefficient of Refractive Index
1.37×10^{-5} K^{-1} at 5.461 μm
1.4×10^{-5} K^{-1} at 5.791 μm

Dispersion
(See dispersion equation, Table 6.2.)

Dielectric Constant
10.55 for 10^2–10^8 Hz at 298 K (parallel)
8.6 for 10^2–10^{10} Hz at 298 K (perpendicular)

Melting Temperature
2300 K

Thermal Conductivity
35.1 W/(m·K) at 300 K (parallel)
33.0 W/(m·K) at 300 K (perpendicular)
5.8 W/(m·K) at 773 K (parallel)

Thermal Expansion
5.6×10^{-6} K^{-1} at 293 K (parallel)
5.0×10^{-6} K^{-1} at 293 K (perpendicular)

Specific Heat
0.18 [cal/(g·K)] at 298 K

Hardness
1370 kg/mm^2
(Knoop number with 1000-g load)

Bandgap
9.9 eV

Solubility
Insoluble in H_2O

Elastic Moduli
Elastic stiffness
$c_{11} = 495$ GPa
$c_{12} = 160$ GPa
$c_{44} = 146$ GPa
$c_{13} = 115$ GPa
$c_{14} = -23$ GPa
$c_{33} = 497$ GPa
Elastic compliance
$s_{11} = 2.38$ TPa^{-1}
$s_{12} = -0.70$ TPa^{-1}
$s_{44} = 7.03$ TPa^{-1}
$s_{13} = -0.38$ TPa^{-1}
$s_{14} = 0.49$ TPa^{-1}
$s_{33} = 2.19$ TPa^{-1}
Young's modulus
344.5 GPa
Shear modulus
148.14 GPa
Bulk modulus
207 GPa at 273 K
Poisson's ratio
0.27

Notes: Since this material is very hard (often used as an abrasive), it must be ground and polished with diamond or boron carbide abrasive; the techniques are therefore difficult and costly. Sapphire has a very high thermal conductivity at liquid nitrogen temperatures and below, and so it can be used as a substrate for cooled cells. Sapphire is not hygroscopic, but it is slightly soluble in acids and alkalies. It is available to 8-in diameters from both Crystal Systems and Saphikon, Inc.

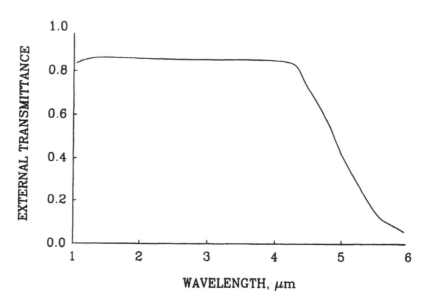

Figure 6.4   The external transmittance of sapphire at 293 K, thickness 8 mm. [Adapted from U. P. Oppenheim and U. Even, *J. Opt. Soc. Am. 52*: 1078 (1962).]

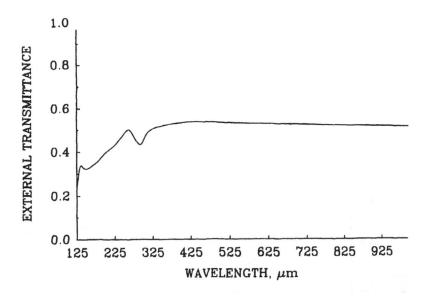

Figure 6.5 The long-wavelength external transmittance of sapphire, thickness 1.0 mm; sample with the optical axis parallel to the face. [Adapted from data from E. V. Loewenstein, *J. Opt. Soc. Am. 51*: 108 (1961).]

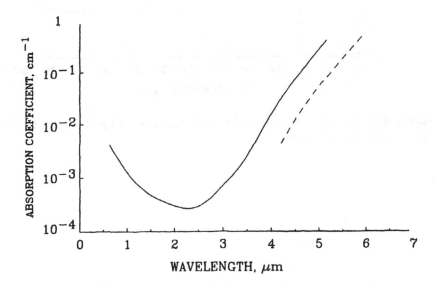

Figure 6.6 Absorption coefficient of sapphire at 1473 K (dashed line) and 296 K (solid line). [Adapted from I. H. Malitson, *J. Opt. Soc. Am. 52*: 1377 (1962) and D. A. Burch, *J. Opt. Soc. Am. 55*: 625 (1965).]

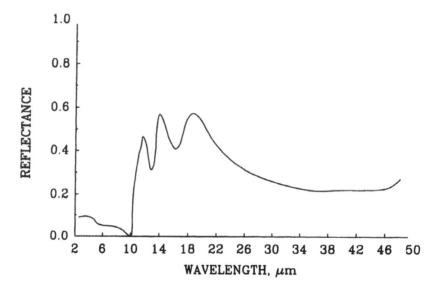

Figure 6.7 The reflectance of sapphire, thickness 2 mm. [Adapted from D. E. Mc-Carthy, *Appl. Opt. 2;* 293 (1963).]

Table 6.2 Sapphire (Synthetic): Refractive Index for the Ordinary Ray

λ (μm)	n	λ (μm)	n
0.26520	1.83360	1.39506	1.74888
0.28035	1.82427	1.52952	1.74660
0.28936	1.81949	1.6932	1.74368
0.29673	1.81595	1.70913	1.74340
0.30215	1.81351	1.81307	1.74144
0.3130	1.80906	1.9701	1.73833
0.33415	1.80184	2.1526	1.73444
0.34662	1.79815	2.24929	1.73231
0.361051	1.79450	2.32542	1.73057
0.365015	1.79358	2.4374	1.72783
0.39064	1.78826	3.2439	1.70437
0.404656	1.78582	3.2668	1.70356
0.433834	1.78120	3.3026	1.70231
0.546071	1.77078	3.3303	1.70140
0.576960	1.76884	3.422	1.69818
0.579066	1.76871	3.5070	1.69504
0.64385	1.76547	3.7067	1.68746
0.706519	1.76303	4.2553	1.66371
0.85212	1.75885	4.954	1.62665
0.89440	1.75796	5.1456	1.61514
1.01398	1.75547	5.349	1.60202
1.12866	1.75339	5.419	1.59735
1.36728	1.74936	5.577	1.58638

Dispersion Equation:

$$n^2 - 1 = \sum_{11} \frac{A_i \lambda^2}{\lambda^2 - \lambda_i^2}$$

where at 24°C

$\lambda_1^2 = 0.00377588$ $A_1 = 1.023798$
$\lambda_2^2 = 0.0122544$ $A_2 = 1.058264$
$\lambda_3^2 = 321.3616$ $A_3 = 5.280792$

Temperature coefficients of index:
Temperature coefficients of refractive index dn/dT were determined from the differences between the indexes at 19 and 24°C. The results indicate that the coefficient is positive and decreases from about 20×10^{-6} per Celsius degree at the short wavelengths to about 10×10^{-6} per Celsius degree near 4 μm. The average value of 13×10^{-6} per Celsius degree for the visible region was determined from additional measurements made at 17, 24, and 31°C on a Wild precision spectrometer.

Note: E.V. Loewenstein, *J. Opt. Soc. Am. 51*:108 (1961) gives 3.14 ± 0.13 for n_O and 3.61 ± 0.14 for n_E from 167 to 500 μm.
Source: I.H. Malitson, *J. Opt. Soc. Am. 52*:1377 (1962).

Ammonium Dihydrogen Phosphate (ADP), $NH_4H_2PO_4$

Molecular Weight
115.04

Specific Gravity
1.803 [g/cm^3] at 293 K

Transmission
Long-wavelength limit 1.7 μm
Short-wavelength limit 0.125 μm
(See Fig. 6.8.)

Reflection Loss
10.5% for two surfaces at 0.7 μm

Refractive Index
(See Table 6.3.)

Dispersion
(See dispersion equation, Table 6.3.)

Melting Temperature
463 K

Thermal Conductivity
0.71162 W/(m·K) at 315 K (parallel)
1.2558 W/(m·K) at 313 K (perpendicular)

Thermal Expansion
39.3×10^{-6} K^{-1} at 297–407 K

Solubility
22.7 g/(100 g H_2O) at 273 K, insoluble in alcohol and acetone.

Elastic Moduli
Elastic stiffness
$c_{11} = 67.3$ GPa
$c_{12} = 5.0$ GPa
$c_{13} = 19.8$ GPa
$c_{33} = 33.7$ GPa
$c_{44} = 8.57$ GPa
$c_{66} = 6.02$ GPa
Elastic compliance
$s_{11} = 18.3$ TPa^{-1}
$s_{12} = 2.2$ TPa^{-1}
$s_{13} = -12.0$ TPa^{-1}
$s_{33} = 43.7$ TPa^{-1}
$s_{44} = 117$ TPa^{-1}
$s_{66} = 166$ TPa^{-1}

Notes: ADP can be cut but is easily damaged. It is water-soluble, has a low resistance to thermal shock, and can be polished with alcohol. An electro-optic material, it is useful as a modulator.

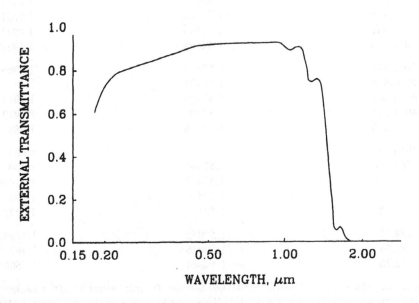

Figure 6.8. The external transmittance of ADP, thickness 1.65 mm. [Adapted from D. E. McCarthy, *Appl. Opt.* 6: 1896 (1967).]

Table 6.3 Refractive Index of ADP

λ $(10^{-4}\ \mu m)$	n_O	n_E
2138.560	1.62598	1.56738
2288.018	1.60785	1.55138
2536.519	1.58688	1.53289
2967.278	1.56462	1.51339
3021.499	1.56270	1.51163
3125.663	1.55917	1.50853
3131.545	1.55897	1.50832
3341.478	1.55300	1.50313
3650.146	1.54615	1.49720
3654.833	1.54608	1.49712
3662.878	1.54592	1.49698
3906.410	1.54174	. . . . .
4046.561	1.53969	1.49159
4077.811	1.53925	1.49123
4358.350	1.53578	1.48831
4916.036	. . . . .	1.48390
5460.740	1.52662	1.48079
5769.590	1.52478	1.47939
5790.654	1.52466	1.47930
6328.160	1.52195	1.47727
10 139.75	1.50835	1.46895
11 287.04	1.50446	1.46704
11 522.76	1.50364	1.46666

The accuracy of the data is believed to be ±0.00003 or better. The index values for ADP are adapted from F. Zernike, Jr., *J. Opt. Soc. Am. 54*: 1215 (1964), and 55, 210E (1965). Index values of ADP for 0.4860, 0.5890, and 0.6560 μm are given in the *International Critical Tables*, vol. VII, pp. 19, 27, McGraw-Hill Book Company, New York, 1930. In addition, Zernike lists many computed index values in air, using the following dispersion equation

$$n^2 = \frac{A + B\nu^2}{1 - \nu^2/C} + \frac{D}{E - \nu^2}$$

where $\nu = 1/\lambda$ in cm^{-1}.

Constants of the dispersion equation for ADP:

	Extraordinary ray	Ordinary ray
A	2.163510	2.302842
B	9.616676 × 10^{-11}	1.1125165 × 10^{-10}
C	7.698751 × 10^9	7.5450861 × 10^9
D	1.479974 × 10^6	3.775616 × 10^6
E	2.500000 × 10^5	2.500000 × 10^5

Aluminum Nitride, AlN

Molecular Weight
40.99

Specific Gravity
3.26 [g/cm^3]

Crystal Class
Hexagonal

Absorption Coefficient
(See Fig. 6.9)

Refractive Index
2.20 at 0.589 μm

Dielectric Constant
9.1 at zero frequency

Melting Temperature
3070 K

Thermal Expansion
(polycrystalline)
0.3×10^{-6} K^{-1} at 75 K
2.6×10^{-6} K^{-1} at 300 K
6.5×10^{-6} K^{-1} at 1600 K

Thermal Conductivity
320 W/(m·K) at 300 K
42 W/(m·K) at 1000 K

Specific Heat
0.016 [cal/(g·K)] at 276 K
0.284 [cal/(g·K)] at 1150 K

Debye Temperature
950 K

Hardness
1200 kg/mm^2

Bulk Modulus
220 GPa

Notes: AlN is difficult to grow. It has also been made from nitridation of AlCl$_3$ powders, but difficulty in densifying the powders results in light scattering. Interest in AlN remains because of its high thermal conductivity and low thermal expansion, which make it very resistant to thermal shock. Its high hardness and strength also make it an attractive material.

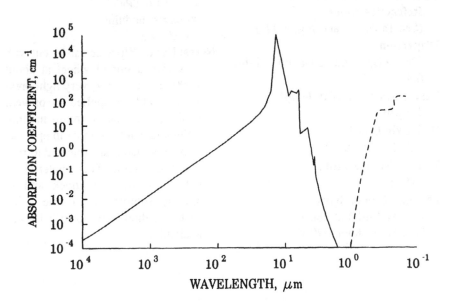

Figure 6.9. Optical absorption coefficient at room temperature of single crystals of AlN. (Adapted from G. A. Slack, General Electric Report No. 79CRD071, 1979.)

Barium Fluoride, BaF$_2$

Molecular Weight
175.36
Specific Gravity
4.89 [g/cm^3]
Crystal Class
Cubic
Transmission
Long-wavelength limit 15 μm
Short-wavelength limit 0.15 μm
(See Fig. 6.10.)
Absorption Coefficient
Adolf Meller sample
0.14 × 10^{-3} cm^{-1} at 2.7 μm
0.20 × 10^{-3} cm^{-1} at 3.8 μm
Optovac sample
1.8 × 10^{-3} cm^{-1} at 2.7 μm
2.0 × 10^{-3} cm^{-1} at 3.8 μm
Reflection Loss
7.7% for two surfaces at 0.6 μm
Refractive Index
(See Table 6.4.)
Temperature Coefficient of Refractive Index
(See Table 6.5 and Fig. 6.11.)
Dispersion
(See dispersion equation, Table 6.4.)
Zero Dispersion Wavelength
2.3 μm
Dielectric Constant
7.33 for 2 × 10^6 Hz
Melting Temperature
1550 K
Thermal Conductivity
11.7 W/(m·K) at 286 K
10.5 W/(m·K) at 370 K

Thermal Expansion
6.7 × 10^{-6} K^{-1} at 75 K
19.9 × 10^{-6} K^{-1} at 300 K
24.7 × 10^{-6} K^{-1} at 500 K
Specific Heat
0.096 [cal/(g·K) at 274 K
Debye Temperature
282 K
Hardness
82 kg/mm^2
(Knoop number with 500-g load)
Bandgap
9.1 eV
Solubility
0.17 g/(100 g H$_2$O)
Elastic Moduli
Elastic stiffness
$c_{11} = 90.7$ GPa
$c_{12} = 41.0$ GPa
$c_{44} = 25.3$ GPa
Elastic compliance
$s_{11} = 15.2$ TPa^{-1}
$s_{12} = -4.7$ TPa^{-1}
$s_{44} = 39.6$ TPa^{-1}
Young's modulus
53.05 GPa

Notes: Barium fluoride can be cut with a Norton diamond wheel at about 4000 rpm, but very, very light pressure should be applied to prevent cleavage. It has a high melting point and is less suitable in applications in which it is subjected to mechanical stress. This material is subject to fracture shock and is slightly soluble in sulfuric acid. Plates of diameters up to 3 in. are available.

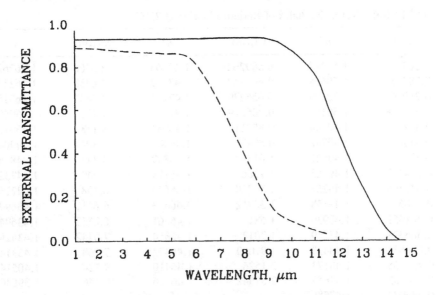

Figure 6.10. The external transmittance of barium fluoride at 300 K (solid line), thickness 8.0 mm; and at 1273 K (dashed line), thickness 7.6 mm. [Adapted from U. P. Oppenheim and A. Goldman, *J. Opt. Soc. Am.* **54**: 127 (1964).]

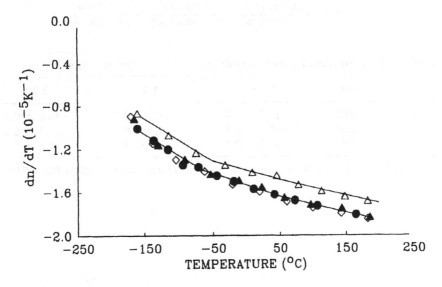

Figure 6.11 The temperature coefficient of barium fluoride as a function of temperature for wavelengths 0.6328 μm (●), 1.15 μm (◇), 3.39 μm (▲), and 10.6 μm (△). [Adapted from A.J. Glass and A.H. Guenther (Eds.), *Laser Induced Damage in Optical Materials*, 1977, p. 78.)

Table 6.4   Refractive Index of Barium Fluoride at 25°C

λ (μm)	n	λ (μm)	n	λ (μm)	n
0.2652	1.51217	0.589262	1.47443	2.5766	1.46262
0.28035	1.50668	0.643847	1.47302	2.6738	1.46234
0.28936	1.50390	0.656279	1.47274	3.2434	1.46018
0.296728	1.50186	0.706519	1.47177	3.422	1.45940
0.30215	1.50044	0.85211	1.46984	5.138	1.45012
0.3130	1.49782	0.89435	1.46942	5.3034	1.44904
0.32546	1.49521	1.01398	1.46847	5.343	1.44878
0.334148	1.49363	1.12866	1.46779	5.549	1.44732
0.340365	1.49257	1.36728	1.46673	6.238	1.44216
0.34662	1.49158	1.52952	1.46613	6.6331	1.43899
0.361051	1.48939	1.681	1.46561	6.8559	1.43694
0.366328	1.48869	1.7012	1.46554	7.0442	1.43529
0.404656	1.48438	1.97009	1.46472	7.268	1.43314
0.435835	1.48173	2.1526	1.46410	9.724	1.40514
0.486133	1.47855	2.32542	1.46356	10.346	1.39636
0.546074	1.47586				

Source: Adapted from I. H. Malitson: *J. Opt. Soc. Am. 54*:628 (1964).
Malitson also reports several computed index values, using the following dispersion equation:

$$n^2 - 1 = \frac{0.643356\lambda^2}{\lambda^2 - (0.057789)^2} + \frac{0.506762\lambda^2}{\lambda^2 - (0.10968)^2} + \frac{3.8261\lambda^2}{\lambda^2 - (46.3864)^2}$$

Table 6.5   Experimental Values of dn/dT (10^{-6} K^{-1}) for Barium Fluoride

Crystal	T (K)	0.6328	1.15	3.39
BaF$_2$	310	-16.7 ± 0.4	-17.1 ± 0.5	-16.8 ± 0.4
	330	-17.3 ± 0.3	-17.5 ± 0.5	-17.3 ± 0.5
	350	-17.9 ± 0.3	-17.8 ± 0.5	-17.6 ± 0.4

Source: H. G. Lipson, Y. F. Tsay, B. Bendow, and P. A. Ligon, *Appl. Opt. 15*(10):2353 (1976).

Barium Titanate, BaTiO$_3$

Molecular Weight
232.96
Specific Gravity
5.90 [g/cm^3]
Crystal Class
Orthorhombic 203–278 K
Tetragonal 278–393 K
Cubic above ~393 K
Transmission
Long-wavelength limit 6.9 μm
Short-wavelength limit <0.5 μm
(See Fig. 6.12.)
Reflection Loss
29.0% for two surfaces, in the infrared.
Refractive Index
(See Fig. 6.13.)
Dispersion
(See dispersion equation, Table 6.6.)
Dielectric Constant
1240–1100 for 10^2–10^8 Hz at 298 K
Melting Temperature
1870 K
Thermal Conductivity
1.34 W/(m·K) at 293 K
1.05 W/(m·K) at 403 K
Thermal Expansion
19 × 10^{-6} K^{-1} at 283–343 K
Solubility
Insoluble in H$_2$O
Specific Heat
0.077 [cal/(g·K)] at 175 K
0.103 [cal/(g·K)] at 273 K
0.140 [cal/(g·K)] at 1800 K
Hardness
200–580 kg/mm^2 (Vickers)

Elastic Moduli
Elastic stiffness
(Cubic)
$c_{11} = 214$ GPa
$c_{12} = 140$ GPa
$c_{44} = 119$ GPa
Elastic compliance
(cubic)
$s_{11} = 9.8$ TPa^{-1}
$s_{12} = -3.9$ TPa^{-1}
$s_{44} = 8.4$ TPa^{-1}
(Tetragonal with constant electric field)
$c_{11} = 275$ GPa
$c_{12} = 179$ GPa
$c_{13} = 152$ GPa
$c_{33} = 165$ GPa
$c_{44} = 54.4$ GPa
$c_{66} = 113$ GPa
(tetragonal with constant electric field)
$s_{11} = 8.05$ TPa^{-1}
$s_{12} = -2.35$ TPa^{-1}
$s_{13} = -5.24$ TPa^{-1}
$s_{33} = 15.7$ TPa^{-1}
$s_{44} = 18.4$ TPa^{-1}
$s_{66} = 8.84$ TPa^{-1}
Bulk modulus
161.92 GPa
Shear modulus
126.09 GPa
Young's modulus
33.76 GPa

Notes: Barium titanate, a ferroelectric material, is well known for its electrical properties. Because of its transmission properties, it has applications as immersion lenses for infrared detectors.

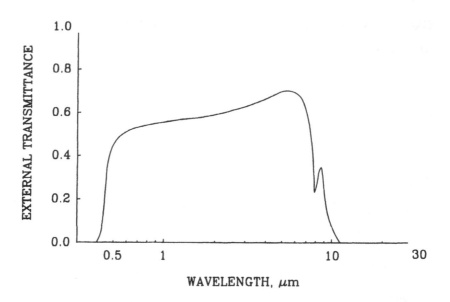

Figure 6.12 The external transmittance of barium titanate, thickness 0.25 mm. [Adapted from A. F. Iatsenko, *Soviet Phys.-Tech. Phys. 2;* 2257 (1957).]

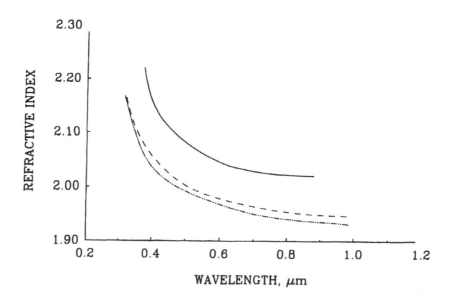

Figure 6.13 The refractive index of sputter-deposited barium titanate. The dot-dot-dash, dash, and solid lines are for deposition at 300, 500, and 600 K respectively. [Adapted from *J. Appl. Phys. 48*: 1748 (April 1977).]

Table 6.6 Refractive Indexes of Melt-Grown Barium Titanate at Room Temperature

λ (μm)	n_O	n_E
0.4579	2.5637	2.4825
0.4765	2.5355	2.4605
0.4880	2.5206	2.4487
0.5145	2.4917	2.4255
0.5321	2.4760	2.4128
0.6328	2.4164	2.3637
1.0642	2.3379	2.2970
2.1284	2.2947	2.2593

Dispersion equation:

$$n_o^2 - 1 = \frac{4.239\lambda^2}{\lambda^2 - (0.2229)^2}$$

$$n_E^2 - 1 = \frac{4.0854\lambda^2}{\lambda^2 - (0.2087)^2}$$

where λ is in μm.

Note: Coherence lengths measure in μm at $\lambda_1 = 1.058$ μm: $l_{15} = 1.57$. $l_{31} = 2.90$, $l_{33} = 2.07$.

Source: M.J. Weber, Ed., *Handbook of Laser Science and Technology*, Vol. III, *Optical Materials*; Part 1, CRC Press, Boca Raton, Fla., 1988, p. 158

Boron Nitride, BN

Molecular Weight
24.82
Specific Gravity
2.25 [g/cm^3]
Crystal Class
Cubic
Transmission
(See Fig. 6.14.)
Reflectivity
(See Fig. 6.15.)
Refractive Index
2.117 at 0.5893 μm
Melting Temperature
3500 K

Thermal Conductivity
1300 W/(m·K) at 300 K
Thermal Expansion
0.1 × 10^{-6} K^{-1} at 75 K
1.2 × 10^{-6} K^{-1} at 300 K
6.8 × 10^{-6} K^{-1} at 1200 K
Specific Heat
0.35 [cal/(g·K)] at 473 K
Debye Temperature
600 K
Bandgap
8 eV (indirect)
Solubility
Insoluble in H$_2$O

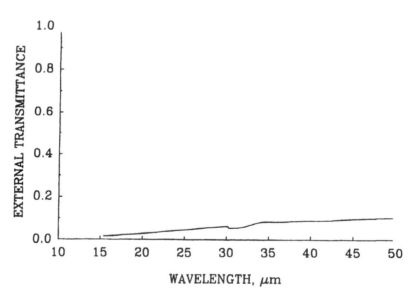

Figure 6.14 The external transmittance of boron nitride, thickness 6.0 mm. [Adapted from D.E. McCarthy, (no date).]

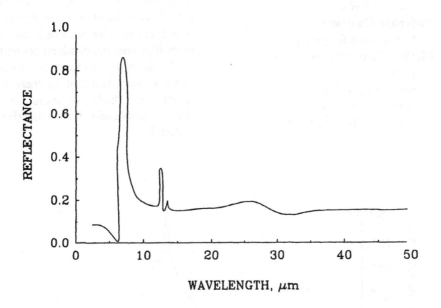

Figure 6.15. The reflectance of boron nitride, thickness 6.0 mm. [Adapted from D.E. McCarthy (no date).]

214   Browder et al.

Boron Phosphide, BP

Molecular Weight
41.78
Specific Gravity
2.97 [g/cm^3]
Crystal Class
Cubic
Absorption Coefficient
(See Fig. 6.16.)
Refractive Index
3.1 at 0.589 μm
Dielectric Constant
11.0 at zero frequency
Melting Temperature
3000 K
Thermal Expansion
3.0 × 10^{-6} K^{-1} at 300 K
5.4 × 10^{-6} K^{-1} at 1000 K

Thermal Conductivity
350 W/(m·K) at 300 K
45 W/(m·K) at 1000 K
Debye Temperature
985 K
Hardness
3600 kg/mm^2
Bulk Modulus
200 GPa
Notes: BP is difficult to grow. It has been made by both vapor transport growth and chemical vapor deposition. Interest in BP remains because of its high thermal conductivity and moderate thermal expansion, which make it very resistant to thermal shock. Its high hardness and strength also make it an attractive material.

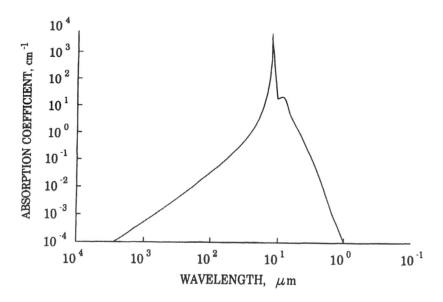

Figure 6.16 Optical absorption coefficient at room temperature of single crystals of BP. (Adapted from G.A. Slack, General Electric Report No. 79CRD071, 1979.)

Cadmium Fluoride, CdF$_2$

Molecular Weight
150.41

Specific Gravity
6.64 [g/cm^3]

Crystal Class
Cubic

Transmission
(See Fig. 6.17.)

Temperature Coefficient of Refractive Index
-1.10×10^{-5} K^{-1} at 0.4579 μm 293 K
-1.15×10^{-5} K^{-1} at 1.15 μm 293 K
-1.12×10^{-5} K^{-1} at 3.39 μm 293 K

Melting Temperature
1370 K

Thermal Expansion
27×10^{-6} K^{-1} at 293–333 K

Hardness
3 (Mohs; estimate)

Bandgap
6 eV

Solubility
4.4 g/(100 g H$_2$O)

Elastic Moduli
Elastic stiffness
$c_{11} = 184$ GPa
$c_{12} = 67$ GPa
$c_{44} = 21.8$ GPa
Elastic compliance
$s_{11} = 6.74$ TPa^{-1}
$s_{12} = -1.80$ TPa^{-1}
$s_{44} = 45.9$ TPa^{-1}

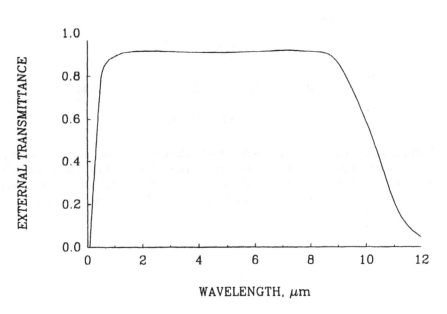

Figure 6.17 The external transmittance of cadmium fluoride; thickness 5.0 mm. [Adapted from D.A. Jones et al., *Proc. Phys. Soc. B65*: 906 (1952).]

Cadmium Phosphide, Cd₃P₂

Molecular Weight
   399.15
Specific Gravity
   5.60 [g/cm³]

Transmission
   (See Fig. 6.18.)
Reflectivity
   (See Fig. 6.19.)
Refractive Index
   (See Fig. 6.20.)

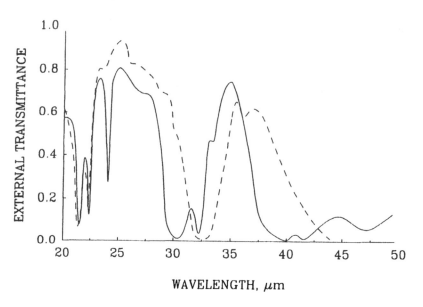

Figure 6.18 The external transmittance of cadmium phosphide, thickness 30 μm. The crystals were cut in the form of plates approximately 4 mm × 15 mm with the C axis lying in the plane of the plates. *Note*: Solid curve indicates perpendicular to C axis, and dashed curve indicates parallel to C axis. [Adapted from I.S. Gorban' et al., *Zh. Prikl. Spektrosk.* 25(5), 935–937 (1976).]

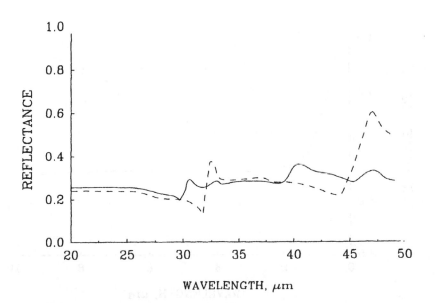

Figure 6.19 The reflectance of cadmium phosphide, thickness 30 μm. The crystals were cut in the form of plates approximately 4 mm × 15 mm with the *C* axis lying in the plane of the plates. *Note*: Solid curve indicates perpendicular to *C* axis, and dashed curve indicates parallel to *C* axis. [Adapted from I.S. Gorban' et al., *Zh. Prikl. Spektrosk.* *25*(5): 935–937 (1976).]

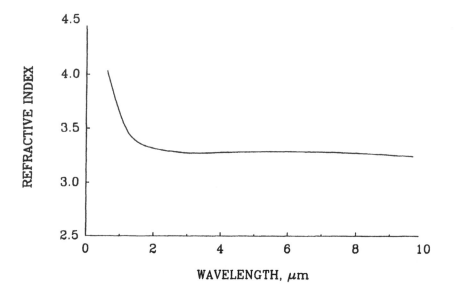

Figure 6.20 The refractive index of amorphous cadmium phosphide layers. [Adapted from I.V. Potykevich, V. K. Maksimov, V. V. Borshch, and A. I. Krivotenko, *Opt. Spektrosk.* 52: 175–177 (1982).]

Cadmium Selenide, CdSe

Molecular Weight
191.36

Specific Gravity
5.63–5.81 [g/cm^3]

Crystal Class
Hexagonal, cubic

Transmission
(See Fig. 6.21.)

Absorption Coefficient
0.032 cm^{-1} at 10.6 μm

Refractive Index
(See Table 6.7.)

Dispersion
(See dispersion equation, Table 6.7.)

Melting Temperature
>1620 K

Thermal Conductivity
9.0 W/(m·K) at 300 K

Thermal Expansion
1.4 × 10^{-6} K^{-1} at 80 K
7.3 × 10^{-6} K^{-1} at 320 K

Debye Temperature
230 K

Bandgap
1.8 eV

Hardness
44–90 kg/mm^2
(Knoop number)

Solubility
Insoluble in H$_2$O

Elastic Moduli
Elastic stiffness
(hexagonal)
$c_{11} = 74.1$ GPa
$c_{12} = 45.2$ GPa
$c_{13} = 38.9$ GPa
$c_{33} = 84.3$ GPa
$c_{44} = 13.4$ GPa
(cubic)
$c_{11} = 66.7$ GPa
$c_{12} = 46.3$ GPa
$c_{44} = 22.3$ GPa

Elastic compliance
(hexagonal)
$s_{11} = 23.2$ TPa^{-1}
$s_{12} = -11.2$ TPa^{-1}
$s_{13} = -5.5$ TPa^{-1}
$s_{33} = 16.9$ TPa^{-1}
$s_{44} = 74.7$ TPa^{-1}
(cubic)
$s_{11} = 34.8$ TPa^{-1}
$s_{12} = -14.2$ TPa^{-1}
$s_{44} = 44/$TPa^{-1}

Notes: The surface of cadmium selenide takes a mirrorlike polish but scratches easily.

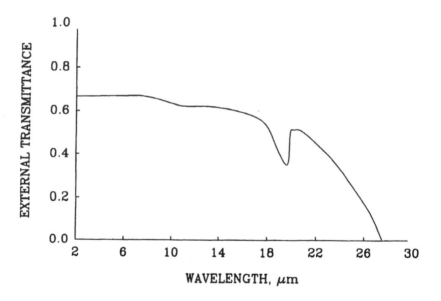

Figure 6.21 The external transmittance of cadmium selenide, thickness 1.67 mm. [Adapted from D. E. McCarthy, *Appl. Opt. 4*: 317 (1965).]

Table 6.7 Measured Refractive Indices of Cadmium Selenide

λ (μm)	n_O	n_E
1.0139	2.5481	2.5677
1.1287	2.5246	2.5444
1.3673	2.4971	2.5170
1.5295	2.4861	2.5059
1.7109	2.4776	2.4974
2.3253	2.4627	2.4823
3.00	2.4553	2.4748
4.00	2.4500	2.4694
5.00	2.4404	2.4657
6.00	2.4434	2.4625
7.00	2.4398	2.4586
8.00	2.4367	2.4552
9.00	2.4333	2.4514
10.00	2.4294	2.4475
11.00	2.4252	2.4430
12.00	2.4204	2.4379

Source: R.L. Harbst and R.L. Byer, *Appl. Phys. Lett. 19*(12):529 (1971).
Sellmeier formula [B. Tatian, *Appl. Opt. 23*(24):4484 (1984)]:

$$n^2 = 1 + \sum_{j=1}^{k} \frac{A_j \lambda^2}{\lambda^2 - B_j^2}$$

$A_1 = 3.97478769E+00$ $B_1 = 2.24269842E-01$
$A_2 = 2.66808089E-01$ $B_2 = 4.66937848E-01$
$A_3 = 7.40772832E-04$ $B_3 = 5.09151386E-01$
$A_4 = -3.22573381E+01$ $B_4 = 9.89949494E+01$

Cadmium Sulfide, CdS

Molecular Weight
144.48
Specific Gravity
4.82 [g/cm^3] at 293 K
Crystal Class
Hexagonal, cubic
Transmission
Long-wavelength limit 16 μm
Short-wavelength limit 0.52 μm
(See Fig. 6.22.)
Absorption Coefficient
0.032 cm^{-1} at 10.6 μm
0.01 cm^{-1} at 3 μm
Reflection Loss
12.9% for two surfaces at 1.5 μm
Emissivity
(See Fig. 6.23.)
Refractive Index
(See Table 6.8.)
Dispersion
(See dispersion equation, Table 6.8.)
Dielectric Constant
9.02–9.35 for 10^4 Hz
Melting Temperature
1670 K
Thermal Conductivity
15.91 W/(m·K) at 287 K
Thermal Expansion
2.1×10^{-6} K^{-1} at 77–298 K (parallel)
4×10^{-6} K^{-1} at 77–298 K (perpendicular)
3.6×10^{-6} K^{-1} at 323–773 K (parallel)
5×10^{-6} K^{-1} at 323–773 K (perpendicular)
Specific Heat
0.0882 [cal/(g·K)] at 273 K
0.0964 [cal/(g·K)] at 1200 K
Hardness
122±4 kg/mm^2
[Knoop number with 25-g load (parallel)]
Bandgap
2.4 eV (parallel)
Solubility
Insoluble in H$_2$O
Elastic Moduli
Elastic stiffness
(hexagonal)
$c_{11} = 84.32$ GPa
$c_{12} = 52.12$ GPa
$c_{13} = 46.38$ GPa
$c_{33} = 93.79$ GPa
$c_{44} = 14.89$ GPa
(cubic)
$c_{11} = 77$ GPa
$c_{12} = 54$ GPa
$c_{44} = 24$ GPa
Elastic compliance
(hexagonal)
$s_{11} = 23.2$ TPa^{-1}
$s_{12} = -11.2$ TPa^{-1}
$s_{13} = -5.5$ TPa^{-1}
$s_{33} = 16.9$ TPa^{-1}
$s_{44} = 74.7$ TPa^{-1}
(cubic)
$s_{11} = 31$ TPa^{-1}
$s_{12} = -13$ TPa^{-1}
$s_{44} = 42$ TPa^{-1}

Notes: Cadmium sulfide is easily cut, ground, lapped, and polished but is relatively soft. It has negligible water solubility but can be dissolved in acids. Cadmium sulfide's crystal structure is cubic if it has been chemically precipitated and hexagonal if grown from the vapor phase. Natural-crystal cadmium sulfide is called greenockite. Commercial compensators and retardation plates constructed from this material are available for use in the infrared. Finally, cadmium sulfide is a piezoelectric crystal.

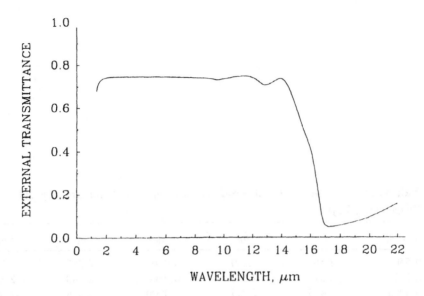

Figure 6.22 The external transmittance of single-crystal cadmium sulfide, thickness 3.02 mm (solid curve). [Adapted from A.B. Francis and A.L. Carlson, *J. Opt. Soc. Am.* *50*: 118 (1960).]

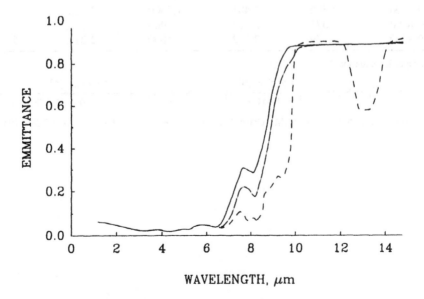

Figure 6.23 Spectral emittance of cadmium sulfide at 473 K (solid line), 373 K (dot-dash line), and 77 K (dashed line), thickness 5.1 mm. [Adapted from Stierwalt, Navy Ordnance Lab Report 667, Corona, Calif., 1966.]

Table 6.8 Refractive Index of Hexagonal Cadmium Sulfide for Ordinary and Extraordinary Rays

λ (μm)	n_O	n_E	λ (μ)	n_O	n_E
0.5500	2.565	2.597	0.9000	2.353	2.358
0.5750	2.518	2.545	0.9500	2.340	2.352
0.6000	2.483	2.511	1.0000	2.338	2.341
0.6250	2.458	2.478	1.0500	2.332	2.338
0.6500	2.438	2.459	1.1000	2.325	2.333
0.6750	2.421	2.437	1.1500	2.322	2.330
0.7000	2.407	2.425	1.2000	2.317	2.322
0.7500	2.386	2.403	1.2500	2.316	2.320
0.8000	2.371	2.383	1.3000	2.315	2.319
0.8500	2.359	2.372	1.4000	2.311	2.314

Dispersion equation:

$$n_O^2 = 5.235 + \frac{1.819 \times 10^7}{\lambda^2 - 1.651 \times 10^7} \qquad n_E^2 = 5.239 + \frac{2.076 \times 10^7}{\lambda^2 - 1.651 \times 10^7}$$

Source: Czyzak, S.J., Baker, W.M., Crane, R.C., and Howe, J.B., *J. Opt. Soc. Am.* 47:240 (1957).

Cadmium Telluride, CdTe

Molecular Weight
240.02

Specific Gravity
6.20 [g/cm^3]

Crystal Class
Cubic

Transmission
Long-wavelength limit 15 μm
Short-wavelength limit 0.9 μm
(See Fig. 6.24.)

Absorption Coefficient
Single crystal
1.0×10^{-3} cm^{-1} at 10.6 μm
Polycrystalline
3.5×10^{-3} cm^{-1} at 10.6 μm

Reflection Loss
32% for two surfaces at 10 μm

Refractive Index
(See Tables 6.9 and 6.10.)

Temperature Coefficient of Refractive Index
4.9×10^{-5} K^{-1} at 10 μm
5.1×10^{-5} K^{-1} at 15 μm

Dispersion
(See dispersion equations, Table 6.10.)

Dielectric Constant
11.0 for 1–10^5 Hz (5.5×10^{13} carriers/cm^3)

Melting Temperature
1320 K

Thermal Conductivity
6.3 W/(m·K) at 300 K

Thermal Expansion
5.9×10^{-6} K^{-1} at 300 K

Specific Heat
0.056 [cal/(g·K)] at 300 K

Hardness
56 kg/mm^2 (Knoop number)

Bandgap
1.5 eV

Solubility
Insoluble in H$_2$O

Elastic Moduli
Elastic stiffness
$c_{11} = 53.51$ GPa
$c_{12} = 36.81$ GPa
$c_{44} = 19.94$ GPa
Elastic Compliance
$s_{11} = 42.7$ TPa^{-1}
$s_{12} = -17.4$ TPa^{-1}
$s_{44} = 49.5$ TPa^{-1}
Young's modulus
36.52 GPa
Rupture modulus
0.0406 GPa
Bulk modulus
25 GPa

Notes: Generally, the same cutting and polishing techniques that are used for silicon and germanium can be used for cadmium telluride. It has important linear and nonlinear optical properties in the infrared for laser optics. Polycrystalline cadmium telluride is produced by Eastman Kodak Co. under the trade name Irtran VI.

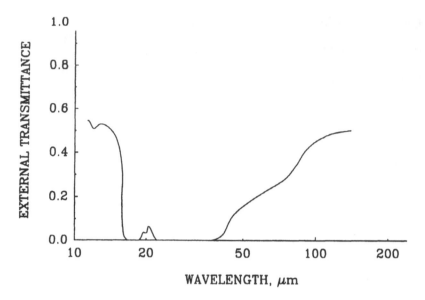

Figure 6.24 External transmittance of high-resistivity cadmium telluride at 300 K; thickness 2.26 mm; [111] single-crystal plate optically polished. Resistivity 0.35 Mohm-cm. [Adapted from C. J. Johnson et al., *Appl. Opt. 8*: 1668 (1969).]

Table 6.9 Refractive Index of Cadmium Telluride at 300 K

Frequency (cm^{-1})	n	Frequency (cm^{-1})	n
11,074.19	2.910	45.938	3.2733
10,000.00	2.839	45.089	3.2720
6666.60	2.742	42.608	3.2628
5000.00	2.713	41.778	3.2597
4000.00	2.702	40.937	3.2574
3333.33	2.695	40.095	3.2550
2857.10	2.691	34.194	3.2360
2500.00	2.688	33.305	3.2371
2000.00	2.684	30.752	3.2294
1666.60	2.681	25.550	3.2203
1428.60	2.679	22.941	3.2133
1250.00	2.677	20.306	3.2136
1111.10	2.674	18.563	3.2096
1000.00	2.672	15.929	3.2062

Source: Left-hand columns, D.T.F. Marple, *J. Appl. Phys. 35*:539 (1964) and L.S. Ladd, *Infrared Phys. 6*:145 (1966). Right-hand columns, E.J. Danielewicz and P.D. Coleman, *Appl. Opt. 13*(5):1166 (1974).

Table 6.10   Refractive Index of Cadmium Telluride as a Function of Wavelength

λ (μm)	n		
	20 K	80 K	300 K
6	2.65607	2.65925	2.68198
8	2.65139	2.65459	2.67730
10	2.64661	2.64956	2.67242
12	2.64081	2.64408	2.66677
14	2.63413	2.63734	2.66020
16	2.62645	2.62981	2.65253
18	2.61789	2.62065	2.64366
20	2.60768	2.61039	2.63343
22	2.59604	2.59866	2.62177

Sellmeier equation:

$$n^2 - 1 = \frac{A_1\lambda^2}{\lambda^2 - \lambda_1^2} + \frac{A_2\lambda^2}{\lambda^2 - \lambda_2^2}$$

Herzberger equation:

$$n = A + BL + CL^2 + D\lambda^2 + E\lambda^4 \qquad L = (\lambda^2 - 0.028)^{-1}$$

The coefficients obtained by a least-square technique are as follows (λ in micrometers).

Two-term Sellmeier equation coefficients:

Temp.	A_1	A_2	λ_1^2	λ_2^2
300 K	6.1977889	3.2243821	0.1005326	5279.518
80 K	6.0756642	2.8743304	0.1053945	4773.944
20 K	6.0599879	3.7564378	0.1004272	6138.789

Herzberger equation coefficients:

Temp.	A	B	C	$10^4 D$	$10^8 E$
300 K	2.6825805	0.1321326	−0.2774556	−1.1169905	−2.9967524
80 K	2.660743	0.1127150	0.1835710	−1.1263461	−3.0545993
20 K	2.6558958	0.2065638	−2.0528983	−1.0980731	−3.0241469

Source: A.G. DeBell, E.L. Derniak, J. Harvey, J. Nissley, J. Palmer, A. Selvarajan, and W.L. Wolfe, *Appl. Opt.* *18*(18):3114 (1979).

Calcium Carbonate (Calcite), CaCO₃

Molecular Weight
100.09
Specific Gravity
2.7102 [g/cm^3]
Crystal Class
Hexagonal
Transmission
Long-wavelength limit 5.5 μm
Short-wavelength limit 0.3 μm
(See Fig. 6.25.)
Reflection Loss
10.6% for two surfaces for the ordinary ray, at 2 μm
(See Fig. 6.26.)
Refractive Index
(See Table 6.11.)
Temperature Coefficient of Refractive Index
(See Table 6.12.)
Dielectric Constant
8.0 for 10^4 Hz at 290–294 K (parallel)
8.5 for 10^4 Hz at 290–294 K (perpendicular)
Melting Temperature
1612 K at 102.5 atm
Thermal Conductivity
15.8 W/(m·K) at 8 K (perpendicular)
5.526 W/(m·K) at 273 K (parallel)
4.646 W/(m·K) at 273 K (perpendicular)
3.56 W/(m·K) at 374 K (perpendicular)
Thermal Expansion
25 × 10^{-6} K^{-1} at 273 K (parallel)
-5.8×10^{-6} K^{-1} at 273 K (perpendicular)
Specific Heat
0.203 [cal/(g·K)] at 273 K
Hardness
3 (Mohs)
Solubility
0.0014 g/(100 g H₂O)
Elastic Moduli
Elastic stiffness
$c_{11} = 137.1$ GPa
$c_{12} = 45.6$ GPa
$c_{13} = 45.1$ GPa
$c_{14} = -20.8$ GPa
$c_{33} = 79.7$ GPa
$c_{44} = 34.2$ GPa
Elastic compliance
$s_{11} = 11.4$ TPa^{-1}
$s_{12} = -4.0$ TPa^{-1}
$s_{13} = -4.5$ TPa^{-1}
$s_{14} = 9.5$ TPa^{-1}
$s_{33} = 17.4$ TPa^{-1}
$s_{44} = 41.4$ TPa^{-1}
Young's modulus
72.35 GPA (parallel)
88.19 GPa (perpendicular)
Bulk modulus
129.53 GPa

Notes: Calcite's birefringent properties are well known, and although they are important academically and in polarizing prisms, they are disadvantageous in many instrument applications. Calcite is a form of calcium carbonate but is also known to exist as Iceland spar, oriental alabaster, and onyx.

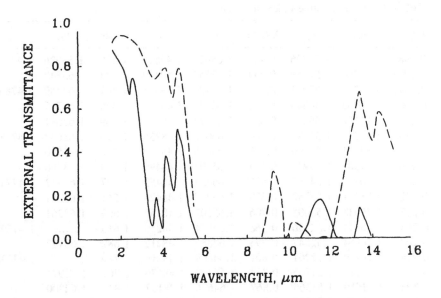

Figure 6.25 The external transmittance of calcite for the ordinary ray (——) and for the extraordinary ray (----), thickness 1 mm. [Adapted from R. E. Nysander, *Phys. Rev.* 28: 291 (1909).]

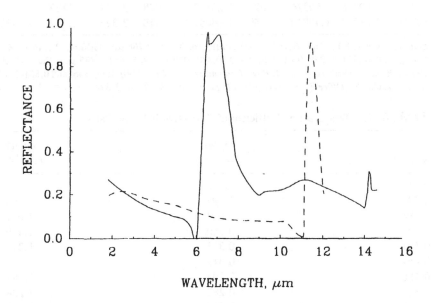

Figure 6.26 Reflectance of calcite for the ordinary ray (——) and the extraordinary ray (----). [Adapted from R. E. Nysander, *Phys. Rev.* 28: 291 (1909).]

Table 6.11   Refractive Index of Calcite

λ (μm)	n_O	n_E	λ (μm)	n_O	n_E	λ (μm)	n_O	n_E
0.198	......	1.57796	0.410	1.68014	1.49640	1.229	1.63926	1.47870
0.200	1.90284	1.57649	0.434	1.67552	1.49430	1.273	1.63849	
0.204	1.88242	1.57081	0.441	1.67423	1.49373	1.307	1.63789	1.47831
0.208	1.86733	1.56640	0.508	1.66527	1.48956	1.320	1.63767	
0.211	1.85692	1.56327	0.533	1.66277	1.48841	1.369	1.63681	
0.214	1.84558	1.55976	0.560	1.66046	1.48736	1.396	1.63637	1.47789
0.219	1.83075	1.55496	0.589	1.65835	1.48640	1.422	1.63590	
0.226	1.81309	1.54921	0.643	1.65504	1.48490	1.479	1.63490	
0.231	1.80233	1.54541	0.656	1.65437	1.48459	1.497	1.63457	1.47744
0.242	1.78111	1.53782	0.670	1.65367	1.48426	1.541	1.63381	
0.257	1.76038	1.53005	0.706	1.65207	1.48353	1.609	1.63261	
0.263	1.75434	1.52736	0.768	1.64974	1.48259	1.615	......	1.47695
0.267	1.74864	1.52547	0.795	1.64886	1.48216	1.682	1.63127	
0.274	1.74139	1.52261	0.801	1.64869	1.48216	1.749	......	1.47638
0.291	1.72774	1.51705	0.833	1.64772	1.48176	1.761	1.62974	
0.303	1.71959	1.51365	0.867	1.64676	1.48137	1.849	1.62800	
0.312	1.71425	1.51140	0.905	1.64578	1.48098	1.909	......	1.47573
0.330	1.70515	1.50746	0.946	1.64480	1.48060	1.946	1.62602	
0.340	1.70078	1.50562	0.991	1.64380	1.48022	2.053	1.62372	
0.346	1.69833	1.50450	1.042	1.64276	1.47985	2.100	......	1.47492
0.361	1.69316	1.50224	1.097	1.64167	1.47948	2.172	1.62099	
0.394	1.68374	1.49810	1.159	1.64051	1.47910	3.324	......	1.47392

Source: Martens, F.F., *Ann. Physik 6*, 603 (1901), for 0.198–0.768 μm; Gifford, W., *Proc. Roy. Soc. (Lond) 70*:329 (1902), for 0.226, 0.303, 0.330, 0.361, 0.706, and 0.795 μm; Carvallo, A., *Compt. Rend. 126*:950 (1898) and *J. Phys. Radium, ser. 3,9*:465 (1900), for 0.346, and 0.801–2.172 μm; Smakula, A., *Office Tech. Serv. (OTS) Rept.* 111053, 1952, for 3.324 μm.

Table 6.12   Temperature Coefficient of Refractive Index of Calcite

λ (μm)	dn_O/dT $(10^{-5}/°C)$	dn_E/dT $(10^{-5}/°C)$
0.211	2.150	
0.231	1.397	2.198
0.298	0.604	1.641
0.361	0.360	1.449
0.441	0.325	1.318
0.467	0.319	
0.480	0.305	1.287
0.508	0.287	1.234
0.589	0.240	1.213
0.643	0.208	1.185

Source: Mitchell, F.J., *Ann. Physik 4*(7):772 (1902).

Calcium Fluoride, CaF$_2$

Molecular Weight
78.08

Specific Gravity
3.180 [g/cm^3]

Crystal Class
Cubic

Transmission
Long-wavelength limit 12 μm
Short-wavelength limit 0.13 μm
(See Fig. 6.27.)

Absorption Coefficient
Adolf Meller sample:
2.2×10^{-3} cm^{-1} at 2.7 μm
2.1×10^{-3} cm^{-1} at 3.8 μm
Harshaw sample:
5.2×10^{-3} cm^{-1} at 2.7 μm
0.27×10^{-3} cm^{-1} at 3.8 μm
Hughes R.L. sample:
0.49×10^{-3} cm^{-1} at 2.7 μm
0.17×10^{-3} cm^{-1} at 3.8 μm
Optovac sample
0.78×10^{-3} cm^{-1} at 2.7 μm
0.59×10^{-3} cm^{-1} at 3.8 μm
Raytheon sample:
0.50×10^{-3} cm^{-1} at 2.7 μm
0.60×10^{-3} cm^{-1} at 3.8 μm
0.42×10^{-3} cm^{-1} at 5.25 μm

Reflection Loss
5.6% for two surfaces at 4 μm

Refractive Index
(See Table 6.13.)

Temperature Coefficient of Refractive Index
(See Table 6.13 and 6.14 and Fig. 6.28.)

Dispersion
(See dispersion equation, Table 6.13.)

Zero Dispersion Wavelength
1.7 μm

Dielectric Constant
6.76 for 10^5 Hz

Melting Temperature
1630 K

Thermal Conductivity
39.0 W/(m·K) at 83 K
10 W/(m·K) at 273 K
7.99 W/(m·K) at 373 K

Thermal Expansion
4.8×10^{-6} K^{-1} at 75 K
18.9×10^{-6} K^{-1} at 300 K
36.6×10^{-6} K^{-1} at 900 K

Specific Heat
0.204 [cal/(g·K)] at 273 K
0.306 [cal/(g·K)] at 1700 K

Debye Temperature
510 K

Hardness
160–178 kg/mm^2
(Knoop number with 500-g load in $\langle 110 \rangle$ and $\langle 100 \rangle$ directions)

Bandgap
10 eV

Solubility
0.0017 g/(100 g H$_2$O) at 299 K

Elastic Moduli
Elastic stiffness
$c_{11} = 165$ GPa
$c_{12} = 46$ GPa
$c_{44} = 33.9$ GPa
Elastic compliance
$s_{11} = 6.94$ TPa^{-1}
$s_{12} = -1.53$ TPa^{-1}
$s_{44} = 29.5$ TPa^{-1}
Young's modulus
75.79 GPa
Rupture modulus
0.0365 GPa
Shear modulus
33.76 GPa
Bulk modulus
83.03 GPa at 273 K

Poisson's ratio
0.28

Notes: Calcium fluoride cuts nicely on a diamond saw but is fragile on a diamond fine-grinding wheel. It is difficult to grind; very light "cuts" are therefore recommended. Furthermore, it is practically insoluble in water, acids, and bases but soluble in ammonium solutions and sulfuric acid. This material takes an excellent polish. It is used for prisms, IR cell windows, and as a laser host material. Diameters up to 10 in. are available.

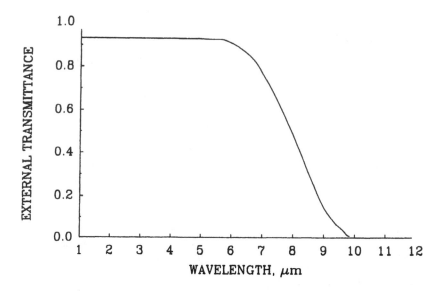

Figure 6.27 The external transmittance of calcium fluoride, thickness 10 mm. [Adapted from R. A. Smith, F. E. Jones, and R. P. Chasmar, *The Detection and Measurement of Infrared Radiation*, Oxford University Press, New York, 1957, p. 341.]

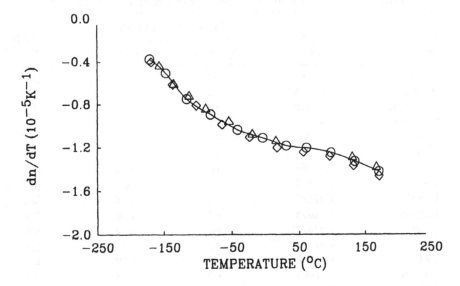

Figure 6.28   The temperature coefficient of calcium fluoride for wavelengths 0.6328 μm (○), 1.15 μm (◇), and 3.39 μm (△). [Adapted from A. J. Glass and A. H. Guenther (Eds.), *Laser Induced Damage in Optical Materials*, 1977, p. 78.]

Table 6.13   Refractive Index of Calcium Fluoride and Residuals × 10^{-5} at 24°C; Also Thermal Coefficient of Index for Mean Temperature of 19°C

λ (μm)	Computed index	Measured differences		$-dn/dt$ × 10^{-6}/°C
		Synthetic	Natural	
0.228803	1.47635	−2	+ 1	6.2
0.24827	1.46793	+3	+ 5	7.0
0.2537	1.46602	+9	+12	7.5
0.26520	1.46233	−1	0	8.1
0.28035	1.45828	−1	+ 1	8.4
0.296728	1.45467	−2	0	8.8
0.334148	1.44852	−1	+ 2	9.2
0.34662	1.44694	−3	0	9.4
0.365015	1.44490	−4	0	9.6
0.4046563	1.44151	−3	+ 1	9.8
0.4358342	1.43949	−3	0	10.0
0.4861327	1.43703	−4	0	10.2
0.546074	1.43494	−3	+ 1	10.4
0.589262	1.43381	−2	+ 2	10.4

(continued)

Table 6.13 Refractive Index of Calcium Fluoride and Residuals $\times\ 10^{-5}$ at 24°C; Also Thermal Coefficient of Index for Mean Temperature of 19°C (Continued)

λ (μm)	Computed index	Measured differences Synthetic	Measured differences Natural	$-dn/dt$ $\times 10^{-6}/°C$
0.643847	1.43268	−2	+ 3	10.4
0.6562793	1.43246	−2	+ 1	10.4
0.6678149	1.43226	−1	+ 1	10.5
0.7065188	1.43167	−2	+ 2	10.5
0.767858	1.43088	−2	+ 2	10.6
0.85212	1.43002	−1	+ 4	10.6
0.8944	1.42966	0	+ 1	10.6
1.01398	1.42879	−2	0	10.5
1.3622	1.42691	+1	+ 8	10.0
1.39506	1.42675	+1	+ 6	9.9
1.52952	1.42612	+4	+ 4	9.6
1.7012	1.42531	+2	+ 4	9.4
1.81307	1.42478	0	+ 9	9.1
1.97009	1.42401	+3	+ 3	8.9
2.1526	1.42306	−1	+ 1	8.7
2.32542	1.42212	+3	+ 4	8.5
2.4374	1.42147	0	+ 2	8.5
3.3026	1.41561	0	+ 3	8.2
3.422	1.41467	+2	+ 2	8.1
3.5070	1.41398	−1	+ 2	8.0
3.7067	1.41229	+2	+ 2	7.8
4.258	1.40713	+4	+ 4	7.5
5.01882	1.39873	+1	+ 5	7.3
5.3034	1.39520	+3	+ 3	7.2
6.0140	1.38539	+5	+5	7.0
6.238	1.38200	−6	0	7.0
6.63306	1.37565	0	+1	6.9
6.8559	1.37186	−8	+2	6.7
7.268	1.36443	+2	+7	6.5
7.4644	1.36070	+5	+6	6.4
8.662	1.33500	−4	+3	6.0
9.724	1.30756	+1	+5	5.6

Dispersion equation:

$$n^2 - 1 = \sum_j \frac{A_j \lambda^2}{\lambda^2 - \lambda_j^2}$$

Constants of the dispersion equation at 24°C:

$\lambda_1 = 0.050263605 \qquad \lambda_1^2 = 0.002526430 \qquad A_1 = 0.5675888$

$\lambda_2 = 0.1003909 \qquad \lambda_2^2 = 0.01007833 \qquad A_2 = 0.4710914$

$\lambda_3 = 34.649040 \qquad \lambda_3^2 = 1200.5560 \qquad A_3 = 3.8484723$

Source: I.H. Malitson, Appl. Opt. 2(11):1103 (1936).

Table 6.14 Experimental Values of Temperature Coefficient of Refractive Index of Calcium Fluoride

	λ (μm)		
T (K)	0.6328	1.15	3.39
310	-11.8 ± 0.7	-12.0 ± 0.5	-11.5 ± 0.7
330	-12.4 ± 0.5	-12.6 ± 0.5	-12.0 ± 0.5
350	-12.7 ± 0.5	-13.1 ± 0.5	-12.4 ± 0.5

Source: H.G. Lipson, Y.F. Tsay, B. Bendow, and P.A. Ligor, *Appl. Opt. 15*(10):2352 (1976).

Calcium Lanthanum Sulfide, CaLa$_2$S$_4$ (90 La$_2$S$_3$:10 CaS)

Specific Gravity
  4.61 [g/cm^3]
Crystal Class
  Cubic
Transmission
  (See Fig. 6.29.)
Absorption Coefficient
  0.1 cm^{-1} at 10.6 μm
Thermal Conductivity
  1.7 W/(m·K) at 300 K
Thermal Expansion
  14.7 × 10^{-6} K^{-1} at 293–673 K
Hardness
  570 kg/mm^2
  (Knoop number)

Elastic Moduli
  Young's modulus
  95 GPa
Notes: Calcium lanthanum sulfide has the Th$_3$P$_4$ cubic crystal structure over a wide range of compositions. It is being investigated for use in windows and domes in the IR band from 8 to 14 μm. Windows are fabricated from powders compressed in a mold and sintered in an H$_2$S atmosphere until a 95% density is reached, with the remaining porosity eliminated by a hot isostatic pressing process. (Private communication, R. W. Tustison, Raytheon.)

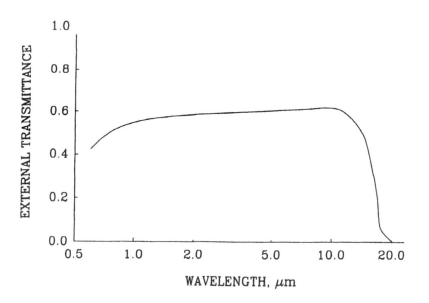

Figure 6.29 The external transmittance of calcium lanthanum sulfide, thickness 2.5 mm. [Adapted from PBN-88-1441, Raytheon Company Research Division, Lexington, Mass., 1985.]

Cesium Bromide, CsBr

Molecular Weight
212.83

Specific Gravity
4.44 [g/cm^3]

Crystal Class
Cubic

Transmission
Long-wavelength limit 55 μm
Short-wavelength limit 0.22 μm
(See Figs. 6.30 and 6.31.)

Reflection Loss
11.6% for two surfaces at 10 μm

Refractive Index
(See Table 6.15.)

Temperature Coefficient of Refractive Index
7.9×10^{-5} K^{-1} at 297–304 K

Dispersion
(See dispersion equation, Table 6.15.)

Zero Dispersion Wavelength
5.3 μm

Dielectric Constant
6.51 for 2×10^6 Hz at 298 K

Melting Temperature
909 K

Thermal Conductivity
0.94 W/(m·K) at 273 K
0.776 W/(m·K) at 367 K

Thermal Expansion
47×10^{-6} K^{-1} at 273 K

Specific Heat
0.063 [cal/(g·K)] at 293 K

Debye Temperature
148K

Hardness
19.5 kg/mm^2
(Knoop number with 200-g load)

Bandgap
6.9 eV

Solubility
124.3 g/(100 g H$_2$O) at 298 K

Elastic Moduli
Elastic stiffness
$c_{11} = 30.7$ GPa
$c_{12} = 8.4$ GPa
$c_{44} = 7.49$ GPa
Elastic compliance
$s_{11} = 36.9$ TPa^{-1}
$s_{12} = -7.9$ TPa^{-1}
$s_{44} = 134$ TPa^{-1}
Young's modulus
15.85 GPa
Rupture modulus
0.0165 GPa

Notes: Like other alkali halide materials, cesium bromide is hygroscopic and must be used in a dry atmosphere. It is also soluble in alcohol and in many acids.

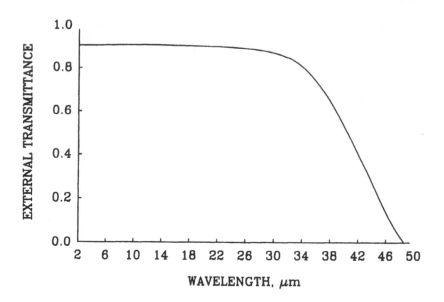

Figure 6.30 The external transmittance of cesium bromide, thickness 10 mm. [Adapted from D. E. McCarthy, *Appl. Opt. 2*: 591 (1963).]

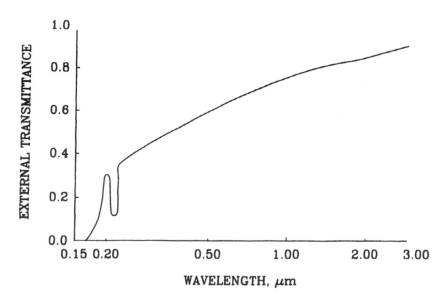

Figure 6.31 The short-wavelength external transmittance of cesium bromide, thickness 10 mm. [Adapted from D. E. McCarthy, *Appl. Opt. 6*: 1896 (1967).]

Table 6.15   Refractive Index of Cesium Bromide as a Function of Wavelength

λ (μm)	n 300 K	n 80 K
2	1.67041	1.68626
3	1.66891	1.68481
4	1.66806	1.68391
6	1.66669	1.68252
7	1.66575	1.68160
8	1.66479	1.68068
10	1.66253	1.67826
12	1.65977	1.67540
13	1.65822	1.67389
14	1.65652	1.67218
16	1.65276	1.66829
18	1.64844	1.66389
20	1.64357	1.65897
22	1.63812	1.65323
24	1.63205	1.64692
26	1.62538	1.64005
28	1.61768	1.63259
30	1.60965	1.62431

Source: A. Selvarajan, J.L. Swedberg, A.G. DeBell, *Appl. Opt. 18*(18):3116 (1979).
Dispersion equation [W.S. Rodney and R.J. Spindler, *J. Res. NBS 51*:123 (1953)]:

$$n^2 = 5.640752 - 0.000003338\lambda^2 + \frac{0.0018612}{\lambda^2} + \frac{41,110.49}{\lambda^2 - 14,390.4} + \frac{0.0290764}{\lambda^2 - 0.024964}$$

The average temperature coefficient of refractive index for two samples of different origin is given by Rodney and Spindler as 7.9×10^{-5} per °C.

Cesium Iodide, CsI

Molecular Weight
259.83
Specific Gravity
4.51 [g/cm^3]
Crystal Class
Cubic
Transmission
Long-wavelength limit 70 μm
Short-wavelength limit 0.24 μm
(See Figs. 6.32 and 6.33.)
Reflection Loss
13.6% for two surfaces at 10 μm
Refractive Index
(See Table 6.16.)
**Temperature Coefficient of
Refractive Index**
(See Fig. 6.34.)
Dispersion
(See dispersion equation, Table 6.16.)
Dielectric Constant
5.65 for 10^6 Hz at 298 K
Melting Temperature
894 K
Thermal Conductivity
1.1 W/(m·K) at 298 K
0.95 W/(m·K) at 360 K
Thermal Expansion
39.1 × 10^{-6} K^{-1} at 75 K

48.3 × 10^{-6} K^{-1} at 293 K
74.7 × 10^{-6} K^{-1} at 850 K
Specific Heat
0.048 [cal/(g·K)] at 293 K
0.073 [cal/(g·K)] at 1150 K
Debye Temperature
93.6 K
Bandgap
6.2 eV
Solubility
44 g/(100 g H$_2$O) at 273 K
Elastic Moduli
Elastic stiffness
$c_{11} = 24.5$ GPa
$c_{12} = 6.6$ GPa
$c_{44} = 6.31$ GPa
Elastic compliance
$s_{11} = 46.1$ TPa^{-1}
$s_{12} = -9.7$ TPa^{-1}
$s_{44} = 158$ TPa^{-1}
Young's modulus
5.30 GPa

Notes: Cesium iodide is hygroscopic. It is mechanically stable and evidences negligible flow or change of shape with time. It is available in ingots of diameters up to 5.5 in. Cesium iodide is soft and difficult to polish. It is used for prisms and windows in spectrophotometers.

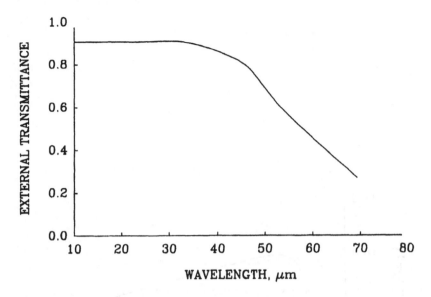

Figure 6.32 The external transmittance of cesium iodide; long-wavelength portion, thickness 5 mm. [Adapted from E. K. Plyler and N. Acquista, *J. Opt. Soc. Am. 43*: 978 (1953) and *48*: 668 (1958).]

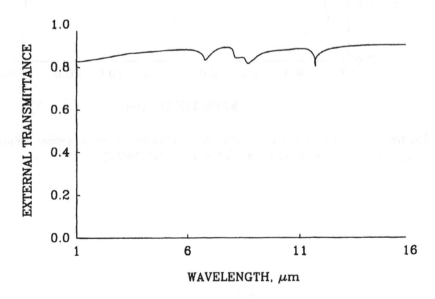

Figure 6.33 The external transmittance of cesium iodide; short-wavelength portion, thickness 3 mm. [Adapted from E. K. Plyler and F. R. Phelps, *J. Opt. Soc. Am. 42*: 432 (1952).]

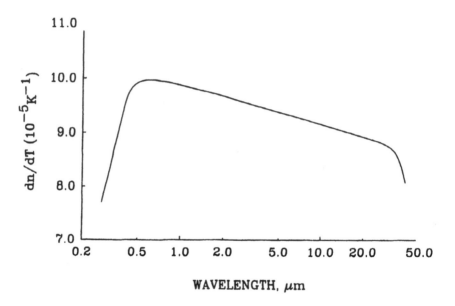

Figure 6.34 The temperature coefficient of refractive index of cesium iodide. [Adapted from W. S. Rodney, *J. Opt. Soc. Am. 45*: 987 (1955).]

Table 6.16 Refractive Index of Cesium Iodide as a Function of Wavelength

λ (μm)	n 300 K	80 K
2	1.74585	1.76317
3	1.74385	1.76141
4	1.74297	1.76037
6	1.74174	1.75924
7	1.74116	1.75868
8	1.74068	1.75811
10	1.73920	1.75661
12	1.73746	1.75483
13	1.73659	1.75389
14	1.73552	1.75285
16	1.73324	1.75068
18	1.73059	1.74778
20	1.72764	1.74476
22	1.72440	1.74144
24	1.72079	1.73777
26	1.71668	1.73351
28	1.71247	1.72915
30	1.70775	1.72438

Dispersion equation:

$$n^2 - 1 = \sum_i \frac{K_i \lambda^2}{\lambda^2 - \lambda_i^2}$$

where the constants for $i = 1, \ldots, 5$ are

i	λ_i^2	K_i
1	0.00052701	0.34617251
2	0.02149156	1.0080886
3	0.032761	0.28551800
4	0.044944	0.39743178
5	25,921.0	3.3605359

Cuprous Chloride, CuCl

Molecular Weight
99.0
Specific Gravity
4.14 [g/cm^3]
Crystal Class
Cubic
Transmission
(See Fig. 6.35.)
Absorption Coefficient
(See Fig. 6.36.)
Refractive Index
(See Fig. 6.37.)
Dielectric Constant
10.0 for 3×10^6 Hz at 293 K
Melting Temperature
695 K
Thermal Expansion
10×10^{-6} K^{-1} at 313–413 K

Hardness
2–2.5 (Mohs)
Bandgap
3.3 eV (excitonic)
Solubility
0.0061 g/(100 g H$_2$O)
Elastic Moduli
Elastic stiffness
$c_{11} = 45.5$ GPa
$c_{12} = 36.3$ GPa
$c_{44} = 14.5$ GPa
Elastic compliance
$s_{11} = 76.1$ TPa^{-1}
$s_{12} = -33.8$ TPa^{-1}
$s_{44} = 69.3$ TPa^{-1}

Notes: Cuprous chloride is hygroscopic. Since it transmits to 20 μm, it serves as an important optical modulator material in the infrared.

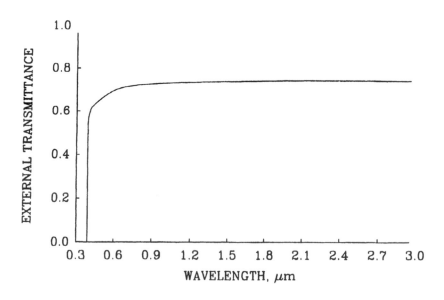

Figure 6.35 The external transmittance of cuprous chloride, thickness 9.1 mm. [Adapted from D. E. McCarthy, *Appl. Opt. 6*: 1896 (1967).]

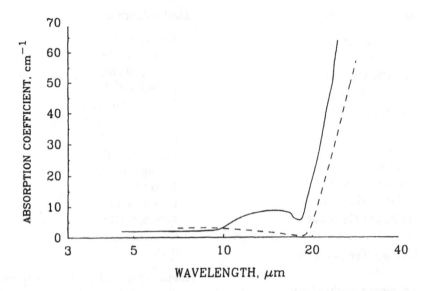

Figure 6.36   Absorption coefficient of cuprous chloride at 300 K for a 90-μm sample (——) and a 200 μm sample (----). [Adapted from P. Alonas et al., *Appl. Opt.* 8: 2557 (1969).]

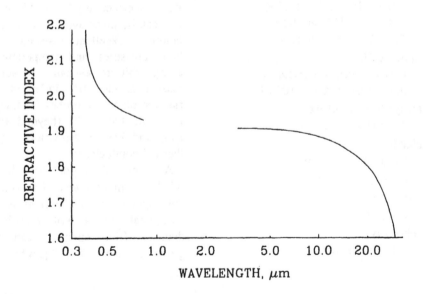

Figure 6.37   Refractive index of cuprous chloride at 300 K. [Adapted from P. Alonas et al., *Appl. Opt.* 8: 2558 (1969).]

Diamond II A, C

Molecular Weight
12.01

Specific Gravity
3.51 [g/cm^3]

Crystal Class
Cubic

Transmission
(See Fig. 6.38.)

Refractive Index
(See Table 6.17.)

Dielectric Constant
5.68 for 1.6×10^6 Hz at 299 K

Melting Temperature
3770 K

Thermal Conductivity
12,000 W/(m·K) at 80 K
2600 W/(m·K) at 273 K

Thermal Expansion
-0.1×10^{-6} K^{-1} at 25 K
0.8×10^{-6} K^{-1} at 293 K
5.8×10^{-6} K^{-1} at 1600 K

Specific Heat
0.124 [cal/(g·K)] at 298 K
0.439 [cal/(g·K)] at 1050 K

Debye Temperature
1860 at 273–1073 K

Hardness
5700–10400 kg/mm^2
(Knoop number)

Bandgap
5.47 eV

Solubility
Insoluble in H_2O

Elastic Moduli
Elastic stiffness
$c_{11} = 1040$ GPa
$c_{12} = 170$ GPa
$c_{44} = 550$ GPa
Elastic compliance
$s_{11} = 1.01$ TPa^{-1}
$s_{12} = -0.14$ TPa^{-1}
$s_{44} = 1.83$ TPa^{-1}
Young's modulus
1050 GPa
Bulk modulus
440–590 GPa
Poisson's ratio
0.16

Notes: There are four main types of diamond: Ia and Ib, which contain dissolved nitrogen (most natural diamonds are type Ia); IIb, which contains dissolved boron; and IIa, which is effectively free of nitrogen and can be made synthetically. Diamond's excellent resistance to high pressures and temperatures along with its resistance to acids make it an ideal IR window material for planetary probes. It transmits into the far-IR (beyond 100 μm) and has an extremely large thermal conductivity.

A chemical vapor deposition (CVD) form of diamond is being developed by various workers as an IR optical material. Polycrystalline diamond films currently can be grown at rates of 1–200 μm/hr.

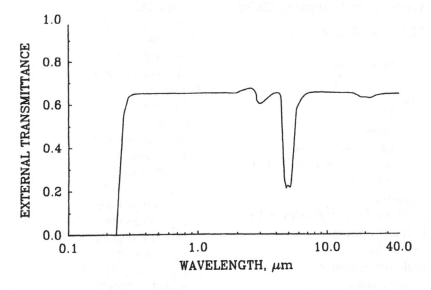

Figure 6.38 The external transmittance of diamond IIA, thickness 1.0 mm. [Adapted from Oriel Corporation catalog.]

Table 6.17 Refractive Index of Diamond IIA

Wavenumber (cm^{-1})	λ (μm)	n
4000	2.5	2.382
3000	3.33	2.3818
2000	5.0	2.3809
1900	5.26	2.3809
1800	5.55	2.3808
1700	5.88	2.3808
1600	6.25	2.3807
1500	6.66	2.3807
1400	7.14	2.3806
1300	7.69	2.3806
1200	8.33	2.3806
1100	9.09	2.3805
1000	10.0	2.3805
943.4	10.6	2.3805
900	11.11	2.3805
880	11.36	2.3805
840	11.90	2.3805
825	12.12	2.3805
800	12.15	2.3805

Source: Oriel Corporation catalog.

Gallium Antimonide, GaSb

Molecular Weight
191.48

Specific Gravity
5.619 [g/cm^3]

Crystal Class
Cubic

Transmission
Long-wavelength limit ~2.5 μm
Short-wavelength limit 2 μm
(See Fig. 6.39.)

Reflection Loss
49.7% for two surfaces at 4 μm

Refractive Index
(See Table 6.18.)

Dielectric Constant
15.69 static
14.44 optical

Melting Temperature
993 K

Thermal Conductivity
43.953 W/(m·K) at 300 K

Thermal Expansion
6.9×10^{-6} K^{-1}

Specific Heat
0.019 [cal/(g·K)]

Hardness
469 kg/mm^2
(Knoop number)

Bandgap
0.72 eV

Solubility
Insoluble in H$_2$O

Elastic Moduli
Elastic stiffness
$c_{11} = 88.49$ GPa
$c_{12} = 40.37$ GPa
$c_{44} = 43.25$ GPa
Elastic compliance
$s_{11} = 15.8$ TPa^{-1}
$s_{12} = -4.95$ TPa^{-1}
$s_{44} = 23.2$ TPa^{-1}
Young's modulus
63.32 GPa
Bulk modulus
56.43 GPa

Notes: Gallium antimonide is similar in composition and preparation to indium antimonide.

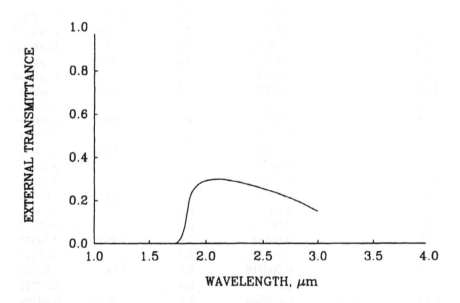

Figure 6.39 The external transmittance of gallium antimonide, thickness 0.66 mm. [Adapted from D. F. Edwards, University of Michigan, Willow Run Laboratories, private communication, 1959.]

Table 6.18 Refractive Index and Extinction Coefficient of Gallium Antimonide*

λ (μm)	n	k	λ (μm)	n	k
1.49		0.0970	1.97		0.000110
1.51		0.0945	2.00	3.789	0.000108
1.53		0.0905			0.0000987
1.55		0.0867	2.03		0.000109
1.56		0.0852	2.07		0.000113
1.57		0.0816	2.1	3.780	
1.58		0.0790	2.2	3.764	
1.59		0.0768	2.3	3.758	
1.60		0.0739	2.4	3.755	0.000143
1.61		0.0710	2.5	3.749	0.000165
1.62		0.0680	2.8		0.000265
1.63		0.0649	3.0	3.898	0.000365
1.64		0.0614	3.4		0.000666
1.65		0.0582	3.5	3.861	0.00746
1.66		0.0547	3.7		0.000925
1.68		0.0510	4.0	3.833	0.00126
1.69		0.0470	4.5		0.00188
1.70		0.0406	5.0	3.824	0.00253
1.71		0.0251	5.4		0.00313
1.72		0.00748	5.8		0.00366
1.73		0.00668	6.0	5.824	0.00394
		0.00214	6.2		0.00422
1.74		0.00305	6.7		0.00490
1.75		0.00183	7.0	3.843	0.00533
1.76		0.00124	7.4		0.00590
1.77		0.000918	8.0	3.843	0.00668
		0.00123	8.4		0.00721
1.80	3.820	0.000540	9.0	3.843	0.00799
		0.000551	9.5		0.00863
1.82		0.000356	10.0	3.843	0.00926
		0.000355	10.6		0.00995
1.84		0.000196	11.1		0.0106
1.85		0.000252	12.0	3.843	0.0116
1.88		0.000200	12.4		0.0121
		0.000141	12.8		0.0126
1.90	3.802		13.4		0.0141
1.91		0.000141	14.0	3.861	0.0140
1.94		0.000118	14.9	3.880	

*Two or more values for one wavelength are from different sources; for identification of all sources the original book should be consulted.

Source: B.O. Seraphin and H.E. Bennett, in *Semiconductors and Semimetals* (R.K. Willardson and A.C. Beer, Eds.), Academic, New York, 1967, p. 525.

Gallium Arsenide, GaAs

Molecular Weight
144.63

Specific Gravity
5.32 [g/cm^3]

Crystal Class
Cubic

Transmission
Long-wavelength limit 16 μm
Short-wavelength limit 1 μm
(See Figs. 6.40 and 6.41.)

Absorption Coefficient
(See Table 6.19.)

Reflection Loss
45% for two surfaces at 12 μm

Refractive Index
(See Table 6.20.)

**Temperature Coefficient of
Refractive Index**
1.47×10^{-4} K^{-1} at 10 μm

Zero Dispersion Wavelength
6.3 μm

Dielectric Constant
12.85 (300 K) static
10.88 (300 K) high frequency

Melting Temperature
1511 K

Thermal Conductivity
55 W/(m·K) at 300 K

Thermal Expansion
0.9×10^{-6} K^{-1} at 75 K
5.7×10^{-6} K^{-1} at 300 K
7.3×10^{-6} K^{-1} at 1000 K

Specific Heat
0.076 [cal/(g·K)] at 273 K

Debye Temperature
360 K

Hardness
721 kg/mm^2
(Knoop number)

Bandgap
1.4 eV

Solubility
Insoluble in H$_2$O

Elastic Moduli
Elastic stiffness
$c_{11} = 118$ GPa
$c_{12} = 53.5$ GPa
$c_{44} = 59.4$ GPa
Elastic compliance
$s_{11} = 11.75$ TPa^{-1}
$s_{12} = -3.66$ TPa^{-1}
$s_{44} = 16.8$ TPa^{-1}
Young's modulus
82.68 GPa
Bulk modulus
75.5 GPa
Poisson's ratio
0.31

Notes: Gallium arsenide finds broad use as an optical material. Single crystals can be grown from the melt by the Czochralski method. Semi-insulating polycrystalline gallium arsenide has been produced by a horizontal Bridgman process to sizes of 8 × 12 × 0.7 in. at Texas Instruments Inc. and is being studied as a material for windows and domes (Klocek et al., *SPIE Proc.* **929**, 1988). Gallium arsenide is being studied for use in photonic switching and waveguides in integrated optics [Alferness and Leonberger, *Optics News*, **14**(2), 46 (1988)].

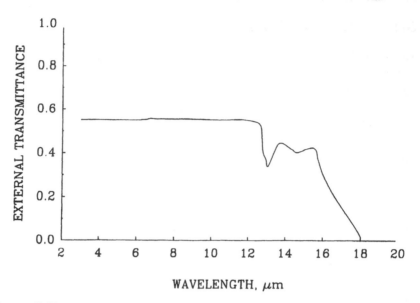

Figure 6.40 The external transmittance of gallium arsenide, thickness 6.48 mm. [Adapted from P. Klocek, L. Stone, M. Boucher, and C. DeMilo, *SPIE Proc. 929*: 65–78 (1988).]

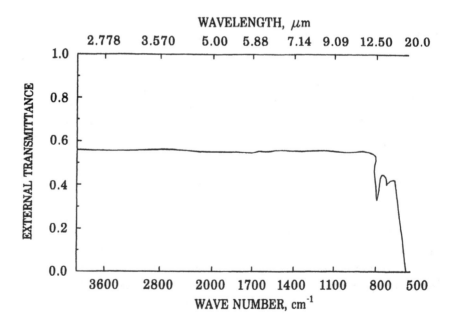

Figure 6.41 The external transmittance of gallium arsenide, thickness 0.5 mm. [Adapted from D. E. McCarthy, *Appl. Opt. 7*: 1997 (1968).]

Table 6.19   Absorption Coefficient of Gallium Arsenide

Wavenumber	λ	Abs. coeff. (cm^{-1})		
(cm^{-1})	(μm)	25°C	200°C	400°C
4,000	2.50	<0.01	<0.01	<0.01
3,000	3.33	<0.01	<0.01	<0.01
2,000	5.00	<0.01	<0.01	0.01
1,900	5.26	<0.01	<0.01	0.01
1,800	5.55	<0.01	<0.01	0.02
1,700	5.88	<0.01	<0.01	0.02
1,600	6.25	<0.01	<0.01	0.02
1,500	6.67	<0.01	<0.01	0.03
1,400	7.14	<0.01	<0.01	0.04
1,300	7.69	<0.01	<0.01	0.03
1,200	8.33	<0.01	<0.01	0.02
1,100	9.09	<0.01	<0.01	0.02
1,000	10.00	<0.01	<0.01	0.05
943.4	10.60	<0.01	<0.01	0.06
900	11.11	<0.01	<0.01	0.09
850	11.76	<0.01	0.02	0.14
825	12.12			
800	12.50	0.08	0.12	0.26

Source: P. Klocek et al., *Proc. 3rd DoD Symp. ElectroMagnetic Windows*, GACIAC PR 89–03, 147 1989.

Table 6.20   Refractive Index of Gallium Arsenide

Wavenumber (cm^{-1})	λ (μm)	Refractive index
4000	2.5	3.3240
3000	3.33	3.310
2000	5.0	3.2975
1900	5.26	3.296
1800	5.55	3.295
1700	5.88	3.294
1600	6.25	3.292
1500	6.66	3.290
1400	7.14	3.288
1300	7.69	3.286
1200	8.33	3.283
1100	9.09	3.280
1000	10.0	3.274
943.4	10.6	3.270
825	12.12	3.262

Source: P. Klocek et al., *SPIE Proc.* 929:65–78 (1988).

Gallium Phosphide, GaP

Molecular Weight
  100.70
Specific Gravity
  4.13 [g/cm^3]
Crystal Class
  Cubic
Transmission
  Long-wavelength limit 11 μm
  Short- wavelength limit 0.6 μm
  (See Fig. 6.42.)
Absorption Coefficient
  (See Table 6.21.)
Reflection Loss
  40% for two surfaces at 2 μm
  (See Fig. 6.43.)
Refractive Index
  (See Table 6.22.)
Temperature Coefficient of
  Refractive Index
  1.00×10^{-4} K^{-1} at 10 μm
Dielectric Constant
  9.036 ± 0.2 at optical frequencies
  10.182 ± 0.2 at zero frequency
Melting Temperature
  1740 K
Thermal Conductivity
  110 W/(m·K) at 300 K
Thermal Expansion
  5.3×10^{-6} K^{-1} at 300 K
  6.0×10^{-6} K^{-1} at 850 K
Specific Heat
  0.20 [cal/(g·K)] at 400 K

Debye Temperature
  446 K
Hardness
  845 kg/mm^2
  (Knoop number)
Bandgap
  2.2 eV (indirect)
Solubility
  Insoluble in the H$_2$O
Elastic Moduli
  Elastic stiffness
  $c_{11} = 142$ GPa
  $c_{12} = 63$ GPa
  $c_{44} = 71.6$ GPa
  Elastic compliance
  $s_{11} = 9.60$ TPa^{-1}
  $s_{12} = -2.93$ TPa^{-1}
  $s_{44} = 14.0$ TPa^{-1}
  Young's modulus
  102.6 GPa
  Poisson's ratio
  0.31

Notes: Gallium phosphide has been produced by a liquid encapsulated Czochralski technique to 2 in. in diameter and 3 in. in length by Texas Instruments, Inc. Due to the broad transmission, hardness, low thermal expansion, and high thermal conductivity of gallium phosphide, it could be useful for windows and domes.

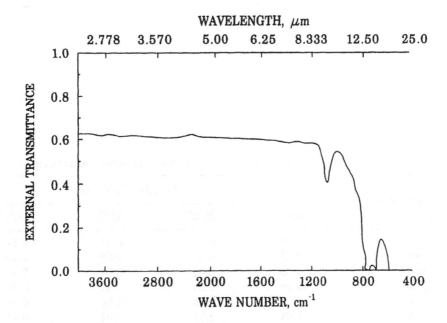

Figure 6.42 The external transmittance of gallium phosphide, thickness 4.67 mm. [Adapted from P. Klocek, L. Stone, M. Boucher, and C. DeMilo, *SPIE Proc. 929*: 65–78 (1988).]

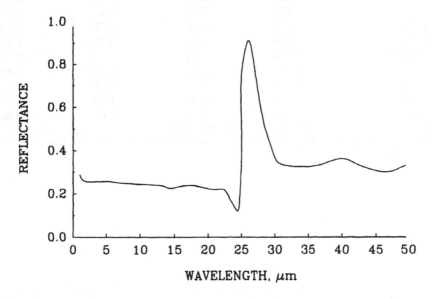

Figure 6.43 The reflectance of gallium phosphide, thickness 1.0 mm. [Adapted from D. E. McCarthy (no date).]

Table 6.21 Absorption Coefficient of Gallium Phosphide Versus Wavelength

Wavenumber (cm^{-1})	λ (μm)	Abs. Coeff. (cm^{-1})		
		25°C	200°C	400°C
4,000	2.50	0.11	0.11	0.14
3,000	3.33	0.07	0.06	0.10
2,000	5.00	0.03	0.04	0.07
1,900	5.26	0.05	0.04	0.07
1,800	5.55	0.05	0.04	0.08
1,700	5.88	0.05	0.04	0.07
1,600	6.25	0.05	0.04	0.08
1,500	6.67	0.06	0.05	0.09
1,400	7.14	0.08	0.09	0.14
1,300	7.69	0.08	0.08	0.13
1,200	8.33	0.09	0.11	0.20
1,100	9.09	0.66	0.77	0.94
1,000	10.00	0.21	0.30	0.67
943.4	10.60	0.44	0.70	1.11
900	11.11	0.62	1.10	1.88
850	11.76	1.08	1.96	3.36
825	12.12	1.44	2.45	4.37
800	12.50	2.53	3.66	5.60

Source: P. Klocek et al., *Proc. 3rd DoD Symp. ElectroMagnetic Windows*, GACIAC PR 89–03, 147, 1989.

Table 6.22 Refractive Index of Gallium Phosphide

Wavenumber (cm^{-1})	λ (μm)	Refractive index
4000	2.5	3.028
3000	3.33	3.010
2000	5.0	3.004
1900	5.26	3.003
1800	5.55	3.001
1700	5.88	2.999
1600	6.25	2.997
1500	6.66	2.994
1400	7.14	2.991
1300	7.69	2.988
1200	8.33	2.981
1100	9.09	2.975
1000	10.0	2.964
943.4	10.6	2.956
900	11.11	2.949
880	11.36	2.946
840	11.90	2.938
800	12.15	2.933

Source: P. Klocek et al., *SPIE Proc. 929*:65–78 (1988).

Gallium Selenide, GaSe

Molecular Weight
148.68
Specific Gravity
5.03 [g/cm^3]
Crystal Class
Hexagonal
Transmission
Long-wavelength limit ~18 μm
Short-wavelength limit ~1 μm
(See Fig. 6.44.)
Absorption Coefficient
0.45 cm^{-1} at 1.06 μm
0.1 cm^{-1} at 1.9 μm
Reflectivity
(See Figs. 6.45 and 6.46.)
Refractive Index
(See Fig. 6.47.)

Dispersion
(See Table 6.23.)
Temperature Coefficient of
Refractive Index
(See Fig. 6.48.)
Dielectric Constant
7.443 (high-frequency limit)
Melting Temperature
1233 K
Notes: Single crystals of gallium selenide can be grown by the Bridgeman–Stockborger method. The material crystallizes in a hexagonally layered structure forming a uniaxial crystal with the optical axis perpendicular to the layers. The nonlinear nature of this material has been used to generate coherent tunable infrared radiation.

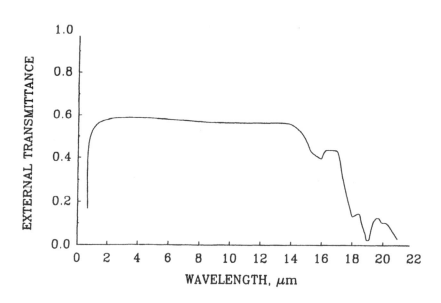

Figure 6.44 The external transmittance of gallium selenide, thickness 0.4 mm. [Adapted from V.L. Cardetta and A.M. Mancini, *Opt. Commun.* 25: 257 (1978).]

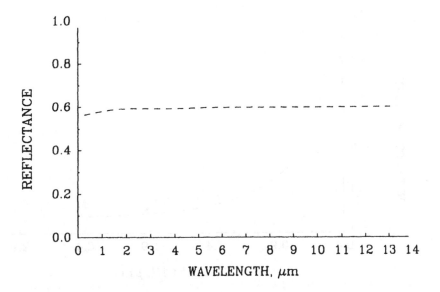

Figure 6.45   The reflectance of gallium selenide. [Adapted from V.L. Cardetta and A.M. Mancini, *Opt. Commun. 25*: 257 (1978).]

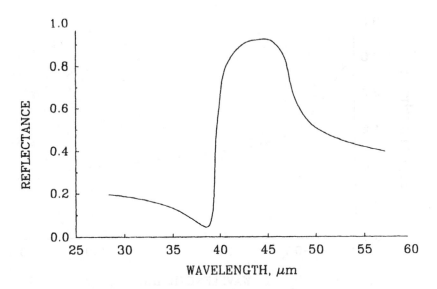

Figure 6.46   Calculated reflectance of gallium selenide at 300 K. [Adapted from N. Piccioli et al., *Appl. Opt. 16*: 5 (1977).]

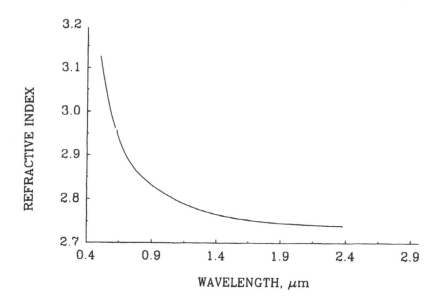

Figure 6.47 Ordinary refractive index of gallium selenide at 300 K. [Adapted from G. Antonioli, D. Bianchi, and P. Franzosi, *Appl. Opt. 18*: 3848 (1979).]

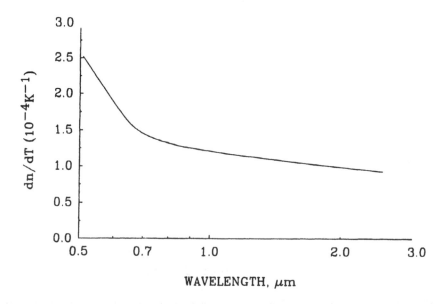

Figure 6.48 Temperature coefficient of the refractive index of gallium selenide. [Adapted from G. Antonioli, D. Bianchi, and P. Franzosi, *Appl. Opt. 18*: 3848 (1979).]

Table 6.23 Coefficients of the Refractive Index Dispersion Representation $n(h\nu)$ $= A + Bh\nu + C(h\nu)^3$ for Gallium Selenide

T (K)	A	B (eV^{-1})	C (eV^{-3})
300	2.716	0.04053	0.01986
275	2.714	0.04055	0.01963
250	2.711	0.04057	0.01940
225	2.709	0.04059	0.01916
200	2.707	0.04078	0.01888
175	2.704	0.04099	0.01859
150	2.702	0.04116	0.01831
125	2.700	0.04139	0.01812
100	2.699	0.04162	0.01793
75	2.697	0.04186	0.01774

Source: G. Antonioli, D. Bianchi, and P. Franzosi, *Appl. Opt.* *18*(22):3848 (1979).

Germanium, Ge

Molecular Weight
72.60

Specific Gravity
5.35 [g/cm^3]

Crystal Class
Cubic

Transmission
Long-wavelength limit 23 μm
Short-wavelength limit 1.8 μm
(See Fig. 6.49.)

Absorption Coefficient
1.3×10^{-3} cm^{-1} at 2.7 μm
1.35×10^{-3} cm^{-1} at 3.8 μm
(Cal. Tech. sample)
(See Fig. 6.50.)

Reflection Loss
52.9% for two surfaces at 10 μm
(See Fig. 6.51.)

Refractive Index
(See Table 6.24.)

**Temperature Coefficient of
Refractive Index**
(See Table 6.25.)

Dispersion
(See dispersion equation, Table 6.24.)

Dielectric Constant
16.6 for 9.37×10^9 Hz at 300 K
15.8 for 10^6 Hz at 77 K

Melting Temperature
1210 K

Thermal Conductivity
165.8 W/(m·K) at 125 K
59 W/(m·K) at 293 K
43.95 W/(m·K) at 400 K

Thermal Expansion
2.4×10^{-6} K^{-1} at 100 K
6.1×10^{-6} K^{-1} at 298 K
8.0×10^{-6} K^{-1} at 1200 K

Specific Heat
0.074 [cal/(g·K)] at 273–373 K
0.095 [cal/(g·K)] at 853 K

Debye Temperature
370 K

Hardness
800 kg/mm^2
(Knoop number)

Bandgap
0.67 eV

Solubility
Insoluble in H$_2$O

Elastic Moduli
Elastic stiffness
$c_{11} = 129$ GPa
$c_{12} = 48$ GPa
$c_{44} = 67.1$ GPa
Elastic compliance
$s_{11} = 9.73$ TPa^{-1}
$s_{12} = -2.64$ TPa^{-1}
$s_{44} = 14.9$ TPa^{-1}
Young's modulus
102.66 GPa
Shear modulus
67.04 GPa
Bulk modulus
77.86 GPa
Poisson's ratio
0.278

Notes: Germanium can be used as an optical material both as a single crystal and in polycrystalline form. It is hard and brittle at room temperature. Germanium has zero cold-water solubility but can be dissolved in aqua regia and hot sulfuric acid. Its transmission is very temperature sensitive, becoming opaque near 100°C.

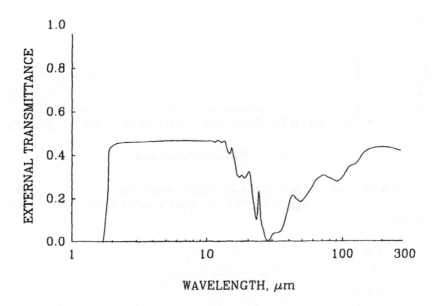

Figure 6.49 The external transmittance of polycrystalline, n-type germanium, thickness 2 mm. [Adapted from Exotic Materials Inc., *Infrared Phys. 5*: (1965).]

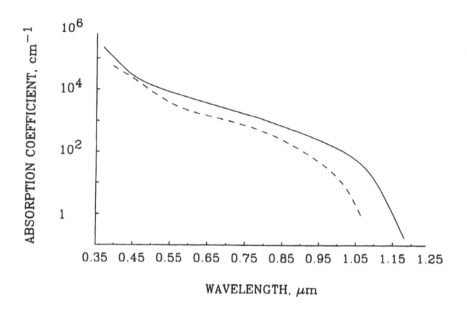

Figure 6.50 The absorption coefficient of germanium, thickness 6 mm. At 300 K (——) and 77 K (----).[Adapted from W.C. Dash and R. Newman, *Phys. Rev. 99*: 1151 (1955).]

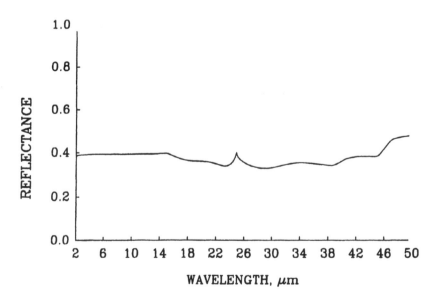

Figure 6.51 The reflectance of germanium, thickness 1.6 mm. [Adapted from D. E. McCarthy, *Appl. Opt. 2*: 591 (1963).]

Table 6.24 Refractive Index of Germanium

λ (μm)	297 K	275 K	204 K	94 K
2.554	4.06230	4.05659	4.02528	3.98859
2.652	4.05754	4.05201	4.01955	3.98462
2.732	4.05310	4.04725	4.01511	3.98052
2.856	4.04947	4.04338	4.01139	3.97720
2.958	4.04595	4.03957	4.00796	3.97390
3.090	4.04292	4.03649	4.00485	3.97100
4.120	4.02457	4.01732	3.98662	3.95334
5.190	4.01617	4.00853	3.97820	3.94536
8.230	4.00743	3.99933	3.96934	3.93720
10.270	4.00571	3.99729	3.96745	3.93597
12.360	4.00627	3.99607	3.96625	3.94026

Source: H.W. Icenogle, B.C. Platt, and W.L. Wolfe, *Appl. Opt. 15*(10), 2349 (1976).
Dispersion equation:

$$n^2 = 1 + \sum_{j=1}^{K} \frac{A_j \lambda^2}{\lambda^2 - B_j \lambda^2}$$

$A_1 = 1.47587446E + 1$ $B_1 = 4.34303403E - 1$

$A_2 = 2.35256294E - 1$ $B_2 = 1.26245893$

$A_3 = -2.48822748E + 1$ $B_3 = 1.30200000E + 3$

Source: B. Tatian, *Appl. Opt. 23*(24), 4481 (1984).

Table 6.25 Temperature Coefficient of the Refractive Index of Germanium

λ (μm)	dn/dT $(°C)^{-1}$
1.934	5.919×10^{-4}*
2.174	5.286×10^{-4}*
2.246	5.251×10^{-4}*
2.401	5.037×10^{-4}*
2.554–12.1	$\sim 3.96 \times 10^{-4}$†

†From H.W. Icenogle, B.C. Platt, and W.L. Wolfe, *Appl. Opt. 15* (10): (1976).
*From D.H. Rank, H.E. Bennett, and D.C. Crouemeyer, *J. Opt. Soc. Am. 44*:12 (1954).

Indium Antimonide, InSb

Molecular Weight
  237
Specific Gravity
  5.78 [g/cm^3]
Crystal Class
  Cubic
Transmission
  Long-wavelength limit 16 μm
  Short-wavelength limit 7 μm
  (See Fig. 6.52.)
Absorption Coefficient
  0.009 cm^{-1} at 10.6 μm
  (See Fig. 6.53.)
Reflection Loss
  53.2% for two surfaces at 10 μm
Refractive Index
  (See Table 6.26.)
Temperature Coefficient of
  Refractive Index
  [Values are for $(1/n)(dn/dt)$.]
  $(3.9 \pm 0.4) \times 10^{-5}$ K^{-1} at 77–400
  K
Dielectric Constant
  17.88 static
  15.68 optical
Melting Temperature
  796 K
Thermal Conductivity
  35.58 W/(m·K) at 293 K
Thermal Expansion
  -0.06×10^{-6} K^{-1} at 10 K

4.89×10^{-6} K^{-1} at 270 K
5.04×10^{-6} K^{-1} at 300 K
Specific Heat
  0.0496 [cal/(g·K)] at 273 K
  0.0562 [cal/(g·K)] at 1050 K
Debye Temperature
  200 K
Hardness
  225 kg/mm^2
  (Knoop number)
Bandgap
  0.17 at 293 K
Solubility
  Insoluble in H$_2$O
Elastic Moduli
  Elastic stiffness
  $c_{11} = 64.72$ GPa
  $c_{12} = 32.65$ GPa
  $c_{44} = 30.71$ GPa
  Elastic compliance
  $s_{11} = 24.6$ TPa^{-1}
  $s_{12} = -8.65$ TPa^{-1}
  $s_{44} = 33.2$ TPa^{-1}
  Young's modulus
  42.79 GPa
  Bulk modulus
  43.27 GPa
Notes: Indium antimonide is soft and
  brittle. To cut it, one may use a di-
  amond wheel with care. Indium an-
  timonide has a large Faraday effect.

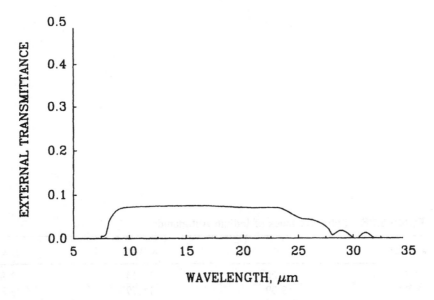

Figure 6.52   The external transmittance of indium antimonide, thickness 1.00 mm. [Adapted from D. E. McCarthy, *Appl. Opt.* 7: 1997 (1968).]

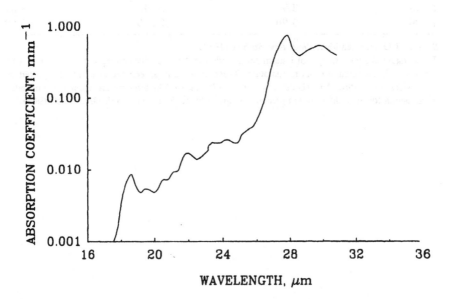

Figure 6.53   Absorption coefficient for indium antimonide at 90 K. [Adapted from R. K. Willardson and A. C. Beer, *Semiconductors Semimetals 3*: 54 (1967).]

Table 6.26 Refractive Index of Indium Antimonide

λ (μm)	n	λ (μm)	n
7.87	4.00	15.13	3.88
8.00	3.99	15.79	3.87
9.01	3.96	16.96	3.86
10.06	3.95	17.85	3.85
11.01	3.93	18.85	3.84
12.06	3.92	19.98	3.82
12.98	3.91	21.15	3.81
13.90	3.90	22.20	3.80

Source: R.G. Breckenridge, *Phys. Rev. 96*:571 (1954).
These values are for a sample of indium antimonide that has a purity corresponding to 2.0×10^{16} carriers/cm^3, measured at room temperature. These data are in agreement with values reported by T. S. Moss, *Proc. Phys. Soc. (Lond.), Ser. B 70:* 776 (1954). The temperature dependence of index of refraction for three different temperatures is given by R. F. Potter, *Appl. Opt. 5*: 35 (1966).

Indium Arsenide, InAs

Molecular Weight
189.73

Crystal Class
Cubic

Transmission
Long-wavelength limit >7.0 μm
Short-wavelength limit 3.8 μm
(See Fig. 6.54.)

Absorption Coefficient
(See Fig. 6.55.)

Reflection Loss
47% for two surfaces at 6 μm

Refractive Index
(See Table 6.27.)

Temperature Coefficient of Refractive Index
[Values are for $(1/n)(dn/dt)$.]
$(6.7–10.3) \times 10^{-5} \, K^{-1}$ (theoretical)

Dielectric Constant
14.55 static
11.8 optical

Melting Temperature
1215 K

Thermal Conductivity
50 W/(m·K) at 280 K

Thermal Expansion
$2.0 \times 10^{-6} \, K^{-1}$ at 42 K
$3.4 \times 10^{-6} \, K^{-1}$ at 77 K
$5.3 \times 10^{-6} \, K^{-1}$ at 273 K

Specific Heat
0.06 [cal/(g·K)] at 273 K
0.068 [cal/(g·K)] at 1200 K

Hardness
330 kg/mm^2 (Knoop number)

Bandgap
0.36 eV

Solubility
Insoluble in H_2O

Elastic Moduli
Elastic stiffness
$c_{11} = 84.4$ GPa
$c_{12} = 46.4$ GPa
$c_{44} = 39.6$ GPa
Elastic compliance
$s_{11} = 19.4$/TPa
$s_{12} = -6.8$/TPa
$s_{44} = 25.3$/TPa
Bulk modulus
57.9 GPa

Notes: Indium arsenide is useful as an IR detector material.

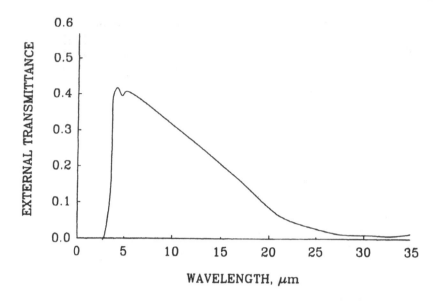

Figure 6.54 The external transmittance of indium arsenide, thickness 0.17 mm. [Adapted from D. E. McCarthy, *Appl. Opt.* 7: 1997 (1968).]

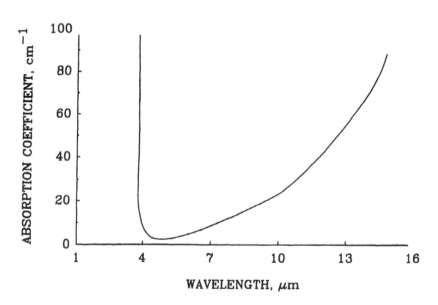

Figure 6.55 Absorption coefficient of indium arsenide, resistivity 1 mohm-cm, apparently with a considerable impurity contribution in the sample measured. [Adapted from F. Oswald and R. Schade, *Z. Naturforsch. 9a*: 611 (1954).]

Table 6.27 Refractive Index and Extinction Coefficient of Indium Arsenide

λ (μm)	n	k	λ (μm)	n	k	λ (μm)	n	k
0.049	1.139	0.168	0.248	1.987	2.647	1.38	3.516	0.047
0.051	1.135	0.195	0.259	2.288	3.086	1.68	....	0.200
0.054	1.135	0.207	0.264	2.617	3.264	1.80	....	0.185
0.056	1.133	0.215	0.269	3.060	3.274	2.00	.....	0.168
0.059	1.31	0.222	0.222	3.800	2.735	2.07	....	0.162
0.062	1.125	0.225	0.310	3.678	1.508	2.25	....	0.149
0.064	1.120	0.224	0.335	3.359	1.340	2.50	....	0.133
0.067	1.110	0.215	0.344	3.271	1.363	2.76	....	0.119
0.070	1.047	0.189	0.354	3.227	1.411	3.00	....	0.102
0.077	0.948	0.272	0.387	3.484	1.547	3.35	....	0.064
0.082	0.894	0.336	0.413	3.331	1.787	3.40	....	0.052
0.089	0.829	0.426	0.443	3.817	1.954	3.44	....	0.037
0.095	0.766	0.563	0.451	3.980	1.865	3.50	....	0.018
0.103	0.745	0.727	0.459	4.087	1.748	3.65	....	0.002
0.108	0.751	0.830	0.468	4.119	1.644	3.74	3.52	
0.112	0.755	0.905	0.477	4.192	1.618	4.00	3.51	
0.123	0.835	1.071	0.496	4.489	1.452	5.00	3.46	
0.136	0.890	1.260	0.517	4.558	1.047	6.67	3.45	
0.153	0.967	1.552	0.563	4.320	0.554	10.0	3.42	
0.172	1.184	1.889	0.620	4.101	0.348	14.3	3.39	
0.180	1.332	1.998	0.689	3.934	0.231	16.7	3.38	
0.188	1.483	2.020	0.775	3.800	0.157	20.0	3.35	
0.195	1.583	2.120	0.885	3.696	0.100	25.0	3.26	
0.211	1.782	2.011	1.03	3.613	0.076	33.3	2.95	
0.225	1.765	2.202	1.24	3.548	0.051			

Source: B.O. Seraphin and H.E. Bennett, in *Semiconductors and Semimetals*, Vol. 3 (R.K. Willardson and A.C. Beer, Eds.), Academic, New York, 1967, p. 535.

Indium Phosphide, InP

Molecular Weight
145.80
Crystal Class
Cubic
Transmission
Long-wavelength limit 14 μm
Short-wavelength limit 1 μm
(See Fig. 6.56.)
Absorption Coefficient
(See Fig. 6.57.)
Reflection Loss
43.2% for two surfaces at 12 μm
(See Fig. 6.58.)
Refractive Index
(See Table 6.28.)
**Temperature Coefficient of
Refractive Index**
[Values are for $(1/n)(dn/dt)$.]
2.7×10^{-5} K^{-1}
Dielectric Constant
12.35 static
9.52 optical
Melting Temperature
1330 K

Thermal Conductivity
70 W/(m·K)
Thermal Expansion
4.5×10^{-6} K^{-1}
Specific Heat
0.073 [cal/(g·K)] at 273 K
Debye Temperature
420 K
Hardness
430 kg/mm^2
(Knoop number)
Bandgap
1.35 eV
Elastic Moduli
Elastic stiffness
$c_{11} = 102$ GPa
$c_{12} = 58$ GPa
$c_{44} = 46$ GPa
Elastic compliance
$s_{11} = 16.4$ TPa^{-1}
$s_{12} = -5.9$ TPa^{-1}
$s_{44} = 21.7$ TPa^{-1}
Notes: Indium phosphide can be prepared by fusing together the pure elements and using a Kyropoulos crystal-growing technique.

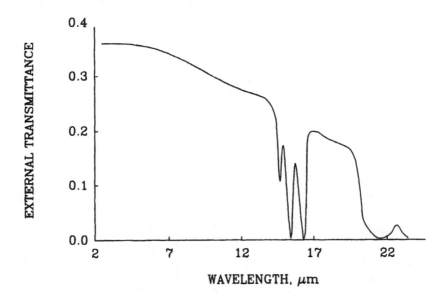

Figure 6.56 The external transmittance of indium phosphide, thickness 1.0 mm. [Adapted from D. E. McCarthy, *Appl. Opt. 7*: 1997 (1968).]

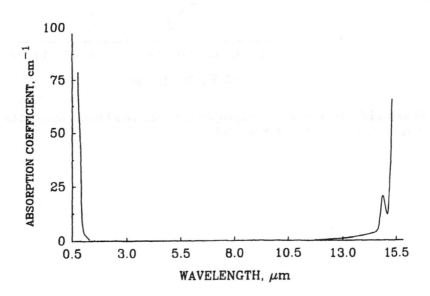

Figure 6.57 Absorption coefficient of indium phosphide, no impurity data given. [Adapted from F. Oswald and R. Schade, *Z. Naturforsch. 9a*: 611 (1954).]

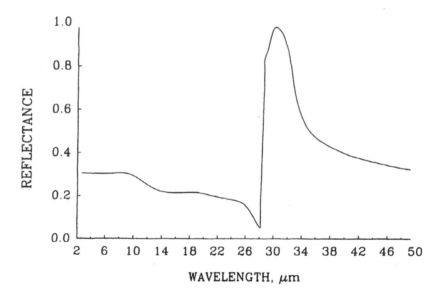

Figure 6.58   The reflectance of indium phosphide, thickness 1.0 mm. [Adapted from D. E. McCarthy, *Appl. Opt. 7*: 1997 (1968).]

Table 6.28   Refractive Index and Extinction Coefficient of Indium Phosphide

λ (μm)	n	k	λ (μm)	n	k	λ (μm)	n	k
0.062	0.793	0.494	0.197	1.526	1.991	0.945	3.374	0.000967
0.064	0.815	0.499	0.200	1.525	1.982	0.950	3.369	0.000527
0.065	0.834	0.493	0.203	1.508	2.005	0.953	.....	0.000281
0.067	0.843	0.487	0.207	1.500	2.063	0.955	3.364	
0.069	0.846	0.477	0.210	1.516	2.130	0.957	.....	0.000145
0.071	0.840	0.469	0.214	1.544	2.191	0.960	3.359	0.0000739
0.073	0.824	0.454	0.217	1.573	2.267	0.965	3.355	0.0000392
0.075	0.785	0.457	0.221	1.616	2.349	0.968	.....	0.0000246
0.077	0.742	0.491	0.225	1.668	2.442	0.970	3.351	
0.079	0.719	0.529	0.229	1.737	2.553	0.972	.....	0.0000166
0.083	0.695	0.574	0.234	1.834	2.675	0.975	3.346	0.0000113
0.085	0.675	0.645	0.238	1.960	2.801	1.00	3.327	
0.089	0.688	0.706	0.243	2.132	2.982	1.10	3.268	
0.092	0.701	0.765	0.248	2.451	3.166	1.20	3.231	
0.095	0.726	0.820	0.253	2.885	3.144	1.30	3.205	
0.099	0.754	0.861	0.258	3.335	3.039	1.40	3.186	
0.103	0.771	0.899	0.264	3.729	2.635	1.50	3.172	
0.108	0.781	0.946	0.269	3.849	2.117	1.60	3.161	
0.113	0.793	0.996	0.275	3.655	1.691	1.70	3.152	
0.118	0.797	1.056	0.282	3.473	1.549	1.80	3.145	
0.124	0.806	1.154	0.288	3.347	1.468	1.90	3.139	
0.125	0.820	1.172	0.295	3.248	1.415	2.00	3.134	
0.126	0.832	1.185	0.302	3.162	1.389	5.00	3.08	
0.128	0.840	1.198	0.310	3.105	1.392	6.00	3.07	
0.129	0.847	1.210	0.318	3.054	1.401	7.00	3.07	
0.130	0.852	1.225	0.326	3.027	1.440	8.00	3.06	
0.132	0.859	1.237	0.335	3.024	1.489	9.00	3.06	
0.133	0.861	1.253	0.344	3.047	1.550	10.00	3.05	
0.135	0.865	1.269	0.354	3.082	1.622	12.00	3.05	0.000527
0.136	0.868	1.287	0.364	3.192	1.747	13.08	.....	0.000667
0.138	0.872	1.304	0.375	3.441	1.857	14.00	3.04	0.000886
0.139	0.874	1.324	0.387	3.835	1.804	14.40	.....	0.00128
0.141	0.875	1.346	0.399	4.100	1.439	14.85	3.03	0.00300
0.142	0.877	1.375	0.413	4.083	1.056	15.00	.....	0.00371
0.144	0.885	1.403	0.427	3.982	0.816	15.24	.....	0.00525
0.146	0.894	1.433	0.443	3.833	0.670	15.32	.....	0.00617
0.148	0.909	1.458	0.459	3.754	0.599	15.45	.....	0.00626
0.149	0.919	1.486	0.477	3.675	0.531	15.52	.....	0.00563
0.151	0.934	1.512	0.496	3.621	0.480	15.74	.....	0.00522
0.153	0.947	1.539	0.517	3.567	0.430	15.85	.....	0.00613

(*continued*)

Table 6.28 Refractive Index and Extinction Coefficient of Indium Phosphide (Continued)

λ (μm)	n	k	λ (μm)	n	k	λ (μm)	n	k
0.155	0.960	1.566	0.539	3.521	0.389	16.00	....	0.00712
0.157	0.973	1.594	0.563	3.472	0.358	16.14	....	0.00746
0.159	0.984	1.627	0.590	3.450	0.334	16.21	....	0.00667
0.161	1.000	1.664	0.620	3.430	0.298	16.28	....	0.00516
0.163	1.022	1.700	0.652	3.410	0.253	16.39	....	0.00333
0.165	1.046	1.736	0.689	.....	0.206	16.55	....	0.00231
0.167	1.072	1.771	0.729	.....	0.176	17.00	....	0.00177
0.170	1.100	1.812	0.775	.....	0.163	18.00	....	0.00181
0.172	1.136	1.847	0.826	.....	0.140	18.93	....	0.00232
0.175	1.174	1.882	0.885	.....	0.0906	19.51	....	0.00384
0.177	1.215	1.915	0.921	.....	0.0571	19.62	....	0.00473
0.180	1.261	1.941	0.925	3.396	0.0355	19.78	....	0.00602
0.182	1.307	1.966	0.928	.....	0.0204	20.00	....	0.00794
0.185	1.354	1.986	0.930	3.390	0.0109	20.19	....	0.00949
0.188	1.402	2.004	0.935	3.385	0.00590	20.34	....	0.0108
0.191	1.453	2.010	0.940	3.379	0.00318	20.42	....	0.0115
0.194	1.496	2.008	0.942	.....	0.00171	20.57	....	0.0130

Source: B.O. Seraphin and H.E. Bennett, in *Semiconductors and Semimetals* (R.K. Willardson and A.C. Beer, Eds.), Academic, New York, 1967, p. 529.

Lanthanum Fluoride, LaF$_3$

Molecular Weight
195.9

Specific Gravity
5.94 [g/cm^3]

Crystal Class
Hexagonal

Transmission
(See Figs. 6.59 and 6.60.)

Reflectivity
(See Fig. 6.61.)

Refractive Index
(See Table 6.29 and Fig. 6.62.)

Dispersion
(See dispersion equation, Table 6.29.)

Melting Temperature
1760 K

Thermal Conductivity
5.1 W/(m·K) at 300 K

Thermal Expansion
11×10^{-6} K^{-1} at 288 K (parallel)
17×10^{-6} K^{-1} at 288 K (perpendicular)

Specific Heat
0.121 [cal/(g·K)]

Hardness
4.5 (Mohs)

Bandgap
6.6 eV

Solubility
Insoluble in H$_2$O

Notes: Lanthanum fluoride exhibits a high degree of thermal and chemical stability. In a 100°C oven for 100 hr, it is insoluble in water and sodium hydroxide and has negligible solubility in sulfuric, hydrochloric, and nitric acids.

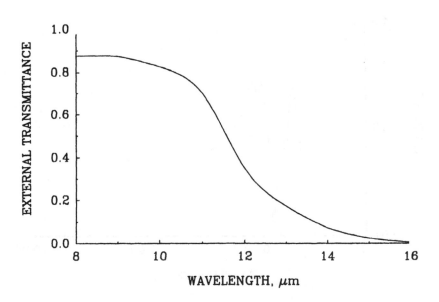

Figure 6.59 The external transmittance of lanthanum fluoride, thickness 11.7 mm. [Adapted from J. B. Mooney, *Infrared Phys.* 6: 153 (1966).]

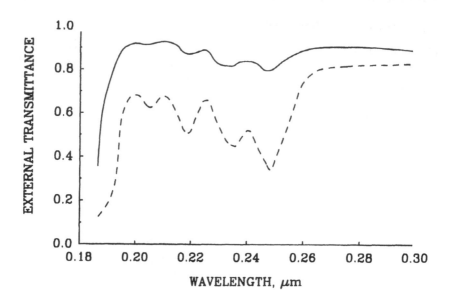

Figure 6.60 The ultraviolet external transmittance of lanthanum fluoride for two thicknesses: (———) 1.3 mm, optical axis perpendicular to light; (---) 8 mm, optical axis parallel to light. [Adapted from M. P. Wirick, *Appl. Opt. 5*: 1966 (1966).]

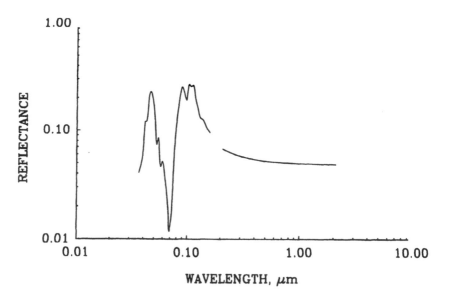

Figure 6.61 The reflectance for the basal plane of hexagonal lanthanum fluoride. [Adapted from D. W. Lynch, *Handbook of Laser Science and Technology*, Vol. IV, *Optical Materialsz:* Part 2, M.J. Weber, Ed., CRC Press, Boca Raton, 1986, p. 214.]

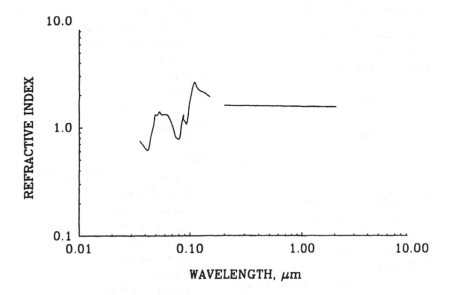

Figure 6.62 Refractive index for the basal plane of hexagonal lanthanum fluoride. [Adapted from D.W. Lynch, *Handbook of Laser Science and Technology,* Vol. IV, Part 2, *Optical Materials,* M.J. Weber, Ed., CRC Press, Boca Raton, Fla., 1986, p. 214.]

Table 6.29 Refractive Index of Lanthanum Fluoride

λ (μm)	n_E	n_O
0.25365	1.64866	1.65587
0.31315	1.61803	
0.36633	1.61803	
0.40465	1.61184	1.61797
0.43583	1.60950	1.61664
0.54607	1.60223	1.60597

Source: Adapted from Wirick, M. P., *Appl. Opt. 5*: 1966 (1966).
Dispersion equations:

$$n_E = 1.58330 + \frac{77.850}{\lambda - 1346.5} \qquad n_o = 1.57376 + \frac{153.137}{\lambda - 686.2}$$

A mean value between n_O and n_E of about 1.58 between 0.8 and 2.0 μm is reported by J. B. Mooney, *Infrared Phys. 6*: 153 (1966).

Lead Fluoride, PbF$_2$

Molecular Weight
  245.21
Specific Gravity
  8.24 [g/cm^3]
Crystal Class
  Cubic
Transmission
  Long-wavelength limit ~17 μm
  Short-wavelength limit 0.25 μm
  (See Fig. 6.63.)
Absorption Coefficient
  0.018 cm^{-1} at 4 μm
  0.1 cm^{-1} at 9.5 μm
  (See Fig. 6.64.)
Reflection Loss
  13.8% for two surfaces at 10 μm
Refractive Index
  (See Table 6.30.)
Zero Dispersion Wavelength
  3.3 μm
Dielectric Constant
  13 for 1×10^6 Hz
Melting Temperature
  1100 K
Thermal Expansion
  29×10^{-6} K^{-1} at 283 K

Specific Heat
  0.072 [cal/(g·K)] at 282 K
Debye Temperature
  219 K
Hardness
  200 kg/mm^2
  (Knoop number)
Bandgap
  5.0 eV
Solubility
  0.064 g/(100 g H$_2$O) at 293 K
Elastic Moduli
  Elastic stiffness
  $c_{11} = 91$ GPa
  $c_{12} = 46$ GPa
  $c_{44} = 23$ GPa
  Elastic compliance
  $s_{11} = 15.34$ TPa^{-1}
  $s_{12} = -4.9$ TPa^{-1}
  $s_{44} = 44.0$ TPa^{-1}

Notes: Lead fluoride has a high index of refraction and specific gravity, which are the reasons for its research interests. Because it is difficult to make of very high optical quality and its low strength and hardness, it is not widely used as an optical component.

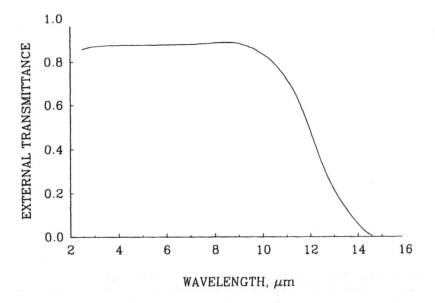

Figure 6.63   The external transmittance of lead fluoride, thickness 5 mm. [Adapted from Optovac, Inc., Bulletin 50, P. V-17, North Brookfield, Mass., 1969.]

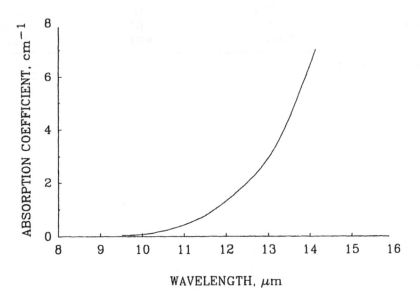

Figure 6.64   The absorption coefficient of lead fluoride. [Adapted from Optovac Inc., Bulletin 50, p. V-17, North Brookfield, Mass., 1969.]

Table 6.30 Refractive Index of Single-Crystal Lead Fluoride

λ (μm)	n	λ (μm)	n
0.30	1.93665	0.90	1.74455
0.40	1.81804	1.00	1.74150
0.50	1.78220	3.00	1.72363
0.60	1.76489	5.00	1.70805
0.70	1.75502	7.00	1.68544
0.80	1.74879	9.00	1.65504

Source: Optovac Inc.

Lead Selenide, PbSe

Molecular Weight
286.17

Specific Gravity
8.10 [g/cm^3] at 288 K

Crystal Class
Cubic

Absorption Coefficient
(See Fig. 6.65.)

Refractive Index
(See Table 6.31.)

Dielectric Constant
206 (low-frequency limit)
22.9 (high-frequency limit)

Melting Temperature
1338 K

Thermal Conductivity
4.2 W/(m·K)

Thermal Expansion
7.7 × 10^{-6} K^{-1} at 303 K

Specific Heat
0.029 [cal/(g·K)] at 25 K
0.042 [cal/(g·K)] at 240 K

Debye Temperature
144 K

Bandgap
0.27 eV

Solubility
Insoluble in H_2O

Notes: Lead selenide can be used in single-crystal or thin-film form. Crystals of lead selenide can be grown by a variety of methods, the most widely used being the Bridgman–Stockman method. Single crystals can be prepared by pulling from the melt (Czochralski technique) or by slow evaporation in a vacuum onto a substrate of the same crystal structure. Also, lead selenide has been useful as an IR detector. Diode lasers of this material have been reported to emit radiation at 6.5 μm.

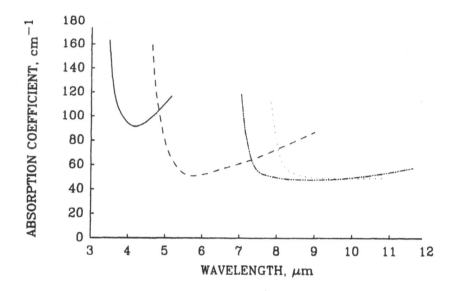

Figure 6.65 The absorption coefficient of lead selenide for several temperatures, thickness 0.68 mm, purity unknown. Temperatures: (——) 590 K, (– – –) 290 K, (—·—) 77 K, (···) 20 K. [Adapted from A. F. Gibson, *Proc. Phys. Soc. 65B*: 378 (1952).]

Table 6.31 Refractive Index of Lead Selenide at 77 K and 300 K

λ (μm)	n 300 K	E (eV)	n 77 K
5.4	4.866	0.11	5.04
6.2	4.828	0.12	5.06
7.0	4.797	0.13	5.10
8.0	4.784	0.14	5.16

Source: B. Jensen and A. Torabi, *IEEE J. Quant. Electronics QE-20*(6): 619 (1984).

Lead Sulfide, PbS

Molecular Weight
  239.28
Specific Gravity
  7.5 [g/cm^3]
Crystal Glass
  Cubic
Transmission
  Long-wavelength limit 7 μm
  Short-wavelength limit 3 μm
  (See Fig. 6.66.)
Absorption Coefficient
  (See Fig. 6.67.)
Reflection Loss
  54% for two surfaces at 3 μm)
Refractive Index
  (See Table 6.32.)
Temperature Coefficient of
  Refractive Index
  6×10^{-4} K^{-1} at 293–573 K
Dielectric Constant
  17.9 for 10^6 Hz at 298 K
Melting Temperature
  1387 K
Thermal Conductivity
  0.67 W/(m·K)

Thermal Expansion
  18×10^{-6} K^{-1}
Specific Heat
  0.050 [cal/(g·K)] at 273 K
Debye Temperature
  194 K
Bandgap
  0.41 eV
Elastic Moduli
  Elastic stiffness
  $c_{11} = 127$ GPa
  $c_{12} = 29.8$ GPa
  $c_{44} = 24.8$ GPa

Notes: Lead sulfide occurs as natural crystals (called galena). It is also available as synthetic crystals and in thin-film form. Crystals of lead sulfide can be grown by a variety of methods, the most widely used technique being the Bridgman–Stockbarger method. Single crystals can be produced by slow evaporation in a vacuum onto a substrate of the same crystal structure. Also, lead sulfide has been useful as an IR detector. Diode lasers of this material have been reported to emit radiation at 4.27 μm.

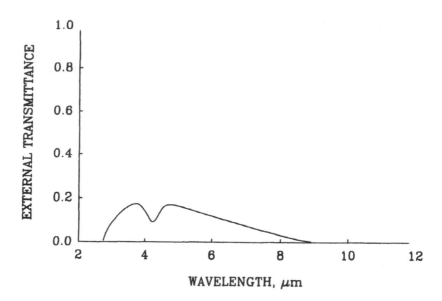

Figure 6.66 The external transmittance of natural crystalline lead sulfide, thickness unknown; no detectable impurity by X-ray analysis. [Adapted from S. S. Ballard, K. A. McCarthy, and W. L. Wolfe, Optical Materials for Infrared Instrumentation, Univ. Michigan Rept. 2389-11S, 101, 1959.]

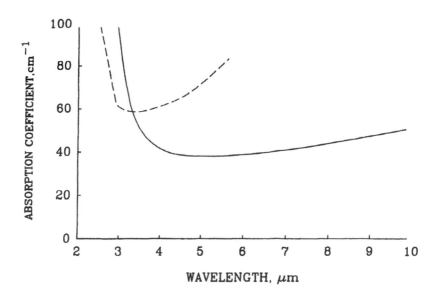

Figure 6.67 Absorption of *n*-type lead sulfide at 576 K (– – –) and 290 K (——), purity unknown. [Adapted from A.F. Gibson, *Proc. Phys. Soc. (Lond.) 65B*: 378 (1952).

Table 6.32 Refractive Index of Lead Sulfide at 77 K and 300 K

λ (μm)	n 300 K	E (eV)	n 77 K
3.10	4.30	0.15	4.25
3.54	4.20	0.20	4.35
4.13	4.13	0.25	4.40
4.96	4.10	0.27	4.48

Source: B. Jensen and A. Torabi, *IEEE J. Quant. Electronics QE-20*(6): 619 (1984).

Lead Telluride, PbTe

Molecular Weight
  334.79
Specific Gravity
  8.16 [g/cm^3]
Crystal Class
  Cubic
Transmission
  (See Fig. 6.68.)
Refractive Index
  (See Table 6.33.)
Dielectric Constant
  380 (low-frequency limit)
  28.5 (high-frequency limit)
Melting Temperature
  1190 K
Thermal Conductivity
  2.4 W/(m·K) at 273 K
Thermal Expansion
  9.0×10^{-6} K^{-1} at 303 K
Specific Heat
  0.013 [cal/(g·K)] at 25 K
  0.036 [cal/(g·K)] at 240 K
Debye Temperature
  128 K
Bandgap
  0.34 eV

Solubility
  Insoluble in H$_2$O
Elastic Moduli
  Elastic stiffness
  $c_{11} = 107.5$ GPa
  $c_{12} = 8.1$ GPa
  $c_{44} = 13.2$ GPa
  Elastic compliance
  $s_{11} = 9.42$ TPa^{-1}
  $s_{12} = -0.66$ TPa^{-1}
  $s_{44} = 75.9$ TPa^{-1}

Notes: Lead telluride can be used in synthetic single-crystal or thin-film form. Crystals of lead telluride can be grown by a variety of methods, the most widely used being the Bridgman–Stockman method. Single crystals can be prepared by pulling from the melt (Czochralski technique) or by slow evaporation in a vacuum onto a substrate of the same crystal structure. Also, lead telluride has been useful as an IR detector. Diode lasers of this material have been reported to emit radiation at 6.5 μm.

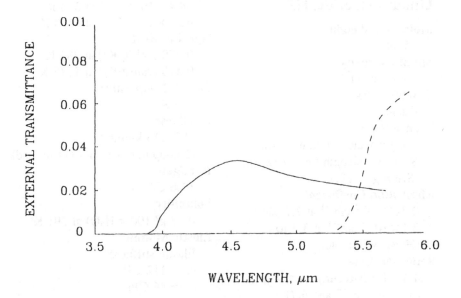

Figure 6.68 The external transmittance of lead telluride for 296 K (– – –) and 79 K (———), thickness 0.11 mm. [Adapted from M. A. Clark and R. J. Cashman, *Phys. Rev.* *85*: 1043 (1952).]

Table 6.33 Refractive Index of Lead Telluride at 77 K and 300 K

λ (μm)	*n* 300 K	E (eV)	*n* 77 K
4.5	5.878	0.12	5.95
5.4	5.798	0.14	6.00
7.0	5.734	0.16	6.05
8.0	5.723	0.18	6.12

Source: B. Jensen and A. Torabi, *IEEE J. Quant. Electronics QE-20*(6), 619 (1984).

Lithium Fluoride, LiF

Molecular Weight
25.94

Specific Gravity
2.635 [g/cm^3]

Crystal Class
Cubic

Transmission
Long-wavelength limit 9 μm
Short-wavelength limit 0.12 μm
(See Fig. 6.69.)

Absorption Coefficient
0.74×10^{-3} cm^{-1} at 2.7 μm
2.1×10^{-3} cm^{-1} at 3.8 μm
(Harshaw sample)

Reflection Loss
4.4% for two surfaces at 4 μm
(See Figs. 6.70 and 6.71.)

Refractive Index
(See Table 6.34.)

**Temperature Coefficient of
Refractive Index**
(See Fig. 6.72.)

Dispersion
(See dispersion equation, Table 6.34.)

Zero Dispersion Wavelength
1.3 μm

Dielectric Constant
9.0 for 10^2–10^{10} Hz at 298 K
9.1 for 10^2–10^{10} Hz at 353 K

Melting Temperature
1140 K

Thermal Conductivity
11.3 W/(m·K) at 314 K

Thermal Expansion
5.5×10^{-6} K^{-1} at 75 K

34.4×10^{-6} K^{-1} at 300 K
66.7×10^{-6} K^{-1} at 1100 K

Specific Heat
0.370 [cal/(g·K)] at 283 K
0.598 [cal/(g·K)] at 1175 K

Debye Temperature
732 K

Hardness
102–113 kg/mm^2
(Knoop number with 600-g load)

Bandgap
13.6 eV

Solubility
0.27 g/(100 g H_2O) at 291 K

Elastic Moduli
Elastic stiffness
$c_{11} = 112$ GPa
$c_{12} = 46$ GPa
$c_{44} = 63.5$ GPa
Elastic compliance
$s_{11} = 11.6$ TPa^{-1}
$s_{12} = -3.35$ TPa^{-1}
$s_{44} = 15.8$ TPa^{-1}
Young's modulus
64.77 GPa
Rupture modulus
137.8 GPa
Shear modulus
55.12 GPa
Bulk modulus
62.0 GPa

Notes: When lithium fluoride is grown in a vacuum, the absorption at 2.8 μm attributed to the H–F band disappears. Cylindrical castings of diameter 6 in. are available. Lithium fluoride is only slightly soluble in water but can be dissolved in acids.

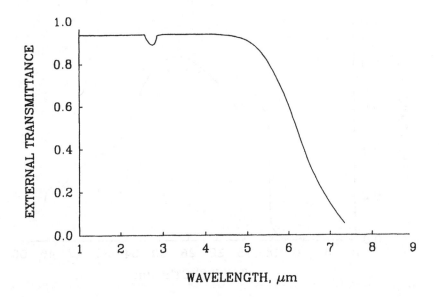

Figure 6.69 The external transmittance of lithium fluoride, thickness 12 mm. [Adapted from R. A. Smith, F. E. Jones, and R. P. Chasmar, *The Detection and Measurement of Infrared Radiation*, Oxford University Press, New York, 1957.]

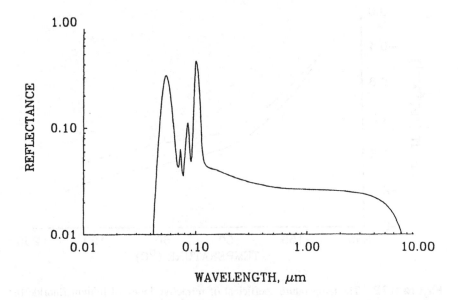

Figure 6.70 The reflectance for lithium fluoride calculated for normal incidence. No effect from a second surface was considered. [Adapted from M.J. Weber (Ed.), *Handbook of Laser Science and Technology*, Vol. IV, Part 2, CRC, Boca Raton, Fla., 1986, p. 212.]

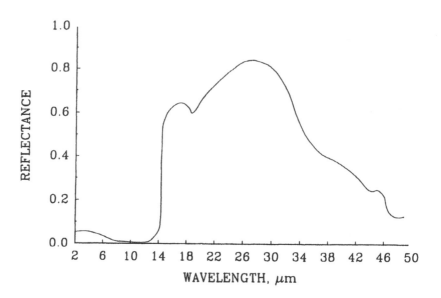

Figure 6.71   The long-wavelength reflectance of lithium fluoride, thickness 5 mm. [Adapted from D. E. McCarthy, *Appl. Opt. 2*: 591 (1963).]

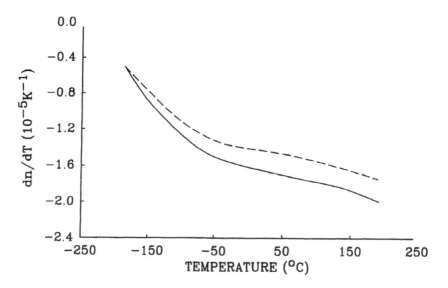

Figure 6.72   The temperature coefficient of refractive index of lithium fluoride for wavelengths 3.39 μm (− − −) and 1.15 μm (——). [Adapted from A.J. Glass and A.H. Guenther (Eds.), *Laser Induced Damage in Optical Materials*, 1977, p. 80.]

Table 6.34 Refractive Index of Lithium Fluoride

λ (μm)	n	λ (μm)	n	λ (μm)	n
0.1935	1.4450	0.366	1.40121	4.50	1.33875
0.1990	1.4413	0.391	1.39937	5.00	1.32661
0.2026	1.4390	0.4861	1.39480	5.50	1.31287
0.2063	1.4367	0.50	1.39430	6.00	1.29745
0.2100	1.4346	0.80	1.38896	6.91	1.260
0.2144	1.4319	1.00	1.38711	7.53	1.239
0.2194	1.4300	1.50	1.38320	8.05	1.215
0.2265	1.4268	2.00	1.37875	8.60	1.190
0.231	1.4244	2.50	1.37327	9.18	1.155
0.254	1.41792	3.00	1.36660	9.79	1.109
0.280	1.41188	3.50	1.35868		
0.302	1.40818	4.00	1.34942		

Data at a temperature of 20°C for wavelengths 0.193–0.231 μm are taken from Z. Gyulai, *Z. Physik* *46*: 84 (1927); at 20°C for 0.254–0.486 μm are taken from H. Harting, *Sitzber. Deut. Akad. Wiss. Berlin IV*: 1–25 (1948); at 23.6°C for 0.50–6.0 μm from L. W. Tilton and E. K. Plyler, *J. Res. NBS 47*: 25 (1951); at 18°C for 6.91–9.79 μm from H. W. Hohls, *Ann. Physik 29*: 433 (1937). The data for the four spectral regions reported here fit together to within a few parts in the fourth decimal place.

For computational purposes, a dispersion equation for the wavelength range 0.5–6.0 μm is given by M. Herzberger and C. D. Salzberg, *J. Opt. Soc. Am. 52*: 420 (1962):

$$n = A + BL + CL^2 + D\lambda^2 + E\lambda^4$$

where $L = \dfrac{1}{\lambda^2 - 0.028}$

$A = 1.38761$
$B = 0.001796$
$C = -0.000041$
$D = -0.0023045$
$E = -0.00000557$

Lithium Niobate, $LiNbO_3$

Molecular Weight
147.9
Specific Gravity
4.644 [g/cm^3]
Crystal Class
Trigonal
Transmission
(See Fig. 6.73.)
Refractive Index
(See Tables 6.35 and 6.36 and Fig.
6.74.)
Temperature Coefficient of
Refractive Index
(See Table 6.36.)
Dispersion
(See dispersion equations, Table
6.35.)
Dielectric Constant
29
Melting Temperature
1530 K
Thermal Conductivity
5.6 W/(m·K) at 300 K
Thermal Expansion
4.1×10^{-6} K^{-1} at 300 K (parallel)
15×10^{-6} K^{-1} at 300 K (perpendicular)
Specific Heat
0.15 [cal/(g·K)]
Hardness
~5 (Mohs)

Bandgap
4.0 eV
Solubility
Insoluble in H_2O
Elastic Moduli
Elastic stiffness
$c_{11} = 202$ GPa
$c_{12} = 55$ GPa
$c_{13} = 71$ GPa
$c_{14} = 8.3$ GPa
$c_{33} = 242$ GPa
$c_{44} = 60.1$ GPa
Elastic compliance
$s_{11} = 5.81$ TPa^{-1}
$s_{12} = -1.15$ TPa^{-1}
$s_{13} = -1.36$ TPa^{-1}
$s_{14} = -0.96$ TPa^{-1}
$s_{33} = 4.94$ TPa^{-1}
$s_{44} = 16.9$ TPa^{-1}

Notes: Lithium niobate is an important electrooptic and ferroelectric material. Among electrooptic materials, the photorefractive effects have been investigated most extensively in lithium niobate. Single crystals of the material can be grown from a stoichiometric melt. Lithium niobate is an important material in integrated optics because of its favorable combination of optical and electrooptical properties. See *Optics News, 14*(2), 1988.

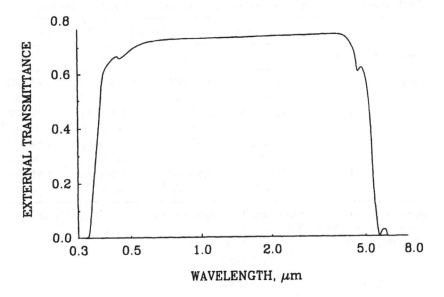

Figure 6.73   The external transmittance of lithium niobate, thickness 6 mm. [Adapted from K. Nassau et al., *J. Phys. Chem. Solids 27*: 989–996 (1966).]

Table 6.35   Refractive Indexes of Congruently Melting Lithium Niobate at 20°C

λ (μm)	n_O	n_E
0.43584	2.39276	2.29278
0.54608	2.31657	2.22816
0.63282	2.28647	2.20240
1.1523	2.2273	2.1515
3.3913	2.1451	2.0822

Sellmeier dispersion equation:

$$n_o^2 = 4.9048 - \frac{0.11768}{0.04750 - \lambda^2} - 0.027169\,\lambda^2$$

$$n_E^2 = 4.5820 - \frac{0.099169}{0.044432 - \lambda^2} - 0.021950\,\lambda^2$$

where λ is μm.

Source: D. S. Smith, H. D. Riccus, and R. P. Edwin, *Opt. Commun. 17*(3): 333 (1976).

Table 6.36 Temperature-Dependent Refractive Indexes of Lithium Niobate

λ (μm)	Polarization	$n = n_o [1 + a_1 T + a_2 T^2 + a_3 T^3 + a_4 T^4$			
		a_1	a_2	a_3	a_4
0.633	O	1.898802×10^{-6}	1.814362×10^{-8}	$-2.545860 \times 10^{-11}$	1.851642×10^{-14}
	E	1.714651×10^{-5}	5.330995×10^{-8}	$-7.023143 \times 10^{-11}$	5.590826×10^{-14}
3.39	O	1.302416×10^{-7}	1.031300×10^{-8}	$-1.316482 \times 10^{-11}$	8.880433×10^{-15}
	E	1.384239×10^{-5}	4.564947×10^{-8}	$-5.744349 \times 10^{-11}$	4.912069×10^{-14}

Source: D. S. Smith, H. D. Riccus, and R. P. Edwin, *Opt. Commun. 17*(3): 333 (1976).
n_0 = refractive index at 0°C
T = temperature

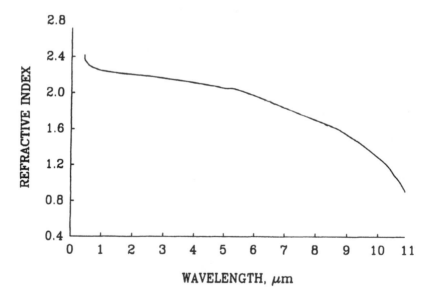

Figure 6.74 The refractive index of lithium niobate. [Adapted from A. S. Barker, Jr. and R. Loudon, *Phys. Rev. 158*: 433–445 (1967).]

Magnesium Aluminate (Spinel), $MgAl_2O_4$

Molecular Weight
356.74
Specific Gravity
3.55 [g/cm^3]
Crystal Class
Cubic
Transmission
Long-wavelength limit 6 μm
Short- wavelength limit 0.21 μm
(See Fig. 6.75.)
Refractive Index
(See Table 6.37.)
Dielectric Constant
8.0–9.0
Melting Temperature
2320 K
Thermal Conductivity
58.5 W/(m·K) at 80.49 K
13.8 W/(m·K) at 308 K
8.5 W/(m·K) at 773 K
Thermal Expansion
5.9×10^{-6} K^{-1} at 313 K
11.7×10^{-6} K^{-1} at 2000 K
Specific Heat
0.26 [cal/(g·K)] at 441 K

Hardness
1140 kg/mm^2 (Knoop number with 1000-g load)
Solubility
Insoluble in H_2O
Elastic Moduli
Elastic stiffness
$c_{11} = 282$ GPa
$c_{12} = 154$ GPa
$c_{44} = 154$ GPa
Elastic compliance
$s_{11} = 5.80$ TPa^{-1}
$s_{12} = -2.05$ TPa^{-1}
$s_{44} = 6.49$ TPa^{-1}

Notes: Spinel can be grown by the flame fusion (Verneuil) process. Spinel is somewhat softer than sapphire and is thus more easily worked for optical purposes. For applications where optical isotropy is desired, the advantages of cubic spinel over trigonal sapphire are obvious. An absorption band at 2.8 μm may be observed due to entrapped water. The optical properties of spinel are discussed by K. A. Wickersheim and R. A. Lefever, *J. Opt. Soc. Am.* **50**:831 (1960). Also see data sheet for Raytran spinel (Raytheon).

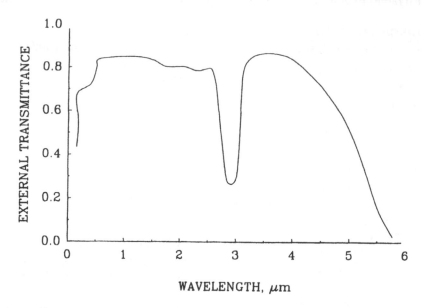

Figure 6.75   The external transmittance of spinel, thickness 5.45 mm. [Adapted from G. Calingaert et al., *Trans. Soc. Automot. Eng. 39*: 448 (1936).]

Table 6.37   Refractive Index of Spinel

λ (μm)	n
0.4861	1.736
0.5893	1.727
0.6563	1.724

Source: Linde Air Products Company technical data sheets.

Magnesium Aluminate (Raytran Spinel), $MgAl_2O_4$

Specific Gravity
3.57 [g/cm³]

Crystal Class
Cubic

Transmission
(See Fig. 6.76.)

Absorption Coefficient
(See Fig. 6.77.)

Refractive Index
1.715 at 0.589 μm
1.62 at 4 μm

Melting Point
2408 K

Thermal Conductivity
14.6 W/(m·K) at 300 K

Thermal Expansion
6.5×10^{-6} K^{-1} at 303–473 K

Specific Heat
0.248 [cal/(g·K)]

Hardness
1645 kg/mm² (Knoop number)

Elastic Moduli
Young's modulus
193 GPa
Poisson's ratio
0.26

Notes: Raytran spinel is an extremely durable material that is fabricated by using a Raytheon-developed powder-processing technique. The material has the isotropic cubic crystal structure and has better transmission than sapphire in the 3–5-μm range. Windows up to 5 in. in diameter and 0.5 in. thick and domes up to 5 in. in diameter can be made. Rods and tubes are also available. See also Spinel. Spinel is available from Alpha Optical Systems, Inc.

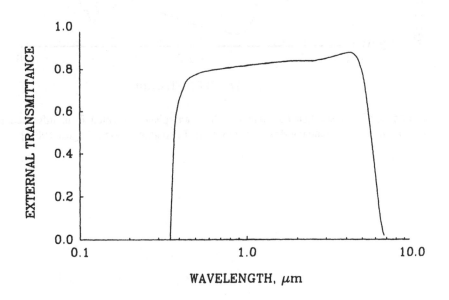

Figure 6.76 The external transmittance of Raytran spinel, thickness 2.58 mm. [Adapted from PBN-84-745, Raytheon Company Research Division, Lexington, Mass., August 1987.]

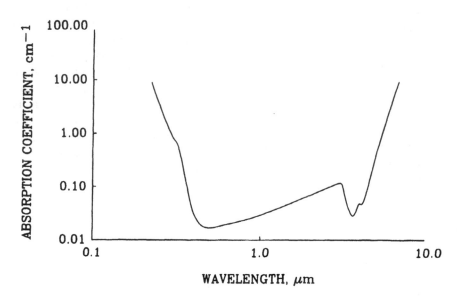

Figure 6.77 Absorption coefficient of Raytran spinel. [Adapted from information given by Raytheon Company Research Division, Lexington, Mass., August 1987.]

Magnesium Fluoride, MgF$_2$

Molecular Weight
62.31
Specific Gravity
3.18 [g/cm^3]
Crystal Class
Tetragonal
Transmission
Long-wavelength limit 7.5 μm
Short-wavelength limit 0.11 μm
(See Fig. 6.78.)
Absorption Coefficient
40.0×10^{-3} cm^{-1} at 2.7 μm
5.4×10^{-3} cm^{-1} at 3.8 μm
(See Fig. 6.79.)
Reflection Loss
4.8% for two surfaces in the visible
Refractive Index
(See Table 6.38 and Fig. 6.80.)
Temperature Coefficient of Refractive Index
1.9×10^{-6} K^{-1} at 0.7065 μm for ordinary ray
1.0×10^{-8} K^{-1} at 0.7065 μm for extraordinary ray
Dispersion
(See dispersion equation, Table 6.37.)
Zero Dispersion Wavelength
1.4 μm
Dielectric Constant
4.87 (parallel)
5.45 for 95×10^3–42×10^6 Hz (perpendicular)
Melting Temperature
1528 K
Thermal Conductivity
21 W/(m·K) at 300 K
Thermal Expansion
14×10^{-6} K^{-1} at 310 K (parallel)
8.9×10^{-6} K^{-1} at 310 K (perpendicular)
Specific Heat
0.24 [cal/(g·K)] at 298 K
0.362 [cal/(g·K) at 1700 K
Hardness
415 kg/mm^2
(Knoop number)
Bandgap
10.8 eV
Solubility
Insoluble in H$_2$O
Elastic Moduli
Elastic stiffness
$c_{11} = 140.2$ GPa
$c_{12} = 89.5$ GPa
$c_{44} = 56.7$ GPa
$c_{13} = 62.9$ GPa
$c_{33} = 204.7$ GPa
$c_{66} = 95.7$ GPa
Elastic compliance
$s_{11} = 12.5$ TPa^{-1}
$s_{12} = -7.2$ TPa^{-1}
$s_{44} = 17.6$ TPa^{-1}
$s_{13} = -1.61$ TPa^{-1}
$s_{33} = 5.9$ TPa^{-1}
$s_{66} = 10.4$ TPa^{-1}

Notes: Magnesium fluoride in single-crystal form occurs in nature as the mineral sellaite. It is a hard material with high thermal and mechanical shock resistance. This material is birefringent, has a low refractive index, and is used as an IR polarizer. Diameters up to 6 in. are available. As a polycrystalline optical material, it has been commercially produced as Irtran 1 by Eastman Kodak Co. In this form magnesium fluoride apparently does not exhibit double refraction. (See also data sheet on Irtran 1.)

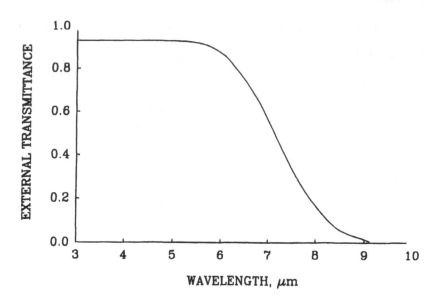

Figure 6.78   The external transmittance of magnesium fluoride, thickness 10.0 mm. [Adapted from Optovac, Inc., Bulletin 50, North Brookfield, Mass., 1969, p. V-19.]

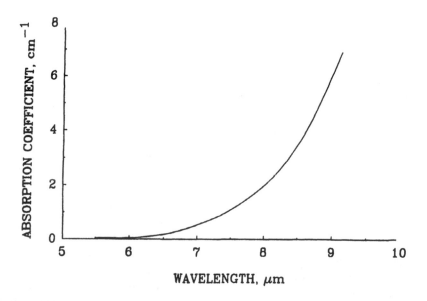

Figure 6.79   Absorption coefficient of magnesium fluoride. [Adapted from Optovac, Inc., Bulletin 50, North Brookfield, Mass., 1969, p. V-19.]

Table 6.38 Refractive Index of Magnesium Fluoride at 19.5°C

λ (μm)	n_E	n_O
0.21386	1.42897	1.41566
0.22675	1.42251	1.40942
0.24827	1.41615	1.40329
0.25763	1.41382	1.40106
0.27528	1.40967	1.39707
0.29673	1.40592	1.39345
0.31315	1.40364	1.39124
0.33415	1.40116	1.38889
0.36501	1.39834	1.38614
0.40466	1.39567	1.38359
0.43584	1.39048	1.38208
0.46782	1.39276	1.38082
0.47999	1.39232	1.38040
0.50858	1.39142	1.37954
0.54607	1.39043	1.37859
0.58756	1.38955	1.37774
0.64385	1.38858	1.37682
0.66781	1.38823	1.37649
0.79476	1.38679	1.37512
1.0830	1.38465	1.37307

Dispersion equation:
$n_O = 1.36957 + 0.0035821 (\lambda - 0.14925)^{-1}$
$n_E = 1.38100 + 0.0037415 (\lambda - 0.14947)^{-1}$
Note that the wavelength must be given in angstroms in this equation.
Source: W. L. Wolfe and G. J. Zissis, *Infrared Handbook*, Office of Naval Research, Dept. of the Navy, Washington, D.C., 1978, p. 7–68.

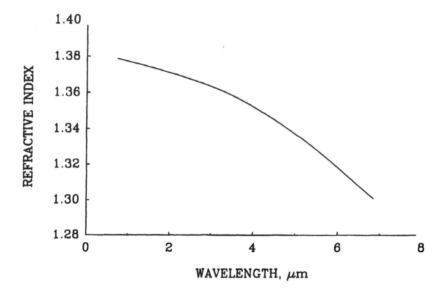

Figure 6.80 The refractive index for magnesium fluoride. [Adapted from Optovac, Inc., Bulletin 50, North Brookfield, Mass., 1969, p. V-19.]

Magnesium Fluoride (Irtran 1), MgF_2

Specific Gravity
3.18 [g/cm^3]
Transmission
Long-wavelength limit ~8.5 μm
Short-wavelength limit ~1 μm
(See Fig. 6.81.)
Absorption Coefficient
0.31 cm^{-1} at 3 μm
0.19 cm^{-1} at 4 μm
0.24 cm^{-1} at 5 μm
Reflection Loss
47.3% for two surfaces at 4.3 μm
(See Fig. 6.82.)
Refractive Index
(See Table 6.39.)
Dispersion
(See dispersion equation, Table 6.39.)
Dielectric Constant
5.1 for 8.5–12.0 × 10^9 Hz
Melting Temperature
1528 K
Thermal Conductivity
14.65 W/(m·K) at 329 K
12.98 W/(m·K) at 381 K
10.88 W/(m·K) at 452 K
Thermal Expansion
0.06 × 10^{-6} K^{-1} at 70 K
7.13 × 10^{-6} K^{-1} at 200 K
10.4 × 10^{-6} K^{-1} at 300 K
Specific Heat
0.12 [cal/(g·K)] at 298 K

Hardness
575 kg/mm^2
(Knoop number)
Solubility
Insoluble in H_2O
Elastic Moduli
Young's modulus
114.37 GPa
Rupture modulus
1.50 GPa
Poisson's ratio
0.25–0.36 (est.)

Notes: Irtran is the trademark for infrared-transmitting optical materials manufactured by Eastman Kodak Company. Irtran I is a hot-pressed polycrystalline compact of magnesium fluoride. Finished blanks in diameters as large as 8 in. and domes up to 9 in. in diameter and 4.5 in. in height have been made. Virtually any shape that can be cut, ground, or polished from glass can also be similarly produced from this material. However, the material cannot be heat-softened, as is done in glass-blowing or forming. The material is not soluble in water or dilute acids or bases. It is resistant to thermal shock and will withstand atmospheric weathering about as well as conventional glasses.

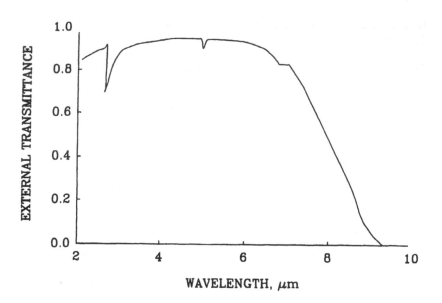

Figure 6.81 The external transmittance of Kodak Irtran 1 material, thickness 2.0 mm. [Adapted from W. L. Wolfe and G. J. Zissis, *The Infrared Handbook*, 1978.]

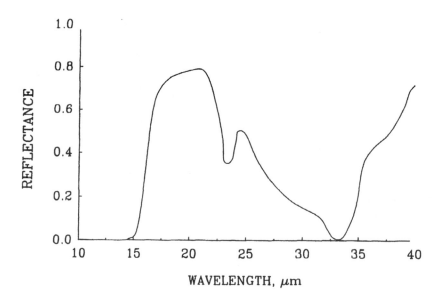

Figure 6.82 The reflectance of Kodak Irtran 1 material, thickness 2.0 mm. [Adapted from D. E. McCarthy, *Appl. Opt. 2*: 594 (1963).]

Table 6.39   Refractive Indices of Irtran 1

λ (μm)	n	λ (μm)	n
1.0000	1.3778	6.0000	1.3179
1.2500	1.3763	6.2500	1.3122
1.5000	1.3749	6.5000	1.3063
1.7500	1.3735	6.7500	1.3000
2.0000	1.3720	7.0000	1.2934
2.2500	1.3702	7.2500	1.2865
2.5000	1.3683	7.5000	1.2792
2.7500	1.3663	7.7500	1.2715
3.0000	1.3640	8.0000	1.2634
3.2500	1.3614	8.2500	1.2549
3.5000	1.3587	8.5000	1.2460
3.7500	1.3558	8.7500	1.2367
4.0000	1.3526	9.0000	1.2269
4.2500	1.3492	9.2500	
4.5000	1.3455	9.5000	
4.7500	1.3416	9.7500	
5.0000	1.3374	10.0000	
5.2500	1.3329	11.0000	
5.5000	1.3282	12.0000	
5.7500	1.3232	13.0000	

Index of refraction values were experimentally determined at selected wavelengths between 1 and 10 μm. Coefficients of an interpolation formula were established and reduced by least-squares methods, and the values computed. All values beyond 10 μm are extrapolated. Herzberger dispersion equation:

$$n = n_o + \frac{b}{\lambda^2 - 0.028} + \frac{c}{(\lambda^2 - 0.028)^2} + d\lambda^2 + e\lambda^4$$

where n = listed index of refraction
n_O = a measured value of index used as a base for modification by the coefficient terms. λ = wavelength in micrometers.
For Irtran 1,

n_o = 1.3776955

b = 1.3515529 × 10^{-3}

c = 2.1254394 × 10^{-4}

d = −1.5041172 × 10^{-3}

e = −4.4109708 × 10^{-6}

Source: Kodak IRTRAN Infrared Optical Materials, Kodak Publ. U-72, Eastman Kodak Co., Rochester, N.Y.

Magnesium Oxide, MgO

Molecular Weight
40.32

Specific Gravity
3.58 [g/cm^3]

Crystal Class
Cubic

Transmission
Long-wavelength limit 8.5 μm
Short-wavelength limit 0.25 μm
(See Fig. 6.83.)

Absorption Coefficient
0.05 cm^{-1} at 5.5 μm

Reflection Loss
11.6% for two surfaces at 4 μm

Refractive Index
(See Table 6.40.)

Temperature Coefficient of Refractive Index
(See Table 6.41.)

Dispersion
(See dispersion equation, Table 6.40.)

Zero Dispersion Wavelength
1.6 μm

Dielectric Constant
9.65 for 10^2–10^8 Hz at 298 K

Melting Temperature
3070 K

Thermal Conductivity
450 W/(m·K) at 80 K
59 W/(m·K) at 300 K
43.1 W/(m·K) at 400 K

Thermal Expansion
1.0×10^{-6} K^{-1} at 75 K
10.5×10^{-6} K^{-1} at 293 K
16.5×10^{-6} K^{-1} at 1600 K

Specific Heat
0.21 [cal/(g·K)] at 273 K

Debye Temperature
946 K

Hardness
692 kg/mm^2
(Knoop number with 600-g load perpendicular to cleavage planes)

Bandgap
7.8 eV

Solubility
Insoluble in H$_2$O; soluble in acids and ammonia salts

Elastic Moduli
Elastic stiffness
$c_{11} = 294$ GPa
$c_{12} = 93$ GPa
$c_{44} = 155$ GPa
Elastic compliance
$s_{11} = 4.01$ TPa^{-1}
$s_{12} = -0.96$ TPa^{-1}
$s_{44} = 6.47$ TPa^{-1}
Young's modulus
248.73 GPa
Shear modulus
154.34 GPa
Bulk modulus
154.34 GPa
Poisson's ratio
0.18

Notes: Magnesium oxide, also called periclase, is a cubic crystal of fairly high hardness and high melting point. Although less hard than sapphire, it may be pressed against metal gaskets to form a leaktight seal. This method, though not always successful, has been used for sealing magnesium oxide windows to absorption cells at high temperatures. The crystal can be cut on a disk grinder with no lubricant. Hard work with an aluminum oxide finishing cloth is necessary to get a smooth finish. It can also be used without polishing if a perfect cleavage of the single crystal has been obtained. Some specimens show little O–H absorption, probably due to water. The polished surfaces of optical components of magnesium oxide can be protected from attack by atmospheric moisture with evapo- rated coatings of silicon monoxide. Magnesium oxide has a slippage plane that may affect the mechanical strength of certain optical components.

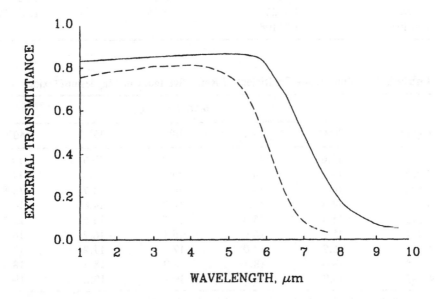

Figure 6.83 The external transmittance of magnesium oxide, thickness 5.5 mm, at 298 K (——) and 1273 K (– – –). [Adapted from U. P. Oppenheim and A. Goldman, *J. Opt. Soc. Am. 54*: 127 (1964).]

Table 6.40 Refractive Index of Magnesium Oxide at 23.3°C

λ (μm)	n	λ (μm)	n
0.36117	1.77318	1.97009	1.70885
0.365015	1.77186	2.24929	1.70470
1.01398	1.72259	2.32542	1.70350
1.12866	1.72059	3.3033	1.68526
1.36728	1.71715	3.5078	1.68055
1.52952	1.71496	4.258	1.66039
1.6932	1.71281	5.138	1.63138
1.7092	1.71258	5.35	1.62404
1.81307	1.71108		

Table 6.41 Temperature Coefficient of Refractive Index of Magnesium Oxide

λ (μm)	dn/dT (10^{-6}/°C)				
	20°C	25°C	30°C	35°C	40°C
7.679	13.6	13.7	13.8	13.9	14.0
7.065	14.1	14.2	14.3	14.4	14.5
6.678	14.4	14.5	14.6	14.7	14.8
6.563	14.5	14.6	14.7	14.8	14.9
5.893	15.3	15.4	15.5	15.6	15.7
5.461	15.9	16.0	16.1	16.2	16.3
4.861	16.9	17.0	17.1	17.2	17.3
4.358	18.0	18.1	18.2	18.3	18.4
4.047	18.9	19.0	19.1	19.2	19.3

Source: R. E. Stephens and I. H. Malitson, *J. Res. NBS 49*: 249 (1952).

Magnesium oxide (Magnesia), MgO

Molecular Weight
40.31
Specific Gravity
3.585 [g/cm^3]
Crystal Class
Cubic
Transmission
(See Figs. 6.84 and 6.85.)
Refractive Index
(See Table 6.42.)
Melting Temperature
3053 K ± 20 K
Thermal Conductivity
40.6 W/(m·K) at 298 K
36.4 W/(m·K) at 373 K
8.36 W/(m·K) at 1273 K

Thermal Expansion
8.0×10^{-6} K^{-1} at 373 K
10.3×10^{-6} K^{-1} at 673 K
12.8×10^{-6} K^{-1} at 973 K
Specific Heat
0.24 [cal/(g·K)]
Hardness
910 microvickers for (100 facet and 500-g load)
Notes: The crystalline magnesia is processed from high-purity calcinated magnesia arc furnaces at temperatures around 3073 K. Magnesia is distributed by Advanced Composite Materials Corporation. The cyrstal is transparent in the visible and also exhibits good transmission in the ultraviolet and infrared. (See also preceding data sheet magnesium oxide.)

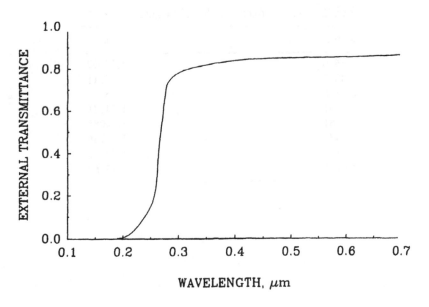

Figure 6.84 The external transmittance of magnesia, thickness 10 mm. [Adapted from Advanced Composite Materials Corporation, Greer, S.C.]

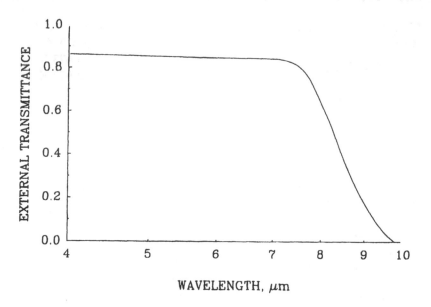

Figure 6.85 The external transmittance of magnesia, thickness 10 mm. [Adapted from Advanced Composite Materials Corporation, Greer, S.C.]

Table 6.42 Refractive Index of Magnesia at 24.5°C

λ (μm)	n
0.3650	1.77197
0.4047	1.76111
0.4358	1.75478
0.4861	1.74721
0.5461	1.74085
0.5876	1.73756
0.6563	1.73341
1.0140	1.72267

Potassium Aluminosilicate (Mica), $KAl_3Si_3O_{10}(OH)_2$

Specific Gravity
2.78 [g/cm^3]

Crystal Class
Monoclinic

Transmission
(See Fig. 6.86.)

Refractive Index
(See Table 6.43.)

Dielectric Constant
5.4 for 10^2–3×10^9 Hz at 299 K

Melting Temperature
1473–1573 K

Thermal Conductivity
0.25–0.59 W/(m·K)

Thermal Expansion
27×10^{-6} K^{-1} at 293 K

Specific Heat
0.208 [cal/(g·K)] at 293–373 K

Solubility
Insoluble in H_2O

Elastic Moduli
Elastic stiffness
$c_{11} = 178$ GPa
$c_{12} = 42.4$ GPa
$c_{13} = 14.5$ GPa
$c_{33} = 54.9$ GPa
$c_{44} = 12.2$ GPa
Elastic compliance
$s_{11} = 6.0$ TPa^{-1}
$s_{12} = -1.3$ TPa^{-1}
$s_{13} = -1.2$ TPa^{-1}
$s_{33} = 18.9$ TPa^{-1}
$s_{44} = 81.9$ TPa^{-1}

Notes: Mica is a birefringent material often used for quarter-wave plates. In addition to muscovite, mica also occurs as biotite, phlogopite, and fluorphogopite. Mica has been used for large-aperture wave plates in high-power laser systems. It exhibits dichroism in the infrared.

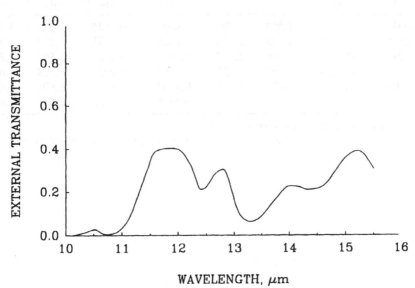

Figure 6.86 The external transmittance of muscovite mica, thickness 8 μm. [Adapted from J. M. Serratosa and A. Hildalgo, *Appl. Opt. 3*: 315 (1964).]

Table 6.43 Refractive Indices of Muscovite Mica

λ_F (μm)	λ_S (μm)	n_F	n_S	λ_F (μm)	λ_S (μm)	n_F	n_S
$t = 5.24$ μm thick				$t = 20.82$ μm thick			
0.6665	0.6675	1.590	1.592	0.6960	0.6985	1.598	1.594
0.6188	0.6210	1.594	1.600	0.6316	0.6336	1.593	1.598
0.5573	0.5590	1.595	1.600	0.5539	0.5555	1.596	1.601
0.5082	0.5090	1.600	1.603	0.4935	0.4950	1.600	1.605
0.4538	0.4555	1.602	1.608	0.4600	0.4615	1.602	1.607
0.4320	0.4330	1.608	1.611	0.4310	0.4326	1.604	1.610
$t = 13.91$ μm thick				$t = 48.68$ μm thick			
0.6910	0.6930	1.590	1.594	0.6914	0.6935	1.591	1.596
0.5995	0.6010	1.595	1.599	0.6110	0.6125	1.594	1.598
0.5293	0.5308	1.598	1.603	0.5470	0.5487	1.596	1.601
0.4740	0.4754	1.602	1.606	0.4958	0.4971	1.599	1.603
0.4300	0.4308	1.607	1.611	0.4667	0.4680	1.601	1.606
				0.4408	0.4425	1.603	1.609

λ_F and λ_S are wavelengths of fast and slow rays.
Source: Adapted from M. A. Jeppeson and A. M. Taylor, *J. Opt. Soc. Am. 56:* 451 (1966).

Potassium Bromide, KBr

Molecular Weight
119.01
Specific Gravity
2.75 [g/cm^3]
Crystal Class
Cubic
Transmission
Long-wavelength limit 40 μm
Short-wavelength limit 0.23 μm
(See Fig. 6.87.)
Absorption Coefficient
1.2×10^{-4} cm^{-1} at 2.7 μm
1.6×10^{-4} cm^{-1} at 3.8 μm
Reflection Loss
8.4% for two surfaces at 10 μm
(See Fig. 6.88.)
Refractive Index
(See Table 6.44.)
Temperature Coefficient of Refractive Index
(See Fig. 6.89.)
Dispersion
(See dispersion equation, Table 6.44.)
Zero Dispersion Wavelength
4.2 μm
Dielectric Constant
4.9 for 10^2–10^{10} Hz at 298 K
4.97 for 10^2–10^{10} Hz at 360 K
Melting Temperature
1000 K
Thermal Conductivity
14.0 W/(m·K) at 80 K
4.8 W/(m·K) at 319 K
4.80 W/(m·K) at 372 K
Thermal Expansion
25.0×10^{-6} K^{-1} at 75 K
38.5×10^{-6} K^{-1} at 300 K
72.4×10^{-6} K^{-1} at 1000 K
Specific Heat
0.104 [cal/(g·K)] at 273 K
Debye Temperature
174 K
Hardness
7 kg/mm^2 (Knoop number with 200-g load in ⟨100⟩ direction)
Bandgap
7.6 eV
Solubility
53.48 g/(100 g H$_2$O) at 273 K
Elastic Moduli
Elastic stiffness
$c_{11} = 34.5$ GPa
$c_{12} = 5.5$ GPa
$c_{44} = 5.1$ GPa
Elastic compliance
$s_{11} = 30.3$ TPa^{-1}
$s_{12} = -4.2$ TPa^{-1}
$s_{44} = 196$ TPa^{-1}
Young's modulus
26.87 GPa
Rupture modulus
0.0039 GPa
Shear modulus
5.08 GPa
Bulk modulus
15.02 GPa

Notes: Potassium bromide is grown in the same manner as sodium chloride and is hygroscopic. It is soluble in alcohol and glycerin. Potassium bromide is used in IR prisms and windows and in laser windows. Diameters as large as 8 in. are available.

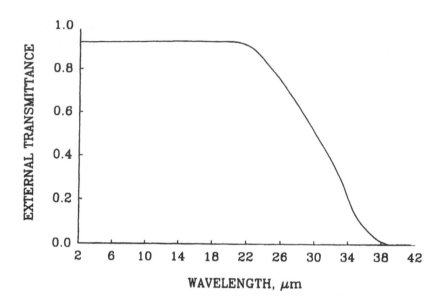

Figure 6.87 The external transmittance of potassium bromide, thickness 5 mm. [Adapted from D. E. McCarthy, *Appl. Opt. 2*: 591 (1963).]

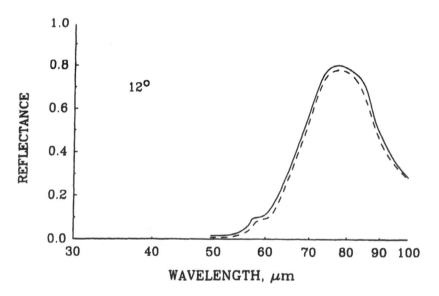

Figure 6.88 The reflectance of plane-polarized light by potassium bromide crystal at an incident angle of 12° (——) perpendicular, (----) parallel. [Adapted from A. Mitsuishi, *J. Opt. Soc. Am. 50*: 433 (1960).]

Table 6.44 Refractive Index of Potassium Bromide at 22°C

λ (μm)	n	λ (μm)	n	λ (μm)	n	λ (μm)	n
0.404656	1.589752	1.01398	1.54408	6.238	1.53288	17.40	1.50390
0.435835	1.581479	1.12866	1.54258	6.692	1.53225	18.16	1.50076
0.486133	1.571791	1.36728	1.54061	8.662	1.52903	19.01	1.49703
0.508582	1.568475	1.7012	1.53901	9.724	1.52695	19.91	1.49288
0.546074	1.563928	2.44	1.53733	11.035	1.52404	21.18	1.48655
0.587562	1.559965	2.73	1.53693	11.862	1.52200	21.83	1.48311
0.643847	1.555858	3.419	1.53612	14.29	1.51505	23.86	1.47140
0.706520	1.552447	4.258	1.53523	14.98	1.51280	25.14	1.46324

Source: R. E. Stephens, E. K. Plyler, W. S. Rodney, and R. J. Spindler, *J. Opt. Soc. Am. 43*: 111–112 (1953).
Dispersion equation:

$$n^2 = 2.361323 - 0.000311497\lambda^2 - 0.000000058613\lambda^4 + \frac{0.007676}{\lambda^2} + \frac{0.0156569}{\lambda^2 - 0.0324}$$

The average value of the temperature coefficient of refractive index is given as 4.0×10^{-5} per °C.

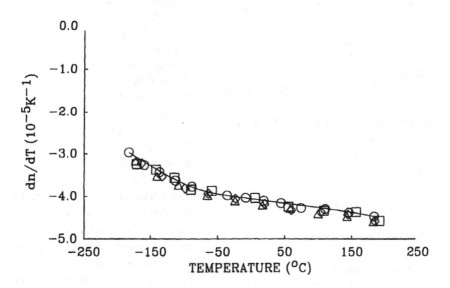

Figure 6.89 The temperature coefficient of refractive index of potassium bromide for wavelengths 0.6328 μm (○), 1.15 μm (△), 3.39 μm (◇), and 10.6 μm (□). [Adapted from A. J. Glass and A. H. Guenther (Eds.), *Laser Induced Damage in Optical Materials*, 1977, p. 79.]

Potassium Chloride, KCl

Molecular Weight
74.55
Specific Gravity
1.984 [g/cm^3]
Crystal Class
Cubic
Transmission
Long-wavelength limit 30 μm
Short- wavelength limit 0.21 μm
(See Fig. 6.90.)
Absorption Coefficient
Harshaw sample:
1.5×10^{-3} cm^{-1} at 2.7 μm
0.33×10^{-3} cm^{-1} at 3.8 μm
Adolf Meller:
1×10^{-3} cm^{-1} at 2.7 μm
0.33×10^{-3} cm^{-1} at 3.8 μm
Reflection Loss
6.8% for two surfaces at 10 μm
(See Fig. 6.91.)
Refractive Index
(See Table 6.45.)
Dispersion
(See dispersion equation, Table 6.44.)
Temperature Coefficient of Refractive Index
(See Fig. 6.92.)
Zero Dispersion Wavelength
3.3 μm
Dielectric Constant
4.64 for 10^6 Hz at 302 K
Melting Temperatures
1050 K
Thermal Conductivity
33.5 W/(m·K) at 85 K
6.7 W/(m·K) at 315 K
3.85 W/(m·K) at 460 K
Thermal Expansion
20.7×10^{-6} K^{-1} at 75 K

36.6×10^{-6} K^{-1} at 300 K
62.1×10^{-6} K^{-1} at 1000 K
Specific Heat
0.162 [cal/(g·K)] at 273 K
0.182 [cal/(g·K)] at 701 K
Debye Temperature
235 K
Hardness
7.2 kg/mm^2 (Knoop number with 200-g load in ⟨110⟩ direction)
Bandgap
8.5 eV
Solubility
34.7 g/(100 g H$_2$O) at 293 K
Elastic Moduli
Elastic stiffness
$c_{11} = 40.5$ GPa
$c_{12} = 6.9$ GPa
$c_{44} = 6.27$ GPa
Elastic compliance
$s_{11} = 25.9$ TPa^{-1}
$s_{12} = -3.8$ TPa^{-1}
$s_{44} = 159$ TPa^{-1}
Young's modulus
29.63 GPa
Rupture modulus
0.0044 GPa
Shear modulus
6.24 GPa
Bulk modulus
17.36 GPa

Notes: Potassium chloride is grown in the same way as sodium chloride, but sometimes multiple crystals instead of single-crystal ingots result. Its water solubility is only about half that of potassium bromide; it is soluble in alkalies, ether, and glycerin. Diameters as large as 12 in. are available.

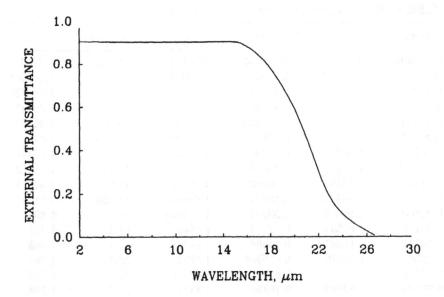

Figure 6.90 The external transmittance of potassium chloride, thickness 10 mm. [Adapted from D. E. McCarthy, *Appl. Opt. 4*: 317 (1965).]

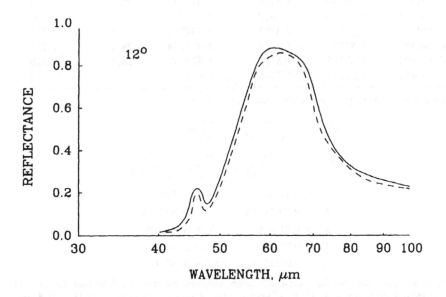

Figure 6.91 The reflectance of plane-polarized light by potassium chloride crystal at an incident angle of 12° (——) perpendicular, (– – –) parallel. [Adapted from A. Mitsuishi, *J. Opt. Soc. Am. 50*: 433 (1960).]

Table 6.45 Refractive Index of Potassium Chloride

λ (μm)	n	λ (μm)	n	λ (μm)	n
0.185409	1.82710	0.410185	1.50907	5.3039	1.470013
0.186220	1.81853	0.434066	1.50503	5.8932	1.468804
0.197760	1.73120	0.441587	1.50390	8.2505	1.462726
0.198990	1.72438	0.467832	1.50044	8.8398	1.460858
0.200090	1.71870	0.486149	1.49841	10.0184	1.45672
0.204470	1.69817	0.508606	1.49620	11.786	1.44919
0.208216	1.68308	0.53383	1.49410	12.965	1.44346
0.211078	1.67281	0.54610	1.49319	14.144	1.43722
0.21445	1.66188	0.56070	1.49218	15.912	1.42617
0.21946	1.64745	0.58931	1.49044	17.680	1.41403
0.22400	1.63612	0.58932	1.490443	18.2	1.409
0.23129	1.62043	0.62784	1.48847	18.8	1.401
0.242810	1.60047	0.64388	1.48777	19.7	1.398
0.250833	1.58979	0.656304	1.48727	20.4	1.389
0.257317	1.58125	0.67082	1.48669	21.1	1.379
0.263200	1.57483	0.76824	1.48377	22.2	1.374
0.267610	1.57044	0.78576	1.483282	23.1	1.363
0.274871	1.56386	0.88398	1.481422	24.1	1.352
0.281640	1.55836	0.98220	1.480084	24.9	1.336
0.291368	1.55140	1.1786	1.478311	25.7	1.317
0.308227	1.54136	1.7680	1.475890	26.7	1.300
0.312280	1.53926	2.3573	1.474751	27.2	1.275
0.340358	1.52726	2.9466	1.473834	28.2	1.254
0.358702	1.52115	3.5359	1.473049	28.8	1.226
0.394415	1.51219	4.7146	1.471122		

Dispersion equations (for the ultraviolet and visible, respectively):

$$n^2 = a^2 + \frac{M_1}{\lambda^2 - \lambda_1^2} + \frac{M_2}{\lambda^2 - \lambda_2^2} - k\lambda^2 - h\lambda^4$$

$$n^2 = b^2 + \frac{M_1}{\lambda^2 - \lambda_1^2} + \frac{M_2}{\lambda^2 - \lambda_2^2} - \frac{M_2}{\lambda^2 - \lambda_3^2}$$

$$a^2 = 2.174967 \qquad k = 0.000513495$$
$$M_1 = 0.008344206 \qquad h = 0.06167587$$
$$\lambda_1^2 = 0.0119082 \qquad b^2 = 3.866619$$
$$M_2 = 0.00698382 \qquad M_3 = 5{,}569.715$$
$$\lambda_2^2 = 0.0255550 \qquad \lambda_3^2 = 3{,}292.47$$

Refractive index data for the wavelength ranges indicated are from the following sources: (1) 0.185409–0.76824 μm at 18°C, F. F. Martens, *Ann. Physik 6:* 619 (1901); (2) 18.2–28.8 μm, H. W. Hohls, *Ann. Physik 29:* 433 (1937); (3) 0.58932–17.680 μm at 15°C, F. Paschen, *Ann. Physik 26:* 120 (1908). Note that the data of Paschen and of Martens overlap in a small region, and both sets are presented. There is less spread between Hohls's data and Paschen's data than there is among Paschen's data in the region where they join. The fit in this region is good (~0.0005). Paschen also presents two dispersion curves that fit the data of Martens to about five parts in the fifth decimal.

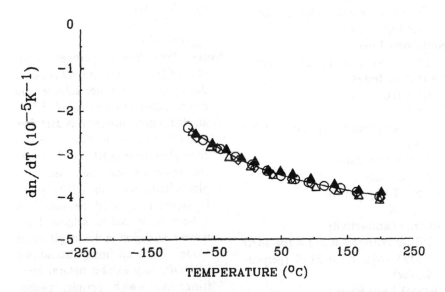

Figure 6.92 The temperature coefficient of refractive index of potassium chloride for wavelengths 0.6328 μm (○), 1.15 μm (△), 3.39 μm (◇), and 10.6 μm (▲). [Adapted from A. J. Glass and A.H. Guenther (Eds.), *Laser Induced Damage in Optical Materials*, 1977, p. 79.]

Potassium Dihydrogen Phosphate (KDP), KH_2PO_4

Molecular Weight
136.09

Specific Gravity
2.338 [g/cm^3]

Crystal Class
Tetragonal

Transmission
Long-wavelength limit 1.7 μm
Short-wavelength limit 0.25 μm
(See Fig. 6.93.)

Reflection Loss
8% for two surfaces at 0.59 μm

Refractive Index
(See Table 6.46.)

Dispersion
(See dispersion equation, Table 6.46.)

Dielectric Constant
21 (parallel)

Melting Temperature
525.6 K

Thermal conductivity
1.2139 W/(m·K) at 312 K (parallel)
1.3395 W/(m·K) at 319K (perpendicular)

Thermal Expansion
42×10^{-6} K^{-1} at 300 K (parallel)
27×10^{-6} K^{-1} at 300 K (perpendicular)

Specific Heat
0.21 [cal/(g·K)] at 306 K

Solubility
33 g/(100 g H_2O); insoluble in alcohol.

Elastic Moduli
Elastic stiffness
$c_{11} = 71.4$ GPa
$c_{12} = -4.9$ GPa
$c_{13} = 12.9$ GPa
$c_{33} = 56.2$ GPa
$c_{44} = 12.7$ GPa
$c_{66} = 6.2$ GPa

Notes: Potassium dihydrogen phosphate (KDP) can be cut but is easily damaged. It is water-soluble and can be polished with alcohol. KDP, an electrooptic material, is useful as a modulator. Potassium dideuterium phosphate or KD*P (the asterisk indicates that deuterium replaces hydrogen in the KDP crystal) is superior to KDP because less voltage is needed to achieve half-wave retardation in the longitudinal mode. There are many isomorphs of KDP; they exhibit natural birefringence, which permits phase-matched second-harmonic generation.

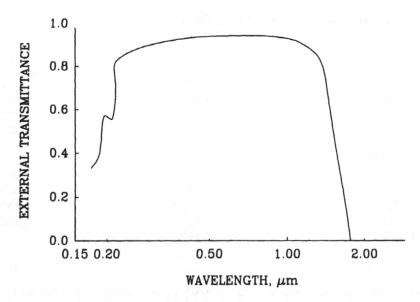

Figure 6.93 The external transmittance of KDP, thickness 1.65 mm. [Adapted from D. E. McCarthy, *Appl. Opt. 6*: 1896 (1967).]

Table 6.46   Refractive Index of KDP

λ (× 10^{-4} μm)	n_O	n_E
2138.560	1.60177	1.54615
2288.018	1.58546	
2446.905	1.57228	
2464.068	1.57105	
2536.519	1.56631	1.51586
2800.869	1.55263	1.50416
2980.628	1.54618	1.49824
3021.499	1.54433	1.49708
3035.781	....	1.49667
3125.663	1.54117	1.49434
3131.545	1.54098	1.49419
3341.478	....	1.48954
3650.146	1.52932	1.48432
3654.833	1.52923	1.48423
3662.878	1.52909	1.48409
3906.410	....	1.48069
4046.561	1.52341	1.47927

(continued)

Table 6.46 Refractive Index of KDP (Continued)

λ ($\times\ 10^{-4}$ μm)	n_O	n_E
4077.811	1.52301	1.47898
4358.350	1.51990	1.47640
4916.036	....	1.47254
5460.740	1.51152	1.46982
5769.580	1.50987	
5790.654	1.50977	1.46858
6328.160	1.50737	1.46685
10 139.75	1.49535	1.46041
11 287.04	1.49205	1.45917
11 522.76	1.49135	1.45893
13 570.70	1.46455	
15 281.00	....	1.45521
15 295.25	....	1.45512

The accuracy of the data is believed to be $\pm$ 0.00003 or better. The indexes for KDP are from F. Zernike, Jr., *J. Opt. Soc. Am. 54:* 1215 (1964); Zernike also reports several absolute index values. In addition, Zernike lists many computed index values in air, using the following dispersion equation:

$$n^2 = \frac{A + Bv^2}{1 - v^2/C} + \frac{D}{E - v^2}$$

where $v = 1/\lambda$ in cm^{-1}.

Constants of the dispersion equation for KDP:

	Extraordinary ray	Ordinary ray
A	2.132668	2.259276
B	8.637494×10^{-11}	1.008056×10^{-10}
C	8.142631×10^{9}	7.726408×10^{9}
D	8.069981×10^{6}	3.251305×10^{4}
E	2.500000×10^{5}	2.500000×10^{5}

Source: F. Zernike, Jr., *J. Opt. Soc. Am. 54:* 1215 (1964).

Potassium Iodide, KI

Molecular Weight
166.02

Specific Gravity
3.13 [g/cm^3]

Crystal Class
Cubic

Transmission
Long-wavelength limit 42 μm
Short-wavelength limit <0.38 μm
(See Fig. 6.94.)

Reflection Loss
10.6% for two surfaces at 12 μm

Refractive Index
(See Table 6.47.)

Temperature Coefficient of Refractive Index
-5.0×10^{-5} K^{-1} at 0.546 μm at 311–363 K

Dielectric Constant
4.94 for 2×10^6 Hz

Melting Temperature
996 K

Thermal Conductivity
2.1 W/(m·K) at 299 K

Thermal Expansion
28.5×10^{-6} K^{-1} at 75 K
40.2×10^{-6} K^{-1} at 293 K
59.0×10^{-6} K^{-1} at 700 K

Specific Heat
0.075 [cal/(g·K)] at 270 K

Debye Temperature
132 K

Bandgap
6.2 eV

Solubility
127.5 g/(100 g H$_2$O) at 273 K

Elastic Moduli
Elastic stiffness
$c_{11} = 27.4$ GPa
$c_{12} = 4.3$ GPa
$c_{44} = 3.7$ GPa
Elastic compliance
$s_{11} = 38.2$ TPa^{-1}
$s_{12} = -5.2$ TPa^{-1}
$s_{44} = 270$ TPa^{-1}
Young's modulus
31.49 GPa
Shear modulus
6.20 GPa

Notes: Potassium iodide is valuable as a prism material, but it is too hygroscopic (being about twice as soluble in water as potassium bromide) and too soft for field use. It is also soluble in alcohol and in ammonia.

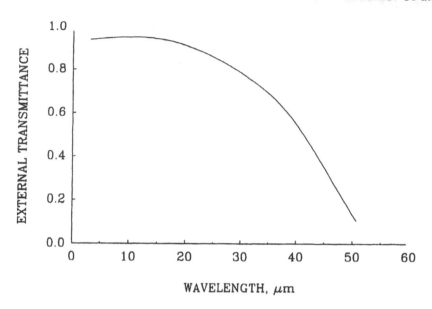

Figure 6.94 The external transmittance of potassium iodide, thickness 0.83 mm. [Adapted from J. Strong, *Phys. Rev. 38*: 1818 (1931).]

Table 6.47 Refractive Index of Potassium Iodide

λ (μm)	n	λ (μm)	n	λ (μm)	n
0.248	2.0548	0.656	1.65809	10.02	1.6201
0.254	2.0105	0.707	1.6537	11.79	1.6172
0.265	1.9424	0.728	1.6520	12.97	1.6150
0.270	1.9221	0.768	1.6494	14.14	1.6127
0.280	1.8837	0.811	1.6471	15.91	1.6085
0.289	1.85746	0.842	1.6456	18.10	1.6030
0.297	1.83967	0.912	1.6427	19	1.5997
0.302	1.82769	1.014	1.6396	20	1.5964
0.313	1.80707	1.083	1.6381	21	1.5930
0.334	1.77664	1.18	1.6366	22	1.5895
0.366	1.74416	1.77	1.6313	23	1.5858
0.391	1.72671	2.36	1.6295	24	1.5819
0.405	1.71843	3.54	1.6275	25	1.5775
0.436	1.70350	4.13	1.6268	26	1.5729
0.486	1.68664	5.89	1.6252	27	1.5681
0.546	1.67310	7.66	1.6235	28	1.5629
0.588	1.66654	8.84	1.6218	29	1.5571
0.589	1.66643				

Source: For wavelengths 0.248–1.083 μm: H. Harting, *Sitzber. Deut. Akad. Wiss. Berlin IV*: 1 (1948). For wavelengths 1.18–29 μm: K. Korth, *Z. Physik. 84*: 677 (1933).

Rubidium Bromide, RbBr

Molecular Weight
165.38

Specific Gravity
3.35 [g/cm^3]

Crystal Class
Cubic

Transmission
Long-wavelength limit ~40 μm
Short-wavelength limit <0.3 μm
(See Figs. 6.95 and 6.96.)

Absorption Coefficient
0.0016 cm^{-1} at 10.6 μm

Refractive Index
1.73 at 0.248 μm
1.60 at 0.351 μm
1.56 at 0.488 μm
1.56 at 0.5461 μm
1.55 at 0.5893 μm
1.55 at 0.6328 μm
1.54 at 1.06 μm
1.53 at 1.55 μm
1.53 at 2.8 μm
1.52 at 10.6 μm

Temperature Coefficient of Refractive Index
-4.0×10^{-5} K^{-1} at 0.288 μm
-4.4×10^{-5} K^{-1} at 0.40 μm
-4.5×10^{-5} K^{-1} at 0.64 μm
-4.5×10^{-5} K^{-1} at 1.15 μm
-4.5×10^{-5} K^{-1} at 3.40 μm
-4.4×10^{-5} K^{-1} at 10.6 μm
-4.3×10^{-5} K^{-1} at 17.0 μm

Melting Temperature
966 K

Thermal Conductivity
12.2 W/(m·K) at 105 K

Thermal Expansion
27.20×10^{-6} K^{-1} at 80 K
36.98×10^{-6} K^{-1} at 270 K

Specific Heat
0.0648 [cal/(g·K)] at
86.8 K
0.074 [cal/(g·K)] at 283 K

Bandgap
7.2 eV

Solubility
98 g/(100 g H$_2$O)

Elastic Moduli
Elastic stiffness
$c_{11} = 31.5$ GPa
$c_{12} = 4.8$ GPa
$c_{44} = 3.82$ GPa
Elastic compliance
$s_{11} = 33.1$ TPa^{-1}
$s_{12} = -4.4$ TPa^{-1}
$s_{44} = 262$ TPa^{-1}

Notes: Rubidium bromide is hygroscopic; care must be used in handling it to preserve the surface polish. The gradual decrease in transmittance at shorter wavelengths is due to surface scattering that is caused by the roughness of the surface and not by absorption or scattering within the material itself.

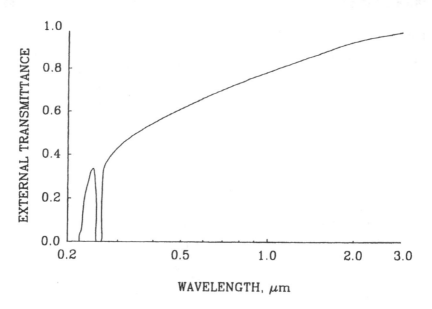

Figure 6.95 The external transmittance of rubidium bromide, thickness 5.31 mm. [Adapted from D.E. McCarthy, *Appl. Opt.* 7: 1243 (1968).]

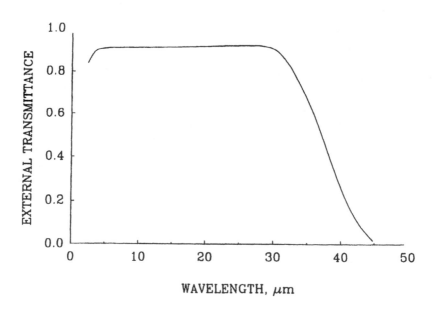

Figure 6.96 The infrared external transmittance of rubidium bromide, thickness 5.3 mm. [Adapted from D.E. McCarthy, *Appl. Opt.* 7: 1997 (1968).]

Rubidium Chloride, RbCl

Molecular Weight
120.92
Specific Gravity
2.80 [g/cm^3]
Crystal Class
Cubic
Transmission
Long-wavelength limit ~35 μm
Short-wavelength limit <0.2 μm
(See Figs. 6.97 and 6.98.)
Absorption Coefficient
0.00092 cm^{-1} at 10.6 μm
Refractive Index
1.60 at 0.248 μm
1.53 at 0.351 μm
1.50 at 0.488 μm
1.50 at 0.5461 μm
1.49 at 0.5893 μm
1.49 at 0.6328 μm
1.48 at 1.06 μm
1.48 at 1.55 μm
1.48 at 2.8 μm
1.46 at 10.6 μm
Temperature Coefficient of
Refractive Index
-3.8×10^{-5} K^{-1} at 0.288 μm
-3.9×10^{-5} K^{-1} at 0.40 μm
-3.9×10^{-5} K^{-1} at 0.64 μm
-3.9×10^{-5} K^{-1} at 1.15 μm
-3.9×10^{-5} K^{-1} at 3.4 μm

-3.8×10^{-5} K^{-1} at 10.6 μm
-3.5×10^{-5} K^{-1} at 17.0 μm
Dielectric Constant
4.68–5.20
Melting Temperature
991 K
Thermal Conductivity
7.6 W/(m·K) at 124 K
Thermal Expansion
21.9 × 10^{-6} K^{-1} at 70 K
36.1 × 10^{-6} K^{-1} at 300 K
Specific Heat
0.10 [cal/(g·K)] at 283 K
Debye Temperature
165 K
Bandgap
8.3 eV
Solubility
77 g/(100 g H$_2$O)
Elastic Moduli
Elastic stiffness
$c_{11} = 36.4$ GPa
$c_{12} = 6.3$ GPa
$c_{44} = 4.7$ GPa
Elastic compliance
$s_{11} = 29.3$ TPa^{-1}
$s_{12} = -4.3$ TPa^{-1}
$s_{44} = 212$ TPa^{-1}
Notes: Rubidium chloride is hygroscopic; care must be used in handling it to preserve the surface polish.

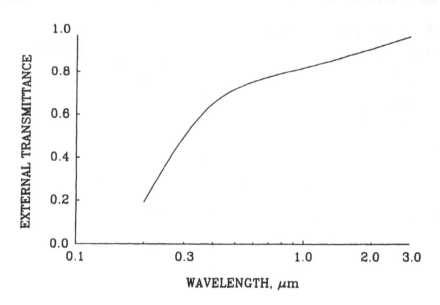

Figure 6.97 The external transmittance of rubidium chloride, thickness 2.1 mm. [Adapted from D.E. McCarthy, *Appl. Opt. 7*: 1243 (1968).]

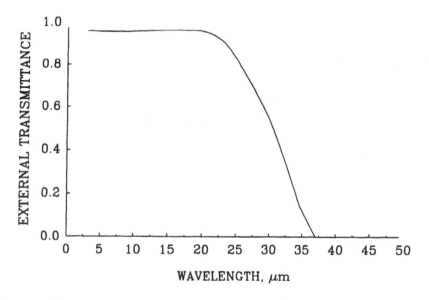

Figure 6.98 The external transmittance of rubidium chloride, thickness 3.0 mm. [Adapted from D.E. McCarthy (no date).]

Rubidium Iodide, RbI

Molecular Weight
212.37

Specific Gravity
3.55 [g/cm^3]

Crystal Class
Cubic

Transmission
Long-wavelength limit ~50 μm
Short-wavelength limit
~0.25 μm
(See Fig. 6.99.)

Refractive Index
2.05 at 0.248 μm
1.73 at 0.351 μm
1.67 at 0.488 μm
1.65 at 0.5461 μm
1.65 at 0.5893 μm
1.64 at 0.6328 μm
1.62 at 1.06 μm
1.62 at 1.55 μm
1.61 at 2.8 μm
1.61 at 10.6 μm

Temperature Coefficient of Refractive Index
-3.7×10^{-5} K^{-1} at 0.288 μm
-5.5×10^{-5} K^{-1} at 0.4 μm
-5.6×10^{-5} K^{-1} at 0.64 μm
-5.6×10^{-5} K^{-1} at 1.15 μm
-5.6×10^{-5} K^{-1} at 3.4 μm
-5.6×10^{-5} K^{-1} at 10.6 μm
-5.5×10^{-5} K^{-1} at 17.0 μm

Melting Temperature
920 K

Thermal Conductivity
9.9 W/(m·K) at 84 K

Thermal Expansion
39×10^{-6} K^{-1} at 293 K

Specific Heat
0.058 [cal/(g·K)] at 283 K

Debye Temperature
103 K

Hardness
1.0 (Mohs)

Bandgap
5.8 eV

Solubility
152 g/(100 g H$_2$O)

Elastic Moduli
Elastic stiffness
$c_{11} = 25.6$ GPa
$c_{12} = 3.7$ GPa
$c_{44} = 2.79$ GPa
Elastic compliance
$s_{11} = 40.5$ TPa^{-1}
$s_{12} = -5.1$ TPa^{-1}
$s_{44} = 358$ TPa^{-1}

Notes: Rubidium iodide is hygroscopic; care must be used in handling it to preserve the surface polish.

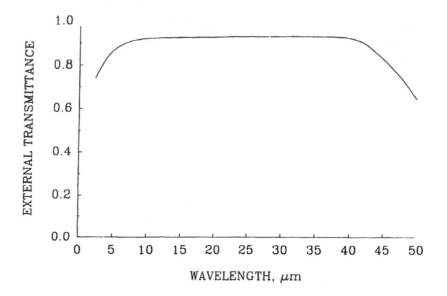

Figure 6.99   The external transmittance of rubidium iodide, thickness 3.91 mm.
[Adapted from D.E. McCarthy (no date).]

Selenium, Se

Molecular Weight
78.96

Specific Gravity
4.82 [g/cm³] (trigonal)
4.26 [g/cm³] (amorphous)

Crystal Class
Trigonal
Amorphous
Hexagonal

Transmission
(See Fig. 6.100.)

Refractive Index
(See Table 6.48.)

Dielectric Constant
6.00 for 10^2–10^{10} Hz at 298 K

Melting Temperature
490 K

Thermal Conductivity
1.3 W/(m·K) at 308 K (trigonal)

Thermal Expansion
20.3×10^{-6} K^{-1} at 195–292 K
48.7×10^{-6} K^{-1} at 273–294 K
45.2×10^{-6} K^{-1} at 478 K

Specific Heat
0.068 [cal/(g·K)] at 85–291 K
0.077 [cal/(g·K)] at 293.5 K
0.131 [cal/(g·K)] at 311 K

Debye Temperature
90 K

Hardness
2.6 (Mohs; trigonal)

Bandgap
1.7 eV (trigonal)

Solubility
Insoluble in H_2O (trigonal)

Elastic Moduli
Elastic stiffness
$c_{11} = 18.6$ GPa
$c_{12} = 7.3$ GPa
$c_{13} = 25.2$ GPa
$c_{14} = 5.6$ GPa
$c_{33} = 76.1$ GPa
$c_{44} = 14.8$ GPa
Elastic compliance
$s_{11} = 131$ TPa^1
$s_{12} = -13$ TPa^{-1}
$s_{13} = -40$ TPa^{-1}
$s_{14} = 56$ TPa^{-1}
$s_{33} = 41$ TPa^{-1}
$s_{44} = 112$ TPa^{-1}

Notes: Selenium can exist in various forms. The trigonal or crystalline form is most stable up to 443 K and then transforms into the hexagonal. The amorphous or vitreous form is fairly stable below 323 K but converts to the trigonal form at higher temperatures. Selenium has been used in the photocopying process and photocells. It is also used as a reflection polarizer in the infrared. It has important linear and nonlinear optical properties in the IR for laser optics.

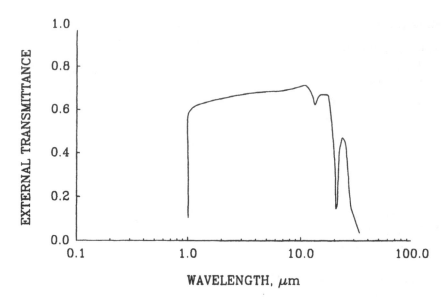

Figure 6.100   The external transmittance of selenium, thickness 1.69 mm. [Adapted from R.S. Caldwell, Special Report on Contract DA 36-039-SC-71131, Purdue University, January 1958.]

Table 6.48   Refractive Index of Selenium at 23°C (±2°C)

λ (μm)	n_O	n_E
1.06	2.790 ± 0.008	3.608 ± 0.008
1.15	2.737 ± 0.008	3.573 ± 0.008
3.39	2.65 ± 0.01	3.46 ± 0.01
10.6	2.64 ± 0.01	3.41 ± 0.01

These values are for single-crystal selenium [from L. Gampel and F. M. Johnson, *J. Opt. Soc. Am.* *59*: 72 (1969)]. For the region 9–23 μm, the index values are 2.78 ± 0.02 for the orindary ray and 3.58 ± 0.02 for the extraordinary ray with no appreciable variation [from R. S. Caldwell and H. Y. Fan, *Phys. Rev. 114*: 664 (1959)]. For amorphous selenium in the region 2.5–15 μm, index values of 2.46–2.38 are referenced by Caldwell and Fan.

Silicon, Si

Molecular Weight
28.09

Specific Gravity
2.32–2.34 [g/cm^3]

Crystal Class
Cubic

Transmission
(See Fig. 6.101.)

Absorption Coefficient
5.5×10^{-3} cm^{-1} 2% at 2.7 m
0.785×10^{-3} cm^{-1} 6% at 3.8 m
(Cal. Tech. sample)
(See Fig. 6.102 and Table 6.49.)

Reflectivity
(See Fig. 6.103.)

Refractive Index
(See Table 6.50.)

Temperature Coefficient of Refractive Index
(Values are for $1/n$ dn/dt) 3.9–4.7×10^{-5} K^{-1}

Dispersion
(See dispersion equation, Table 6.50.)

Dielectric Constant
13 for 9.37×10^9 Hz

Melting Temperature
1690 K

Thermal Conductivity
598.6 W/(m·K) at 125 K
163 W/(m·K) at 313 K
105.1 W/(m·K) at 400 K

Thermal Expansion
-0.5×10^{-6} K^{-1} at 75 K
2.6×10^{-6} K^{-1} at 293 K
4.6×10^{-6} K^{-1} at 1400 K

Specific Heat
0.18 [cal/(g·K)] at 298 K
0.253 [cal/(g·K)] at 1800 K

Debye Temperature
640 K

Hardness
1100 kg/mm^2 (Knoop number)

Bandgap
1.1 eV

Solubility
Insoluble in H$_2$O

Elastic Moduli
Elastic stiffness
$c_{11} = 165$ GPa
$c_{12} = 64$ GPa
$c_{44} = 79.2$ GPa
Elastic compliance
$s_{11} = 7.74$ TPa^{-1}
$s_{12} = -2.16$ TPa^{-1}
$s_{44} = 12.6$ TPa^{-1}
Young's modulus
130.91 GPa
Shear modulus
79.92 GPa
Bulk modulus
101.97 GPa
Poisson's ratio
0.28

Notes: The physical and chemical properties of silicon are very similar to those of germanium. Silicon is hard and can be finished with ordinary polishing equipment. It can be used in single-crystal or polycrystalline form with a minimum of scattering. Optical-grade silicon has high resistance to thermal and mechanical shocks. Polycrystalline silicon is available in n and p types and can be produced in diameters larger than 16 in. Silicon has a zero cold-water solubility but can be dissolved in a mixture of hydrofluoric acid and nitric acid. Silicon can be ground on a Blanchards polisher; diamond curve generators can be used; and normal pitches and polishing compounds are acceptable.

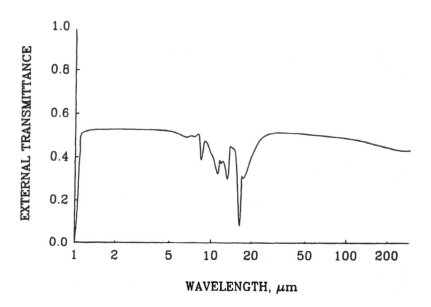

Figure 6.101   The external transmittance of polycrystalline *n*- or *p*-type silicon, thickness 2 mm. [Adapted from Exotic Materials, Inc., *Infrared Phys.* 5 (1965).]

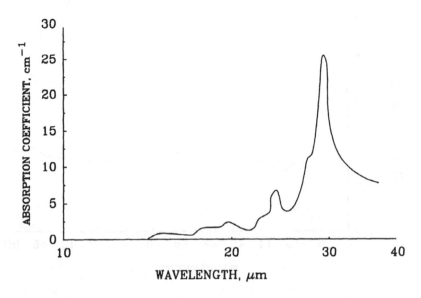

Figure 6.102 Lattice absorption curve for silicon. [Adapted from Laz and Burstein, *Phys. Rev. 97*: 39 (1955).]

Table 6.49 Absorption Coefficient (cm^{-1}) at Room Temperature and 200 °C of Silicon (7000 ohm·cm)

Wavenumber (cm^{-1})	λ (μm)	Room Temperature	200°C
4000	2.5	<0.01	<0.01
3000	3.33	<0.01	<0.01
2000	5.0	<0.01	0.01
1900	5.26	0.003	0.014
1800	5.55	0.005	0.174
1700	5.88	0.011	0.04
1600	6.25	0.012	0.039
1500	6.66	0.028	0.059
1400	7.14	<0.131	0.265
1300	7.69	0.195	0.343
1200	8.33	0.151	0.27
1100	9.09		
1000	10.0	0.354	0.579
943.4	10.6	0.533	0.957
900	11.11	0.893	1.27
800	12.5	0.695	1.40

Source: P. Klocek et al., *SPIE Proc. 929*: 65–78 (1988).

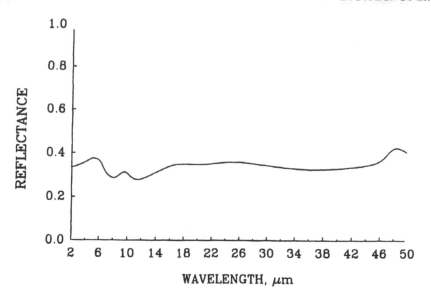

Figure 6.103 The reflectance of silicon, thickness 10 mm. [Adapted from D. E. McCarthy, *Appl. Opt.* 2: 595 (1963).]

Table 6.50 Refractive Index of Silicon at 26°C

λ (μm)	$n^\dagger$	$n^\ddagger$
2.4373	3.4434	3.4408
2.50	3.4424	—
2.7144	3.4393	3.4358
3.00	3.4361	3.4320
3.3033	3.4335	3.4297
3.4188	3.4327	3.4286
3.50	3.4321	3.4284
4.00	3.4294	3.4255
4.258	3.4283	3.4242
4.50	3.4275	3.4236
5.00	3.4261	3.4223
5.50	3.4250	3.4213
6.00	3.4242	3.4202
6.50	3.4236	3.4195
7.00	3.4231	3.4189
7.50	3.4227	3.4186
8.00	3.4224	3.4184
8.50	3.4221	3.4182

(continued)

Table 6.50 Refractive Index of Silicon at 26°C (Continued)

λ (μm)	n†	n‡
9.00	3.4219	—
9.50	3.4217	—
10.00	3.4215	3.4179
10.50	3.4214	—
11.00	3.4213	—
11.04	3.4213	3.4176
11.50	3.4212	—
12.00	3.4211	—
12.50	3.4210	—
13.00	3.4209	—
14.00	3.4208	—
14.50	3.4208	—
15.00	3.4207	—
15.50	3.4207	—
16.00	3.4206	—
17.00	3.4206	—
18.00	3.4205	—
19.00	3.4205	—
20.00	3.4204	—
21.00	3.4204	—
22.00	3.4203	—
23.00	3.4203	—
24.00	3.4202	—
25.00	3.4201	—

†From D. F. Edwards and E. Ochoa *Appl. Opt. 19*(24): 4130 (1980).
‡From C. D. Salzberg and J. J. Villa, *J. Opt. Soc. Am. 47*: 244 (1957).
Disperison equation [D.F. Edwards and E. Ochoa, *Appl. Opt. 19*(24): 4130 (1980)]:

$$n = A + BL + CL^2 + D\lambda^2 + E\lambda^4$$

where L = $1/(\lambda^2 - 0.028)$

$$A = 3.41983$$
$$B = 1.59906 \times 10^{-1}$$
$$C = -1.23109 \times 10^{-1}$$
$$D = 1.26878 \times 10^{-6}$$
$$E = -1.95104 \times 10^{-9}$$

Silicon Carbide, SiC

Molecular Weight
40.1
Specific Gravity
3.217 [g/cm^3]
Crystal Class
Hexagonal
Transmission
(See Fig. 6.104.)
Absorption Coefficient
(See Fig. 6.105.)
Reflectivity
(See Fig. 6.106.)
Refractive Index
(See Fig. 6.107 and Table 6.51.)
Dispersion
(See Fig. 6.108.)
Melting Temperature
3000 K
Thermal Conductivity
3500 W/(m·K) at 85 K
490 W/(m·K) at 300 K
Thermal Expansion
0.09×10^{-6} K^{-1} at 75 K
2.8×10^{-6} K^{-1} at 300 K
6.5×10^{-6} K^{-1} at 1800 K
Specific Heat
0.14 [cal/(g·K)] at 273 K
0.165 [cal/(g·K)] at 298 K
0.350 [cal/(g·K)] at 2800 K

Hardness
2130–2755 kg/mm^2 (Knoop number)
Bandgap
2.86 eV
Solubility
Insoluble in H$_2$O
Elastic Moduli
Elastic stiffness
$c_{11} = 502$ GPa
$c_{12} = 95$ GPa
$c_{44} = 169$ GPa
$c_{13} = 56$ GPa
$c_{33} = 565$ GPa
Elastic compliance
$s_{11} = 2.08$ TPa^{-1}
$s_{12} = -0.37$ TPa^{-1}
$s_{44} = 5.92$ TPa^{-1}
$s_{13} = -0.17$ TPa^{-1}
$s_{33} = 1.8$ TPa^{-1}
Young's modulus
386 GPa
Bulk modulus
273 GPa at 293 K

Notes: These data are for single crystals of hexagonal structure. Silicon carbide is a very hard material with a high melting temperature. Its birefringent characteristics are reported in the *Handbook of Optical Constants of Solids* (Palik, Academic Press, 1985, p. 587).

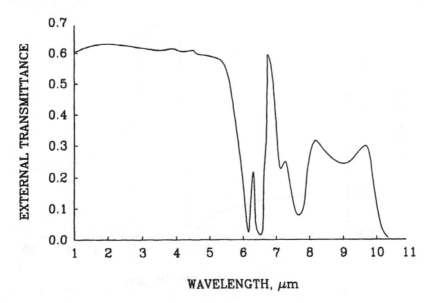

Figure 6.104 The external transmittance of silicon carbide, thickness 0.07 mm. [Adapted from H. G. Lipson, Conf. on Silicon Carbide, Boston, 1959, *Silicon Carbide— A High Temperature Semiconductor*, J. R. O'Connor and J. Smittens, Eds., Pergamon, New York, 1960.]

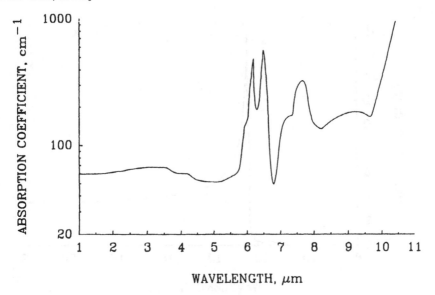

Figure 6.105 The absorption curve for silicon carbide. [Adapted from W.G. Spitzer et al., *Semiconductors and Semimetals*, R. K. Willardson and A.C. Beer, Eds., 1967, Vol. 3, p. 17.]

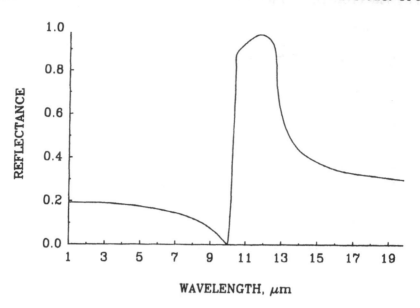

Figure 6.106 The reflectance of silicon carbide. [Adapted from W. G. Spitzer et al., *Phys. Rev. 113*: 127–132 (1959); H. R. Philipp and E. A. Tafi, Conf. on Silicon Carbide, Boston, 1959, *Silicon Carbide—A High Temperature Semiconductor*, J. R. O'Connor and J. Smittens, Eds., Pergamon, New York, 1960.]

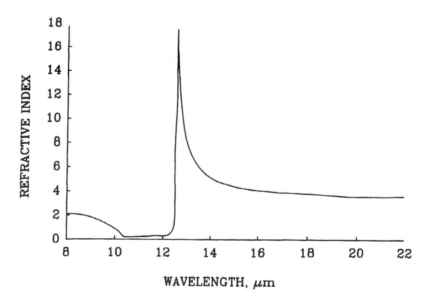

Figure 6.107 The refractive index of silicon carbide. [Adapted from W. G. Spitzer et al., *Phys. Rev. 113*: 127–132 (1959).]

Table 6.51 Refractive Index of Silicon Carbide

λ (μm)	n_O	k_O
0.4358	2.7305	
0.467	2.7074	
0.4959	2.684	1.18×10^{-5}
0.498	2.6870	
0.515	2.6789	
0.5461	2.6631	
0.568	2.6557	
0.5781	2.6511	
0.589	2.6488	
0.5895	2.6475	
0.616	2.6411	
0.6199	2.634	
0.6563	2.6296	
0.691	2.6243	
0.8266	2.598	
1.248	2.573	
2	2.572	3.98×10^{-4}
2.222	2.568	
2.480	2.558	
2.5	2.562	5.17
2.857	2.554	
3.100	2.556	
3.333	2.540	
4.0	2.516	6.37
4.133	2.555	
5.0	2.467	8.75
5.263	2.450	
5.556	2.430	
5.882	2.404	
6.250	2.372	
6.667	2.328	
7.143	2.267	
7.692	2.178	
8.333	2.034	
9.091	1.768	
9.259	1.684	
9.434	1.581	
9.615	1.453	
9.8	1.286	1.40
10.10	0.888	4.5
10.20	0.663	6.7×10^{-2}

(*continued*)

Table 6.51 Refractive Index of Silicon Carbide (Continued)

λ (μm)	n_O	k_O
10.31	0.274	0.18
10.42	0.0872	0.63
10.53	0.0663	0.95
10.64	0.0593	1.21
10.75	0.0569	1.45
10.87	0.0569	1.69
10.99	0.0587	1.93
11.11	0.0621	2.18
11.24	0.0672	2.45
11.36	0.0746	2.75
11.49	0.0850	3.08
11.63	0.0999	3.47
11.76	0.122	3.93
11.90	0.156	4.51
12.05	0.215	5.27
12.12	0.262	5.77
12.20	0.332	6.38
12.27	0.443	7.18
12.35	0.639	8.27
12.42	1.05	9.93
12.50	2.22	12.8
12.58	8.74	18.4
12.66	17.7	6.03
12.74	12.7	1.76
12.82	10.3	0.868
12.90	8.91	0.531
12.99	8.00	0.364
13.07	7.35	0.268
13.16	6.86	0.208
13.33	6.16	0.136
13.51	5.68	9.7×10^{-2}
13.70	5.32	7.4
13.89	5.05	5.8
14.08	4.83	4.7
14.29	4.65	3.9
15.38	4.09	1.8
16.67	3.80	1.1
18.18	3.61	
20.00	3.49	
22.22	3.40	
25.00	3.34	

Source: W. J. Choyke and E. D. Palik, Silicon Carbide, in *Handbook of Optical Constants of Solids* (E. D. Palik, Ed.), Academic, New York, 1985, pp. 593–595.

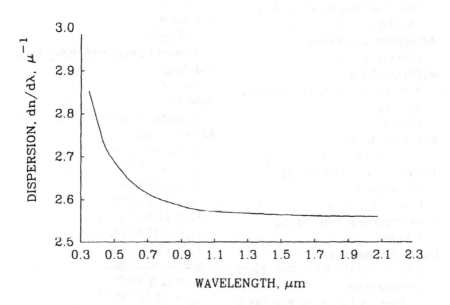

Figure 6.108   The dispersion of silicon carbide versus wavelength. [Adapted from W. J. Choyke and L. Patrick, *J. Opt. Soc. Am. 58*: 377–379 (1968).]

Silicon Dioxide (Crystal Quartz), SiO₂

Molecular Weight
60.06

Specific Gravity
2.635–2.660 [g/cm^3]

Crystal Class
Hexagonal

Transmission
Long-wavelength limit 4.5 μm
Short-wavelength limit 0.4 μm
(See Figs. 6.109–6.111.)

Absorption Coefficient
(See Fig. 6.112.)

Reflection Loss
8.2% for two surfaces for the ordinary ray at 2 μm
(See Fig. 6.113.)

Refractive Index
(See Table 6.52 and Fig. 6.114.)

Temperature Coefficient of Refractive Index
(See Table 6.53.)

Zero Dispersion Wavelength
1.3 μm

Dielectric Constant
4.34 for 3×10^7 Hz at 290–295 K (perpendicular)
4.27 for 3×10^7 Hz at 290–295 K (parallel)

Melting Temperature
1740 K

Thermal Conductivity
720 W/(m·K) at 20 K (parallel)
370 W/(m·K) at 20 K (perpendicular)
20 W/(m·K) at 194 K (parallel)
10 W/(m·K) at 194 K (perpendicular)
10.7 W/(m·K) at 323 K (parallel)
6.2 W/(m·K) at 323 K (perpendicular)

Thermal Expansion
8.0×10^{-6} K^{-1} at 273–353 K (parallel)
13.4×10^{-6} K^{-1} at 273–353 K (perpendicular)

Specific Heat
0.17 [cal/(g·K)] at 273 K

Debye Temperature
470 K

Hardness
741 kg/mm^2
(Knoop number with 500-g load)

Bandgap
8.4 eV

Solubility
Insoluble in H₂O

Elastic Moduli
Elastic stiffness
$c_{11} = 86.75$ GPa
$c_{12} = 6.87$ GPa
$c_{44} = 57.86$ GPa
$c_{13} = 12.6$ GPa
$c_{14} = -17.8$ GPa
$c_{33} = 106.1$ GPa
Elastic compliance
$s_{11} = 12.8$ TPa^{-1}
$s_{12} = -1.74$ TPa^{-1}
$s_{44} = 20.0$ TPa^{-1}
$s_{13} = -1.32$ TPa^{-1}
$s_{14} = 4.48$ TPa^{-1}
$s_{33} = 9.75$ TPa^{-1}
Young's modulus
76.5 GPa (perpendicular)
97.2 GPa (parallel)
Bulk modulus
36.4 GPa at 273 K
Shear modulus
31.14 GPa

Notes: Crystal quartz is an anisotropic, piezoelectric material. Historically, this material was of great interest until synthetic fused silica became available. Quartz crystals crack easily when heated, in contrast to fused silica. It is birefringent and can serve as a retarder (waveplate). See also fused silica (Chapter 7).

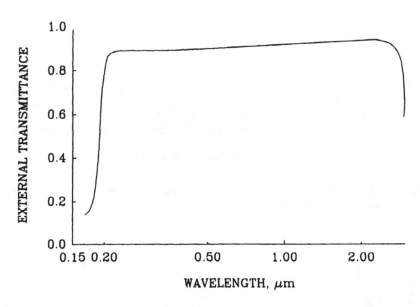

Figure 6.109 The external transmittance of natural crystal quartz, thickness 1.0 mm. [Adapted from D. E. McCarthy, *Appl. Opt. 6*: 1896 (1967).]

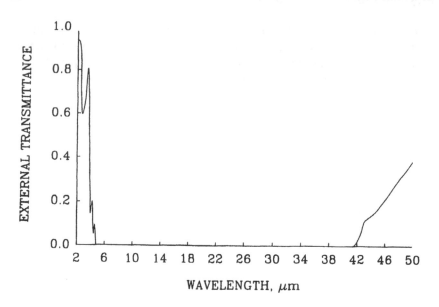

Figure 6.110   The external transmittance of crystal quartz. [Adapted from D.E. Mc-Carthy, *Appl. Opt. 2*: 591 (1963).]

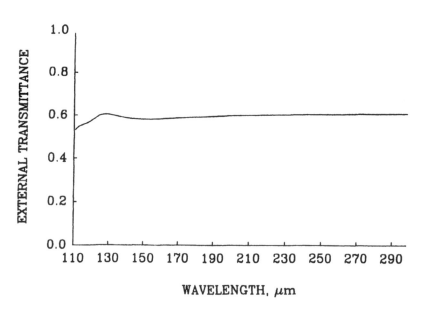

Figure 6.111   The external transmittance of quartz, thickness 0.6 mm. [Adapted from Hadni et al., *Rev. Opt. 38*: 463 (1959).]

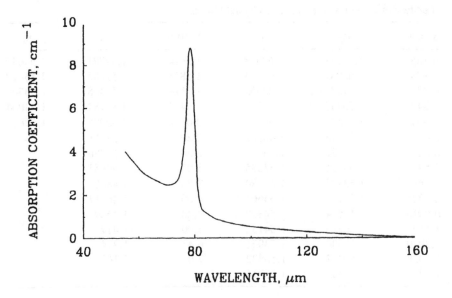

Figure 6.112 Absorption coefficient of crystal quartz. [Adapted from E. E. Russell and E. E. Bell, *J. Opt. Soc. Am. 57:* 341 (1967).]

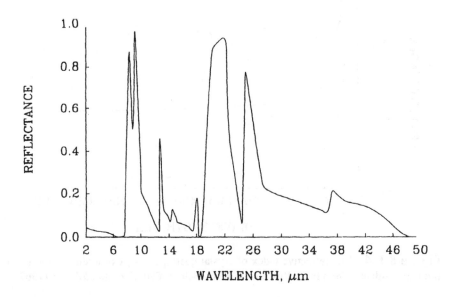

Figure 6.113 The reflectance of crystal quartz (ordinary or extraordinary ray not specified). [Adapted from D.E. McCarthy, *Appl. Opt. 2:* 591 (1963).]

Table 6.52   Refractive Index of Crystal Quartz

λ (μm)	n_O	n_E	λ (μm)	n_O	n_E
0.185	1.65751	1.68988	1.5414	1.52781	1.53630
0.198	1.65087	1.66394	1.6815	1.52583	1.53422
0.231	1.61395	1.62555	1.7614	1.52468	1.53301
0.340	1.56747	1.57737	1.9457	1.52184	1.53004
0.394	1.55846	1.56805	2.0531	1.52005	1.52823
0.434	1.55396	1.56339	2.30	1.51561	
0.508	1.54822	1.55746	2.60	1.50986	
0.5893	1.54424	1.55335	3.00	1.49953	
0.768	1.53903	1.54794	3.50	1.48451	
0.8325	1.53773	1.54661	4.00	1.46617	
0.9914	1.53514	1.54392	4.20	1.4569	
1.1592	1.53283	1.54152	5.00	1.417	
1.3070	1.53090	1.53951	6.45	1.274	
1.3958	1.52977	1.53832	7.0	1.167	
1.4792	1.52865	1.53716			

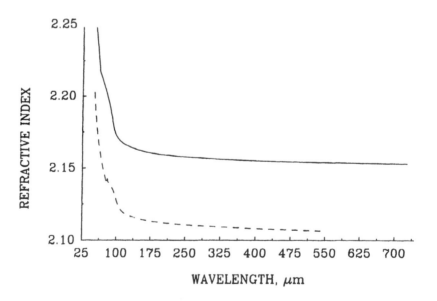

Figure 6.114   The refractive index of crystal quartz, (———) extraordinary and (----) ordinary. [Adapted from E. E. Russell and E. E. Bell, *J. Opt. Soc. Am. 57*: 341 (1967).]

Table 6.53 Temperature Coefficient of Refractive Index of Crystal Quartz

$\lambda(\mu m)$	dn_o/dT $(10^{-5}/°C)$	dn_E/dT $(10^{-5}/°C)$	$\lambda(\mu m)$	dn_o/dT $(10^{-5}/°C$	dn_E/dT $(10^{-5}/°C)$
0.202	+0.321	+0.267	0.298	−0.311	−0.415
0.206	0.253	0.198	0.313	−0.348	−0.450
0.210	0.193	0.143	0.325	−0.352	−0.469
0.214	0.124	0.083	0.340	−0.393	−0.501
0.219	0.074	0.027	0.361	−0.418	−0.521
0.224	0.017	−0.048	0.441	−0.475	−0.593
0.226	−0.008	−0.075	0.467	−0.485	−0.601
0.228	−0.027	−0.093	0.480	−0.499	−0.610
0.231	−0.052	−0.112	0.508	−0.514	−0.616
0.257	−0.186	−0.265	0.589	−0.539	−0.642
0.274	−0.235	−0.323	0.643	−0.549	−0.653
0.288	−0.279	−0.385			

Source: F.J. Micheli, *Ann. Physik* 4:7 (1902).

Silicon Nitride, Si$_3$N$_4$

Molecular Weight
140.28

Specific Gravity
3.24 [g/cm^3]

Crystal Class
Hexagonal

Transmission
Long-wavelength limit 4.6 μm
Short-wavelength limit 0.3 μm

Refractive Index
2.04 at 0.5893 μm

Melting Temperature
>2300 K

Thermal Conductivity
33 W/(m·K) at 300 K

Thermal Expansion
2.1 × 10^{-6} K^{-1} at 293–400 K (parallel)
1.1 × 10^{-6} K^{-1} at 273–400 K (perpendicular)

Specific Heat
0.210 [cal/(g·K)] at 533 K
0.380 [cal/(g·K)] at 1922 K

Hardness
3400 kg/mm^2 (Knoop number)

Bandgap
5.0 eV

Solubility
Insoluble in H$_2$O

Silver Bromide, AgBr

Molecular Weight
187.78

Specific Gravity
6.473 [g/cm^3]

Crystal Class
Cubic

Transmission
Long-wavelength limit ~35 μm
Short-wavelength limit 0.45 μm
(See Fig. 6.115.)

Absorption Coefficient
(See Figs. 6.116 and 6.117.)

Reflectivity
(See Fig. 6.118.)

Refractive Index
2.167 at 9.93 μm
2.162 at 12.67 μm
(See Figs. 6.119 and 6.120.)

Zero Dispersion Wavelength
6.7 μm

Dielectric Constant
13.1 for 10^6 Hz at 300 K

Melting Temperature
705 K

Thermal Conductivity
1.21 W/(m·K) at 273 K
0.711 W/(m·K) at 413 K

Thermal Expansion
19.8 × 10^{-6} K^{-1} at 75 K
33.8 × 10^{-6} K^{-1} at 300 K
35.0 × 10^{-6} K^{-1} at 350 K

Specific Heat
0.07 [cal/(g·K)] at 273 K

Debye Temperature
144 K

Hardness
7 kg/mm^2 (Knoop number)

Bandgap
4.3 eV

Solubility
Insoluble in H$_2$O

Elastic Moduli
Elastic stiffness
c_{11} = 56.3 GPa
c_{12} = 32.8 GPa
c_{44} = 7.25 GPa
Elastic compliance
s_{11} = 31.1 TPa^{-1}
s_{12} = −11.5 TPa^{-1}
s_{44} = 138 TPa^{-1}
Young's modulus
31.97 GPa
Bulk modulus
44.03 GPa

Notes: Silver bromide is relatively inexpensive. It is not hygroscopic and is not as toxic as KRS-5. It is soft and decomposes in ultraviolet light.

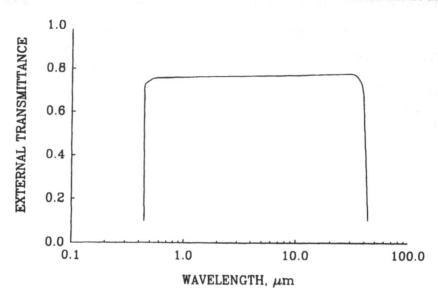

Figure 6.115 The external transmittance of silver bromide, thickness 1 mm. [Adapted from *Laser Focus Buyer's Guide*, 1982, p. 507.]

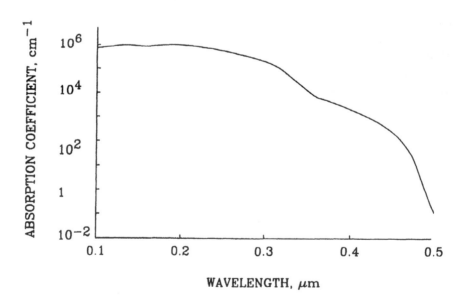

Figure 6.116 The absorption coefficient of silver bromide as a function of wavelength. [Adapted from V. I. Saunders, *J. Opt. Soc. Am. 67*: 6 (1977).]

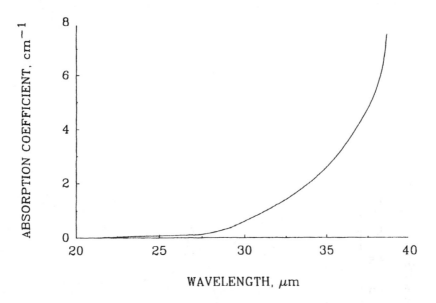

Figure 6.117   The absorption coefficient of silver bromide. [Adapted from G.W. Luckey, *J. Phys. Chem. 57*: 791 (1953).]

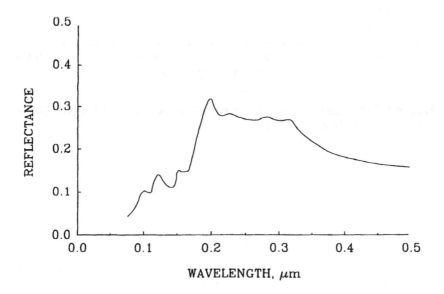

Figure 6.118   The reflectance of silver bromide at room temperature. [Adapted from F. Bassani, R. S. Knox, and W. B. Fowler, *Phys. Rev. 137*: A1217 (1965).]

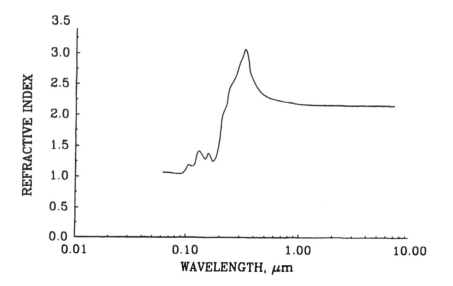

Figure 6.119   The refractive index of silver bromide. [Adapted from J.J. White, *J. Opt. Soc. Am. 62*(2): 212 (1974).]

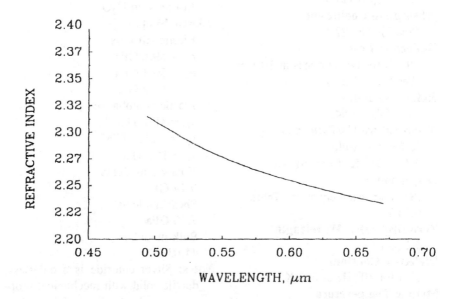

Figure 6.120 The refractive index of silver bromide. [Adapted from H. Schroeder, *Z. Physik 67:* 24–36 (1931).]

Silver Chloride, AgCl

Molecular Weight
143.34

Specific Gravity
5.56 [g/cm^3]

Crystal Class
Cubic

Transmission
Long-wavelength limit ~28 μm
Short-wavelength limit 0.4 μm
(See Fig. 6.121.)

Absorption Coefficient
(See Fig. 6.122.)

Reflection Loss
19.5% for two surfaces at 10 μm
(See Fig. 6.123.)

Refractive Index
(See Table 6.54.)

Temperature Coefficient of Refractive Index
6.1×10^{-5} K^{-1} at 0.61 μm

Dispersion
(See dispersion equation, Table 6.54.)

Zero Dispersion Wavelength
5.1 μm

Dielectric Constant
12.3 for 10^6 Hz at 293 K

Melting Temperature
731 K

Thermal Conductivity
1.12 W/(m·K) at 295 K
1.05 W/(m·K) at 372 K

Thermal Expansion
22.8×10^{-6} K^{-1} at 120 K
31.0×10^{-6} K^{-1} at 302 K
82.3×10^{-6} K^{-1} at 713 K

Specific Heat
0.085 [cal/(g·K)] at 273 K

Debye Temperature
183 K

Hardness
9.5 kg/mm^2
(Knoop number with 200-g load)

Bandgap
5.1 eV

Solubility
Insoluble in H_2O

Elastic Moduli
Elastic stiffness
$c_{11} = 59.6$ GPa
$c_{12} = 36.1$ GPa
$c_{44} = 6.22$ GPa
Elastic compliance
$s_{11} = 31.1$ TPa^{-1}
$s_{12} = -11.7$ TPa^{-1}
$s_{44} = 161$ TPa^{-1}
Young's modulus
0.14 GPa
Shear modulus
7.10 GPa
Bulk modulus
44.03 GPa

Notes: Silver chloride is a colorless, ductile solid with mechanical properties similar to those of lead. It is extremely corrosive to metal and can be fused to glass or silver by a permanent vacuum-type seal. Furthermore, its cold-water solubility is zero, but it is soluble in ammonium hydroxide, sodium thiosulfate (hypo), and potassium cyanide.

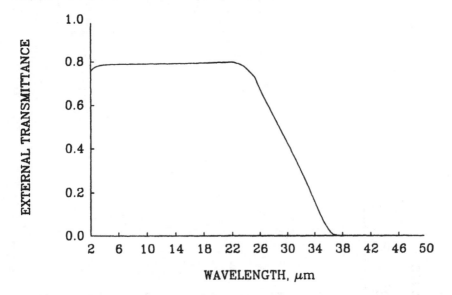

Figure 6.121 The external transmittance of silver chloride, thickness 0.5 mm. [Adapted from D. E. McCarthy, *Appl. Opt. 2*: 591 (1963).]

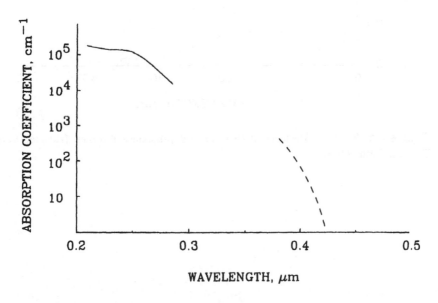

Figure 6.122 The absorption coefficient of silver chloride at 299 K in the ultraviolet range [adapted from S. Tutihasi, *Phys. Rev. 105*: 882 (1957)]; and at 299 K in the visible range [adapted from F. Moser and F. Urbach, *Phys. Rev. 102*: 1519 (1956).]

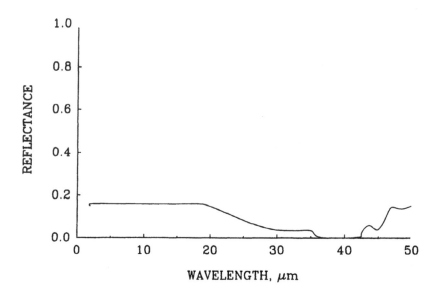

Figure 6.123 The reflectance of silver chloride, thickness 0.5 mm. [Adapted from D.E. McCarthy (1963).]

Table 6.54   Refractive Index of Silver Chloride at 23.9°C

λ (μm)	n	λ (μm)	n	λ (μm)	n	λ (μm)	n
0.5	2.09648	2.3	2.00465	4.0	1.99983	12.5	1.96742
0.6	2.06385	2.4	2.00424	4.5	1.99866	13.0	1.96444
0.7	2.04590	2.5	2.00386	5.0	1.99745	13.5	1.96133
0.8	2.03485	2.6	2.00351	5.5	1.99618	14.0	1.95807
0.9	2.02752	2.7	2.00318	6.0	1.99438	14.5	1.95467
1.0	2.02239	2.8	2.00287	6.5	1.99339	15.0	1.95113
1.1	2.01865	2.9	2.00258	7.0	1.99185	15.5	1.94743
1.2	2.01582	3.0	2.00230	7.5	1.99021	16.0	1.94358
1.3	2.01363	3.1	2.00203	8.0	1.98847	16.5	1.93958
1.4	2.01189	3.2	2.00177	8.5	1.98661	17.0	1.93542
1.5	2.01047	3.3	2.00151	9.0	1.98464	17.5	1.93109
1.6	2.00931	3.4	2.00126	9.5	1.98255	18.0	1.92660
1.7	2.00833	3.5	2.00102	10.0	1.98034	18.5	1.92194
1.8	2.00750	3.6	2.00078	10.5	1.97801	19.0	1.91710
1.9	2.00678	3.7	2.00054	11.0	1.97556	19.5	1.91208
2.0	2.00615	3.8	2.00030	11.5	1.97297	20.0	1.90688
2.1	2.00559	3.9	2.00007	12.0	1.97026	20.5	1.90149
2.2	2.00510						

Dispersion equation:

$$n^2 = 4.00804 - 0.00085111\lambda^2 - 0.0000001976\lambda^4 + \frac{0.079086}{\lambda^2 - 0.04585}$$

Temperature coefficient of index:

$$\frac{dn}{dT} \approx 6.1 \times 10^{-5}/°C \quad \text{at } 0.61 \text{ μm}$$

Source: W.L. Tilton and E.K. Plyler, *Natl. Bur. Stand. J. Res.* 47:25, 1951.

Sodium Chloride, NaCl

Molecular Weight
58.45
Specific Gravity
2.165 [g/cm^3]
Crystal Class
Cubic
Transmission
Long-wavelength limit 18 μm
Short-wavelength limit 0.17 μm
(See Fig. 6.124.)
Absorption Coefficient
7×10^{-6} cm^{-1} at 1.06 μm
Reflection Loss
7.5% for two surfaces at 10 μm
(See Fig. 6.125.)
Refractive Index
(See Table 6.55.)
**Temperature Coefficient of
Refractive Index**
-32.81×10^{-6} K^{-1} at 4.96 μm
-32.41×10^{-6} K^{-1} at 6.4 μm
-25.23×10^{-6} K^{-1} at 8.85 μm
(See Table 6.56.)
Zero Dispersion Wavelength
2.9 μm
Dielectric Constant
5.9 for 10^2 to 2.5×10^{10} Hz at
298 K
6.35–5.97 for 10^2 to 2.5×10^{10} Hz
at 358 K
Melting Temperature
1070 K
Thermal Conductivity
35.1 W/(m·K) at 80 K
6.5 W/(m·K) at 289 K
4.85 W/(m·K) at 400 K
Thermal Expansion
19.0×10^{-6} K^{-1} at 75 K
40.0×10^{-6} K^{-1} at 273 K
69.5×10^{-6} K^{-1} at 1000 K

Specific Heat
0.20 [cal/(g·K)] at 273 K
Debye Temperature
321 K
Hardness
15.2 kg/mm^2 (Knoop number with
200-g load in ⟨110⟩ direction)
18.2 kg/mm^2 (Knoop number with
200-g load in ⟨100⟩ direction)
Bandgap
9.0 eV
Solubility
35.7 g/(100 g H$_2$O) at 273 K
Elastic Moduli
Elastic stiffness
$c_{11} = 49.1$ GPa
$c_{12} = 12.8$ GPa
$c_{44} = 12.8$ GPa
Elastic compliance
$s_{11} = 22.9$ TPa^{-1}
$s_{12} = -4.8$ TPa^{-1}
$s_{44} = 78.3$ TPa^{-1}
Young's modulus
39.96 GPa
Rupture modulus
0.00393 GPa
Shear modulus
8.97 GPa
Bulk modulus
24.32 GPa

Notes: Sodium chloride, the natural
form of which is ordinary rock salt,
polishes easily, and, although hygroscopic, can be protected by
evaporated coatings and plastics;
selenium films have been used suc-
cessfully. It is soluble in glycerin.
Synthetic single crystals can be
grown by the Czochralski method.
Sodium chloride is used for IR
spectrometer window cells and laser windows. Diameters as large as
12 in. are available.

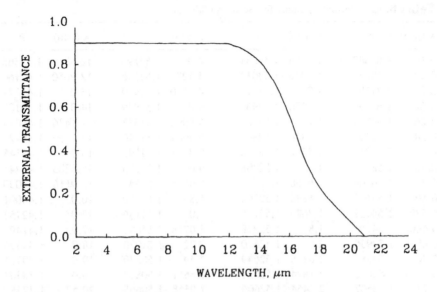

Figure 6.124 The external transmittance of sodium chloride, thickness 10 mm. [Adapted from R. A. Smith, F. E. Jones, and R. P. Chasmar, *The Detection and Measurement of Infrared Radiation*, Oxford University Press, New York, 1957.]

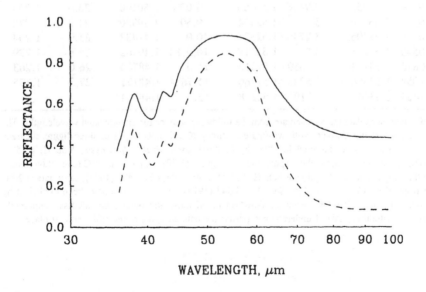

Figure 6.125 Reflectance of plane-polarized light by sodium chloride at incident angle of 52; perpendicular (– – –) and parallel (——). [Adapted from A. Mitsuishi, *J. Opt. Soc. Am. 50*: 433 (1960).]

Table 6.55 Refractive Index for Sodium Chloride

λ (μm)	n	λ (μm)	n	λ (μm)	n	λ (μm)	n
0.19	1.85343	1.1786	1.53031	4.0	1.52190	12.50	1.47568
0.20	1.79073	1.2016	1.53014	4.1230	1.52156	12.9650	1.47160
0.22	1.71591	1.2604	1.52971	4.7120	1.51979	13.0	1.47141
0.24	1.67197	1.3126	1.52937	5.0	1.51899	14.0	1.46189
0.26	1.64294	1.4	1.52888	5.0092	1.51883	14.1436	1.46044
0.28	1.62239	1.4874	1.52845	5.3009	1.51790	14.7330	1.45427
0.30	1.60714	1.5552	1.52815	5.8932	1.51593	15.0	1.45145
0.35	1.58232	1.6	1.52798	6.0	1.51548	15.3223	1.44743
0.40	1.56769	1.6368	1.52781	6.4825	1.51347	15.9116	1.44090
0.50	1.55175	1.6848	1.52764	6.80	1.51200	16.0	1.44001
0.589	1.54427	1.7670	1.52736	7.0	1.51136	17.0	1.42753
0.6400	1.54141	1.8	1.52728	7.0718	1.51093	17.93	1.4149
0.6874	1.53930	2.0	1.52670	7.22	1.51020	18.0	1.41393
0.70	1.53881	2.0736	1.52649	7.59	1.50850	19.0	1.39914
0.7604	1.53682	2.1824	1.52621	7.6611	1.50822	20.0	1.38307
0.7858	1.53607	2.2464	1.52606	7.9558	1.50665	20.57	1.3735
0.80	1.53575	2.3	1.52594	8.0	1.50655	21.0	1.36563
0.8835	1.53395	2.3560	1.52579	8.04	1.5064	21.3	1.352
0.90	1.53366	2.6	1.52525	8.8398	1.50192	22.3	1.3403
0.9033	1.53361	2.6505	1.52512	9.0	1.50105	22.8	1.318
0.9724	1.53253	2.9466	1.52466	9.00	1.50100	23.6	1.299
1.0	1.53216	3.0	1.52434	9.50	1.49980	24.2	1.278
1.0084	1.53206	3.2736	1.52371	10.0	1.49482	25.0	1.254
1.0540	1.53153	3.5	1.52317	10.0184	1.49462	25.8	1.229
1.0810	1.53123	3.5359	1.52312	11.0	1.48783	26.6	1.203
1.1058	1.53098	3.6288	1.52286	11.7864	1.48171	27.3	1.175
1.1420	1.53063	3.8192	1.52238	12.0	1.48004		

Refractive index data for rock salt are from the following sources for the indicated wavelengths: The data for wavelengths given with two-figure accuracy (0.19, 0.50, . . .) or three-figure accuracy (10.0, 11.0, . . .) are reported for 20°C by F. Kohlrausch, *Praktische Physik*, Vol. II, p. 528, Teubner, Leipzig, 1943; the data reported to four figures (1.299) in index at 18°C (even though they are three figures in wavelength) are from H.W. Hohls, *Ann. Physik 29:*433 (1937); other data at 20°C are from W.W. Coblentz, *J. Opt. Soc. Am. 4:*443 (1914). Still more data have been published by Langley, Martens, Paschen, Rubens, Trowbridge, Nichols, and others, but all have apparently measured natural crystals of undetermined purity; the data all agree to the fifth decimal place.

Table 6.56 Temperature Coefficient of Refractive Index of Sodium Chloride at about 60°C

λ (μm)	dn/dT $(10^{-5}/°C)$	λ (μm)	dn/dT $(10^{-5}/°C)$
0.202	3.134	0.589	−3.622
0.206	2.229	0.643	−3.636
0.210	1.570	0.656	−3.652
0.214	0.861	1.1	−3.642
0.219	0.235	1.6	−3.557
0.224	−0.187	2.7	−3.427
0.226	−0.382	3.96	−3.286
0.229	−0.598	4.96	−3.172
0.231	−0.757	6.4	−3.149
0.257	−1.979	8.85	−2.405
0.274	−2.396	10.02	−2.2
0.288	−2.602	11.79	−1.6
0.298	−2.727	12.97	−1.4
0.313	−2.862	14.14	−1.2
0.325	−2.987	14.73	−1.0
0.340	−3.068	15.32	−0.8
0.361	−3.194	15.91	−0.7
0.441	−3.425	17.93	−0.5
0.467	−3.454	20.57	0
0.480	−3.468	22.3	0
0.508	−3.517		

Source: F.J. Micheli, *Ann. Physik 4*:7 (1902) for the region 0.202–0.643 μm; E. Liebreich, *Verhandl. Deut. Physik. Ges. 13*:709 (1911) for the region 0.656–15.91 μm; and H. Rubens and E.F. Nichols, *Weid. Ann. 60*:454 (1897) for the region 17.93–22.3 μm.

Sodium Fluoride, NaF

Molecular Weight
42.00

Specific Gravity
2.558 [g/cm^3]

Crystal Class
Cubic

Transmission
Long-wavelength limit 15 μm
Short-wavelength limit <0.19 μm
(See Fig. 6.126.)

Absorption Coefficient
4.3×10^{-4} cm^{-1} at 2.7 μm
2.7×10^{-4} cm^{-1} at 3.8 μm

Reflection Loss
3.6% for two surfaces at 4 μm

Refractive Index
(See Table 6.57.)

Temperature Coefficient of Refractive Index
-1.6×10^{-5} K^{-1} at 0.546 μm at 18–80 K
-1.6×10^{-5} K^{-1} at 3.5 μm at 18–80 K
-0.7×10^{-5} K^{-1} at 8.5 μm at 18–80 K
(See Fig. 6.127.)

Zero Dispersion Wavelength
1.7 μm

Dielectric Constant
6 for 2×10^6 Hz at 292 K

Melting Temperature
1270 K

Thermal Conductivity
9.2 W/(m·K) at 298 K

Thermal Expansion
9.2×10^{-6} K^{-1} at 75 K
33.5×10^{-6} K^{-1} at 300 K
62.3×10^{-6} K^{-1} at 1200 K

Specific Heat
0.26 [cal/(g·K)] at 273 K

Debye Temperature
492 K

Hardness
60 kg/mm^2 (Knoop number in ⟨100⟩ direction)

Bandgap
10.5 eV

Solubility
4.22 g/(100 g H$_2$O) at 291 K

Elastic Moduli
Elastic stiffness
$c_{11} = 97.0$ GPa
$c_{12} = 24.2$ GPa
$c_{44} = 28.1$ GPa
Elastic compliance
$s_{11} = 11.5$ TPa^{-1}
$s_{12} = -2.3$ TPa^{-1}
$s_{44} = 35.6$ TPa^{-1}
Bulk modulus
62.00 GPa

Notes: Sodium fluoride transmits in about the same region as calcium fluoride and lithium fluoride; it is less useful mechanically because it exhibits cleavage similar to that of sodium chloride. Sodium fluoride has some significance, however, in cases where a remarkably low refractive index is needed.

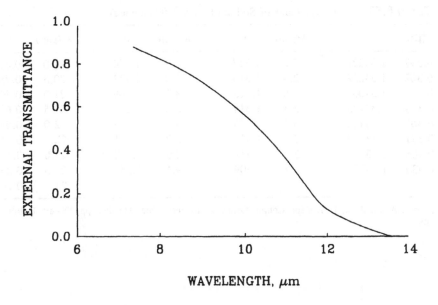

Figure 6.126 The external transmittance of sodium fluoride, 10 mm thickness. [Data of S.S. Ballard, private communication.]

Table 6.57 Refractive Index of Sodium Fluoride

λ(μm)	n	λ(μm)	n	λ (μm)	n	λ (μm)	n
0.186	1.3930	0.486	1.32818	3.9	1.309	9.4	1.251
0.193	1.3854	0.546	1.32640	4.1	1.308	9.8	1.241
0.199	1.3805	0.588	1.32552	4.5	1.305	10.3	1.233
0.203	1.3772	0.589	1.32549	4.7	1.303	10.8	1.222
0.206	1.3745	0.656	1.32436	4.9	1.302	11.3	1.209
0.210	1.3718	0.707	1.32372	5.1	1.301	11.7	1.193
0.214	1.3691	0.720	1.32349	5.3	1.299	12.5	1.180
0.219	1.3665	0.768	1.32307	5.5	1.297	13.2	1.163
0.227	1.3630	0.811	1.32272	5.7	1.295	13.8	1.142
0.231	1.3606	0.842	1.32247	5.9	1.294	14.3	1.118
0.237	1.3586	0.912	1.32198	6.1	1.292	15.1	1.093
0.240	1.35793	1.014	1.32150	6.3	1.290	15.9	1.065
0.248	1.35500	1.083	1.32125	6.5	1.288	16.7	1.034
0.254	1.35325	1.27	1.320	6.7	1.286	17.3	1.000
0.265	1.34999	1.48	1.319	6.9	1.284	18.1	0.963
0.270	1.34881	1.67	1.318	7.1	1.281	18.6	0.924
0.280	1.34645	1.83	1.318	7.3	1.279	19.3	0.881
0.289	1.34462	2.0	1.317	7.5	1.277	19.7	0.838

(*continued*)

Table 6.57 Refractive Index of Sodium Fluoride (Continued)

λ(μm)	n	λ(μm)	n	λ (μm)	n	λ (μm)	n
0.297	1.34328	2.2	1.317	7.7	1.274	20.0	0.82
0.302	1.34232	2.4	1.316	7.9	1.272	20.5	0.75
0.313	1.34062	2.6	1.315	8.1	1.269	21.0	0.70
0.334	1.33795	2.8	1.314	8.3	1.266	21.5	0.65
0.366	1.33482	3.1	1.313	8.5	1.263	22.0	0.55
0.391	1.33290	3.3	1.312	8.7	1.261	22.5	0.45
0.405	1.33194	3.5	1.311	8.9	1.258	23.0	0.33
0.436	1.33025	3.7	1.309	9.1	1.252	23.5	0.25
						24.0	0.24

Source: A. Smakula, U.S. Dept. Comm. Office Tech. Serv. Doc. 111,052. pp. 88–89, October 1952.

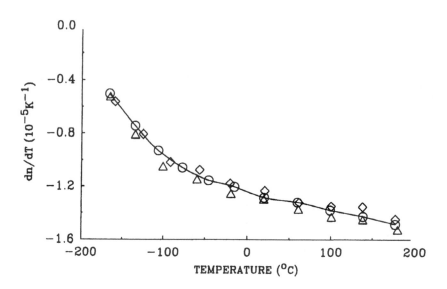

Figure 6.127 The temperature coefficient of refractive index of sodium fluoride for wavelengths 0.6328 μm (○), 1.15 μm (△), and 3.39 μm (◇). [Adapted from A. J. Glass and A. H. Guenther (Eds.), *Laser Induced Damage in Optical Materials*, 1977, p. 80.]

Strontium Barium Niobate, SrBaNb₃O₆

Transmission
(See Fig. 6.128.)
Refractive Index
2.23 at 1 μm (extraordinary)
2.22 at 1 μm (ordinary)

Solubility
<0.005 g/(100 g H₂O)

Notes: Strontium barium niobate is a ferroelectric with promising nonlinear and electrooptical properties.

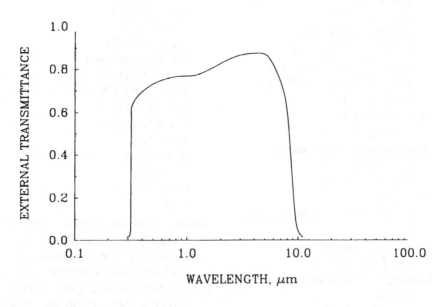

Figure 6.128 The external transmittance of strontium barium niobate, thickness 0.03 mm. [Adapted from *Laser Focus Buyer's Guide, 233* (1978).]

Strontium Fluoride, SrF$_2$

Molecular Weight
   125.63
Specific Gravity
   4.24 [g/cm^3]
Crystal Class
   Cubic
Transmission
   (See Fig. 6.129.)
Absorption Coefficient
   $<10^{-4}$ cm^{-1} at 4 μm
   60×10^{-4} cm^{-1} at 7.5 μm
   600×10^{-4} cm^{-1} at 10.6 μm
   (See Fig. 6.130.)
Refractive Index
   (See Fig. 6.131.)
**Temperature Coefficient of
   Refractive Index**
   (See Fig. 6.132.)
Dispersion
   (See dispersion equation, Table
   6.58.)
Dielectric Constant
   7.69 for 2×10^6 Hz
Melting Temperature
   1190 K
Thermal Conductivity
   1.42 W/(m·K) at 298.2 K
Thermal Expansion
   5.2×10^{-6} K^{-1} at 75 K
   18.4×10^{-6} K^{-1} at 293 K
   31.8×10^{-6} K^{-1} at 1300 K
Specific Heat
   0.130 [cal/(g·K)] at 274 K

Hardness
   130 kg/mm^2 (Knoop number)
Bandgap
   3.4 eV
Solubility
   0.011 g/(100 g H$_2$O) at 273 K
Elastic Moduli
   Elastic stiffness
   $c_{11} = 124$ GPa
   $c_{12} = 45$ GPa
   $c_{44} = 31.7$ GPa
   Elastic compliance
   $s_{11} = 9.89$ TPa^{-1}
   $s_{12} = -2.59$ TPa^{-1}
   $s_{44} = 31.6$ TPa^{-1}
   Young's modulus
   99.91 GPa
   Bulk modulus
   24.65 GPa
   Poisson's ratio
   0.25

Notes: Strontium fluoride is hard and is slightly soluble in sulfuric acid. It is used as a laser host material and as a laser window material at chemical laser wavelengths. Windows made from SrF$_2$ permit large power transmission before they fracture or optically distort the beam. Optovac, Inc. produces this crystal in sizes up to 6 in. in diameter. Also cast polycrystalline strontium fluoride has been produced by Raytheon Research Labs.

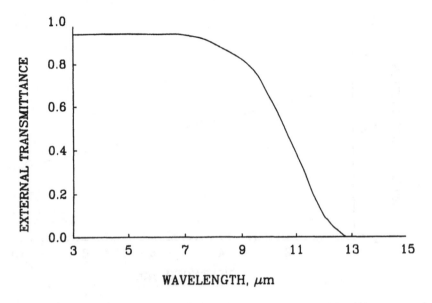

Figure 6.129 The external transmittance of strontium fluoride, thickness 10.0 mm. [Adapted from Optovac, Inc., Bulletin 50, North Brookfield, Mass., 1968, p. V-21.]

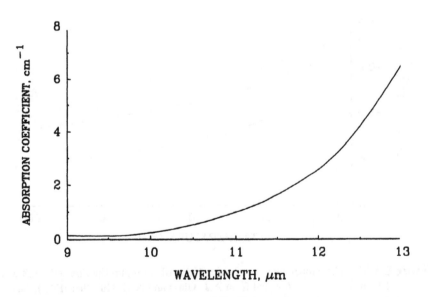

Figure 6.130 Absorption coefficient of strontium fluoride. [Adapted from Optovac Inc. Bulletin 50, North Brookfield, Mass., 1968, p. V-21.]

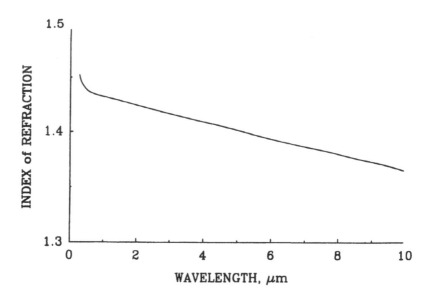

Figure 6.131   The refractive index of strontium fluoride. [Adapted from Optovac Inc. Bulletin 50, North Brookfield, Mass., 1968, p. V-21.]

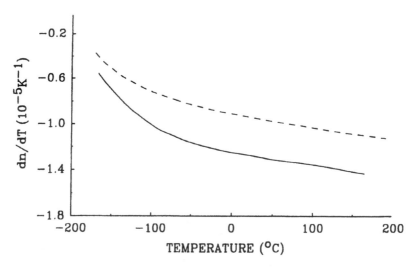

Figure 6.132   The temperature of refractive index of strontium fluoride at 0.6328 μm (----), and 10.6 μm (——). [Adapted from A.J. Glass and A.H. Guenther (Eds.), *Laser Induced Damage in Optical Materials*, (1977).]

Table 6.58 Dispersion Equation for Strontium Fluoride

The refractive index of fusion-cast SrF_2 was measured from 0.2138 to 11.475 μm at 20°C. Fifty-three experimental data points were fitted to a three-term Sellmeier-type dispersion equation,

$$n^2 - 1 = \sum_j \frac{A_j \lambda^2}{\lambda^2 - \lambda_j^2}$$

Constants for this equation to calculate the refractive index of SrF_2 at 20°C are

$$A_1 = 0.67805894 \quad \lambda_1 = 0.05628989$$
$$A_2 = 0.37140533 \quad \lambda_2 = 0.10801027$$
$$A_3 = 3.3485284 \quad \lambda_3 = 39.906666$$

Source: M. J. Dodge, *Laser Induced Damage in Optical Materials*, NBS Spec. Publ. 541, 1978, pp. 55–56.

Strontium Titanate, SrTiO$_3$

Molecular Weight
183.53
Crystal Class
Cubic
Transmission
Long-wavelength limit <7 μm
Short-wavelength limit 0.4 μm
(See Fig. 6.133.)
Absorption Coefficient
6×10^{-3} cm^{-1} at 2.7 μm
20×10^{-3} cm^{-1} at 3.8 μm
(NL Industries sample)
(See Fig. 6.134.)
Reflection Loss
29% for two surfaces at 0.6 μm
(See Fig. 6.135.)
Refractive Index
(See Table 6.59.)
Dielectric Constant
306 for 10^2–10^5 Hz at 298 K
Melting Temperature
2350 K
Thermal Conductivity
18.4 W/(m·K) at 80 K
11.2 W/(m·K) at 300 K

Thermal Expansion
9.4×10^{-6} K^{-1}
Specific Heat
0.13 [cal/(g·K)] at 300 K
1.70 [cal/(g·K)] at 1800 K
Hardness
595 kg/mm^2 (Knoop number)
Bandgap
4.1 eV
Solubility
Insoluble in H$_2$O
Elastic Moduli
Elastic stiffness
$c_{11} = 316$ GPa
$c_{12} = 102$ GPa
$c_{44} = 123$ GPa
Elastic compliance
$s_{11} = 3.75$ TPa^{-1}
$s_{12} = -0.92$ TPa^{-1}
$s_{44} = 8.15$ TPa^{-1}

Notes: Strontium titanate is a ferroelectric material. It is of interest for special applications such as immersion lenses.

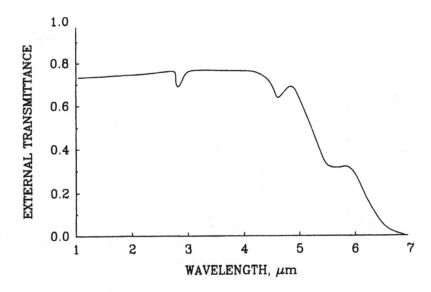

Figure 6.133 The external transmittance of single-crystal strontium titanate at 299 K, thickness 3 mm. [Adapted from C. D. Salzberg, *J. Opt. Soc. Am. 51*: 1149 (1961).]

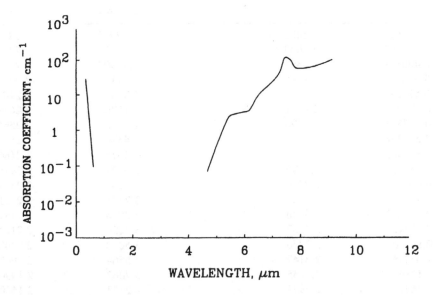

Figure 6.134 Absorption coefficient of synthetic strontium titanate. [Adapted from S. B. Levin et al., *J. Opt. Soc. Am. 45*: 737 (1955).]

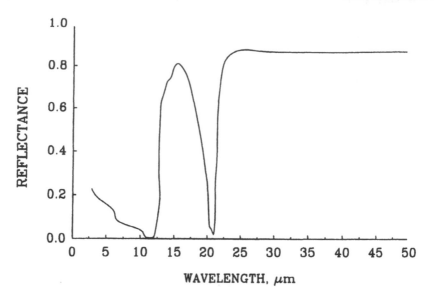

Figure 6.135   The reflectance of strontium titanate, thickness 1.0 mm. [Adapted from D. E. McCarthy (no date).]

Table 6.59   Refractive Index of Strontium Titanate

λ (μm)	n	λ (μm)	n
0.404657	2.6481	1.52952	2.2848
0.435384	2.5680	1.7012	2.2783
0.486132	2.4897	1.81307	2.2744
0.546074	2.4346	1.871	2.2710
0.576960	2.4149	1.918	2.2704
0.579066	2.4137	2.1526	2.2624
0.587562	2.4090	2.3126	2.2564
0.589262	2.4081	2.4374	2.2525
0.643847	2.3837	2.5628	2.2466
0.656279	2.3790	2.6707	2.2438
0.667815	2.3750	2.7248	2.2404
0.706519	2.3634	3.2434	2.2211
0.767858	2.3488	3.3026	2.2181
0.85212	2.3337	3.4226	2.2124
0.89440	2.3276	3.5070	2.2088
1.01398	2.3147	3.5564	2.2063
1.12866	2.3055	3.7067	2.1990
1.3622	2.2921	4.2553	2.1680
1.39506	2.2906	5.138	2.1119
1.517	2.2859	5.3034	2.1004

Source: I.H. Malitson, National Bureau of Standards, 1960.

Tellurium, Te

Atomic Weight
127.61

Specific Gravity
6.24 [g/cm^3] at 293 K

Crystal Class
Trigonal

Transmission
(See Figs. 6.136 and 6.137.)

Refractive Index
(See Table 6.60.)

Temperature Coefficient of Refractive Index
0.3 cm^{-1} at 10.6 μm

Melting Temperature
723 K

Thermal Conductivity
6.3 W/(m·K)

Thermal Expansion
−4.8 × 10^{-6} K^{-1} at 100 K (parallel)
26.0 × 10^{-6} K^{-1} at 100 K (perpendicular)
−2.6 × 10^{-6} K^{-1} at 260 K (parallel)
28.4 × 10^{-6} K^{-1} at 260 K (perpendicular)
−1.52 × 10^{-6} K^{-1} at 303 K (parallel)
27.5 × 10^{-6} K^{-1} at 303 K (perpendicular)

Specific Heat
0.0483 [cal/(g·K)] at 561–646 K

Debye Temperature
153 K

Hardness
18 kg/mm^2 (Knoop number)

Bandgap
0.33 eV

Solubility
Insoluble in H_2O

Elastic Moduli
Elastic stiffness
c_{11} = 34.4 GPa
c_{12} = 9.0 GPa
c_{13} = 24.9 GPa
c_{14} = 13.1 GPa
c_{33} = 70.8 GPa
c_{44} = 32.7 GPalh
Elastic compliance
s_{11} = 53.4 TPa^{-1}
s_{12} = −16.1 TPa^{-1}
s_{13} = −13.6 TPa^{-1}
s_{14} = 26.7 TPa^{-1}
s_{33} = 24.3 TPa^{-1}
s_{44} = 52.1 TPa^{-1}

Notes: Tellurium is an anisotropic material. It exhibits dichroism in the infrared as a single crystal and in the visible in needle form. When tellurium is ground, a conducting layer tends to form on its surface; this layer can be removed by chemical etching, although this does not leave an optical surface. Optical polishing can probably be accomplished without this effect; thin layers of tellurium have been made that have a good optical finish.

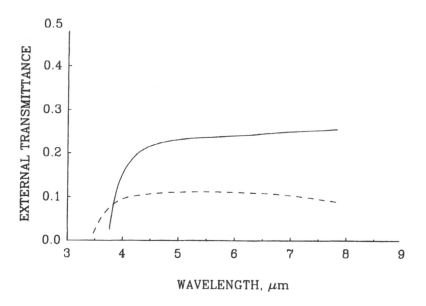

Figure 6.136   The external transmittance of tellurium for two polarizations perpendicular (——), parallel (– – –), thickness 0.85 mm. [Adapted from J.J. Loferski, *Phys. Rev. 93*: 707 (1954) and R.S. Caldwell and H.V. Fan, *Phys. Rev. 114*: 664 (1954).]

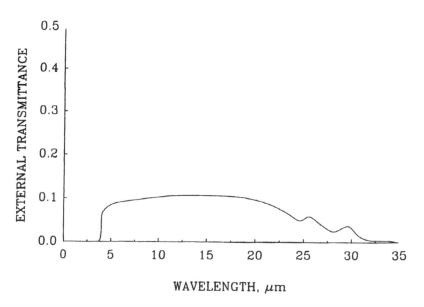

Figure 6.137   The external transmittance of polycrystalline tellurium, thickness 3.5 mm. [Adapted from D.E. McCarthy, *Appl. Opt. 7*: 1997 (1968).]

Table 6.60   Refractive Index of Tellurium

λ (μm)	n_O	n_E	λ (μm)	n_O	n_E
4.0	4.929	6.372	9.0	4.802	6.253
5.0	4.864	6.316	10.0	4.796	6.246
6.0	4.838	6.286	12.0	4.789	6.237
7.0	4.821	6.270	14.0	4.785	6.230
8.0	4.809	6.257			

The data are for single-crystal tellurium [from R. S. Caldwell and H. Y. Fan, *Phys. Rev. 114:* 664 (1959)].

Thallium Bromide, TlBr

Molecular Weight
284.31
Specific Gravity
7.557 [g/cm^3]
Crystal Class
Cubic
Transmission
Long-wavelength limit 40 μm
Short-wavelength limit 0.44 μm
(See Fig. 6.138.)
Reflection Loss
29% for two surfaces at 0.6 μm
Refractive Index
(See Table 6.61.)
Zero Dispersion Wavelength
8.5 μm
Dielectric Constant
30.3 for 10^3–10^7 Hz at 298 K
Melting Temperature
733 K
Thermal Conductivity
0.59 W/(m·K) at 316 K
0.586 W/(m·K) at 343 K
Thermal Expansion
51 × 10^{-6} K^{-1} at 293–333 K
Specific Heat
0.045 [cal/(g·K)] at 293 K

Hardness
11.9 kg/mm^2
(Knoop number with 500-g load in
the ⟨110⟩ and ⟨100⟩ directions)
Bandgap
3.1 eV (excitonic)
Solubility
0.05 g/(100 g H$_2$O)
Elastic Moduli
Elastic stiffness
$c_{11} = 37.6$ GPa
$c_{12} = 14.8$ GPa
$c_{44} = 7.54$ GPa
Elastic compliance
$s_{11} = 34.2$ TPa^{-1}
$s_{12} = -9.6$ TPa^{-1}
$s_{44} = 133$ TPa^{-1}
Young's modulus
29.49 GPa
Shear modulus
7.58 GPa
Bulk modulus
22.46 GPa

Notes: Thallium bromide is soft and difficult to polish. It can be ground a very small amount at a time without cracking or chipping. It bends like lead and is only slightly soluble in water.

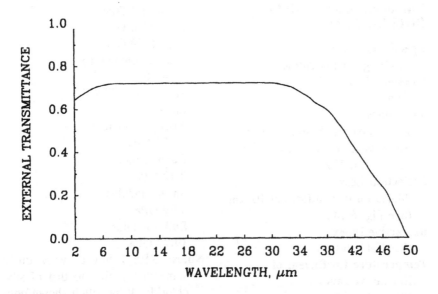

Figure 6.138 The external transmittance of thallium bromide, thickness 2 mm. [Adapted from D. E. McCarthy, *Appl. Opt. 4*: 317 (1965).]

Table 6.61 Refractive Index of Thallium Bromide

λ (μm)	n	λ (μm)	n
0.438	2.652	0.750	2.350
0.546	2.452	9.98	2.338
0.578	2.424	13.95	2.321
0.589	2.418	19.76	2.321
0.650	2.384	24.39	2.321

The indices in the visible region are from T. F. W. Barth, *Am. Mineral. 14:* 358 (1929). The indices in the infrared region were measured at 45°C by D. E. McCarthy, *Appl. Opt. 4:* 878 (1965).

Thallium Bromoiodide (KRS-5), Tl(Br,I)

Specific Gravity
7.371 [g/cm^3] at 289 K

Crystal Class
Cubic

Transmission
Long-wavelength limit 40 μm
Short-wavelength limit 0.5 μm
(See Fig. 6.139.)

Reflection Loss
28.4% for two surfaces at 10 μm
(See Fig. 6.140.)

Refractive Index
(See Table 6.62.)

Temperature Coefficient of Refractive Index
(See Table 6.63.)

Dispersion
(See dispersion equation, Table 6.62.)

Dielectric Constant
32.9–32.5 for 10^2–10^7 Hz at 298 K

Melting Temperature
687 K

Thermal Conductivity
0.544 W/(m·K) at 293 K

Thermal Expansion
60×10^{-6} K^{-1} at 223 K
58×10^{-6} K^{-1} at 293 K
57×10^{-6} K^{-1} at 473 K

Hardness
40.2 kg/mm^2 (Knoop number with 200-g load)

Solubility
0.05 g/(100 g H$_2$O)

Elastic Moduli
Elastic stiffness

$c_{11} = 34.1$ GPa
$c_{12} = 13.6$ GPa
$c_{44} = 5.79$ GPa
Elastic compliance
$s_{11} = 38.0$ TPa^{-1}
$s_{12} = -10.8$ TPa^{-1}
$s_{44} = 173$ TPa^{-1}
Young's modulus
15.85 GPa
Rupture modulus
0.12 GPa
Shear modulus
5.79 GPa
Bulk modulus
19.77 GPa

Notes: KRS-5 has a waxy quality somewhat similar to that of silver chloride. It has often shown polarization properties due to strain birefringence. The thallium salts are toxic, and so KRS-5 should be handled with care. It has a serious tendency to cold-flow and change its shape with time. The refractive index of KRS-5 is not always uniform and sometimes appears as a gradient. If the composition corresponding to the minimum melting point is chosen, the change in refractive index should be less than 10^{-5}. Because of its higher reflection losses and shorter wavelength range, it is not as good a prism material as cesium iodide. KRS-5 is only slightly soluble in water but can be dissolved in alcohol, nitric acid, and aqua regia.

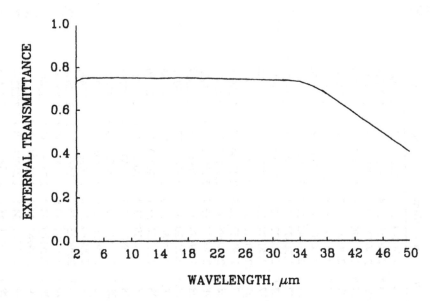

Figure 6.139 The external transmittance of KRS-5, thickness 2 mm. [Adapted from D. E. McCarthy, *Appl. Opt. 2*: 594 (1963).]

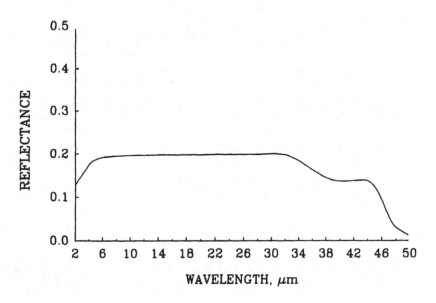

Figure 6.140 The reflectance of KRS-5, thickness 2 mm. [Adapted from D. E. McCarthy, *Appl. Opt. 2*: 591 (1963).]

TABLE 6.62 Refractive Index of Thallium Bromoiodide (KRS-5) at 25°C

λ (μm)	n	λ (μm)	n	λ (μm)	n	λ (μm)	n	λ (μm)	n	λ (μm)	n	λ (μm)	n	λ (μm)	n
0.540	2.68059	1.46	2.40938	5.80	2.37832	15.0	2.35812	22.1	2.33161	26.7	2.30844	31.3	2.28011	35.9	2.24609
0.560	2.64959	1.48	2.40854	6.00	2.37797	15.2	2.35751	22.2	2.33116	26.8	2.30789	31.4	2.27943	36.0	2.24528
0.580	2.62390	1.50	2.40774	6.20	2.37763	15.4	2.35690	22.3	2.33070	26.9	2.30732	31.5	2.27875	36.1	2.24447
0.600	2.60221	1.52	2.40697	6.40	2.37729	15.6	2.35629	22.4	2.33025	27.0	2.30676	31.6	2.27807	36.2	2.24366
0.620	2.58261	1.54	2.40623	6.60	2.37695	15.8	2.35566	22.5	2.32979	27.1	2.30619	31.7	2.27738	36.3	2.24284
0.640	2.56748	1.56	2.40552	6.80	2.37661	16.0	2.35502	22.6	2.32933	27.2	2.30562	31.8	2.27669	36.4	2.24202
0.660	2.55337	1.58	2.40484	7.00	2.37627	16.2	2.35438	22.7	2.32887	27.3	2.30505	31.9	2.27600	36.5	2.24120
0.680	2.54092	1.60	2.40419	7.20	2.37592	16.4	2.35373	22.8	2.32840	27.4	2.30448	32.0	2.27531	36.6	2.24038
0.700	2.52986	1.62	2.40355	7.40	2.37558	16.6	2.35307	22.9	2.32793	27.5	2.30390	32.1	2.27461	36.7	2.23955
0.720	2.51998	1.64	2.40295	7.60	2.37523	16.8	2.35240	23.0	2.32746	27.6	2.30332	32.2	2.27391	36.8	2.23872
0.740	2.51110	1.66	2.40236	7.80	2.37488	17.0	2.35173	23.1	2.32699	27.7	2.30274	32.3	2.27321	36.9	2.23788
0.760	2.50309	1.68	2.40180	8.00	2.37452	17.2	2.35104	23.2	2.32652	27.8	2.30216	32.4	2.27251	37.0	2.23705
0.780	2.49583	1.70	2.40125	8.20	2.37416	17.4	2.35035	23.3	2.32604	27.9	2.30157	32.5	2.27180	37.1	2.23621
0.800	2.48922	1.72	2.40073	8.40	2.37380	17.6	2.34965	23.4	2.32556	28.0	2.30098	32.6	2.27109	37.2	2.23536
0.820	2.48318	1.74	2.40022	8.60	2.37343	17.8	2.34894	23.5	2.32508	28.1	2.30039	32.7	2.27038	37.3	2.23452
0.840	2.47766	1.76	2.39974	8.80	2.37305	18.0	2.34822	23.6	2.32460	28.2	2.29979	32.8	2.26966	37.4	2.23367
0.860	2.47258	1.78	2.39926	9.00	2.37267	18.2	2.34750	23.7	2.32411	28.3	2.29920	32.9	2.26895	37.5	2.23281
0.880	2.46790	1.80	2.39881	9.20	2.37229	18.4	2.34676	23.8	2.32362	28.4	2.29860	33.0	2.26823	37.6	2.23196
0.900	2.46358	1.82	2.39837	9.40	2.37190	18.6	2.34602	23.9	2.32313	28.5	2.29800	33.1	2.26750	37.7	2.23110
0.920	2.45958	1.84	2.39794	9.60	2.37150	18.8	2.34527	24.0	2.32264	28.6	2.29739	33.2	2.26678	37.8	2.23024
0.940	2.45587	1.86	2.30753	9.80	2.37110	19.0	2.34451	24.1	2.32215	28.7	2.29679	33.3	2.26605	37.9	2.22937
0.960	2.45242	1.88	2.39713	10.0	2.37069	19.2	2.34374	24.2	2.32165	28.8	2.29618	33.4	2.26532	38.0	2.22850
0.980	2.44920	1.90	2.39674	10.2	2.37027	19.4	2.34296	24.3	2.32115	28.9	2.29556	33.5	2.26458	38.1	2.22763
1.00	2.44620	1.92	2.30637	10.4	2.36985	19.6	2.34217	24.4	2.32065	29.0	2.29495	33.6	2.26384	38.2	2.22676
1.02	2.44339	1.94	2.39600	10.6	2.36942	19.8	2.34138	24.5	2.32014	29.1	2.29433	33.7	2.26310	38.3	2.22588
1.04	2.44076	1.96	2.39565	10.8	2.36898	20.0	2.34058	24.6	2.31964	29.2	2.29371	33.8	2.26236	38.4	2.22500
1.06	2.43830	1.98	2.39531	11.0	2.36854	20.1	2.34017	24.7	2.31913	29.3	2.29309	33.9	2.26161	38.5	2.22412
1.08	2.43598	2.00	2.39498	11.2	2.36809	20.2	2.33976	24.8	2.31861	29.4	2.29247	34.0	2.26087	38.6	2.22323
1.10	2.43380	2.20	2.39214	11.4	2.36763	20.3	2.33935	24.9	2.31810	29.5	2.29184	34.1	2.26011	38.7	2.22234
1.12	2.43175	2.40	2.38997	11.6	2.36717	20.4	2.33894	25.0	2.31758	29.6	2.29121	34.2	2.25936	38.8	2.22145
1.14	2.42981	2.60	2.38826	11.8	2.36669	20.5	2.33853	25.1	2.31707	29.7	2.29058	34.3	2.25860	38.9	2.22055

1.16	2.42798	2.80	2.38688	12.0	2.36622	20.6	2.33811	25.2	2.31655	29.8	2.28994	34.4	2.25784	39.0	2.21965
1.18	2.42625	3.00	2.38574	12.2	2.36573	20.7	2.33770	25.3	2.31602	29.9	2.28931	34.5	2.25708	39.1	2.21875
1.20	2.42462	3.20	2.38478	12.4	2.36523	20.8	2.33727	25.4	2.31550	30.0	2.28867	34.6	2.25631	39.2	2.21784
1.22	2.42307	3.40	2.38396	12.6	2.36473	20.9	2.33685	25.5	2.31497	30.1	2.28802	34.7	2.25554	39.3	2.21693
1.24	2.42159	3.60	2.38325	12.8	2.36422	21.0	2.33643	25.6	2.31444	30.2	2.28738	34.8	2.25477	39.4	2.21602
1.26	2.42020	3.80	2.38261	13.0	2.36371	21.1	2.33600	25.7	2.31391	30.3	2.28673	34.9	2.25400	39.5	2.21510
1.28	2.41887	4.00	2.38204	13.2	2.36318	21.2	2.33557	25.8	2.31337	30.4	2.28608	35.0	2.25322	39.6	2.21418
1.30	2.41760	4.20	2.38153	13.4	2.36265	21.3	2.33514	25.9	2.31283	30.5	2.28543	35.1	2.25244	39.7	2.21326
1.32	2.41640	4.40	2.38105	13.6	2.36211	21.4	2.33471	26.0	2.31229	30.6	2.28477	35.2	2.25166	39.8	2.21233
1.34	2.41525	4.60	2.38061	13.8	2.36157	21.5	2.33427	26.1	2.31175	30.7	2.28411	35.3	2.25087	39.9	2.21140
1.36	2.41416	4.80	2.38019	14.0	2.36101	21.6	2.33383	26.2	2.31121	30.8	2.28345	35.4	2.25008	40.0	2.21047
1.38	2.41312	5.00	2.37979	14.2	2.36043	21.7	2.33339	26.3	2.31066	30.9	2.28279	35.5	2.24929		
1.40	2.41212	5.20	2.37940	14.4	2.35988	21.8	2.33295	26.4	2.31011	31.0	2.28212	35.6	2.24849		
1.42	2.41117	5.40	2.37903	14.6	2.35930	21.9	2.33251	26.5	2.30956	31.1	2.28145	35.7	2.24769		
1.44	2.41025	5.60	2.37867	14.8	2.35871	22.0	2.33206	26.6	2.30900	31.2	2.28078	35.8	2.24689		

Dispersion equation:

$$n^2 - 1 = \sum_i \frac{K_i \lambda^2}{\lambda^2 - \lambda_i^2}$$

Constants of the dispersion equation at 25°C for Tl(Br.I) (45.7:54.3)

i	λ_i^2	K_i
1	0.0225	1.8293958
2	0.0625	1.6675593
3	0.1225	1.1210424
4	0.2025	0.04513366
5	27.089.737	12.380234

From W. S. Rodney and I. H. Malitson. *J. Opt. Soc. Am. 46*: 956 (1956). who also give data for temperatures of 19 and 31°C. A comparison of these data. taken with the 45.7 to: 54.3 mol % mixed crystal. is made with the older data taken with 42:58 mol % crystal material by L. W. Tilton. E. K. Plyler. and R. E. Stephens. *J. Res. NBS 43*: 81 (1949). The 45.7:54.3 composition. which has the lowest freezing temperature of the binary system. should give the best optical homogeneity.

Table 6.63   Temperature Coefficient of Refractive Index of Thallium Bromoiodide

λ (μm)	dn/dT $(10^{-6}/K)$	λ (μm)	dn/dT $(10^{-6}/K)$
0.577	−254	14	−228
1.1	−240	16	−225
2	−238	20	−217
4	−237	25	−207
6	−237	30	−195
8	−236	35	−175
10	−235	40	−152
12	−232		

Source: A. I. Funai, Lockheed Missiles & Space Co. Rept. LMSC/6-78-68-34, 1968, p. 46, who references W. S. Rodney and I. H. Malitson, *J. Opt. Soc. Am. 46*: 956 (1956).

Thallium Chloride, TlCl

Molecular Weight
238.85

Specific Gravity
7.004 [g/cm^3]

Crystal Class
Cubic

Transmission
Long-wavelength limit 34 μm
Short-wavelength limit 0.44 μm
(See Fig. 6.141.)

Reflection Loss
21.8% for two surfaces at 10 μm
(See Fig. 6.142.)

Refractive Index
(See Table 6.64.)

Zero Dispersion Wavelength
3.5 μm

Dielectric Constant
31.9 for 2×10^6 Hz

Melting Temperature
703 K

Thermal Conductivity
0.75 W/(m·K) at 311–354 K

Thermal Expansion
53×10^{-6} K^{-1} at 293–333 K

Specific Heat
0.052 [cal/(g·K)] at 273 K

Hardness
12.8 kg/mm^2

(Knoop number with 500-g load in $\langle 100 \rangle$ and $\langle 110 \rangle$ directions)

Bandgap
3.6 eV

Solubility
0.32 g/(100 g H$_2$O) at 293 K

Elastic Moduli
Elastic stiffness
$c_{11} = 40.3$ GPa
$c_{12} = 15.5$ GPa
$c_{44} = 7.69$ GPa
Elastic compliance
$s_{11} = 31.6$ TPa^{-1}
$s_{12} = -8.8$ TPa^{-1}
$s_{44} = 130$ TPa^{-1}
Young's modulus
31.69 GPa
Shear modulus
7.58 GPa
Bulk modulus
23.56 GPa

Notes: Thallium chloride can be cut easily on a diamond saw. It is difficult to grind to dimension, because the material flows to the edge and fills the grinding wheel. Thin strips bend like lead. It is soluble in water.

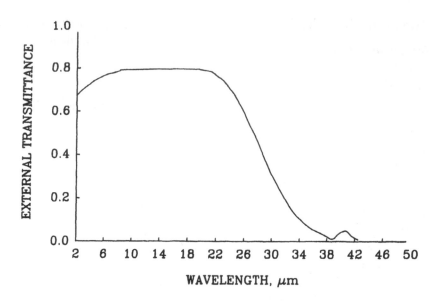

Figure 6.141   The external transmittance of thallium chloride, thickness 1.65 mm.
[Adapted from D. E. McCarthy, *Appl. Opt.* **4**: 317 (1965).]

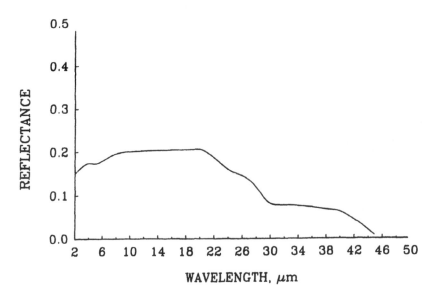

Figure 6.142   The reflectance of thallium chloride, thickness 1.65 mm. [Adapted
from D. E. McCarthy, *Appl. Opt.* **4**: 317 (1965).]

Table 6.64   Refractive Index of Thallium Chloride

λ (μm)	n	λ (μm)	n
0.436	2.400	0.750	2.198
0.546	2.270	10.0	2.193
0.578	2.253	12.47	2.191
0.589	2.247	18.35	2.182
0.650	2.223		

The indices in the visible region are from T. F. W. Barth, *Am. Mineral. 14:* 358 (1929). The indices in the infrared region were measured at 45°C by D. E. McCarthy, *Appl. Opt. 4:* 878 (1965).

Titanium Dioxide (Rutile), TiO_2

Molecular Weight
79.90

Specific Gravity
4.25 [g/cm^3]

Crystal Class
Tetragonal

Transmission
Long-wavelength limit 6.2 μm
Short-wavelength limit ~0.43 μm
(See Fig. 6.143.)

Reflection Loss
30% for two surfaces for the ordinary ray at 2 μm
34.8% for two surfaces for the extraordinary ray at 2 μm
(See Fig. 6.144.)

Refractive Index
(See Table 6.65.)

Zero Dispersion Wavelength
2.8 μm

Dielectric Constant
200–160 for 10^4–10^7 Hz

Melting Temperature
2090 K

Thermal Conductivity
200 W/(m·K) at 4.2 K (parallel)
160 W/(m·K) at 4.2 K (perpendicular)
13.0 W/(m·K) at 273 K (parallel)
9.0 W/(m·K) at 273 K (perpendicular)
12.6 W/(m·K) at 310 K (parallel)
8.8 W/(m·K) at 310 K (perpendicular)

Thermal Expansion
9.2×10^{-6} K^{-1} at 313 K (parallel)
7.1×10^{-6} K^{-1} at 313 K (perpendicular)

Specific Heat
0.17 [cal/(g·K)] at 298 K

Debye Temperature
760 K

Hardness
879 kg/mm^2 (Knoop number with 500-g load)

Bandgap
3.3 eV

Solubility
Insoluble in H_2O; soluble in acids

Elastic Moduli
Elastic stiffness
$c_{11} = 270$ GPa
$c_{12} = 176$ GPa
$c_{44} = 124$ GPa
$c_{13} = 147$ GPa
$c_{33} = 480$ GPa
$c_{66} = 193$ GPa
Elastic compliance
$s_{11} = 6.71$ TPa^{-1}
$s_{12} = -3.92$ TPa^{-1}
$s_{44} = 8.04$ TPa^{-1}
$s_{13} = -0.85$ TPa^{-1}
$s_{33} = 2.60$ TPa^{-1}
$s_{66} = 5.20$ TPa^{-1}

Notes: Rutile is the most common form of titanium dioxide. Anatase is another form of titanium dioxide, with elongated tetragonal structure. It is less stable than rutile and decomposes at high temperatures. There is also another form called brookite, which has rhombic structure and occurs only rarely. Titanium dioxide is used as a coating material with highly desirable properties. It is hard, chemically resistant, and has a high refractive index that is useful for interference applications.

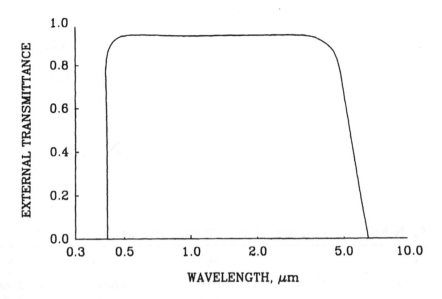

Figure 6.143 The external transmittance of titanium dioxide, thickness 2 mm. [Adapted from Beals and Merker, *Materials in Design Engineering*, Reinhold, New York, 1960.]

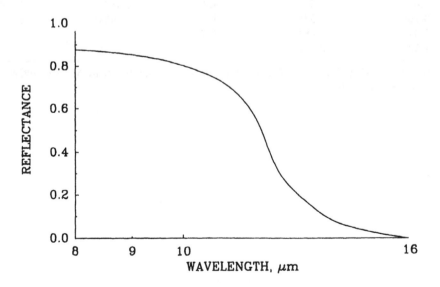

Figure 6.144 Reflectance of titanium dioxide. [Adapted from Spitzer et al., *Phys. Rev. 126*: 1710 (1962).]

Table 6.65   Refractive Index of Rutile (Titanium Dioxide)

λ (μm)	n_O	n_E	λ (μm)	n_O
0.4358	2.853	3.216	2.0000	2.399
0.4916	2.725	3.051	2.5000	2.387
0.4960	2.718	3.042	3.0000	2.380
0.5461	2.652	2.958	3.5000	2.367
0.5770	2.623	2.921	4.0000	2.350
0.5791	2.621	2.919	4.5000	2.322
0.6907	2.555	2.836	5.0000	2.290
0.7082	2.548	2.826	5.5000	2.200
1.0140	2.484	2.747		
1.5296	2.454	2.710		

Source: D. E. Gray, Coordinating Ed., *American Institute of Physics Handbook*, McGraw-Hill, New York, 1972, p. 6–50.

Yttrium Oxide, Y_2O_3

Molecular Weight
225.81

Specific Gravity
5.01 [g/cm³]

Crystal Class
Cubic

Transmission
(See *next data sheet*, Raytran yttria)

Reflectivity
(See Fig. 6.145.)

Dielectric Constant
13 at 293 K

Melting Temperature
2650 K

Thermal Conductivity
164 W/(m·K) at 93 K
27 W/(m·K) at 300 K

Thermal Expansion
6.4×10^{-6} K^{-1} at 300 K

Specific Heat
0.11 [cal/(g·K)] at 300 K
0.137 [cal/(g·K)] at 1700 K

Hardness
875 kg/mm² (Knoop number)

Bandgap
5.6 eV

Solubility
Insoluble in H_2O

Notes: Yttrium oxide can be used as a multilayer film material. Also see next data sheet, Raytran yttria (Raytheon).

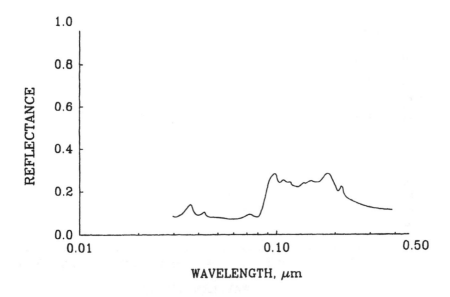

Figure 6.145 The reflectance of yttrium oxide. [Adapted from Y. Nigara, *Jpn. J. Appl. Phys. 7*: 404 (1968).]

Yttrium Oxide (Raytran Yttria) Y_2O_3

Specific Gravity
5.03 [g/cm^3]
Crystal Class
Cubic
Transmission
(See Figs. 6.146 and 6.147.)
Absorption Coefficient
(See Fig. 6.148.)
Refractive Index
(See Fig. 6.149.)
Melting Temperature
2723 K
Thermal Conductivity
14.0 W/(m·K) at 300 K
Thermal Expansion
5.8×10^{-6} K^{-1} at 293–473 K
Specific Heat
0.110 [cal/(g·K)]

Hardness
650 kg/mm^2
(Knoop number)
Elastic Moduli
Young's modulus
164 GPa
Poisson's ratio
0.29
Notes: Raytran yttria is a durable polycrystalline material that is fabricated by using a Raytheon-developed powder-processing technique. The material has the isotropic cubic crystal structure and transmits in the 3–5-μm range even at elevated temperatures. Windows up to 5 in. in diameter and 0.5 in. thick and domes up to 5 in. in diameter can be made. Rods and tubes are also available.

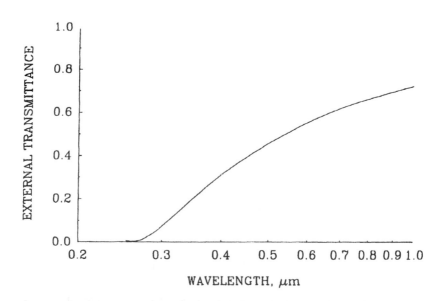

Figure 6.146   The external transmittance of yttria, thickness 3.2 mm. [Adapted from PBN-87-1411, Raytheon Company Research Division, Lexington, Mass., August 1987.]

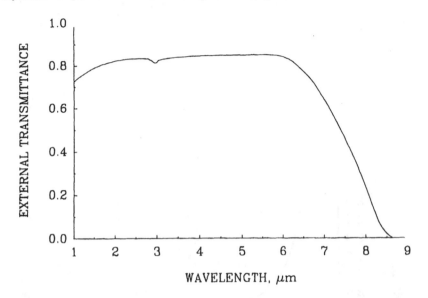

Figure 6.147 The external transmittance of yttria, thickness 3.2 mm. [Adapted from PBN-87-1411, Raytheon Company Research Division, Lexington, Mass., August 1987.]

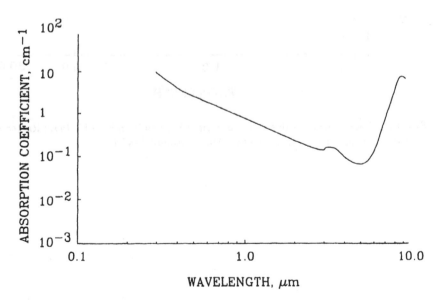

Figure 6.148 Absorption coefficient of yttria. [Adapted from Raytheon Company Research Division, Lexington, Mass., August 1987.]

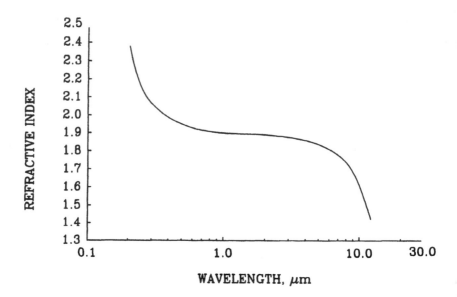

Figure 6.149   The refractive index of yttria. [Adapted from PBN-87-1412; Raytheon Company Research Division, Lexington, Mass., August 1987.]

Yttrium Oxide, Lanthanum Doped, Y_2O_3-La

Specific Gravity
5.13 [g/cm^3]

Crystal Class
Cubic

Transmission
(See Fig. 6.150.)

Absorption Coefficient
(See Fig. 6.151.)

Refractive Index
(See Fig. 6.152.)

Temperature Coefficient of Refractive Index
50×10^{-6} K^{-1} at 0.357 μm at 273–1673 K
32×10^{-6} K^{-1} at 3.39 μm at 273–1673 K

Dielectric Constant
12.2 for 10^5 Hz

Melting Temperature
2737 K

Thermal Conductivity
6.02 W/(m·K) at 296 K
4.28 W/(m·K) at 573 K
3.95 W/(m·K) at 973 K

Thermal Expansion
8.12×10^{-6} K^{-1} at 298–1273 K

Specific Heat
0.1078 [cal/(g·K)] at 325 K
0.1339 [cal/(g·K)] at 975 K

Hardness
730 kg/mm^2 (Knoop number)

Bandgap
5.5 eV

Elastic Moduli
Elastic stiffness
$c_{11} = 229$ GPa
$c_{12} = 102$ GPa
$c_{44} = 63.7$ GPa
Elastic compliance
$s_{11} = 6.01$ TPa^{-1}
$s_{12} = -1.85$ TPa^{-1}
$s_{44} = 16.7$ TPa^{-1}
Young's modulus
166.5 GPa
Rupture modulus
0.166 ± 0.034 GPa
Bulk modulus
144.5 GPa
Poisson's ratio
0.308

Notes: GTE lanthanum-doped yttria, an IR window material, has a long-wavelength transmission limit of 9 μm, which is among the longest for oxides. It is fabricated from powders by the transient solid second-phase sintering technique.

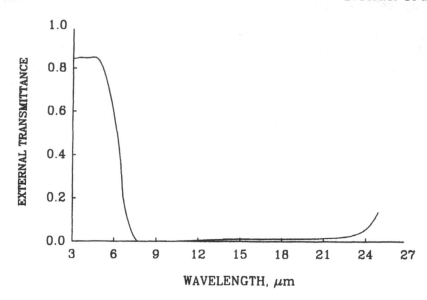

Figure 6.150   The external transmittance of lanthanum-doped yttria, thickness 1 mm. [Adapted from GTE Laboratories Incorporated, Waltham, Mass.]

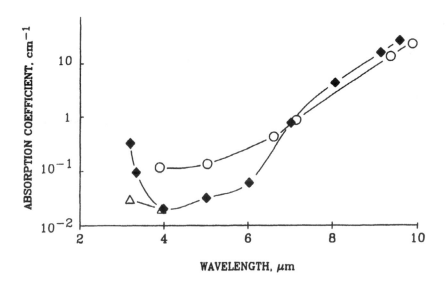

Figure 6.151   The absorption coefficients of lanthanum-doped yttria showing three stages of development: first stage (○), second stage (◆), and third stage (△). [Adapted from GTE Laboratories Incorporated, Waltham, Mass.]

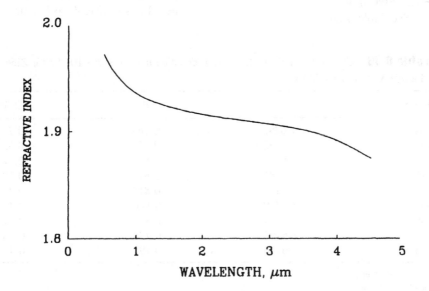

Figure 6.152   The refractive index of lanthanum-doped yttria. [Adapted from GTE Laboratories Incorporated, Waltham, Mass.]

Zinc Chloride, ZnCl$_2$

Molecular Weight
  136.28
Specific Gravity
  2.91 [g/cm^3]
Transmission
  Long-wavelength limit ~15 μm
  Short-wavelength limit ~0.4 μm
Refractive Index
  (See Table 6.66.)

Thermal Conductivity
  0.313 W/(m·K) at 562.9 K
Specific Heat
  0.007 [cal/(g·K)] at 76 K
  0.123 [cal/(g·K)] at 271 K
  0.128 [cal/(g·K)] at 346 K
Notes: Zinc chloride is very hygroscopic. This accounts for the lack of physical property data, because this material is so difficult to handle.

Table 6.66 Constants for Refractive Index Equation $n = a - b \times 10^{-4} t$ for Zinc Chloride Melt at 370–600°C

λ (nm)	a	b	$10^3 \delta$*
434.1	1.6395	0.763	0.25
460	1.6362	0.811	0.55
486.2	1.6316	0.813	0.32
510	1.6281	0.811	0.39
530	1.6259	0.818	0.45
560	1.6219	0.800	0.43
589.3	1.6194	0.805	0.41
620	1.6162	0.789	0.38
656.3	1.6136	0.792	0.44

*δ is standard error
Cauchy's relation:

$$n(\lambda, t) = P + \frac{Q}{\lambda^2} + \frac{R}{\lambda^4} + \left(P_t + \frac{Q_t}{\lambda^2} + \frac{R_t}{\lambda^4} \right) t$$

For the refractive index of ZnCl$_2$, λ in nanometers and t in °C,

P =	1.5846
$10^{-3} Q$ =	13.9390
$10^{-9} R$ =	−0.6679
$10^4 P_t$ =	−0.5868
$10^{-1} Q_t$ =	−1.2363
$10^{-6} R_t$ =	1.6739
$10^4 \epsilon$ =	4.06

(ε is standard error.)

Source: Y. Iwadate, K. Kawamura, K. Murakami, K. Igarashi, and J. Mochinaga, *J. Chem. Phys.* 77(12): 6177 (1982).

Zinc Selenide, ZnSe

Molecular Weight
144.33

Specific Gravity
5.42 [g/cm^3]

Crystal Class
Cubic, hexagonal

Transmission
(See Fig. 6.153 for CVD ZnSe.)

Absorption Coefficient
0.03 cm^{-1} at 10.6 μm

Refractive Index
2.440 at 3 μm
2.405 at 10 μm

Temperature Coefficient of
Refractive Index
4.8×10^{-5} K^{-1} at 4 μm

Melting Temperature
1790 K

Thermal Conductivity
19 W/(m·K) at 300 K

Thermal Expansion
-2.1×10^{-6} K^{-1} at 25 K
7.0×10^{-6} K^{-1} at 293 K
10.9×10^{-6} K^{-1} at 800 K

Specific Heat
0.0090 [cal/(g·K)] at 24 K

Debye Temperature
246 K

Hardness
137 kg/mm^2 (Knoop number)

Bandgap
2.7 eV

Solubility
0.001 g/(100 g H$_2$O) at 298 K

Elastic Moduli
Elastic stiffness
(cubic)
$c_{11} = 85.0$ GPa
$c_{12} = 50.2$ GPa
$c_{44} = 40.7$ GPa
(hexagonal)
$c_{11} = 107$ GPa
$c_{12} = 45$ GPa
$c_{13} = 35$ GPa
$c_{33} = 116$ GPa
$c_{44} = 25.0$ GPa
Elastic compliance
(cubic)
$s_{11} = 21.1$ TPa^{-1}
$s_{12} = -7.8$ TPa^{-1}
$s_{44} = 24.6$ TPa^{-1}
(hexagonal)
$s_{11} = 11.8$ TPa^{-1}
$s_{12} = -4.1$ TPa^{-1}
$s_{13} = -2.3$ TPa^{-1}
$s_{33} = 10.0$ TPa^{-1}
$s_{44} = 40.0$ TPa^{-1}
Young's modulus
70.97 GPa
Bulk modulus
40 GPa
Poisson's ratio
0.28

Notes: Zinc selenide is useful as a film material. See also data sheet for CVC zinc selenide.

Zinc Selenide (CVD Raytran), ZnSe

Molecular Weight
144.33

Specific Gravity
5.27 [g/cm^3]

Transmission
(See Fig. 6.153.)

Absorption Coefficient
100×10^{-3} cm^{-1} at 0.6328 μm
1.5×10^{-3} cm^{-1} at 2.7 μm
1.5×10^{-3} cm^{-1} at 3.8 μm
0.5×10^{-3} cm^{-1} at 10.6 μm
(See Fig. 6.154.)

Refractive Index
(See Table 6.67.)

Temperature Coefficient of Refractive Index
(See Fig. 6.155.)

Dispersion
(See dispersion equation, Table 6.67.)

Thermal Conductivity
18 W/(m·K) at 300 K

Thermal Expansion
5.6×10^{-6} K^{-1} at 173 K
7.1×10^{-6} K^{-1} at 273 K
8.3×10^{-6} K^{-1} at 473 K

Specific Heat
0.081 [cal/(g·K)] at 296 K

Hardness
105 kg/mm^2 (Knoop number with 100-g load)

Elastic Moduli
Young's modulus
70.3 ± 2.8 GPa
Poisson's ratio
0.28 ± 0.01

Notes: CVD Raytran ZnSe (Raytheon) is an improved form of high-clarity polycrystalline zinc selenide produced by a chemical vapor deposition process. Material is available in sizes up to 30 in. × 40 in. and thicknesses up to 1.5 in. Similar material is also available from CVD Inc.

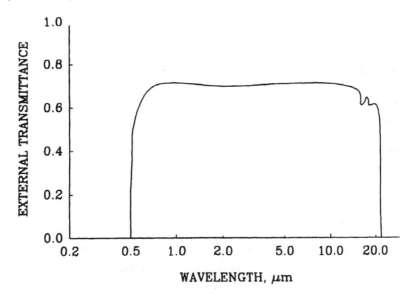

Figure 6.153 The external transmittance of zinc selenide, thickness 6 mm. [Adapted from catalog sheet Zinc Selenide Infrared and Laser Optics, Ptr. Optics Corp., Waltham, Mass.]

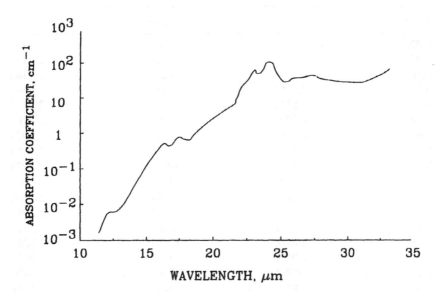

Figure 6.154 Absorption coefficient of zinc selenide. [Adapted from Perry A. Miles, *Appl. Opt. 16*: 2892 (1977).]

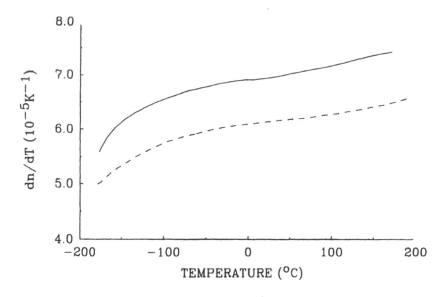

Figure 6.155 The temperature coefficient of refractive index zinc selenide (CVD) as a function of temperature. [Adapted from A. J. Glass and A. H. Guenther (Eds.), *Laser Induced Damage in Optical Materials*, 1977, p. 82.]

Table 6.67   Refractive Index of Zinc Selenide

λ (μm)	n	λ (μm)	n
0.42		7.40	2.4201
0.46		7.80	2.4183
0.50		8.20	2.4163
0.54	2.6754	8.60	2.4143
0.58	2.6312	9.00	2.4122
0.62	2.5994	9.40	2.4100
0.66	2.5755	9.80	2.4077
0.70	2.5568	10.20	2.4053
0.74	2.5418	10.60	2.4028
0.78	2.5295	11.00	2.4001
0.82	2.5193	11.40	2.3974
0.86	2.5107	11.80	2.3945
0.90	2.5034	12.20	2.3915
0.94	2.4971	12.60	1.3883
0.98	2.4916	13.00	2.3850
1.00	2.4892	13.40	2.3816
1.40	2.4609	13.80	2.3781
1.80	2.4496	14.20	2.3744
2.20	2.4437	14.60	2.3705

(continued)

Table 6.67   Refractive Index of Zinc Selenide (Continued)

λ (μm)	n	λ (μm)	n
2.60	2.4401	15.00	2.3665
3.00	2.4376	15.40	2.3623
3.40	2.4356	15.80	2.3579
3.80	2.4339	16.20	2.3534
4.20	2.4324	16.60	2.3487
4.60	2.4309	17.00	2.3438
5.00	2.4295	17.40	2.3387
5.40	2.4281	17.80	2.3333
5.60	2.4266	18.20	2.3278
6.20	2.4251		
6.60	2.4235		
7.00	2.4218		

Source: Data courtesy of Marilyn J. Dodge, Optical Physics Division, National Bureau of Standards.
Dispersion equation (Klein, Raytheon):

$$n^2 - 1 = \sum_i \frac{A_i \lambda^2}{\lambda^2 - \lambda_i^2} \qquad (0.54 \ \mu m \leq \lambda \leq 18.2 \ \mu m)$$

$$A_1 = 4.46395 \qquad \lambda_1 = 0.20108 \ \mu m$$
$$A_2 = 0.46132 \qquad \lambda_2 = 0.39211 \ \mu m$$
$$A_3 = 2.88289 \qquad \lambda_3 = 47.04759 \ \mu m$$

Zinc Sulfide, ZnS

Molecular Weight
  97.43
Specific Gravity
  3.98 [g/cm^3] (alpha)
  4.10 [g/cm^3] (beta)
Crystal Class
  Hexagonal (alpha)
  Cubic (beta)
Transmission
  (See Fig. 6.156.)
Absorption Coefficient
  0.15 cm^{-1} at 10.6 μm
Refractive Index
  (See Tables 6.68 and 6.69.)
Temperature Coefficient of
  Refractive Index
  4.6×10^{-5} K^{-1} at 1.15 μm
  4.2×10^{-5} K^{-1} at 3.39 μm
  4.1×10^{-6} K^{-1} at 10.6 μm
Dispersion
  (See dispersion equations, Table
  6.69.)
Dielectric Constant
  8.3 for 1×10^{12} Hz for beta
  7.9 for 4×10^{8} Hz for beta
Melting Temperature
  2100 K
Thermal Conductivity
  27 W/(m·K) at 300 K
Thermal Expansion
  1.9×10^{-6} K^{-1} at 80 K
  6.7×10^{-6} K^{-1} at 300 K
Specific Heat
  0.099 [cal/(g·K)] at 196 K
  0.120 [cal/(g·K)] at 273 K
Debye Temperature
  315 K
Hardness
  178 kg/mm^2 (Knoop number for
  beta form)
Bandgap
  3.9 eV (beta form)

Solubility
  Beta is insoluble in H$_2$O
Elastic Moduli
  Elastic stiffness
  (beta)
  $c_{11} = 101$ GPa
  $c_{12} = 64.4$ GPa
  $c_{44} = 44.3$ GPa
  Elastic compliance
  (alpha)
  $c_{11} = 122$ GPa
  $c_{12} = 58$ GPa
  $c_{13} = 42$ GPa
  $c_{33} = 138$ GPa
  $c_{44} = 28.7$ GPa
  Elastic compliance
  (beta)
  $s_{11} = 19.7$ TPa^{-1}
  $s_{12} = -7.6$ TPa^{-1}
  $s_{44} = 22.6$ TPa^{-1}
  (alpha)
  $s_{11} = 11.0$ TPa^{-1}
  $s_{12} = -4.5$ TPa^{-1}
  $s_{13} = -2.0$ TPa^{-1}
  $s_{33} = 8.6$ TPa^{-1}
  $s_{44} = 34.8$ TPa^{-1}
  Young's modulus
  66.14 GPa
  Bulk modulus
  80.41 GPa at 273 K
  Poisson's ratio
  0.29

Notes: Zinc sulfide occurs in a
zincblende (cubic) structure (beta)
called sphalerite and a wurtzite (hexagonal) structure (alpha). Natural
crystal zinc sulfide occurs in the cubic (beta) form, and this form can
also be produced by chemical vapor
deposition. Alpha zinc sulfide is optically anisotropic. Zinc sulfide has
applications as a multilayer film material in the visible spectral range.

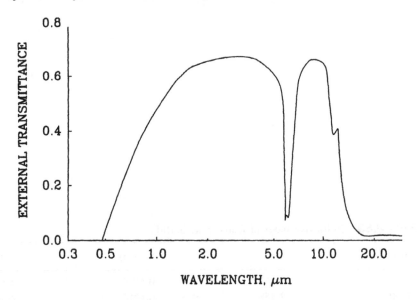

Figure 6.156   The external transmittance of zinc sulfide, thickness 15 mm. [Adapted from A. J. Glass and A. H. Guenther (Eds.), *Laser Induced Damage in Optical Materials*, 1977.]

Table 6.68   Refractive Indices of Hexagonal Zinc Sulfide

λ (μm)	n_E	n_O	λ (μm)	n_E	n_O
0.3600	2.709	2.705	0.5000	2.425	2.421
0.3750	2.640	2.637	0.5250	2.407	2.402
0.4000	2.564	2.560	0.5500	2.392	2.386
0.4100	2.544	2.539	0.5750	2.378	2.375
0.4200	2.525	2.522	0.6000	2.368	2.363
0.4250	2.514	2.511	0.6250	2.358	2.354
0.4300	2.505	2.502	0.6500	2.350	2.346
0.4400	2.488	2.486	0.6750	2.343	2.339
0.4500	2.477	2.473	0.7000	2.337	2.332
0.4600	2.463	2.459	0.8000	2.328	2.324
0.4700	2.453	2.448	0.9000	2.315	2.310
0.4750	2.449	2.445	1.0000	2.303	2.301
0.5800	2.443	2.438	1.2000	2.294	2.290
0.4900	2.433	2.428	1.4000	2.288	2.285

Source: T. M. Bieniewski and S. J. Czyzak, *J. Opt. Soc. Am. 53*: 496 (1963).

Table 6.69 Refractive Index of Cubic Zinc Sulfide

λ (μm)	n	λ (μm)	n
0.4400	2.488	0.6000	2.359
0.4600	2.458	0.6500	2.346
0.4800	2.435	0.7000	2.334
0.5000	2.414	0.9000	2.306
0.5250	2.395	1.0500	2.293
0.5500	2.384	1.2000	2.282
0.5750	2.375	1.4000	2.280

Source: S. J. Czyzak, W. M. Baker, R. C. Crane, and J. B. Howe, *J. Opt. Soc. Am. 47*: 240 (1957). These values are for synthetic cubic zinc sulfide. Refractive index values in the wavelength region 0.365–1.53 μm for natural cubic zinc sulfide are given by J. R. DeVore, *J. Opt. Soc. Am. 41*, 416 (1951).

For computational purposes, Czyzak et al. give a dispersion equation:

$$n^2 = 5.131 + \frac{1.275 \times 10^7}{\lambda^2 - 0.732 \times 10^7}$$

Zinc Sulfide (CVD Raytran Multispectral Grade), ZnS

Specific Gravity
4.09 [g/cm^3]

Transmission
(See Figs. 6.157 and 6.158.)

Absorption Coefficient
1.1×10^{-3} cm^{-1} at 2.7 μm
0.6×10^{-3} cm^{-1} at 3.8 μm
11×10^{-3} cm^{-1} at 9.27 μm
200×10^{-3} cm^{-1} at 10.6 μm

Refractive Index
(See Fig. 6.159.)

Dispersion
(See Fig. 6.159.)

Dielectric Constant
8.393 ± 0.023 (low-frequency limit)
5.105 ± 0.033 (high-frequency limit)

Thermal Conductivity
27 W/(m·K) at 298 K

Thermal Expansion
6.5×10^{-6} K^{-1} at 208–473 K
7.85×10^{-6} K^{-1} at 293–773 K

Specific Heat
0.124 [cal/(g·K)] at 298 K

Hardness
160 kg/mm^2 (Knoop number with 100-g load)

Elastic Moduli
Young's modulus
87.6 ± 0.7 GPa
Poisson's ratio
0.318 ± 0.001

Notes: Raytran multispectral-grade ZnS (Raytheon) behaves like polycrystalline intrinsic cubic zinc sulfide. It is produced by a chemical vapor deposition process. Flat plates as large as 6 in. × 9 in. × 0.5 in. are available. Larger pieces or nonplanar shapes are also available upon request from Raytheon.

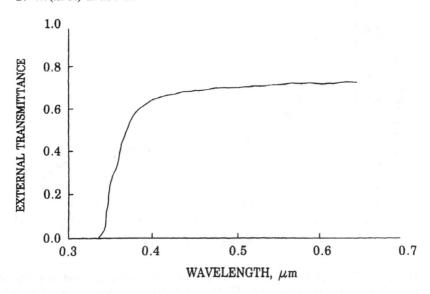

Figure 6.157 External transmittance of multispectral Raytran ZnS, thickness = 0.09 cm (Raytheon data).

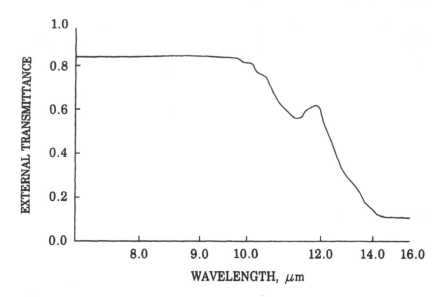

Figure 6.158   External transmittance of multispectral Raytran ZnS, thickness = 1.37 cm (Raytheon data).

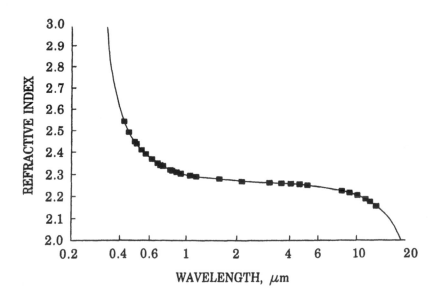

Figure 6.159   Refractive index of HIP-treated CVD-grown ZnS and its dispersion (National Physical Laboratory data). (■) Measured (M. Debenham); (——) Calculated (C. Klein, Raytheon Co.)

Zinc Sulfide (CVD Raytran Standard Grade), ZnS

Specific Gravity
4.08 [g/cm³]

Transmission
(See Figs. 6.160 and 6.161.)

Absorption Coefficient
0.072 ± 0.022 cm⁻¹ at 8 μm
0.079 ± 0.013 cm⁻¹ at 9 μm
0.096 ± 0.011 cm⁻¹ at 10 μm
0.393 ± 0.013 cm⁻¹ at 11 μm
0.400 ± ± 0.016 cm⁻¹ at 12 μm
(See Fig. 6.162.)

Refractive Index
(See Fig. 6.163 and Table 6.70.)

Temperature Coefficient of Refractive Index
(See Fig. 6.164.)

Thermal Conductivity
17 W/(m·K) at 296 K

Thermal Expansion
4.6×10^{-6} K⁻¹ at 173 K
6.6×10^{-6} K⁻¹ at 273 K
7.7×10^{-6} K⁻¹ at 473 K

Specific Heat
0.112 [cal/(g·K)] at 296 K

Hardness
230 kg/mm² (Knoop number with 100-g load)

Elastic Moduli
Young's modulus
74.5±3.5 GPa
Poisson's ratio
0.29±0.01

Notes: Raytran ZnS (Raytheon) is a high-optical-quality material produced by a chemical vapor deposition process. Large and complex shapes of the material can be produced.

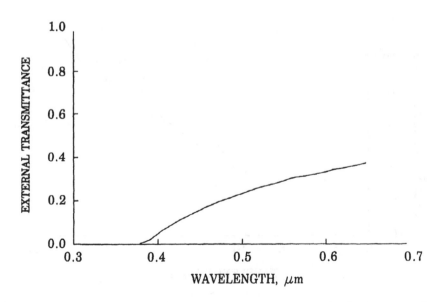

Figure 6.160 External transmittance of standard Raytran ZnS, thickness = 0.09 cm (Raytheon data).

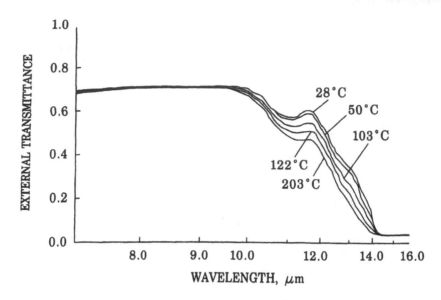

Figure 6.161 External transmittance of standard Raytran ZnS, at elevated temperatures, thickness = 0.668 cm (Raytheon data).

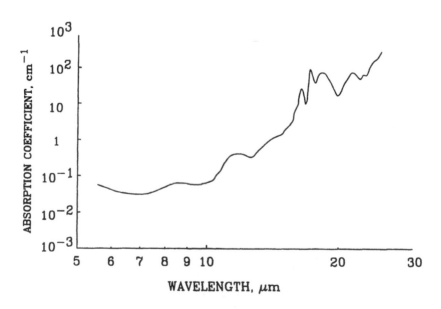

Figure 6.162 Absorption coefficient of chemical-vapor-deposited zinc sulfide. [Adapted from C. Klein et al., Research Division, Raytheon Company, Waltham, Mass.]

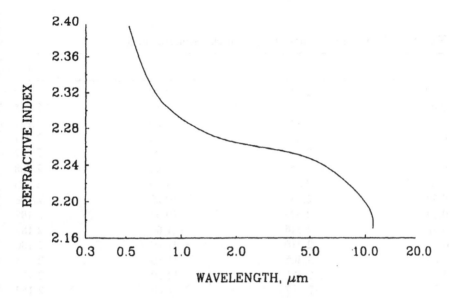

Figure 6.163 The refractive index of CVD zinc sulfide (logarithmic scale). Data near 22°C were calculated from the dispersion equation. [Adapted from M. J. Dodge, article on refractive properties of CVD zinc sulfide, National Bureau of Standards, Washington, D.C.]

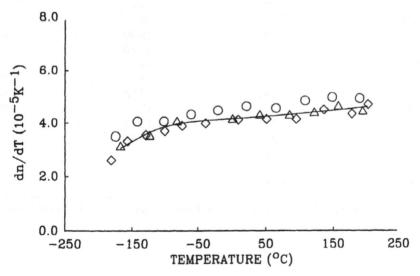

Figure 6.164 Temperature coefficient of the refractive index of zinc sulfide for wavelengths 1.15 μm (O), 3.39 μm (△), and 10.6 μm (◇). [Adapted from A. J. Glass and A. H. Guenther (Eds.), *Laser Induced Damage in Optical Materials*, 81 (1977).]

Table 6.70 Refractive Index of Zinc Sulfide, Standard Grade

λ (μm)	n	λ (μm)	n
0.42	2.516	7.40	2.228
0.46	2.458	7.80	2.225
0.50	2.419	8.20	2.221
0.54	2.391	8.60	2.217
0.58	2.371	9.00	2.212
0.62	2.355	9.40	2.208
0.66	2.342	9.80	2.203
0.70	2.332	10.20	2.198
0.74	2.323	10.60	2.192
0.78	2.316	11.00	2.186
0.82	2.310	11.40	2.180
0.86	2.305	11.80	2.173
0.90	2.301	12.20	2.167
0.94	2.297	12.60	2.159
0.98	2.294	13.00	2.152
1.00	2.292	13.40	2.143
1.40	2.275	13.80	2.135
1.80	2.267	14.20	2.126
2.20	2.263	14.60	2.116
2.60	2.260	15.00	2.106
3.00	2.257	15.40	2.095
3.40	2.255	15.80	2.084
3.80	2.253	16.20	2.072
4.20	2.251	16.60	2.059
4.60	2.248	17.00	2.045
5.00	2.246	17.40	2.030
5.40	2.244	17.80	2.015
5.80	2.241	18.20	1.998
6.20	2.238		
6.60	2.235		
7.00	2.232		

Source: Data courtesy of Marilyn J. Dodge, Optical Physics Division, National Bureau of Standards.

Zinc Sulfide (Irtran 2), ZnS

Specific Gravity
4.09 [g/cm^3]

Transmission
Long-wavelength limit 14.5 μm
Short-wavelength limit 1 μm
(See Fig. 6.165.)

Absorption Coefficient
(See Table 6.71.)

Reflectivity
(See Fig. 6.166.)

Refractive Index
(See Table 6.72.)

Dispersion
(See dispersion equation, Table 6.72.)

Dielectric Constant
8.0 for $(8.5-12.0) \times 10^9$ Hz

Melting Temperature
2103 K

Thermal Conductivity
15.49 W/(m·K) at 327 K
13.81 (w/(m·K) at 379 K
10.88 W/(m·K) at 447 K

Thermal Expansion
0.53×10^{-6} K^{-1} at 70 K
5.26×10^{-6} K^{-1} at 200 K
6.9×10^{-6} K^{-1} at 300 K

Hardness
355 kg/mm^2 (Knoop number)

Solubility
Insoluble in H$_2$O

Elastic Moduli
Young's modulus
96.46 GPa
Rupture modulus
0.971 GPa
Poisson's ratio
0.25–0.36

Notes: Irtran is the trademark for infrared-transmitting optical materials manufactured by Eastman Kodak Company. Irtran 2 is a hot-pressed polycrystalline compact of zinc sulfide. Finished blanks are available in diameters as large as 8 in. and as small as 1 in. Domes up to 9 in. in diameter and 4.5 in. in height are also available. Virtually any shape that can be cut, ground, or polished from glass can also be similarly produced from this material. However, the material cannot be heat-softened, as is done in glass-blowing or forming. The material is not soluble in water or dilute acids or bases. It is resistant to thermal shock and will withstand atmospheric weathering about as well as conventional glasses.

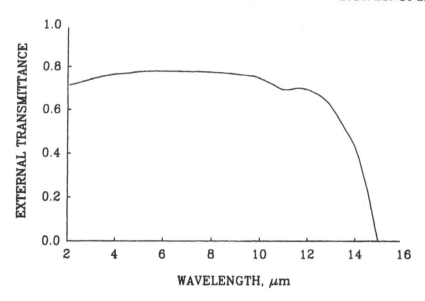

Figure 6.165   The external transmittance of Kodak Irtran 2 material, thickness 2.0 mm. [Adapted from W. L. Wolfe and G. J. Zissis, *The Infrared Handbook*, 1978.]

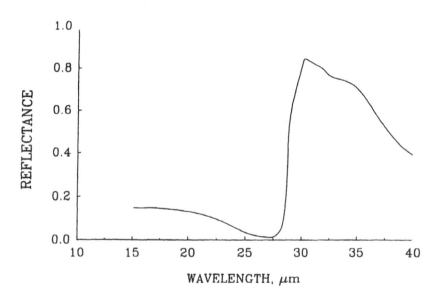

Figure 6.166   The reflectance of Kodak Irtran 2 material. [Adapted from D.E. Mc-Carthy, *Appl. Opt. 2*: 594 (1963).]

Table 6.71 Absorption Coefficient for Kodak Irtran 2

λ (μm)	Absorption coefficient
2	0.39
3	0.22
4	0.12
5	0.07
6	0.06
7	0.07
8	0.07
9	0.08
10	0.11
10.6	0.29
11	0.45
12	0.47

Source: Eastman Kodak Co.

Table 6.72   Refractive Index of Irtran 2

λ (μm)	n
1.0000	2.2907
1.2500	2.2777
1.5000	2.2706
1.7500	2.2662
2.0000	2.2631
2.2500	2.2608
2.5000	2.2589
2.7500	2.2573
3.0000	2.2558
3.2500	2.2544
3.5000	2.2531
3.7500	2.2518
4.0000	2.2504
4.2500	2.2491
4.5000	2.2477
4.7500	2.2462
5.0000	2.2447
5.2500	2.2432
5.5000	2.2416
5.7500	2.2399
6.0000	2.2381
6.2500	2.2363
6.5000	2.2344
6.7500	2.2324
7.0000	2.2304
7.2500	2.2282
7.5000	2.2260
7.7500	2.2237
8.0000	2.2213
8.2500	2.2188
8.5000	2.2162
8.7500	2.2135
9.0000	2.2107
9.2500	2.2078
9.5000	2.2048
9.7500	2.2018

(*continued*)

Table 6.72 Refractive Index of Irtran 2 (Continued)

λ (μm)	n
10.0000	2.1986
11.0000	2.1846
12.0000	2.1688
13.0000	2.1508

Index of refraction values were experimentally determined at selected wavelengths between 1 and 10 μm. Coefficients of an interpolation formula were established and reduced by least-squares methods, and the values computed. All values beyond 10 μm are extrapolated. Herzberger dispersion equation:

$$n = n_o + \frac{b}{\lambda^2 - 0.028} + \frac{c}{(\lambda^2 - 0.028)^2} + d\lambda^2 + e\lambda^4$$

$$n_o = 2.2569735$$
$$b = 3.2640935 \times 10^{-2}$$
$$c = 6.0314637 \times 10^{-4}$$
$$d = -5.2705532 \times 10^{-4}$$
$$e = -6.0428638 \times 10^{-7}$$

Source: Eastman Kodak Co.

Zirconium Dioxide, ZrO$_2$

Molecular Weight
123.22
Specific Gravity
6.27 [g/cm^3]
Crystal Class
Monoclinic

Refractive Index
(See Fig. 6.167.)
Notes: Several crystalline forms of zirconia are known. A stable monoclinic form found in nature is baddeleyite. See cubic zirconia (yttria doped) for more information.

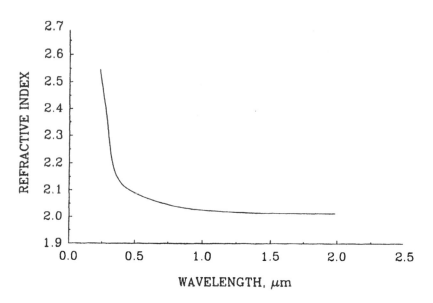

Figure 6.167   The refractive index of zirconium oxide. [Adapted from N. M. Bashara and Y.-K. Peng, *Appl. Opt. 19*(18): 3248 (1980).]

Zirconium Dioxide, Yttrium Oxide Doped, ZrO_2-Y_2O_3

Crystal Class
Cubic

Absorption Coefficient
(See Fig. 6.168.)

Refractive Index
(See Table 6.73.)

Temperature Coefficient of Refractive Index
0.72×10^{-5} K^{-1} at 0.8 μm
0.62×10^{-5} K^{-1} at 1.69 μm

Dispersion
(See dispersion equation, Table 6.73.)

Dielectric Constant
12.0 for $(1.75-2) \times 10^6$ Hz

Melting Temperature
~3000 K

Thermal Conductivity
10.5 W/(m·K) at 260 K

Thermal Expansion
8.8×10^{-6} K^{-1} at 293 K

Specific Heat
0.10 [cal/(g·K)] at 273 K
0.145 [cal/(g·K)] at 1850 K

Hardness
7.5–8.5 (Mohs)

Bandgap
5.0 eV

Solubility
Insoluble in H_2O

Elastic Moduli (doped with 8% Y_2O_3)
Elastic stiffness
$c_{11} = 394$ GPa
$c_{12} = 91$ GPa
$c_{44} = 56$ GPa
Elastic compliance
$s_{11} = 2.78$ TPa^{-1}
$s_{12} = -0.52$ TPa^{-1}
$s_{44} = 17.9$ TPa^{-1}

Notes: This stabilized form of zirconium dioxide, called cubic zirconia, can be produced synthetically at room temperature by doping with Y_2O_3, MgO_2, or CaO. This form is best known for its role in jewelry. Its refractive index and resistance to abrasion and could be useful for refractometer prisms (Wood and Nassau, *Appl. Opt.* **21**(16),2978, 1982).

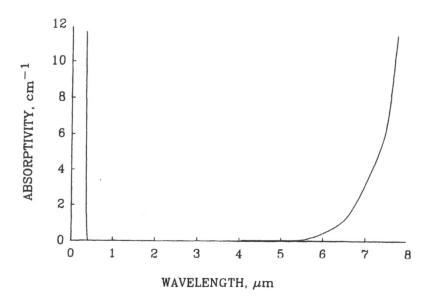

Figure 6.168 The absorptivity of cubic zirconia as a function of wavelength. [Adapted from D. L. Wood and K. Nassau, *Appl. Opt. 21*: 16 (1982).] A correction to the apparent absorption for the 0.172-cm thick samples was made for reflection losses from the two surfaces of the plate according to the simplified relation:

$$\alpha t = \log \frac{I_o}{I} - 2 \log \left| 1 - \left| \frac{n - 1}{n + 1} \right|^2 \right|$$

where α is the absorptivity in cm^{-1}, t is the sample thickness in cm, I_o is the incident radiation intensity, I is the intensity of radiation passing through the sample, and n is the index of refraction at the relevant wavelengths.

Table 6.73 Refractive Indices of Cubic Zirconia at 25°C

λ (μm)	n
0.361051	2.25364
0.365015	2.24990
0.365483	2.24947
0.366308	2.24871
0.388865	2.23052
0.404656	2.21986
0.435833	2.20290
0.447148	2.19783
0.467815	2.18956
0.479991	2.18523
0.492193	2.18120
0.501567	2.17835
0.508582	2.17639
0.546073	2.16694
0.576959	2.16066
0.579066	2.16026
0.587566	2.15878
0.601033	2.15649
0.621287	2.15334
0.643847	2.15026
0.658651	2.14833
0.662865	2.14781
0.667815	2.14723
0.672328	2.14669
0.687045	2.14503
0.697329	2.14394
0.706518	2.14303
0.722853	2.14144
0.727995	2.14097
0.728135	2.14098
0.760901	2.13812
0.794411	2.13559
0.852110	2.13192
0.894350	2.12959
0.917224	2.12842
1.002439	2.12476
1.012360	2.12437
1.013975	2.12425
1.039460	2.12338
1.083030	2.12184
1.128660	2.12038
1.14800	2.11967

(continued)

Table 6.73 Refractive Indices of Cubic Zirconia at 25°C
(Continued)

λ (μm)	n
1.188900	2.11861
1.197730	2.11817
1.357021	2.11439
1.35890	2.11435
1.367531	2.11421
1.395055	2.11358
1.40800	2.11319
1.45950	2.11194
1.529582	2.11072
1.68800	2.10747
1.69272	2.10749
1.71281	2.10716
1.80000	2.10534
1.8600	2.10425
1.9650	2.10219
2.05809	2.10047
2.2280	2.09707
2.3710	2.09419
2.3890	2.09381
3.2700	2.07295
3.3000	2.07203
3.4250	2.06869
3.5050	2.06633
4.2250	2.04282
4.2710	2.04127
5.0790	2.00916
5.1200	2.00732
5.1350	2.00664

The index values at 25°C were fitted to a three-term Sellmeier equation
[L. E. Sutton and O. N. Stavroudis, *J. Opt. Soc. Am. 51:* 901 (1961)]
by a nonlinear least-squares fitting program using the relation

$$n^2 - 1 = \sum_{i=1}^{3} \frac{A_i \lambda^2}{(\lambda^2 - L_i^2)}$$

The constants A_i and L_i are given and λ is the wavelength for which the
index is n. Calculated values from this formula are also given for each
wavelength in Table I, and the residuals between calculated and measured values are given in the last column of the table.

$$A_1 = 2.117788 \quad L_1 = 0.166739$$
$$A_2 = 1.347091 \quad L_2 = 0.062543$$
$$A_3 = 9.452943 \quad L_3 = 24.320570$$

Source: D.L. Wood and K. Nassau, *Appl. Opt. 21*(16): 2978 (1982).

BIBLIOGRAPHY

The Infrared Handbook, Wolfe, W.L. and Zissis, G.J., Eds., IRIA Center, Environmental Research Institute of Michigan, Office of Naval Research, U.S. Government Printing Office, Washington, D.C., 1985.

Handbook of Chemistry and Physics, Weast, R.C. and Astle, M.J., Eds., CRC Press, Boca Raton, Fla., 1982.

Handbook of Laser Science and Technology. Volume IV Optical Materials: Part 2, Weber, M.J., Ed., CRC Press, Boca Raton, Fla., 1986.

Musikant, S., *Optical Materials, An Introduction to Selection and Application,* Marcel Dekker, Inc., New York, 1985.

Thermophysical Properties of Matter, Touloukian, Y.S., and Ho, C.Y., Eds., Plenum Press, New York, 1977.

American Institute of Physics Handbook, Gray, D.E., Ed., McGraw-Hill, New York, 1972.

Moses, A.J., *Handbook of Electronic Materials, Volume 1: Optical Materials Properties,* Plenum Publishing Corp., New York, 1971.

Laser-Induced Damage in Optical Materials: NBS Special Publication 435 (1975), 638 (1983), and 688 (1985), Washington, D.C.

Ballard, S., McCarthy, K., and Wolfe, W., "Optical Materials for Infrared Instrumentation," IRIA Center, Report #2389-11-S, The University of Michigan, Willow Run Laboratories, 1959.

Handbook of Optical Constants of Solids, Palik, E.D., Ed., Academic Press, Orlando, 1985.

Landolt-Bornstein, Numerical Data and Functional Relationships in Science and Technology, Hellwege, K.H., Ed., Springer-Verlag, Berlin, New Series, 1959.

Dielectric Materials and Applications, von Hipple, A.R., Ed., John Wiley and Son, Inc., New York, 1954.

Cardona, M., *Semiconductors and Semimetals,* Beer, A.C., and Willardson, R.K., Eds., Vol. 3, Academic Press, New York, 1967.

Savage, J.A., *Infrared Optical Materials and Their Antireflection Coatings,* Adam Hilger Ltd., Bristol, 1985.

Handbook of Optics, Driscoll, W.G., Ed., McGraw-Hill, New York, 1978.

Lange's Handbook of Chemistry, Dean, J.A., Ed., McGraw-Hill, New York, 1979.

Slack, G.A., Thermal Conductivity of Nonmetallic Crystals, in *Solid State Physics,* Vol. 34, Ehrenreich, H., Seitz, F., and Turnbull, D., Eds., Academic Press, New York, 1979.

Also see references for Chapter 5 (Appendix, pp. 593–598).

7

Physical Properties of Glass Infrared Optical Materials

James Steve Browder   *Jacksonville University, Jacksonville, Florida*

Stanley S. Ballard   *University of Florida, Gainesville, Florida*

Paul Klocek   *Texas Instruments, Inc., Dallas, Texas*

After an optical material has been chosen for a particular application, the engineer may wish to know more about several of its physical properties. All available data are given in Chapters 6 and 7.

The materials are arranged in alphabetical order, as in Tables 5.1 and 5.2 of Chapter 5. The sources of the data on the individual data sheets in Chapters 6 and 7 are identified in the comparison tables of Chapter 5. The sources for the tables and figures in Chapters 6 and 7 are given in the table footnotes and figure legends. For some of the materials, a great deal of data are given, including transmission curves, refractive index tables, and so on. For others, few data were found in our literature search.

Arsenic Trisulfide Glass, As₂S₃

Molecular Weight
248.04
Specific Gravity
3.43 [g/cm³]
Transmission
Long-wavelength limit ~11 μm
Short-wavelength limit 0.6 μm
(See Fig. 7.1.)
Absorption Coefficient
(See Fig. 7.2.)
Reflection Loss
28.5% for two surfaces, at a
wavelength 10 μm
Refractive Index
(See Table 7.1.)
Temperature Coefficient of Refractive Index
(See Fig. 7.3.)
Dispersion
(See dispersion equation, Table 7.1.)
Dielectric Constant
8.1 at 10^3–10^6 Hz
Glass Transition
470 K

Thermal Conductivity
0.1674 W/(m·K) at 313 K
Thermal Expansion
2.462×10^{-5} K^{-1} at 306–438 K
Specific Heat
0.109 at 270 K
Hardness
109 kg/mm² (Knoop number with 100-g load)
Solubility
Insoluble
Elastic Moduli
Young's modulus
15.85 GPa
Rupture modulus
0.017 GPa
Shear modulus
6.48 GPa

Notes: The properties of arsenic trisulfide glass vary with different batches. It is quite soft and brittle and can be pressed, sawed, ground, and polished rather easily. Various compositions of arsenic sulfide and arsenic selenide have been made with a range of similar physical properties.

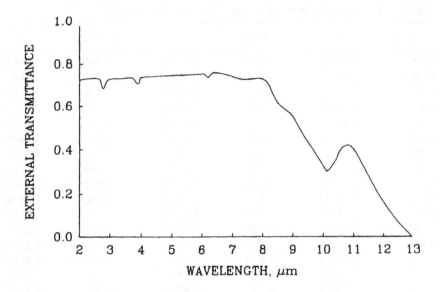

Figure 7.1. The external transmittance of arsenic trisulfide, thickness 6.4 mm. [Adapted from J.A. Savage and S. Nielson, *Infrared Phys.* 5:195 (1965).]

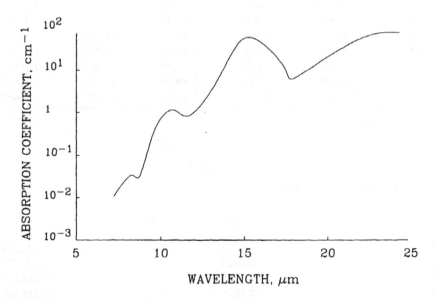

Figure 7.2 The absorption coefficient for arsenic trisulfide glass. [Adapted from R.E. Howard et al., in *Optical Properties of Highly Transparent Solids,* S.S. Mitra and B. Bendon (Eds.), Plenum Publishing Corp., New York, 1975, p. 277.]

Table 7.1 Refractive Index of Arsenic Trisulfide Glass at 25°C

λ (μm)	n	λ (μm)	n	λ (μm)	n
0.560	2.68689	1.800	2.43009	7.000	2.39899
0.580	2.65934	2.000	2.42615	7.200	2.39806
0.600	2.63646	2.200	2.42318	7.400	2.39709
0.620	2.61708	2.400	2.42086	7.600	2.39610
0.640	2.60043	2.600	2.41898	7.800	2.39508
0.660	2.58594	2.800	2.41742	8.000	2.39403
0.680	2.57323	3.000	2.41608	8.200	2.39294
0.700	2.56198	3.200	2.41491	8.400	2.39183
0.720	2.55195	3.400	2.41386	8.600	2.39068
0.740	2.54297	3.600	2.41290	8.800	2.38949
0.760	2.53488	3.800	2.41200	9.000	2.38827
0.780	2.52756	4.000	2.41116	9.200	2.38700
0.800	2.52090	4.200	2.41035	9.400	2.38570
0.820	2.51483	4.400	2.40956	9.600	2.38436
0.840	2.50928	4.600	2.40878	9.800	2.38298
0.860	2.50418	4.800	2.40802	10.000	2.38155
0.880	2.49949	5.000	2.40725	10.200	2.38007
0.900	2.49515	5.200	2.40649	10.400	2.37855
0.920	2.49114	5.400	2.40571	10.600	2.37698
0.940	2.48742	5.600	2.40493	10.800	2.37536
0.960	2.48396	5.800	2.40414	11.000	2.37369
0.980	2.48074	6.000	2.40333	11.200	2.37196
1.000	2.47773	6.200	2.40250	11.400	2.37018
1.200	2.45612	6.400	2.40166	11.600	2.36833
1.400	2.44357	6.600	2.40079	11.800	2.36643
1.600	2.43556	6.800	2.39991	12.000	2.36446

Dispersion equation:

$$n^2 - 1 = \sum_{i=1}^{i=5} \frac{K_i \lambda^2}{\lambda^2 - \lambda_i^2}$$

Constants for the dispersion equation at 25°C:

i	λ_i^2	K_i
1	0.0225	1.8983678
2	0.0625	1.9222979
3	0.1225	0.8765134
4	0.2025	0.1188704
5	750	0.9569903

Source: I.H. Malitson, W.S. Rodney, and T. A. King, *J. Opt. Soc. Am.* 48:633 (1958).

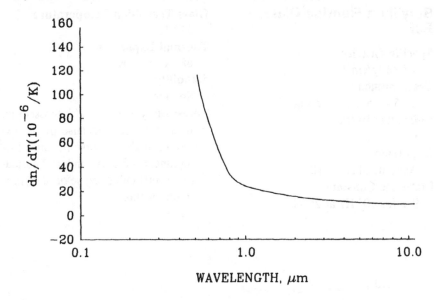

Figure 7.3.   The temperature coefficient of refractive index for arsenic trisulfide glass. [Adapted from I.H. Malitson, W.S. Rodney, and T.A. King, *J. Opt. Soc. Am*: *48*, 633 (1958).]

Beryllium Fluoride Glass, BeF$_2$

Specific Gravity
  1.982 [g/cm^3]
Transmission
  0.15–4.5 μm (See Fig. 7.4.)
Refractive Index
  n_D = 1.27
Dispersion
  Abbe number = 107
Dielectric Constant
  9.7 for 10^6 Hz at 25°C

Glass Transition Temperature
  523 K
Thermal Expansion
  68 × 10^{-7} K^{-1}
Solubility
  Soluble
Notes: BeF$_2$ is toxic and hygroscopic. It is of interest because of its low index, high dispersion, and low nonlinear refractive index. Fabrication is difficult because of its hygroscopic nature.

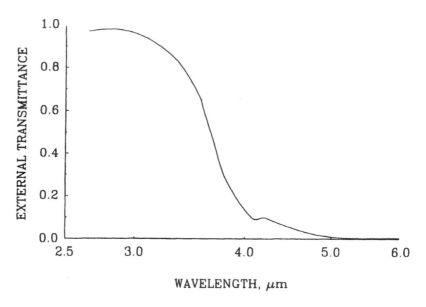

Figure 7.4 The external transmission of beryllium fluoride glass, thickness 5mm. [Adapted from C.M. Baldwin and J.D. Mackenzie, *J. Non-Cryst. Solids 31*:443 (1979).

BZYbT Fluoride Glass, BaF$_2$·ZnF$_2$·YbF$_3$·ThF$_4$ (15:28.3:28.3:28.3)

Specific Gravity
6.43 [g/cm^3]
Transmission
(See Fig. 7.5.)
Absorption Coefficient
(See Fig. 7.6.)
Refractive Index
1.535 at 0.6 μm
Dispersion
Abbe number = 86
Glass Transition Temperature
617 K

Specific Heat
0.106 [cal/(g·K)] at 298 K
Debye Temperature
430 K
Hardness
276 kg/mm^2 (Vickers)
Elastic Moduli
Young's modulus
70 GPa
Notes: BZYbT fluoride glass exhibits high transparency from the mid-IR to the UV. Various compositions of BZYbT have been made as well as many similar glasses with other cations such as Al^{3+}, Mn^{2+}, and Y^{3+}.

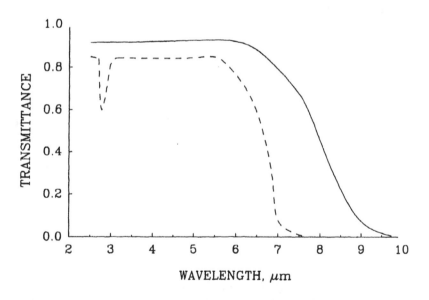

Figure 7.5 The external transmittance of BZYbT fluoride glass comparing InF$_3$ doping of 10% (solid curve) versus AlF$_3$ doping of 10% for samples of 2.4 mm thickness. [Adapted from J. Lucas, D. Tregoat, and G. Fonteneau, abstract 23, Second International Symposium on Halide Glasses (1983).]

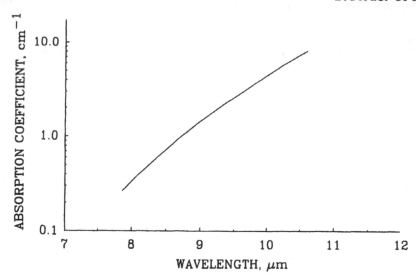

Figure 7.6   Absorption coefficient of BZYbT glass. [Adapted from M. Matecki, Michael Poulain, and Marcel Poulain, abstract 27, Second International Symposium on Halide Glasses (1983).]

Schott IRG 11 Calcium Aluminate Glass

Specific Gravity
3.12 [g/cm^3]

Transmission
(See Fig. 7.7.)

Refractive Index
(See Table 7.2.)

Dielectric Constant
11.5 for 10^6 Hz at 298 K

Thermal Conductivity
1.13 W/(m·K)

Thermal Expansion
8.2 × 10^{-6} K^{-1} at 293–573 K

Specific Heat
0.179 [cal/(g · K)] at 293–373 K

Hardness
608 kg/mm^2 (Knoop number with 200-g load)

Solubility
0.14 g/(100 g H$_2$O) at 293 K

Elastic Moduli
Young's modulus
107.5 GPa
Poisson's ratio
0.284

Notes: IRG 11 is used for IR windows and domes. It is available in disk form up to 200 mm in diameter and 20 mm in thickness. Domes of this material are available with an aperture diameter of up to 150 mm and wall thickness of up to 10 mm. IRG 11 is also available in other dimensions upon request. This glass requires a protective coating and cannot be fused or deformed by reheating.

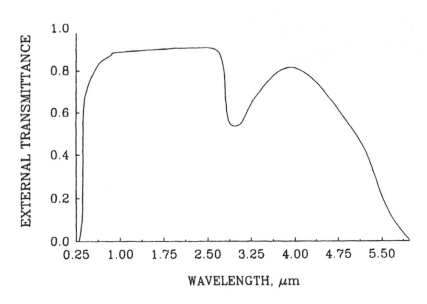

Figure 7.7 External transmittance of calcium aluminate glass (IRG 11), thickness 5 mm. [Adapted from Schott Glass Technologies, Inc., Production Information No. 3112/ie, May 1982.]

Table 7.2 Refractive Index of
calcium aluminate glass (IRG 11)

λ (μm)	n
0.3650	—
0.4047	—
0.4800	1.6926
0.4861	1.6917
0.5461	1.6845
0.5876	1.6809
0.6438	1.6770
0.6563	1.6763
0.7065	1.6741
0.8521	1.6686
1.0140	1.6650
1.5296	1.6581
1.9701	1.6532
2.3254	1.6491
2.674	1.6445
3.303	1.6349
4.258	1.6156
4.586	1.6077

Refractive index deviation from the above values from batch to batch ≤ 0.002. Closer tolerances are possible on request.
Source: Schott Glass Technologies.

Corning 9753: Calcium Aluminosilicate Glass

Specific Gravity
2.798 [g/cm³]
Transmission
(See Fig. 7.8.)
Refractive Index
(See Fig. 7.9)
Dielectric Constant
8.87 for 10^6 Hz at 298 K
9.51 for 10^6 Hz at 773 K
Glass Transition Temperature
1073 K
Thermal Conductivity
~2.3 W/(m·K) at 323 K
~2.5 W/(m·K) at 373 K
~3.5 W/(m·K) at 473 K
~7.1 W/(m·K) at 873 K
Thermal Expansion
5.95×10^{-6} K^{-1} at 298–573 K
7.2×10^{-6} K^{-1} at 298–973 K

Specific Heat
~0.19 [cal/(g·K)] at 323 K
~0.20 [cal/(g·K)] at 373 K
~0.22 [cal/(g·K)] at 473 K
Hardness
601 kg/mm² (Knoop number with 500-g load)
Elastic Moduli
Young's modulus
98.58 GPa
Shear modulus
38.58 GPa
Bulk modulus
74.68 GPa
Poisson's ratio
0.28

Notes: Transparent Cortran code 9753 brand glass transmits in the infrared out to 4.7 μm through a 2-mm section. Its hardness makes it suitable for high-speed airborne applications. Cortran 9753 is now being used on heat-seeking missiles.

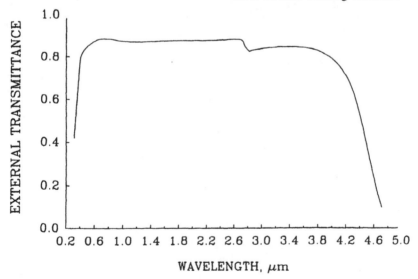

Figure 7.8 External transmittance of calcium aluminosilicate (Corning 9753), thickness 2 mm. [Adapted from Corning Glass Works, Corning Infrared Transmitting Materials data sheet, 1970.]

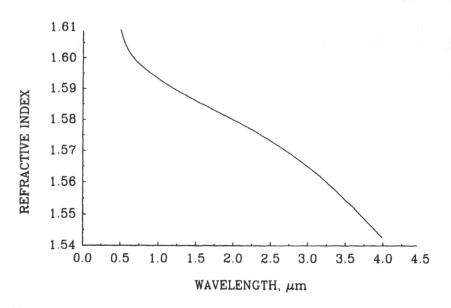

Figure 7.9 The refractive index of calcium aluminosilicate glass (Corning 9753). [Adapted from Corning Glass Works, Corning Infrared Transmitting Materials data sheet, 1970.]

Schott IRG N6 Calcium Aluminosilicate Glass

Specific Gravity
   2.81 [g/cm^3]
Transmission
   (See Fig. 7.10.)
Refractive Index
   (See Table 7.3.)
Dielectric Constant
   9.2 for 10^6 Hz at 298 K
Thermal Conductivity
   1.36 W/(m·K)
Thermal Expansion
   6.3 × 10^{-6} K^{-1} at 293–573 K
Specific Heat
   0.193 [cal/(g·K)] at 293–373 K
Hardness
   623 kg/mm^2 (Knoop number with 200-g load)

Solubility
   0.213 g/(100 g H$_2$O) at 293 K
Elastic Moduli
   Young's modulus
      103.2 GPa
   Poisson's ratio
      0.276
Notes: IRG N6 calcium aluminum silicate glass is used in the production of lenses, IR windows, domes, and other optical elements. Disks up to 200 mm in diameter and up to 50 mm in thickness are available. Domes are available up to 200 mm in aperture diameter and up to 10 mm in wall thickness. This glass can be obtained in other dimensions upon request. IRG N6 cannot be fused or deformed by reheating.

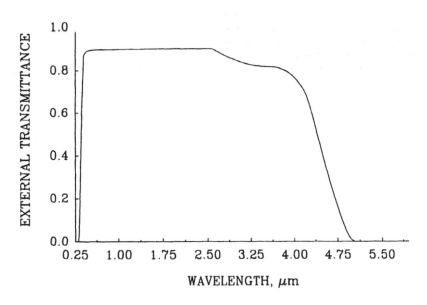

Fig. 7.10 The external transmittance of calcium aluminum silicate glass (Schott IRG N6), thickness 5 mm. [Adapted from Schott Glass Technologies, Inc., Product Information No. 3112/ie, May 1982.]

Table 7.3 Refractive Index of
calcium aluminum silicate Glass
(IRG N6)

λ (μm)	n
0.3650	—
0.4047	1.6069
0.4800	1.5971
0.4861	1.5965
0.5461	1.5915
0.5876	1.5892
0.6438	1.5863
0.6563	1.5857
0.7065	1.5842
0.8521	1.5807
1.0140	1.5777
1.5296	1.5716
1.9701	1.5667
2.3254	1.5620
2.674	1.5567
3.303	1.5451
4.258	1.5209
4.586	—

Refractive index deviation from the above values from batch to batch ≤ 0.002. Closer tolerances are possible on request.
Source: Schott Glass Technologies.

HTF-1 (Ohara Corporation) Fluoride Glass

Specific Gravity
  3.88 [g/cm³]
Transmission
  (See Fig. 7.11.)
Refractive Index
  (See Fig. 7.12.)
Temperature Coefficient of Refractive Index
  $-8.19 \times 10^{-6} \, \text{K}^{-1}$
Glass Transition Temperature
  658 K

Thermal Expansion
  $16.1 \times 10^{-6} \, \text{K}^{-1}$ at 293–373 K
Hardness
  311 kg/mm² (Knoop number)
Solubility
  0.16% weight loss in H_2O
Elastic Moduli
  Young's modulus
    64.2 GPa
  Poisson's ratio
    0.28
Note: HTF-1 is a fluoride glass with broad transparency from the mid-IR to the UV.

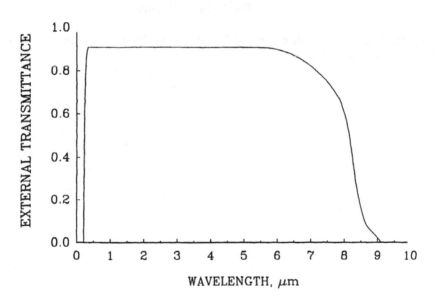

Figure 7.11 The external transmittance of fluoride glass (HTF-1), thickness 5 mm. [Adapted from Ohara Corporation, data sheet.]

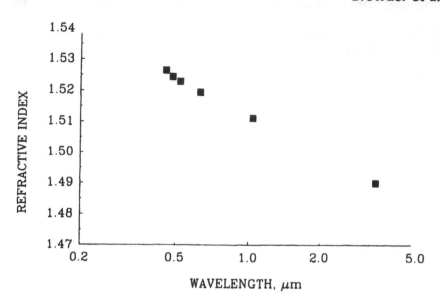

Figure 7.12 The refractive index of fluoride glass (HTF-1). [Adapted from Ohara Corporation, data sheet.]

Schott IRG 9
Fluorophosphate Glass

Specific Gravity
 3.63 [g/cm^3]
Transmission
 (See Fig. 7.13.)
Refractive Index
 (See Table 7.4.)
Dielectric Constant
 10.4 for 10^6 Hz at 298 K
Thermal Conductivity
 0.88 W/(m·K)
Thermal Expansion
 6.1 × 10^{-6} K^{-1} at 293–573 K
Specific Heat
 0.166 [cal/(g·K)] at 293–373 K

Hardness
 346 kg/mm^2 (Knoop number with 200-g load)
Solubility
 0.380 g/(100 g H$_2$O) at 293 K
Elastic Moduli
 Young's modulus
  77.0 GPa
 Poisson's ratio
  0.288

Notes: IRG 9 fluorophosphate glass is used in the production of lenses and other optical elements. Disks up to 120 mm in diameter and up to 30 mm in thickness are available. IRG 9 is also available in other dimensions upon request.

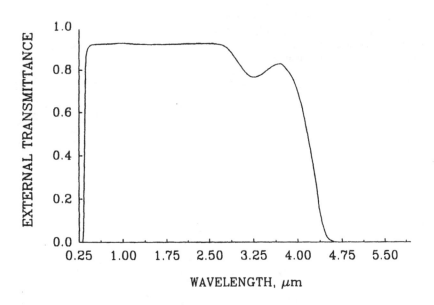

Figure 7.13 The external transmittance of fluorophosphate glass (IRG-9), thickness 5 mm. [Adapted from Schott Glass Technologies, Inc. Product Information No. 3112/ie, May 1982.]

Table 7.4   Refractive Index of
fluorophosphate glass (IRG 9)

λ (μm)	n
0.3650	1.5005
0.4047	1.4961
0.4800	1.4905
0.4861	1.4902
0.5461	1.4875
0.5876	1.4861
0.6438	1.4845
0.6563	1.4842
0.7065	1.4832
0.8521	1.4810
1.0140	1.4793
1.5296	1.4755
1.9701	1.4722
2.3254	1.4692
2.674	1.4658
3.303	1.4583
4.258	—
4.586	—

Refractive index deviation from the above values from batch to batch ≤ 0.002. Closer tolerances are possible on request.
Source: Schott Glass Technologies.

Fused Silica, SiO$_2$

Molecular Weight
60.06
Specific Gravity
2.202 [g/cm^3]
Transmission
Long-wavelength limit 4.5 μm
Short-wavelength limit 0.2 μm
(See Fig. 7.14.)
Absorption Coefficient
(See Fig. 7.15.)
Reflection Loss
6.3% for two surfaces at 2 μm
(See Fig. 7.16.)
Refractive Index
(See Table 7.5 and 7.17.)
Temperature Coefficient of Refractive Index
(See refractive index, Table 7.5.)
Dispersion
(See dispersion equation, Table 7.5.)
Glass Transition Temperature
1273 K
Dielectric Constant
3.78 for 10^2–10^{10} Hz at 300 K
Thermal Conductivity
0.7 W/(m·K) at 20 K

1.2 W/(m·K) at 194 K
1.6 W/(m·K) at 373 K
Thermal Expansion
-0.767×10^{-6} K^{-1} at 80 K
0.42×10^{-6} K^{-1} at 293 K
0.75×10^{-6} K^{-1} at 423 K
Hardness
461 kg/mm^2 (Knoop number with 200-g load)
Solubility
Insoluble in H$_2$O; very slightly soluble in alkalis; soluble in hydrofluoric acid.
Elastic Moduli
Young's modulus
73.03 GPa
Shear modulus
36.38 GPa
Bulk modulus
37 GPa at 293 K
Poisson's ratio
0.17
Notes: Fused silica cuts and grinds very well and is otherwise suitable for many applications. Blanks up to 156 in. in diameter have been successful formed. See also Corning glasses 7940, 7957, and 7958.

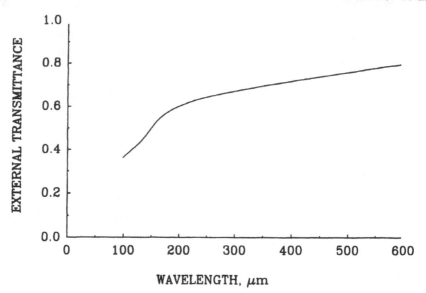

Figure 7.14 The external transmittance of fused silica, thickness 0.56 mm. [Adapted from McCubbin and Sinton, *J. Opt. Soc. Am. 40*:537–539 (1950).]

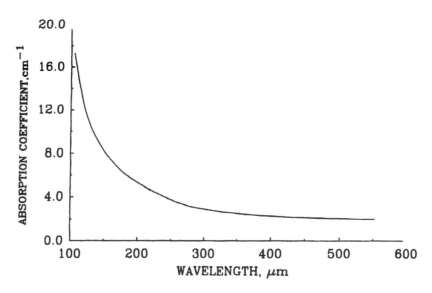

Figure 7.15 The absorption coefficient of fused silica. [Adapted from C.M. Randall and R.D. Rawcliffe, *Appl. Opt. 7*:213 (1968).]

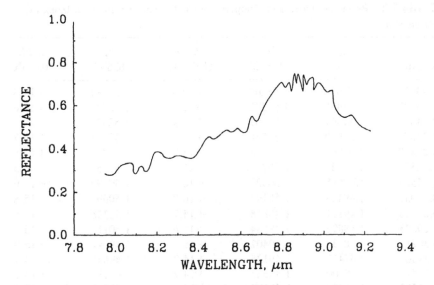

Figure 7.16. The reflectance of fused silica for angle of incidence of 30°. [Adapted from Hardy and Silverman, *Phys. Rev. 37*:176 (1931).]

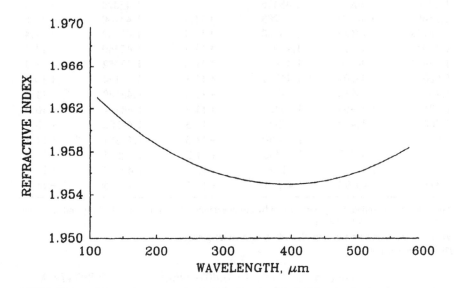

Figure 7.17 The refractive index of fused silica. [Adapted from C.M. Randall and R.D. Rawcliffe, *Appl. Opt. 7*:213 (1968).]

Table 7.5 Refractive Index and Temperature Coefficient of Refractive Index of Fused Silica

λ (μm)	n, 26°C	n, 471°C	dn/dT (10^{-6}/°C)	n, 828°C	dn/dT (10^{-6}/°C)
0.23021	1.52034	1.52908	+19.6	1.53584	+19.3
0.23783	1.51496	1.52332	+18.8	1.52985	+18.6
0.2407	1.51361	1.52201	+18.9	1.52832	+18.3
0.2465	1.50970	1.51774	+18.1	1.52391	+17.7
0.24827	1.50865	1.51665	+18.0	1.52289	+17.8
0.26520	1.50023	1.50763	+16.6	1.51351	+16.5
0.27528	1.49615	1.50327	+16.0	1.50899	+16.0
0.28035	1.49425	1.50143	+16.2	1.50691	+15.8
0.28936	1.49121	1.49818	+15.7	1.50358	+15.4
0.29673	1.48892	1.49584	+15.6	1.50112	+15.2
0.30215	1.48738	1.49407	+15.1	1.49942	+15.0
0.3130	1.48462	1.49126	+14.9	1.49641	+14.7
0.33415	1.48000	1.48633	+14.2	1.49135	+14.1
0.36502	1.47469	1.48089	+14.0	1.48563	+13.6
0.40466	1.46978	1.47575	+13.4	1.48033	+13.2
0.43584	1.46685	1.47248	+12.7	1.47716	+12.9
0.54607	1.46028	1.46575	+12.3	1.47004	+12.2
0.5780	1.45899	1.46429	+11.9	1.46870	+12.1
1.01398	1.45039	1.45562	+11.8	1.45960	+11.5
1.12866	1.44903	1.45426	+11.8	1.45820	+11.4
1.254*	1.44772	1.45283	+11.5	1.45700	+11.6
1.36728	1.44635	1.45140	+11.4	1.45549	+11.4
1.470*	1.44524	1.45031	+11.4	1.45440	+11.4
1.52952	1.44444	1.44961	+11.6	1.45352	+11.3
1.660*	1.44307	1.44799	+11.1	1.45174	+10.8
1.701	1.44230	1.44733	+11.3	1.45140	+11.3
1.981*	1.43863	1.44361	+11.2	1.44734	+10.9
2.262*	1.43430	1.43933	+11.3	1.44306	+10.9
2.553*	1.42949	1.43450	+11.3	1.43854	+11.3
3.00*	1.41995	1.42495	+11.2	1.42877	+11.0
3.245*	1.41353	1.41893	+12.2	1.42243	+11.1
3.37*	1.40990	1.41501	+11.5	1.41915	+11.5

*Wavelength determination by narrow-bandwidth interference filters. From J.H. Wray and J.T. Neu, *J. Opt. soc. Am.* 59:774 (1969).

Values of dn/dT are for Corning No. 7490, ultraviolet grade.

Dispersion equation:

$$n^2 - 1 = \frac{0.6961663\lambda^2}{\lambda^2 - (0.0684043)^2} + \frac{0.4079426\lambda^2}{\lambda^2 - (0.1162414)^2} + \frac{0.8974794\lambda^2}{\lambda^2 - (9.896161)^2}$$

Corning 7940 Fused Silica

Specific Gravity
2.202 [g/cm^3]
Transmission
(See Fig. 7.18)
Refractive Index
(See Table 7.6.)
Temperature Coefficient of Refractive Index
(See Fig. 7.19.)
Dielectric Constant
4.00 for 10^3 Hz at 298 K
4.10 for 10^3 Hz at 673 K
Glass Transition Temperature
1263 K
Thermal Conductivity
1.38 W/(m·K)
Thermal Expansion
-0.6×10^{-6} K^{-1} at 73 K
0.52×10^{-6} K^{-1} at 278–308 K
0.57×10^{-6} K^{-1} at 273–473 K
0.48×10^{-6} K^{-1} at 173–473 K
Specific Heat
0.177 [cal/(g·K)]
Hardness
500 kg/mm^2 (Knoop number with 200-g load)
Elastic Moduli
Young's modulus
73 GPa
Rupture modulus
0.05 GPa
Shear modulus
31 GPa
Bulk modulus
36.9 GPa
Poisson's ratio
0.17

Notes: Corning code 7940 fused silica is a synthetic amorphous silicon dioxide of high purity manufactured by flame hydrolysis. It exhibits virtually no surface clouding or electrical surface leakage when subjected to attack by water, sulfur dioxide, or other atmospheric gases. Rapid corrosion of Corning fused silica occurs only on exposure to hydrofluoric acid and concentrated alkaline solutions, the rate of attack increasing with temperature. Corning 7940 resists discoloration when irradiated with extremely large doses. It is available in sheet form and as circles, squares, rectangles, and items that can be cut from sheets. Corning can also fuse various types and configurations into strong monolithic structures. Corning produces several grades of fused silica that can be formed into solid mirror blanks 200 in. and more in diameter.

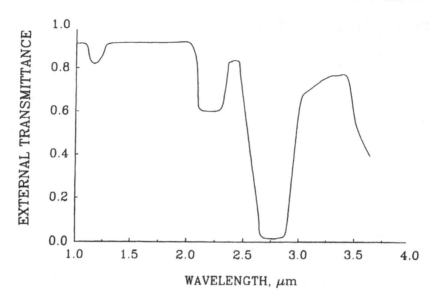

Figure 7.18 The external transmittance of fused silica (Corning 7940). [Adapted from Corning FS7490-12-86 (Rpt.), Goff Communications, Inc., 1986.]

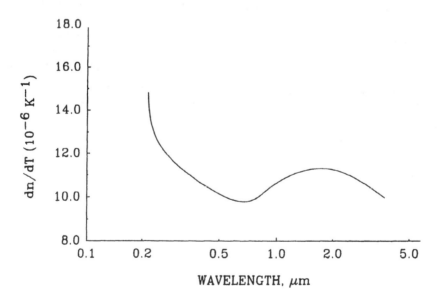

Figure 7.19 The temperature coefficient of the refractive index of fused silica (Corning 7940). [Adapted from Corning FS 7490-12-86 (Rpt.) Goff Communications, Inc., 1986.]

Table 7.6   Refractive Index of Corning Fused Silica No. 7940 (Measured at 20°C)

λ (μm)	n	λ (μm)	n
.155	1.6723	.9	1.45175
.1575	1.65945	.95	1.45107
.16	1.6479	1	1.45042
.165	1.62798	1.3	1.44692
.17	1.61138	1.4	1.44578
.175	1.59734	1.45	1.4452
.18	1.58529	1.5	1.44462
.185	1.57485	1.7	1.44217
.19	1.56572	2	1.43809
.195	1.55766	2.15	1.43581
.2	1.55051	2.25	1.4342
.205	1.54411	2.35	1.4325
.21	1.53836	2.4	1.43163
.215	1.53316	2.6	1.42789
.22	1.52845	2.65	1.4269
.225	1.52416	2.7	1.42588
.23	1.52024	2.75	1.42484
.235	1.51664	2.8	1.42377
.24	1.51333	2.9	1.42156
.242	1.51208	3	1.41925
.245	1.51027	3.1	1.41682
.25	1.50745	3.2	1.41427
.3	1.48779	3.3	1.41161
.32	1.48274	3.4	1.40881
.35	1.47689	3.5	1.40589
.4	1.47012	3.6	1.40282
.45	1.46557	3.7	1.39961
.5	1.46233	3.8	1.39625
.55	1.45991	3.9	1.39272
.6	1.45804	4	1.38903
.65	1.45664	4.15	1.38315
.7	1.45529	4.2	1.3811
.75	1.45424	4.3	1.37684
.8	1.45332	4.4	1.37238
.85	1.4525	4.5	1.3677
		4.6	1.36278

Source: I.H. Malitson, *J. Opt. Soc. Am.*: (1965).

Corning 7957 IR Grade Fused Silica

Specific Gravity
   2.203 [g/cm^3]
Transmission
   (See Fig. 7.20.)
Absorption Coefficient
   1×10^{-5} cm^{-1} at 0.45–1.06 μm
Refractive Index
   1.4633 at 0.486 μm
   1.4586 at 0.589 μm
   1.4566 at 0.656 μm
Dielectric Constant
   3.91 for 10^3 Hz
Glass Transition Temperature
   1266 K
Thermal Conductivity
   1.255 W/(m·K)
Thermal Expansion
   0.39×10^{-6} K^{-1} at 173–473 K
   0.52×10^{-6} K^{-1} at 273–473 K
Specific Heat
   0.18 [cal/(g·K)]

Hardness
   500 kg/mm^2 (Knoop number with 200-g load)
Elastic Moduli
   Young's modulus
      72.4 GPa
   Rupture modulus
      0.050 GPa
   Shear modulus
      31.0 GPa
   Poisson's ratio
      0.16

Notes: Corning's 7957 is a high-purity, amorphous fused silica. It is available in wafers, disks, blocks, rods, bars, domes, and prisms. Disks come in sizes up to 6 in. in diameter and 3 in. thick. Bars are available in sizes up to 14 in. × 3 in. × 3 in. Larger sizes and different forms are available upon request. This material is also available as Corning's 7958, which has the isotropic properties of prism-grade fused silica.

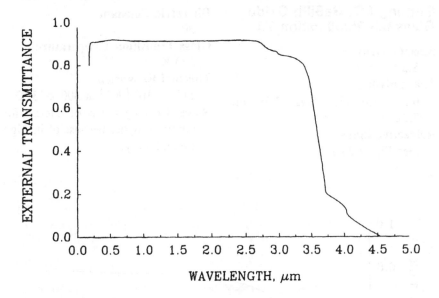

Figure 7.20 The external transmittance of IR grade fused silica (Corning 7957), thickness 1 cm. [Adapted from Corning Glass Works, Corning 7957 data sheet.]

Corning EO: GaBiPb Oxide Glass (25:35:40 cation %)

Specific Gravity
8.2 [g/cm³]
Transmission
0.5–7 μm (See Figs. 7.21 and 7.22.)
Refractive Index
(See Fig. 7.23.)

Dielectric Constant
30
Glass Transition Temperature
573 K
Thermal Expansion
111×10^{-7} K^{-1} at 300–573 K
Notes: Corning EO is of interest for nonlinear optics because of its high susceptibility.

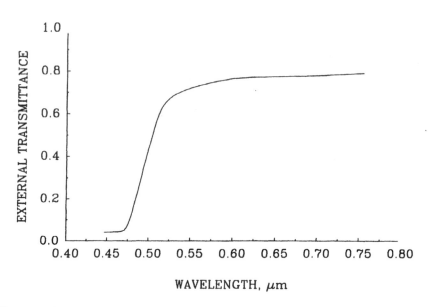

Figure 7.21 The short-wavelength external transmittance of Corning EO glass, thickness 2.0 mm. [Adapted from N.F. Borrelli and W.H. Dumbaugh, *SPIE Proc. 843*: 9 (1987).]

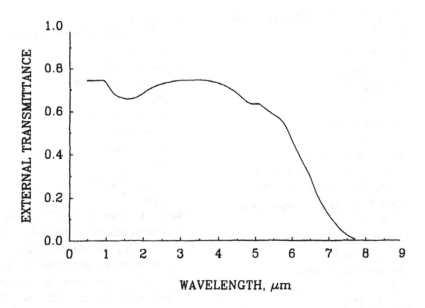

Figure 7.22   The long-wavelength external transmittance of Corning EO glass, thickness 2.0 mm. [Adapted from N.F. Borrelli and W.H. Bumbaugh, *SPIE Proc. 843*:9 (1987).]

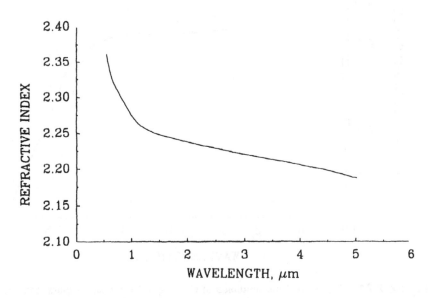

Figure 7.23   The refractive index of Corning EO glass. [Adapted from N.F. Borrelli and W.H. Bumbaugh, *SPIE Proc.* 843:9 (1987).]

Corning 9754 Germanate Glass

Specific Gravity
3.581 [g/cm^3]
Transmission
(See Fig. 7.24.)
Refractive Index
(See Fig. 7.25.)
Dielectric Constant
10.08 for 10^3 Hz at 673 K
10.35 for 10^3 Hz at 823 K
Glass Transition Temperature
970 K
Thermal Conductivity
1.0 W/(m·K)
Thermal Expansion
6.2 × 10^{-6} K^{-1} at 298–573 K
Specific Heat
0.13 [cal/(g·K)]

Hardness
560 kg/mm^2 (Knoop number with 100-g load)
Elastic Moduli
Young's modulus
84.1 GPa
Rupture modulus
0.044 GPa
Shear modulus
35.41 GPa
Poisson's ratio
0.290

Notes: Cortran code 9754 glass is a clear germanate composition material with good transmitting capabilities from the ultraviolet to the infrared. It is resistant to weathering and acid. Corning 9754 is offered on request in plates and molded shapes or polished blanks in sizes up to 4 in.

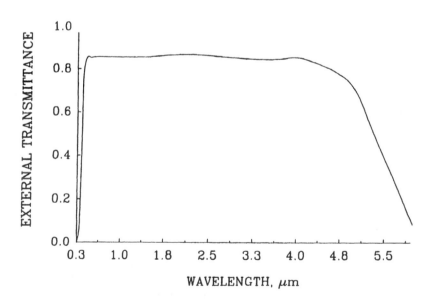

Figure 7.24 The external transmittance of Corning 9754 germanate glass, thickness 1.346 mm. [Adapted from Corning Glass Works, Corning 9754 data sheet (1987).]

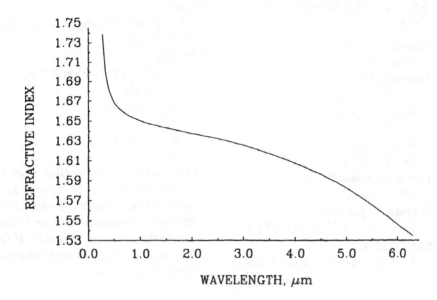

Figure 7.25 The refractive index of Corning 9754 germanate glass. [Adapted from Corning Glass Works, Corning 9754 data sheet (1987).]

Schott IRG 2 Germanate Glass

Specific Gravity
5.00 [g/cm^3]
Transmission
(See Fig. 7.26.)
Refractive Index
(See Table 7.7.)
Dielectric Constant
15.6 for 10^6 Hz at 298 K
Thermal Conductivity
0.91 W/(m·K)
Thermal Expansion
8.8 × 10^{-6} K^{-1} at 293–573 K
Specific Heat
0.108 [cal/(g·K)] at 293–373 K

Hardness
481 kg/mm^2 (Knoop number with 200-g load)
Solubility
0.012 g/(100 g H$_2$O) at 293 K
Elastic Moduli
Young's modulus
95.9 GPa
Poisson's ratio
0.282
Notes: IRG 2 germanate glass is used in the production of lenses and other optical elements. Disks up to 200 mm in diameter and up to 20 mm in thickness are available. IRG 2 is also available in other dimensions upon request.

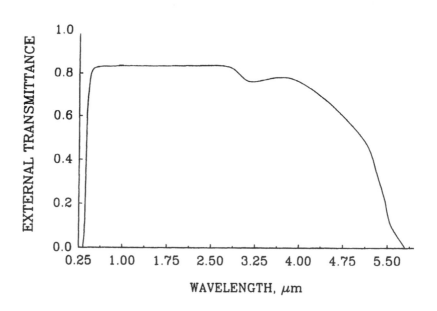

Figure 7.26. The external transmittance of Schott IRG 2 germanate glass, thickness 5 mm. [Adapted from Schott Glass Technologies, Inc. Product Information No. 3112/ie, May 1982.]

Table 7.7 Refractive Index of
Germanate Glass (IRG 2)

λ (μm)	n
0.3650	1.9750
0.4047	1.9462
0.4800	1.9147
0.4861	1.9129
0.5461	1.8988
0.5876	1.8918
0.6438	1.8845
0.6563	1.8832
0.7065	1.8785
0.8521	1.8692
1.0140	1.8630
1.5296	1.8526
1.9701	1.8464
2.3254	1.8414
2.674	1.8362
3.303	1.8253
4.258	1.8041
4.586	1.7954

Refractive index deviation from the above values from batch to batch ≤ 0.002. Closer tolerances are possible on request.
Source: Schott Glass Technologies.

TI-1173 Ge$_{28}$Sb$_{12}$Se$_{60}$ Glass

Specific Gravity
4.67 [g/cm^3]

Transmission
(See Fig. 7.27.)

Absorption Coefficient
0.2×10^{-2} cm^{-1} at 3 μm
1.5×10^{-2} cm^{-1} at 5 μm
1.8×10^{-2} cm^{-1} at 8 μm
1.2×10^{-2} cm^{-1} at 10 μm
1.5×10^{-2} cm^{-1} at 10.6 μm
28×10^{-2} cm^{-1} at 12 μm
59.7×10^{-2} cm^{-1} at 12.8 μm
(See Fig. 7.28.)

Refractive Index
2.626 at 3 μm
2.616 at 5 μm
2.607 at 8 μm
2.600 at 10 μm
2.5976 at 10.6 μm
2.592 at 12 μm
2.5912 at 12.2 μm

Zero Dispersion Wavelength
6.71 μm

Glass Transition Temperature
573 K

Thermal Conductivity
0.301 W/(m·K)

Thermal Expansion
15.0×10^{-6} K^{-1}

Hardness
166 kg/mm^2 (Vickers number)

Solubility
Insoluble in H$_2$O.

Elastic Moduli
Young's modulus
22.1 GPa
Rupture modulus
0.0172 GPa
Poisson's ratio
0.24

Notes: Ge$_{28}$Sb$_{12}$Se$_{60}$ TI-1173 chalcogenide glass is used for infrared optical components because of its transparency in the 8–12-μm region and its relatively good thermal, mechanical, and chemical properties. This glass is insoluble in water and dilute acids but dissolves in alkaline solutions. Various chalcogenide glasses with similar physical properties have been made with a range of Ge$_x$Sb$_y$Se$_z$ compositions.

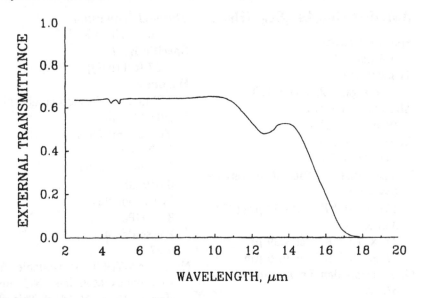

Figure 7.27   The transmittance of TI-1173 glass, thickness 9.08 mm. [Adapted from P. Klocek and L. Colombo, *J. Non. Cryst. Sol.* *93*:1 (1987).]

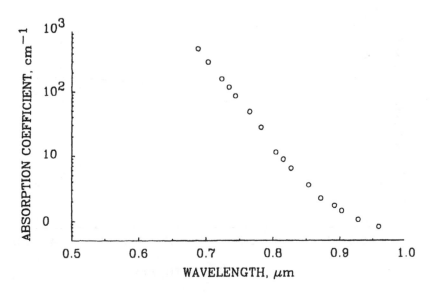

Figure 7.28   The absorption coefficient of TI-1173 glass. [Adapted from P. Klocek and L. Colombo, *J. Non. Cryst. Sol.* *93*:1 (1987).]

AMTIR-1 Ge$_{33}$As$_{12}$Se$_{55}$ Glass

Specific Gravity
4.4 [g/cm^3]
Transmission
(See Figs. 7.29 and 7.30.)
Absorption Coefficient
(See Fig. 7.31.)
Refractive Index
(See Table 7.8.)
Temperature Coefficient of Refractive Index
101×10^{-6} K^{-1} at 1.15 μm (298–338 K)
77×10^{-6} K^{-1} at 3.39 μm
72×10^{-6} K^{-1} at 10.6 μm
Glass Transition Temperature
635 K
Thermal Conductivity
0.25 W/(m·K)

Thermal Expansion
12.0×10^{-6} K^{-1}
Specific Heat
0.07 [cal/(g·K)]
Hardness
170 kg/mm^2 (Knoop number)
Elastic Moduli
Young's modulus
22.0 GPa
Rupture modulus
0.019 GPa
Shear modulus
8.96 GPa
Poisson's ratio
0.27

Notes: AMTIR-1 is available from Amorphous Materials INC. up to sizes of 12 in. × 18 in. This glass is useful from 1 to 12 μm. It is insoluble in water and dilute acids but dissolves in alkalis.

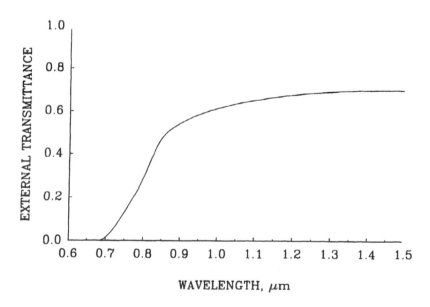

Figure 7.29 The short-wavelength external transmittance of AMTIR-1 (GeAsSe) glass, thickness 1.0 cm. [Adapted from Amorphous Materials Inc. data sheet.]

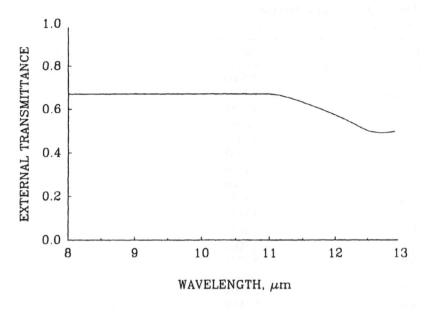

Figure 7.30 The long-wavelength external transmittance of AMTIR-1 (GeAsSe) glass, thickness 0.25 in. [Adapted from Amorphous Materials Inc. data sheet.]

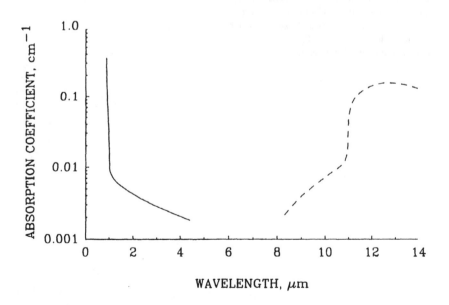

Figure 7.31 Absorption coefficient of AMTIR-1 (GeAsSe) glass; thickness 1 cm (solid curve), thickness 6 cm (dashed curve). [Adapted from Amorphous Materials Inc. data sheet.]

Table 7.8   Refractive Index of
AMTIR-1

λ (μm)	n, at 25°C
1.0	2.6055
1.064	2.5933
1.5	2.5469
2.0	2.5310
2.4	2.5250
3.0	2.5187
4.0	2.5141
5.0	2.5109
6.0	2.5080
7.0	2.5057
8.0	2.5034
9.0	2.5005
10.0	2.4976
11.0	2.4936
12.0	2.4904
13.0	2.4850
14.0	2.4825

Precise infrared refractive index values obtained
using the minimum deviation measurement.
Values are average. Expected batch-to-batch
variation is less than ±0.0010.
Source: Amorphous Materials Inc.

Ge$_{25}$Se$_{75}$ Glass

Transmission
(See Fig. 7.32.)
Absorption Coefficient
(See Fig. 7.33.)
Refractive Index
2.37 at 3 μm
2.36 at 5 μm
2.355 at 10 μm
Glass Transition Temperature
513 K
Zero Dispersion Wavelength
5.93 μm
Thermal Expansion
~22 × 10^{-6} K^{-1}

Hardness
146 kg/mm^2 (Vickers)
Elastic Moduli
Young's modulus
~21.36 GPa (est.)
Poisson's ratio
~0.24 (est.)

Notes: Ge$_{25}$Se$_{75}$ chalcogenide glass is of interest for infrared optical components because of its transparency in the 8–12-μm region and its relatively good thermal, mechanical, and chemical properties. Germanium-selenium glass of various compositions with a range of similar physical properties have been made.

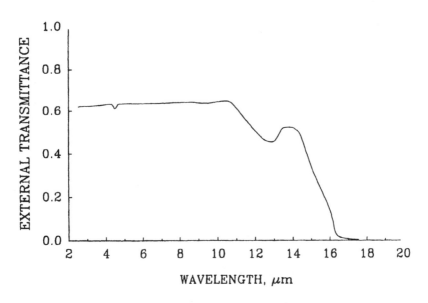

Figure 7.32   The transmittance of Ge$_{25}$Se$_{75}$ glass, thickness 11.8 mm. [Adapted from P. Klocek and L. Colombo, *J. Non Cryst. Sol. 93:*1 (1987).]

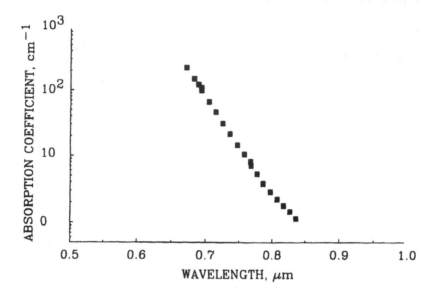

Figure 7.33 The absorption coefficient of $Ge_{25}Se_{25}$ glass. [Adapted from P. Klocek and L. Colombo, *J. Non. Cryst. Sol.* 93:1 (1987).]

Schott IRG 100 Chalcogenide Glass

Specific Gravity
    4.67 [g/cm^3]
Transmission
    (See Fig. 7.34.)
Refractive Index
    (See Table 7.9.)
Temperature Coefficient of Refractive Index
    103 × 10^{-6} K^{-1} at 2.5 μm
    56 × 10^{-6} K^{-1} at 10.6 μm
Softening Temperature
    624 K
Thermal Conductivity
    0.3 W/(m·K) at 293–573 K
Thermal Expansion
    15 × 10^{-6} K^{-1} at 293–573 K
Hardness
    150 kg/mm^2 (Knoop number with 200-g load)

Elastic Moduli
    Young's modulus
        21 GPa
    Poisson's ratio
        0.261

Notes: IRG 100 is a chalcogenide glass with good transmission in the near-IR and mid-IR spectral region between 0.9 and 12 μm. It is suitable for use in manufacturing infrared windows, lenses, and prisms. It is composed of the elements germanium, antimony, and selenium. Schott chalcogenide glass IRG 100 is resistant to attack by water, acids, and organic solvents such as alcohols and acetones. Only alkaline solutions and heavily oxidizing acids attack this glass. Due to its low Knoop hardness, IRG 100 can be ground and polished by conventional optical methods. IRG 100 is, however, sensitive to thermal shock.

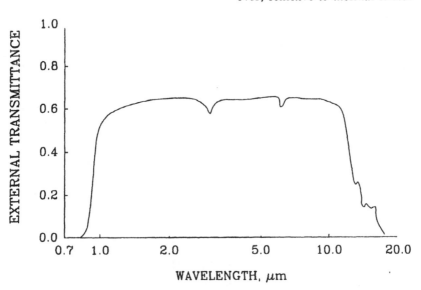

Figure 7.34 The external transmittance of Schott IRG 100 chalcogenide glass composed of germanium, antimony, and selenium, thickness 10 mm. [Adapted from Schott Glass Technologies, Inc. Product Information No. 314.3e/USA, Jan. 1986.]

Table 7.9 Refractive Index of IRG 100

λ (μm)	n	λ (μm)	n
1.0	2.7235	7.5	2.6088
1.5	2.6577	8.0	2.6072
2.0	2.6404	8.5	2.6056
2.5	2.6314	9.0	2.6039
3.0	2.6262	9.5	2.6022
3.5	2.6227	10.0	2.6004
4.0	2.6201	10.5	2.5985
4.5	2.6181	11.0	2.5966
5.0	2.6164	11.5	2.5946
5.5	2.6148	12.0	2.5925
6.0	2.6133	13.0	2.5880
6.5	2.6118	14.0	2.5832
7.0	2.6103		

This index may vary from batch to batch by $\pm 2 \times 10^{-3}$. The optical homogeneity of pieces within one batch is $\pm 1 \times 10^{-4}$.
Source: Schott Glass Technologies.

HBL, Fluoride Glass
$HfF_4BaF_2LaF_3$ (62:33:5)

Specific Gravity
5.78 [g/cm³]

Transmission
(See Fig. 7.35.)

Absorption Coefficient
(See Fig. 7.36.)

Temperature Coefficient of Refractive Index
$-(0.9-1.5) \times 10^{-5} K^{-1}$

Glass Transition Temperature
585 K

Thermal Expansion
$17.7 \times 10^{-6} K^{-1}$

Debye Temperature
510 K

Specific Heat
~0.097 [cal/(g·K)] at 250–550 K
~0.179 [cal/(g·K)] at 625 K

Hardness
228 kg/mm²
(Vickers)

Notes: This glass exhibits a wide range of transparency from the UV to mid-IR and is a candidate for high transparency in the 2–5-μm region. M.G. Drexhage, K.P. Quinlan, C. T. Moynihan, and M. Saleh Boulos, in *Advances in Ceramics*, Vol. 2, *Physics of Fiber Optics*, B. Bendow and S.S. Mitra, Eds., American Ceramics Society, Columbus, Ohio 1981; p. 57).

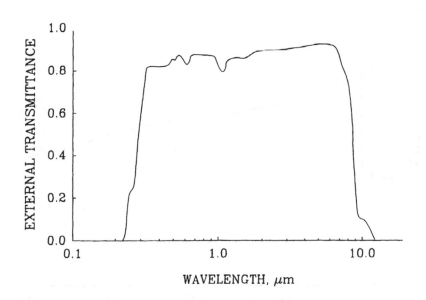

WAVELENGTH, μm

Figure 7.35 The external transmittance of HBL glass, thickness 0.85 mm. [Adapted from M.G. Drexhage, "Fluoride glasses for visible to mid-IR guided wave optics," In *Advances in Ceramics*, American Ceramic Society, Columbus, OH, Vol. 2 (1981).]

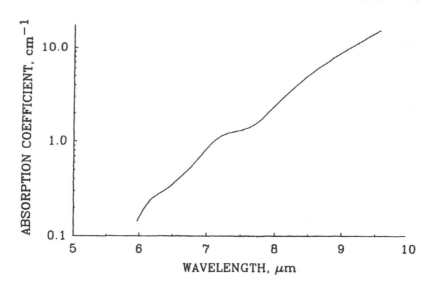

Figure 7.36 Absorption coefficient of HBL glass. [Adapted from M. Matecki, Michel Poulain, and Marcel Poulain, abstract 27, Second International Symposium on Halide Glasses (1983).]

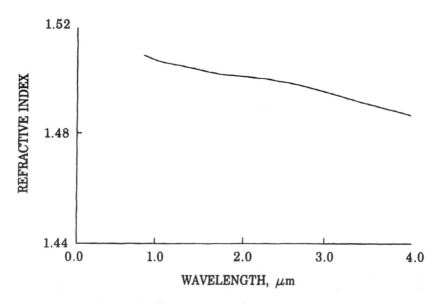

Figure 7.37 Refractive index of HBL glass. [Adapted from M.G. Drexhage, O.H. El-Bayoumi, and C.T. Moynihan, *SPIE Proc. 320*:29 (1982).]

Schott IRG 3 Lanthan Dense Flint Glass

Specific Gravity
 4.47 [g/cm^3]
Transmission
 (See Fig. 7.38.)
Refractive Index
 (See Table 7.10.)
Dielectric Constant
 14.8 for 10^6 Hz at 298 K
Thermal Conductivity
 0.87 W/(m·K)
Thermal Expansion
 8.1×10^{-6} K^{-1} at 293–573 K

Hardness
 541 kg/mm^2 (Knoop number with 200-g load)
Solubility
 0.012 g/(100 g H$_2$O) at 293 K
Elastic Moduli
 Young's modulus
  99.9 GPa
 Poisson's ratio
  0.287

Notes: IRG 3 lanthan dense flint glass is used in the production of lenses and other optical elements. Disks up to 25 mm in diameter and 30 mm in thickness are available. IRG 3 is also available in other dimensions upon request.

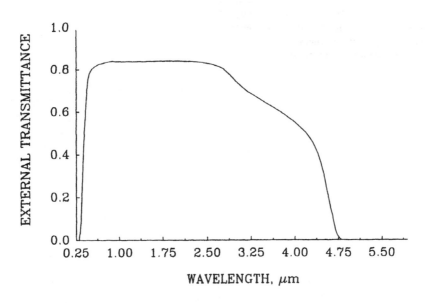

Figure 7.38 The external transmittance of lanthan dense flint glass (IRG 3), thickness 5 mm. [Adapted from Schott Glass Technologies, Inc. Product Information No. 3112/ie, May 1982.]

Table 7.10 Refractive Index of
Lanthan Dense Flint Glass (IRG 3)

λ (μm)	n
0.3650	—
0.4047	1.8925
0.4800	1.8649
0.4861	1.8633
0.5461	1.8510
0.5876	1.8449
0.6438	1.8385
0.6563	1.8373
0.7065	1.8331
0.8521	1.8249
1.0140	1.8193
1.5296	1.8089
1.9701	1.8021
2.3254	1.7963
2.674	1.7900
3.303	1.7764
4.258	1.7491
4.586	1.7375

Refractive index deviation from the above values from batch to batch ≤ 0.002. Closer tolerances are possible on request.
Source: Schott Glass Technologies.

Schott IRG 7 Lead Silicate Glass

Specific Gravity
3.06 [g/cm^3]

Transmission
(See Fig. 7.39.)

Refractive Index
(See Table 7.11.)

Dielectric Constant
6.7 for 10^6 Hz at 298 K

Thermal Conductivity
0.73 W/(m·K)

Thermal Expansion
9.6 × 10^{-6} K^{-1} at 293–573 K

Specific Heat
0.151 [cal/(g·K)] at 293–373 K

Hardness
379 kg/mm^2
(Knoop number with 200-g load)

Solubility
0.171 g/(100 g H$_2$O) at 293 K

Elastic Moduli
Young's modulus
59.7 GPa
Poisson's ratio
0.216

Notes: IRG 7 lead silicate glass is used in the production of lenses and other optical elements. Disks up to 300 mm in diameter and up to 60 mm in thickness and domes up to 300 mm in aperture diameter and 30 mm in wall thickness are available. IRG 7 is also available in other dimensions.

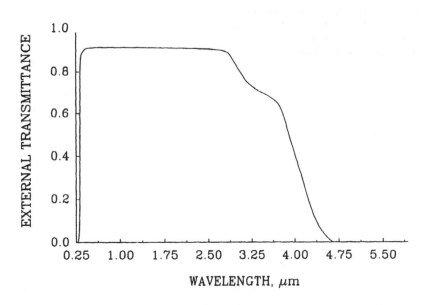

Figure 7.39 The external transmittance of lead silicate glass (IRG 7), thickness 5 mm. [Adapted from Schott Glass Technologies, Inc. Product Information No. 3112/ie, May 1982.]

Table 7.11 Refractive Index of Lead
Silicate Glass (IRG 7)

λ (μm)	n
0.3650	1.5983
0.4047	1.5871
0.4800	1.5743
0.4861	1.5735
0.5461	1.5675
0.5876	1.5644
0.6438	1.5612
0.6563	1.5606
0.7065	1.5585
0.8521	1.5541
1.0140	1.5509
1.5296	1.5442
1.9701	1.5389
2.3254	1.5341
2.674	1.5286
3.303	1.5164
4.258	—
4.586	—

Refractive index deviation from the above values from batch to batch $\leq$ 0.002. Closer tolerances are possible on request.
Source: Schott Glass Technologies.

Corning 7971 ULE Titanium Silicate Glass

Specific Gravity
2.205 [g/cm³]

Transmission
(See Fig. 7.40.)

Refractive Index
1.4892 at 0.486 μm
1.4828 at 0.589 μm
1.4801 at 0.656 μm

Temperature Coefficient of Refractive Index
10.68×10^{-6} K^{-1} at 293–313 K
11.24×10^{-6} K^{-1} at 313–333 K

Dielectric Constant
3.99 for 100 Hz at 298 K
4.00 for 100 Hz at 473 K

Glass Transition Temperature
1163 K

Thermal Conductivity
1.31 W/(m·K)

Thermal Expansion
0.0×10^{-6} K^{-1} at 278–308 K

Specific Heat
0.183 [cal/(g·K)]

Hardness
459 kg/mm²
(Knoop number with 200-g load)

Solubility
Insoluble

Elastic Moduli
Young's modulus
67.52 GPa
Rupture modulus
0.050 GPa
Shear modulus
28.94 GPa
Bulk modulus
34.45 GPa
Poisson's ratio
0.17

Notes: Corning code 7971 ULE glass was developed in the late 1960s to satisfy a need for mirror substrates with a zero expansivity near room temperature. It is "torch-weldable." Its expansion characteristics allow two pieces to be fusion-bonded to form a single piece by heating their mating surfaces with a torch, merging them, and allowing them to cool without the need for controlled-rate cooling in an oven. This property allows the fabrication of honeycomblike cores, ideal for lightweight configurations. ULE glass has been used primarily in applications involving energy reflection. However, newer applications utilizing ULE as a transmitting medium or as a constant-dimension component are developing. Corning ULE titanium silicate exhibits virtually no surface clouding or electrical surface leakage when subjected to attack by water or by sulfur dioxide and other atmospheric gases. Rapid corrosion occurs only on exposure to hydrofluoric acid and concentrated alkaline solutions, the rate of attack increasing with temperature.

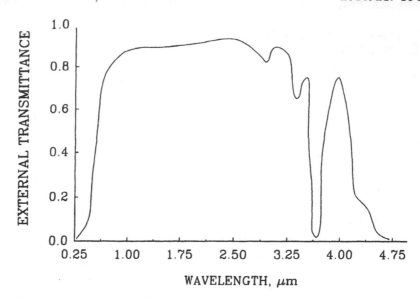

Figure 7.40 The external transmittance of Corning code 7971 ULE™, titanium silicate, thickness 10 mm. [Adapted from Corning Glass Works, Corning Code 7971 data sheet (1985).]

ZBL, Fluoride Glass
ZrF$_4$BaF$_2$LaF$_3$ (62:33:5)

Specific Gravity
  4.79 [g/cm^3]
Transmission
  (See Fig. 7.41.)
Absorption Coefficient
  (See Fig. 7.42.)
Refractive Index
  ~1.523 at 0.5893 μm
  (See Fig. 7.43.)
Glass Transition Temperature
  579 K
Dielectric Constant
  11.8 (high-frequency limit)
Debye Temperature
  540 K

Thermal Expansion
  18.8 × 10^{-6} K^{-1}
Specific Heat
  ~0.125 [cal/(g·K)] at 250–530 K
  ~0.239 [cal/(g·K)] at 610 K
Hardness
  288 kg/mm^2 (Vickers)
Notes: This glass exhibits a wide range of transparency from the UV to the mid-IR and is a candidate for high transparency in the 2–5-μm region. M.G. Drexhage, K.P. Quinlan, C.T. Moynihan, and M. Saleh Boulos, in *Advances in Ceramics*, Vol. 2, *Physics of Fiber Optics*, B. Bendow and S.S. Mitra, Eds., 1981, p. 57).

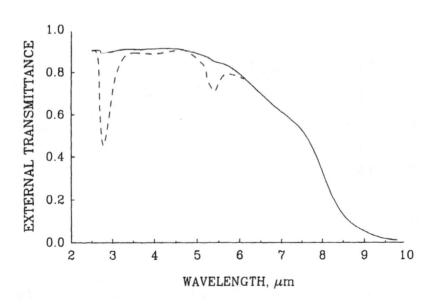

Figure 7.41 The external transmittance of ZBL glass, thickness 2.38 mm, immersed for 0.5 min. in H$_2$O at ambient temperature (solid curve), and for 60 min. in H$_2$O at ambient temperature (dashed curve). [Adapted from A.J. Bruce et al., Rensselaer Polytechnic Institute, abstract P5, Second International Symposium on Halide Glasses (1983).]

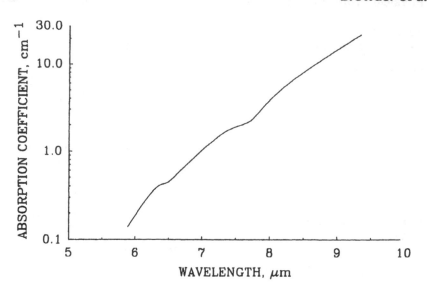

Figure 7.42 Absorption coefficient of ZBL glass. [Adapted from M. Matecki, M. Poulain, and M. Poulain, abstract 27, Second International Symposium on Halide Glasses (1983).]

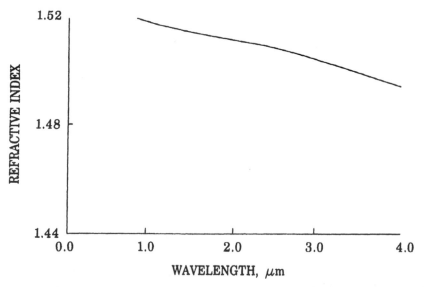

Figure 7.43 The refractive index of a fluorozirconate glass (ZBL). [Adapted from M.G. Drexhage, O.H. El-Bayoumi, and C.T. Moynihan, *SPIE Proc. 320*:29 (1982).]

ZBLA, Fluoride Glass
$ZrF_4BaF_2LaF_3AlF_3$ (57:36:3:4)

Specific Gravity
4.61 [g/cm^3]

Transmission
(See Fig. 7.44.)

Absorption Coefficient
(See Fig. 7.45.)

Refractive Index
$n_D = 1.516$

Glass Transition Temperature
583 K

Thermal Expansion
18.7×10^{-6} K^{-1}

Specific Heat (58:33:5:4)
~0.145 [cal/(g·K)] at 482–540 K
~0.26 [cal/(g·K)] at 590 K

Debye Temperature
580 K

Hardness
267 kg/mm^2
(Vickers)

Elastic Moduli
Elastic stiffness
$c_{11} = 74.7$ GPa
$c_{44} = 25.1$ GPa
Young's modulus
(56:34:6:4)
60.2 GPa
Shear modulus
(56:34:6:4)
24.0 GPa
Bulk modulus
(56:34:6:4)
40.8 GPa
Poisson's ratio
0.250

Notes: ZBLA exhibits high transparency from the mid-IR to the UV. This property makes it a candidate for a wide range of applications from laser windows to infrared fiber optics.

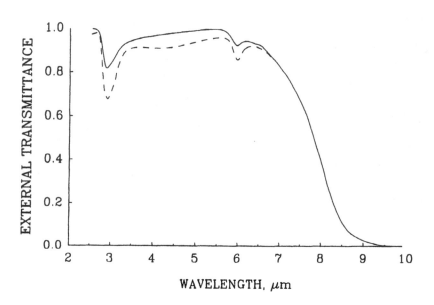

Figure 7.44 The external transmittance of ZBLA glass, diamond paste polished (dashed curve), and cerium oxide polished (solid curve). Both samples were exposed for 20 minutes to 310K deionized water. [Adapted from C.A. Houser and C.G. Pantano, Pennsylvania State University, abstract 15, Second International Symposium on Halide glasses (1983).]

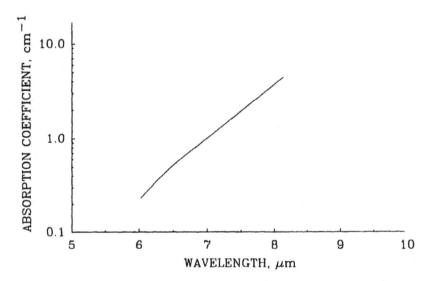

Figure 7.45   Absorption coefficient of ZBLA glass. [Adapted from B. Bendow, abstract 5, Second International Symposium on Halide Glasses (1983).]

8

Optical Thin-Film Coatings

Dale E. Morton *Optic-Electronic Corp., a wholly-owned subsidiary of IMO Industries, Inc., Dallas, Texas*

8.1. INTRODUCTION

Components used in optical systems suffer from losses for many reasons. Energy can be absorbed by the component material. During the fabrication process, imperfections on the surface and inside the material can cause scattering. Refraction occurs as light passes through the boundary between materials and results in reflective losses that depend on the optical properties of the materials. Reflective losses can be as low as 4% per surface for low-index materials and as high as 40% per surface for some high-index materials used in the infrared.

These losses can be reduced selectively over narrow wavelength bands by applying appropriate thin-film coatings or by using sophisticated design techniques to tailor spectral performance over broad wavelength regions. Thin-film coatings that can also be used to enhance system performance include metal mirror coatings, all-dielectric mirror coatings, edge filters, beamsplitters, and bandpass filters. These coatings, when applied to the appropriate optics, allow the system designer to selectively direct various wavelength bands or portions thereof through the system as needed.

This chapter contains a brief discussion of the calculations that can be used to analyze an optical thin-film coating design and how these calculations can be used to design various types of coatings. A final summary will discuss thin-film preparation techniques and the effects they have on the properties of thin-film materials.

8.1.1 Design Analysis: Fresnel Coefficient Approach

What happens when light strikes a coated or uncoated surface can be described by applying Maxwell's equations and the appropriate boundary conditions. Although this concept is straightforward, the calculations are complicated and can be tedious. The advent of high-speed personal computers with sufficient memory has put optical thin-film analysis within the reach of every designer. Although computer systems and software are available that run on all major brands of computers, a basic understanding of these calculations are necessary to anyone who is going to do serious design work. Therefore, two approaches are discussed in this chapter. The first approach, using Fresnel coefficients, is probably easier to understand and is adequate for limited situations. The second approach, the derivation of the matrix, may be more difficult to understand but is mathematically more compact.

Derivation of the matrix is a powerful tool because of the amount of information that can be obtained from it. Consider the situation diagramed in Fig. 8.1, where light passes from medium 1 through medium 2 to medium 3. At each boundary, portions of the light will be reflected and transmitted as well as refracted (bent), depending on the relative optical properties of each medium. At the first boundary, the light reflected back into medium 1 makes the same angle with the normal that the incident ray made. However, the transmitted ray is refracted to another angle depending on the relative refractive index of the two materials according to Snell's law:

$$n_1 \sin \theta_1 = n_2 \sin \theta_2 \tag{1}$$

where n_i = refractive index of medium i.

The transmitted light continues and is incident on the boundary at medium 3, where again a portion is reflected and a portion transmitted. Again, the reflected ray and the incident ray make the same angle with the normal, and the transmitted ray makes an angle to the normal that depends on the relative refractive indices where the subscripts 2 and 3 replace the 1 and 2 in Eq. (1). The ray reflected back into medium 2 continues to reflect back and forth between the boundaries, losing some energy to transmission each time. The relative portions being reflected and transmitted remain the same and depend on the optical properties of each material. Imperfections at the surface also result in energy loss from scattering. This analysis assumes ideal conditions and ignores scattering losses. Assume that the amplitude of the waves reflected back and transmitted forward are given by R_1 and T_1 at the first boundary and by R_2 and T_2 at the second boundary. Also assume that the wave at boundary 1 that is reflected back into medium 2 is given by R_1R (R_1 reverse), and the wave transmitted out is given by $T_1R(T_1$ reverse). Thus, the wave amplitudes reflected and transmitted at each point of incidence for the situation diagramed in Fig. 8.1 are as shown in Fig. 8.2. Then the total wave amplitudes reflected (r) and transmitted (t) are

$$r = R_1 + (T_1 \cdot R_2 \cdot T_1R)e^{-2i\delta} + (T_1 \cdot R_2^2 \cdot R_1R \cdot T_1R)e^{-4i\delta}$$
$$+ (T_1 \cdot R_2^3 \cdot R_1R^2 \cdot T_1R)e^{-6i\delta} + \ldots \tag{2}$$

$$t = (T_1 \cdot T_2)e^{-i\delta} + (T_1 \cdot R_2 \cdot R_1R \cdot T_2)e^{-3i\delta} + (T_1 \cdot R_2^2 \cdot R_1R^2 \cdot T_2)e^{-5i\delta}$$
$$+ (T_1 \cdot R_2^3 \cdot R_1R^3 \cdot T_2)e^{-7i\delta} + \ldots \tag{3}$$

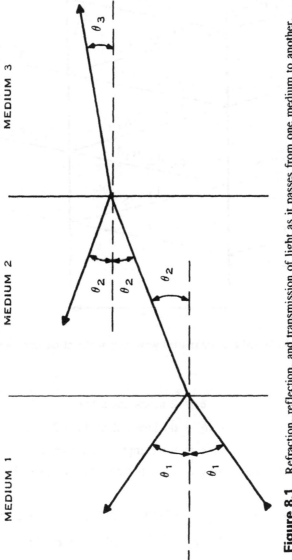

Figure 8.1 Refraction, reflection, and transmission of light as it passes from one medium to another.

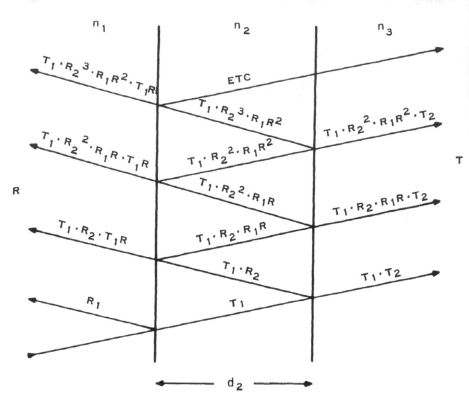

Figure 8.2 Reflected and transmitted wave amplitudes at medium boundaries.

where

$$\delta = 2 \pi n_2 d_2 (\cos \theta_2) / \lambda$$
$$d_2 = \text{thickness of medium 2} \tag{4}$$
$$\lambda = \text{wavelength of the light waves}$$

The reflected and transmitted wave amplitudes are then obtained by summing the above to give

$$r = R_1 + (T_1 \cdot R_2 \cdot T_1 R) e^{-2i\delta} / (1 + R_2 \cdot R_1 R) e^{-2i\delta} \tag{5}$$

$$t = (T_1 \cdot T_2) e^{-i\delta} / [1 + (R_2 \cdot R_1 R) e^{-2i\delta}] \tag{6}$$

The light energy reflected and transmitted is obtained by multiplying the corresponding wave amplitudes by their complex conjugates:

$$R = rr^* \tag{7}$$
$$T = tt^* \tag{8}$$

The reflectance and the transmittance computed by using the above relationships are the ratios of the reflected and transmitted energy, respectively, to the incident energy. The values are always less than or equal to 1. For the remainder of this section, reflectance and transmittance will be used exclusively. The percentage equivalence of these terms, reflection and transmission, are obtained by multiplying the reflectance and/or transmittance by 100.

Substituting Eq. (5) into (7) gives

$$R = \left[R_1 + \frac{(T_1 \cdot R_2 \cdot T_1 R)\, e^{-2i\delta}}{1 + (R_2 \cdot R_1 R)\, e^{-2i\delta}} \right]\left[R_1 + \frac{(T_1 \cdot R_2 \cdot T_1 R)\, e^{2i\delta}}{1 + (R_2 \cdot R_1 R)\, e^{2i\delta}} \right]$$

$$= \left[\frac{R_1 + (R_1 \cdot R_2 \cdot R_1 R + T_1 \cdot R_2 \cdot T_1 R)\, e^{-2i\delta}}{1 + (R_2 \cdot R_1 R)\, e^{-2i\delta}} \right]\left[\frac{R_1 + (R_1 \cdot R_2 \cdot R_1 R + T_1 \cdot R_2 \cdot T_1 R)\, e^{2i\delta}}{1 + (R_2 \cdot R_1 R)\, e^{2i\delta}} \right]$$

$$= \left[\frac{R_1^2 + R_1\,(R_1 \cdot R_2 \cdot R_1 R + T_1 \cdot R_2 \cdot T_1 R)(e^{-2i\delta} + e^{2i\delta}) + (R_1 \cdot R_2 \cdot R_1 R + T_1 \cdot R_2 \cdot T_1 R)^2}{1 + (R_2 \cdot R_1 R)(e^{-2i\delta} + e^{2i\delta}) + R_2^2 \cdot R_1 R\, R^2} \right]$$

$$= \left[\frac{R_1^2 + (R_1 \cdot R_2 \cdot R_1 R + T_1 \cdot R_2 \cdot T_1 R)^2 + 2R_1\,(R_1 \cdot R_2 \cdot R_1 R + T_1 \cdot R_2 \cdot T_1 R)\cos 2\delta}{1 + R_2^2 \cdot R_1\, R^2 + 2R_2 \cdot R_1 R \cos 2\delta} \right] \quad (9)$$

In a similar manner, substituting Eq. (7) into (8) gives

$$T = \left(\frac{T_1 \cdot T_2\, e^{-i\delta}}{1 + (R_2 \cdot R_1 R)\, e^{-2i\delta}} \right)\left(\frac{T_1 \cdot T_2\, e^{i\delta}}{1 + (R_2 R_1 R)\, e^{2i\delta}} \right)$$

$$= \left(\frac{T_1^2 \cdot T_2^2}{1 + (R_2^2 \cdot R_1\, R^2) + 2R_2 \cdot R_1 R \cos 2\delta} \right) \quad (10)$$

Equations (9) and (10) are exact for any situation. For nonabsorbing situations, $T_1 = T_1 R$ and $R_1 = R_1 R$. Since energy is conserved, it follows that

$$T_1 \cdot T_1 R = 1 - R_1^2 \quad (11)$$

Applying Eq. (11) and the equalities from the previous paragraph to Eq. (9) simplifies this expression to

$$R = \frac{[R_1 + [R_1^2 \cdot R_2 + (1 - R_1^2) \cdot R_2]^2 + 2R_1[R_1^2 \cdot R_2 + (1 - R_1^2) R_2]\cos 2\delta}{1 + R_1^2 \cdot R_2^2 + 2R_1 \cdot R_2 \cos 2\delta}$$

$$= \left(\frac{R_1^2 + R_2^2 + 2R_1 \cdot R_2 \cos 2\delta}{1 + R_1^2 \cdot R_2^2 + 2R_1 \cdot R_2 \cos 2\delta} \right) \quad (12)$$

Equation (10) can also be reduced to an expression involving just the Fresnel reflection coefficients:

$$T = \left(\frac{(1 - R_1^2)\,(1 - R_2^2)}{1 + R_1^2 \cdot R_2^2 + 2R_1 \cdot R_2 \cos 2\delta} \right)$$

$$= \left(\frac{1 - (R_1^2 + R_2^2) + R_1^2 \cdot R_2^2}{1 + R_1^2 \cdot R_2^2 + 2R_1 \cdot R_2 \cos 2\delta} \right) \tag{13}$$

Finally, Eq. (12) and (13) can be reduced to

$$R = \left(\frac{a + b \cos c}{d + b \cos c} \right) \tag{14}$$

and

$$T = \left(\frac{d - a}{d + b \cos c} \right) \tag{15}$$

where

$$a = R_1{}^2 + R_2{}^2 \qquad b = 2R_1R_2$$
$$c = 4\pi n_2 d_2 \cos \theta_2/\lambda \qquad d = 1 + R_1{}^2 R_2{}^2$$

and the Fresnel coefficients are

$$R_1 = \frac{n_1 - n_2}{n_1 + n_2} \quad \text{and} \quad R_2 = \frac{n_2 - n_3}{n_2 + n_3}$$

If no film is deposited ($d_2 = 0$), then Eq. (14) can be solved to give

$$R = \left(\frac{n_1 - n_3}{n_3 + n_1} \right)^2 \tag{16}$$

which is the reflectance from the surface of an uncoated substrate (where medium 2 does not exist).

The preceding derivation is exact, and Eqs. (14) and (15) can be readily evaluated to give the reflectance and transmittance of a nonabsorbing film normally incident onto a substrate.

Extending the evaluation to non-normal incidence is only slightly more complicated. The s and p polarizations must be evaluated separately and then averaged together. To do so, replace the refractive index with $n_x \cos \theta_x$ for the s polarization and with $n_x/\cos \theta_x$ for the p polarization when evaluating the Fresnel coefficients.

To extend the above derivation to multiple layers, consider the combination of the first layer down and the substrate as a new medium. This film–substrate combination has an effective index of refraction that can be used to calculate new Fresnel coefficients. To complete this derivation, a phase factor must be calculated. Although not impossible, this derivation is complicated and is tedious to accomplish. Heavens (1955) summarizes several approaches. However, the results of this derivation are valid only for nonabsorbing films. Extending the derivation to include absorbing films requires replacing the refractive index n with the expression $n_x - ik_x$, where k_x is the extinction coefficient of the material. The evaluation of Eqs. (1), (14), and (15) becomes much more tedious because they are now in the complex domain. This was not done in the previous development because of the complexity involved and the amount of space required. Also, an alternative approach exists that provides a more satisfying method to handle both absorbing layers and the extension to multiple layers.

8.1.2. Design Analysis: Matrix Approach

The matrix alternative to Fresnel coefficients uses Maxwell's equations and properties of optical impedance and admittance. Impedance varies continuously through a thin film even though there is a refractive index discontinuity at film boundaries. By equating the tangential components of the electric and magnetic field vectors on both sides of a film boundary and by expressing the change in the admittance as a function of layer depth, even the most complex thin-film structure can be represented in a compact form that is easily programed.

Full derivations of this concept have been presented in numerous publications (e.g., Heavens 1955). One of the better treatments is included in *Optical Thin Films* (MacLeod 1969).

The following development is not restricted to the nonabsorbing, normal incidence case. We will assume only that the thin-film materials are isotropic and homogeneous. Also, film layers will be numbered sequentially as they are deposited, from the substrate out to air. This means that the substrate properties have a 0 subscript, and properties of the medium in which the device is to be used (usually, but not exclusively, air) have the subscript $j + 1$ (where j is the number of thin-film layers).

If the admittance is Y, then the reflectance at the surface of a thin-film assembly is

$$R = \left(\frac{Y - \eta_{j+1}}{Y + \eta_{j+1}}\right)\left(\frac{Y - \eta_{j+1}}{Y + \eta_{j+1}}\right)^* \tag{17}$$

Where the asterisk denotes the complex conjugate. The admittance Y can be obtained as follows:

$$\frac{1}{Y} = \frac{B}{C} = \prod_{x=1}^{i}\begin{bmatrix} \cos \delta_x & i \sin \delta_x/\eta_x \\ i \eta_x \sin \delta_x & \cos \delta_x \end{bmatrix}\begin{bmatrix} 1 \\ \eta_0 \end{bmatrix} \tag{18}$$

where $\delta = 2\pi n_x d_x \cos \theta_x/\lambda$ and the complex refractive index is

$$\eta_x = N_x/\cos \theta_x \quad (p \text{ polarization}) \tag{19}$$

$$\eta_x = N_x \cos \theta_x \quad (s \text{ polarization}) \tag{20}$$

where $N_x = n_x - ik_x$.

The 2×2 matrix must be set up for each film layer and the product evaluated to give the admittance of the thin-film assembly. The transmittance can be calculated in a similar manner:

$$T = \frac{4 \eta_{j+1} \eta_0}{(B + C)(B + C)^*} \tag{21}$$

Since the reflectance and the transmittance must be real, only the real parts of η_{x+1} and η_0 are used in evaluating Eq. (21). The absorbance then becomes $A = 1 - T - R$. If film thickness d_x is 0, then the film matrix becomes the unity matrix and Y is n_{x+1}. Equation (17) then reduces to

$$R = \left(\frac{\eta_{j+1} - \eta_0}{\eta_{j+1} + \eta_0}\right) \tag{22}$$

which is expected, because this is the same as Eq. (16) developed previously for an uncoated surface. Although mathematically elegant and compact, even Eqs. (17)–(21) can become tedious when more than a few layers or a few wavelengths are being calculated for any given design, and most work on these calculations is done with PCs or mainframe computers. However, by considering a few special cases, much can be gained from Eqs. (17)–(21). For example, consider a series of quarter-wave optically thick (QWOT) layers:

$$\frac{1}{Y} = \frac{B}{C} = \cdots \begin{bmatrix} 0 & i/n_3 \\ in_3 & 0 \end{bmatrix} \begin{bmatrix} 0 & i/n_2 \\ in_2 & 0 \end{bmatrix} \begin{bmatrix} 0 & i/n_1 \\ in_1 & 0 \end{bmatrix} \begin{bmatrix} 1 \\ n_0 \end{bmatrix}$$

This reduces to

$$Y = \frac{\cdots n_5^2 \, n_3^2 \, n_1^2}{\cdots n_4^2 \, n_2^2 \, n_0} \tag{23}$$

for an odd number of layers, and to

$$Y = \frac{\cdots n_4^2 \, n_2^2 \, n_0}{\cdots n_3^2 \, n_1^2} \tag{24}$$

for an even number of layers. These expressions can be used to develop simple structures that have good transmission properties [antireflection (AR) coatings] when Y has a value of 1 (impedance matched to air) or good reflection properties when Y has a value very large or very close to 0. Since any thin-film assembly on a substrate can be reduced to its admittance Y, Eq. (17) can be solved for the reflectance as Y is varied close to 1, with the results plotted in

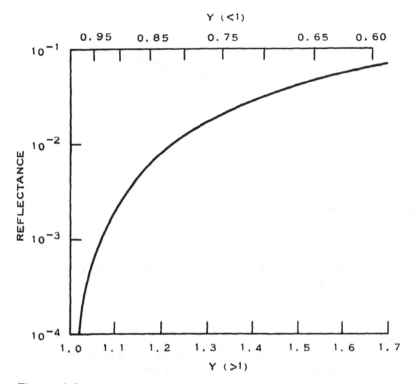

Figure 8.3 Reflectance versus admittance for values near 1. (*Note:* The reflectance is the same for Y and $1/Y$.)

Fig. 8.3. This graph can be used as a quick solution for reflectance once the admittance is known. This is convenient because Eqs. (23) and (24) can be used to determine the admittance of film structures made up of QWOT layers. Figure 8.3 is structured for use with AR-type structures but also can be used to give the reflectance of uncoated glasses, because optical admittance of these glasses is just the refractive index. Another graph (Fig. 8.30) will be developed to use on all-dielectric mirror type structures.

Consider Eq. (23), where, for a single layer, Y has a value of 1. This equation can then be used to solve for an expression that gives the ideal film index for a single-layer antireflection coating as a function of the substrate index. From Eq. (23),

$$1 = n_1^2/n_0 \qquad n_1 = (n_0)^{1/2} \qquad (25)$$

A plot of film indices that perform as ideal single-layer antireflection (SLAR) coatings for various substrate indices is shown in Fig. 8.4.

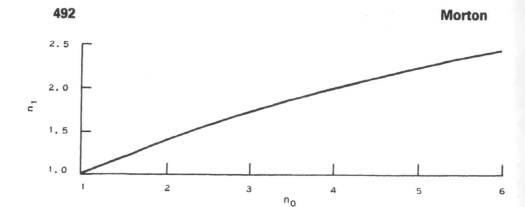

Figure 8.4 n_1 versus n_0 for ideal SLAR coating.

In a similar manner, half-wave optical thickness (HWOT) layers can be represented by the matrix (which is a unity matrix)

$$\begin{bmatrix} 1 & 0 \\ 0 & 1 \end{bmatrix} \qquad (26)$$

Although HWOT layers have no effect in changing the reflectance or transmittance, there can be considerable effect at wavelengths on either side of the central wavelength. Half-wave-thick layers placed between matched all-dielectric quarter-wave reflector stacks form narrow regions of high transmission bounded by high rejection. When properly used in AR coating designs, they act as "achromatizing" layers, which broaden the wavelength band being transmitted. Other uses of half-wave-thick layers as well as those uses mentioned here will be discussed later in this chapter.

8.1.3   Transmittance of a Plate

Before discussing coating designs, one more computational factor should be considered. Equations (14), (15), (17), and (21) give the reflectance or transmittance at one surface of a coated part, not the reflectance or transmittance of the part itself. If the surfaces of the part are parallel to each other (such as a plate), it is possible to set up a multiple bounce situation, and the transmittance of the part will not be the product of the transmittances through the various surfaces. Consider the situation shown in Fig. 8.5, where surface A splits the incident radiation into two portions, T_a and R_a, and surface B splits the incident radiation into two portions, T_b and R_b. Also assume that the material making up

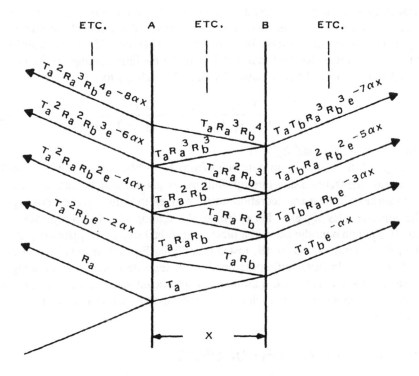

Figure 8.5 Transmittance and reflectance of light incident on a plate.

the part has an absorption coefficient of α. Multiple reflections take place, and the resulting transmittance and reflectance of the part become

$$T = T_a T_b e^{-\alpha x} + T_a T_b R_a R_b \, e^{-3\alpha x} + T_a T_b R_a^2 R_b^2 \, e^{-5\alpha x} + \ldots$$
$$R = T_a^2 R_b e^{-2\alpha x} + T_a^2 R_a R_b^2 \, e^{-4\alpha x} + T_a^2 \, R_a^2 R_b^3 e^{-6\alpha x} \ldots$$

These sums can then be solved to give

$$T = \frac{T_a \, T_b \, e^{-\alpha x}}{1 - R_a \, R_b \, e^{-2\alpha x}} \tag{27}$$

$$R = R_a + \frac{T_a^2 \, R_b \, e^{-2\alpha x}}{1 - R_a \, R_b \, e^{-2\alpha x}} \tag{28}$$

which are equations for the reflectance and transmittance of parts where the various components of the incident energy interact incoherently. Assuming no scattering and a fairly thick parallel plate, these relationships are exact. As the plates become thinner, one must take into account the phase of the interacting light such as was done earlier in deriving Eqs. (14) and (15). Curved surfaces, of course, must be handled individually. If the medium making up the plate is nonabsorbing, then the relationship in Eq. (27) reduces to

$$T = \frac{T_a T_b}{1 - R_a R_b} = \frac{T_a T_b}{T_a + T_b - T_a T_b} \tag{29}$$

since $R_a = 1 - T_a$ and $R_b = 1 - T_b$.

Figure 8.6, a nomogram generated from Eq. (29), can be used to quickly determine the transmission through thin plates given the transmittance through each surface. Similar consideration must be given to using rejection coatings on the opposite sides of plates. For example, if a coating with 40-dB rejection (1×10^{-4}) is put on opposite sides of a plate, and if the faces of the plate are parallel, the rejection will not be 80 dB. instead, applying Eq. (25) results in a rejection of 5×10^{-5}. In fact, putting the coating on separate plates and mounting the plates in series in an optical train is of no significant value unless the plates are tilted at a sufficient angle to each other to prevent multiple reflections from being collected.

8.2. ANTIREFLECTION COATINGS

The ideal AR coating was discussed previously. The refractive index of many materials used in IR systems is high enough to enable the use of existing materials to make efficient SLAR coatings. Examples include the use of ZnS on Ge, SiO on Si, and ThF$_4$ on ZnS. Although the index matches are not exactly correct, they are close enough to be fairly effective and can be entered over a broad wavelength range. Many IR glasses have a fairly low refractive index, but no materials have an index low enough to reduce the reflectance to zero on low-index visible glasses.

Magnesium fluoride with an index of 1.37–1.38 is a durable nonreactive coating material used for single-layer AR coatings in the visible and near-infrared. The most durable coatings are obtained with magnesium fluoride when it is deposited on glass heated to at least 275°C. If deposited at lower temperatures, the coating becomes less dense, lower in index, and soft. At higher temperatures, magnesium fluoride can be deposited in thicknesses to use as SLAR coatings out to about 4.0 μm. At wavelengths greater than 4.0 μm, stresses build up in the coating that cause it to crack and delaminate from the substrate. Recently, ion-assisted deposition (IAD) has been used to deposit magnesium fluoride at lower temperatures, even as low as ambient room temperature. The results have been encouraging and indicate that it may be possible in the future to develop

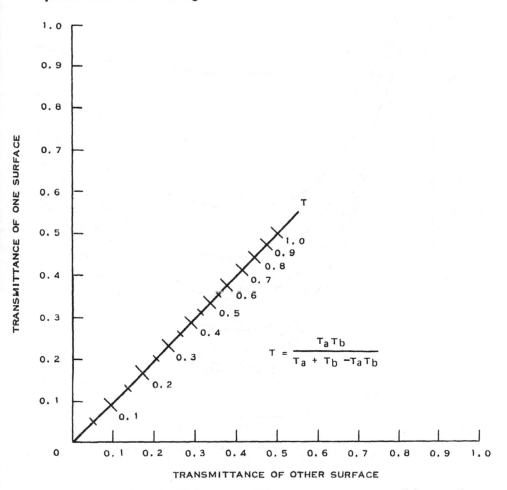

Figure 8.6 Nomogram used to determine transmittance through a thick plate. (Place a straight edge to connect transmittance of each surface, and read transmittance of plate off T line.)

processes involving IAD to make durable magnesium fluoride coating at low substrate temperatures.

Magnesium fluoride is an ideal SLAR coating material for glasses having a refractive index of 1.90. On progressively lower index glasses, the minimum reflectance becomes progressively higher. Figure 8.7 plots the computed reflectance of magnesium fluoride on glasses of various indices from a wavelength of 0.7 to 1.3 λ_0, where λ_0 is the wavelength of the reflectance minimum. Note that these are calculated values. Actual measurements on magnesium fluoride coat-

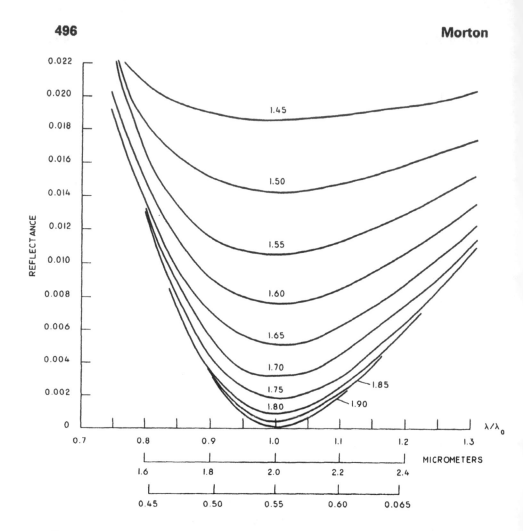

Figure 8.7   Computed reflectance of MgF_2 ($n = 1.38$) glass of indicated refractive index. (Second and third scales can be used to determine reflectance for QWOT films centered at 2.0 and 0.55 μm.)

ings will differ from these values if the deposited film does not have a refractive index sufficiently close to 1.38. However, MgF_2 is a fairly easy material to deposit, and the resulting films usually have a minimum reflectance no more than 0.1–0.2% greater than the computed values.

For non-normal incidence, one must consider the s and p polarizations separately. Figure 8.8 contains plots of the reflectance at the minimum of a QWOT of ThF_4 ($n = 1.50$ in the 4.0-μm region) on glasses of three different indices (2.25, 2.0, and 1.75) for angles of incidence up to 60°. The average reflectance

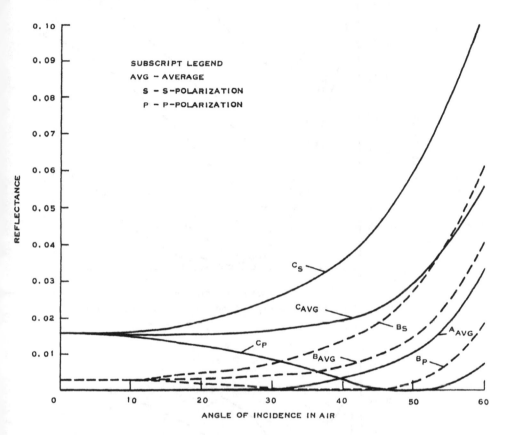

Figure 8.8 R_{min} versus incident angle for single film with 1.50 refractive index on (A) 2.25 index substrate, (B) 2.00 index substrate, and (C) 1.75 substrate.

is shown for each as well as the two polarizations separately when they deviate significantly from the average. The calculation is made using Snell's law to determine the angle the light beam makes to the normal in each material, computing the refractive index for each polarization using Eqs. (19) and (20), and computing the reflectance for the QWOT thickness using Eqs. (17) and (23). The reflectances of the s and p polarizations are then averaged. As can be seen, the reflectance of the s polarization steadily increases, because the film index decreases more rapidly than the higher substrate index. On the other hand, the reflectance of the p polarization steadily decreases until it becomes zero where the relative indices of the film equal the square root of the substrate index and then increases. Not shown in this plot is the fact that, for any given film thickness, the position of the minimum is shifting to a shorter wavelength as well as

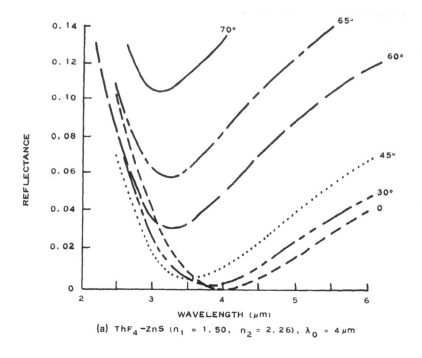

(a) ThF_4-ZnS ($n_1 = 1.50$, $n_2 = 2.26$), $\lambda_0 = 4\,\mu m$

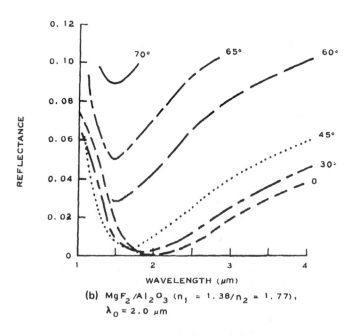

(b) MgF_2/Al_2O_3 ($n_1 = 1.38/n_2 = 1.77$), $\lambda_0 = 2.0\,\mu m$

Figure 8.9 Average reflectance versus wavelength for QWOT films at various angles of incidence.

increasing in reflectance. This is illustrated in Fig. 8.9 for two situations, for ThF_4 on ZnS and for MgF_2 on sapphire (Al_2O_3). Since magnesium fluoride is a common SLAR coating material, a plot of its average reflectance minimum on substrates of indices ranging from 1.46 to 1.79 and at angles of incidence from 30° to 65° is included in Fig. 8.10.

The use of two coating materials gives greater flexibility in achieving low reflectance. Consider the application of Eq. (24) using two materials, and let admittance be 1:

$$Y = 1 = \frac{n_2^2 \, n_0}{n_1^2} \qquad n_1 = n_2 \sqrt{n_0} \qquad (30)$$

Equation (30) can be used to select a film combination that will give a low reflectance at any given wavelength. It would be logical to use magnesium fluoride ($n = 1.38$) for the outer layer and then calculate the index needed for the inner layer. Figure 8.11 is a plot of the inner-layer index versus the substrate index for this situation. Since MgF_2 can be used only to thicknesses of a QWOT of less that 4 μm, another low-index material must be used for two-layer coatings peaked at longer wavelengths. Thorium fluoride is a low-index material often used in IR coatings that are functional in the longer wavelength IR regions. Therefore, Fig. 8.11 also includes a plot of the refractive index of the inner layer that would be needed if the outer layer had an index of 1.55. Because MgF_2 is not very dispersive, the 1.38 index is applicable from 0.4 to beyond 5.0 μm. On the other hand, ThF_4 is very dispersive and has a 1.55 index through the visible region that decreases through the infrared. In the forward-looking infrared (FLIR) region of 8–12 μm, the real part of the index is dropping from about 1.40 to 1.24. The actual index obtained for this material depends on the deposition techniques employed. Another low-index material available for use in the IR region is BaF_2. Its refractive index is about 1.46 in the near-IR region and drops from 1.44 to about 1.37 over the FLIR band. The dashed lines in Fig. 8.11 for outer layers of 1.45 and 1.30 refractive index were included to aid the reader when using this plot for other film combinations.

Figure 8.12 shows the effectiveness of two-layer AR coatings. The plots in this figure show several structures one might select using the information in Fig. 8.11.

Because some of the two-layer AR structures in Fig. 8.12 are very narrow, they are often called V-coats. These narrow structures are those where the impedance matching from the substrate into air is achieved by stepping up to an inner layer with a higher index than the substrate. The broader AR structure involves a straight step down from the substrate to the inner layer, to the outer layer, and then into air. It is also apparent that even the ideal two-layer structure has the problem of finding a material with exactly the right refractive index for

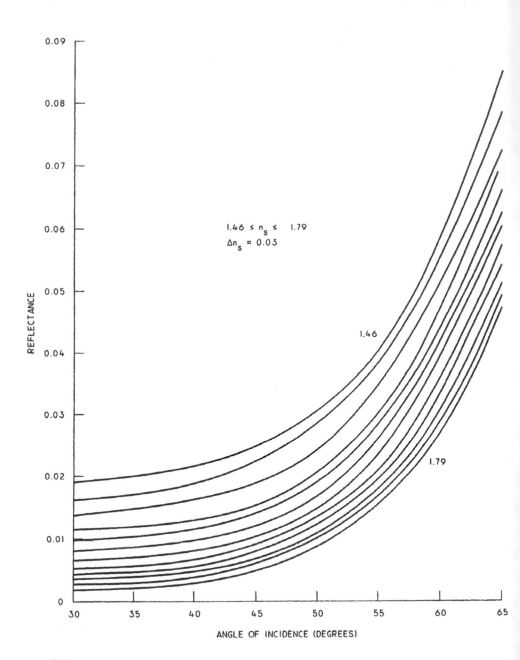

Figure 8.10 Average R_{min} versus incident angle for MgF$_2$ ($n = 1.38$) on substrates of various indices.

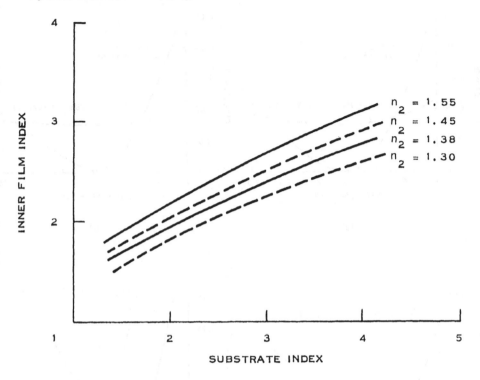

Figure 8.11   Two-layer structures plotting inner layer film index (n_1) needed for ideal AR coating on substrates of various indices.

the inner layer. However, this limitation can be overcome because it is possible to synthesize any film index using film combinations that functionally act as one film of the correct index. This concept of equivalent index structures has been discussed by many, including Herpin (1947), Epstein (1952), Berning (1962), and Thelen (1966), and will be touched on only briefly here. To demonstrate the power of this concept, several examples are included to show the results that can be obtained when the right tools are used.

Three of the examples from Fig. 8.12 were reduced to practice by taking the properties of BaF_2, ThF_4, ZnS, and Ge and replacing the inner QWOT layer with a Herpin structure of the appropriate phase thickness and equivalent index to achieve a good AR structure centered at 10 μm. These results are plotted in Fig. 13. Note that two refractive indices were used for Ge. The lower value of 4.00 is the bulk index for Ge material, whereas the 4.15 value is typical for what is routinely achieved in depositing thin films of Ge.

Two situations are displayed for each example. The one is a wL xH wL structure, and the other is a yH zL yH structure for the Herpin triad, where L and

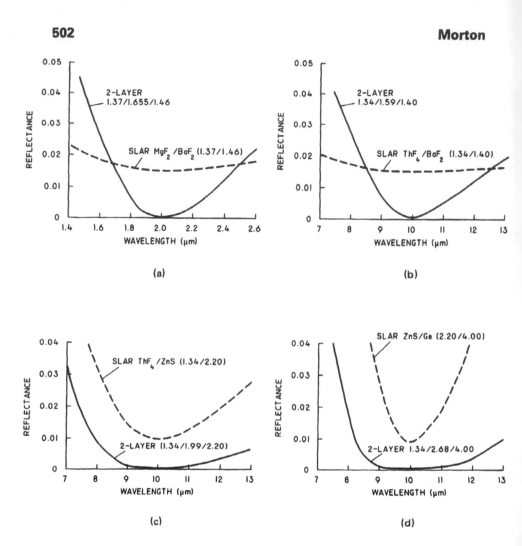

Figure 8.12 Comparison of selected single-layer and two-layer AR coatings. (For the two-layer designs, a real material is chosen and the inner layer is calculated, $n_1 = n_2 \sqrt{n_0}$.)

H are QWOT layers of low- and high-index material; w, x, y, and z are the fractional parts of a QWOT; and $2w + x$ and $2y + z$ should each total 1. (In these examples, the sums may be slightly less than 1 because an optimization routine was used to develop the layer thickness combinations.) These synthesized structures are much more dispersive than most real materials. To show this, Fig. 8.14 presents the index of the structures as a function of the wavelength band 7–13 4µm. In each case, note that the required "ideal" index is achieved

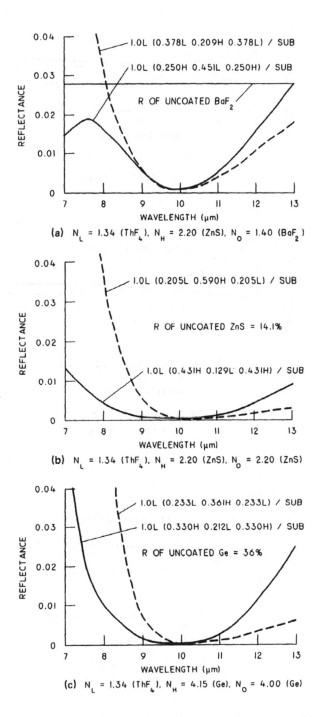

Figure 8.13. Two-layer equivalent AR coatings at 10 μm on various substrates. (Each is shown for LHL and HLH equivalent index profiles as plotted in Fig. 8.14.)

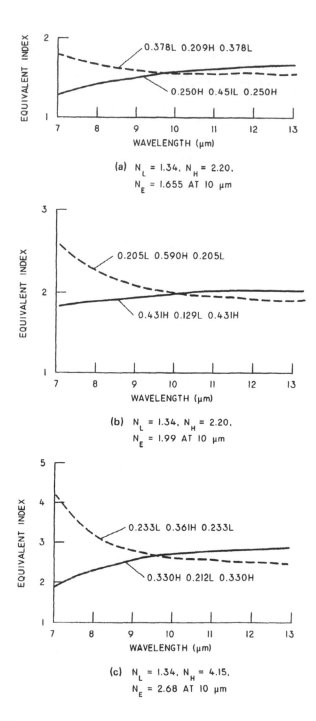

Figure 8.14 Equivalent index profiles to achieve index requirements for AR structures shown in Fig. 8.13.

at 10 μm, which is the design wavelength. Also note that the coating designs no longer consist of only two layers. The actual designs are three or four layers, depending on the layer sequence, because like indices can be coalesced. For example, the wL xH wL of Fig. 8.13a becomes 1.378L 0.209H 0.378L when the adjacent L layers are combined, and the yH zL yH of Fig. 8.13c becomes 1.000L 0.330H 0.212L, where the 0.330H layer of Ge film next to the Ge substrate is ignored on the assumption that the slight difference between the higher film indices is not important.

The previous examples of two-layer AR coating structures are based on the concept of using QWOT film layers. Other types of structures involve non-QWOT layer sequences of the appropriate indices and thicknesses that can be effective in reducing the reflectance to zero at any given wavelength. Early work done by Cox et al. (1954) shows how the Fresnel coefficients can be used to arrive at a numerical solution to this problem using three different types of structures: the quarter–quarter structure already discussed, a quarter–half structure that will be discussed next, and a less-than-QWOT high-index layer followed by a greater-than-QWOT low-index layer now under consideration. Catalan (1962) used the recurrence matrix method to arrive at similar solutions. The first approach is quite cumbersome, and solutions can be achieved only by approximate numerical methods. The latter approach is exact and not too cumbersome when applied to the two-layer structure but increases in complexity if extended to more than two layers.

8.2.1. Vector Design Method Applied to Antireflection Structures

A graphic method exists that is relatively easy to understand and can be extended to any number of layers with little difficulty. Although this method is not exact, the errors are small as long as the Fresnel coefficients at the film boundaries are small. Because this occurs when relatively low index materials are used, this method works well in developing AR coating designs.

Consider the film layer structure shown in Fig. 8.15. The treatment is similar to what has been discussed previously, but the multiple reflections internal in the film structure are being ignored, which is why this method is not exact. Adding up the reflected wave amplitudes results in

$$r = r_a + r_b e^{-2i\Delta_2} + r_c e^{-2i\Delta_1 - 2i\Delta_2}$$

Because the wave amplitudes are vectors, a vector sum must be used to arrive at a solution. This is shown in Fig. 8.16, where all the wave amplitudes are diagramed around a point to ensure that they are properly oriented. (Note that the length of each vector is the value of the Fresnel coefficient, the angle between vectors is 2 × the phase thickness in angular units, and the direction of the vector

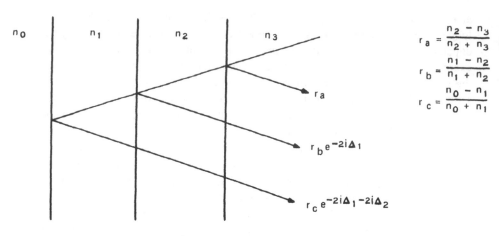

Figure 8.15 Wave amplitude representation of film 2 on film 1 on substrate n_0.

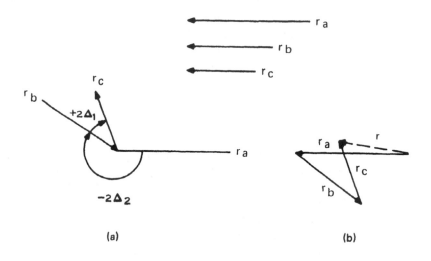

Figure 8.16 Vector sum of wave amplitude representation for $n_{air} < n_2 < n_1 > n_0$; $\Delta_2 > \text{QWOT}$; $\Delta_1 < \text{QWOT}$; (a) vector about origin, (b) vector sum.

is toward the center of the circle if the film boundary is from a higher index material and away from the center if from a lower index material.) Next, the vectors are sequentially summed, and the reflectance is determined from the magnitude of the resulting wave amplitude, $R = rr^*$.

To design an AR coating, one must select the materials to be used, calculate the Fresnel coefficients to determine the vector lengths, and then add them geometrically to give a zero resultant. The layer thicknesses can then be determined by drawing the circle diagram and determining the angles between vectors as one progresses from air to the top layer to the next layer to any subsequent layers and then into the substrate material. Figure 8.17 gives a practical illustration of this technique. The figure also includes a table summarizing the data assumed to set up the situation and the phase thicknesses obtained. Note that,

NO.	MATERIAL	INDEX	r_x	SOLUTION 1 Δ_{xi} (QWOT)	SOLUTION 2 Δ_{xi} (QWOT)
3	AIR	1.00		–	–
			$0.1597 = r_a$		
2	MgF$_2$	1.38		71.72° (0.7967)	108.28° (1.203)
			$0.1953 = r_b$		
1	ZrO$_2$	2.05		153.82° (1.7091)	26.18° (0.291)
			$0.1172 = r_c$		
0	SK–16	1.62			

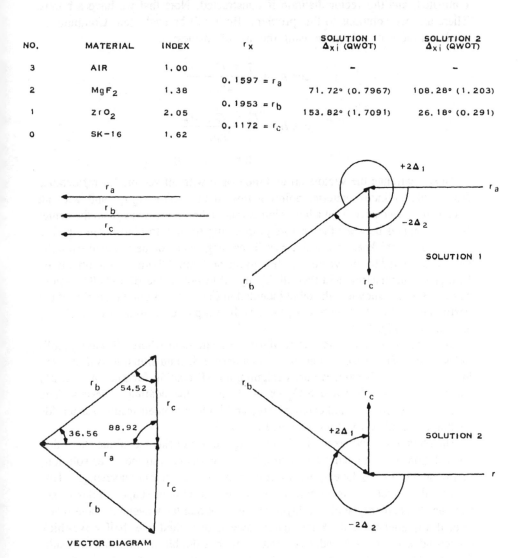

Figure 8.17   Vector solutions for two-layer AR coating of MgF$_2$ and ZrO$_2$ on SK-16 glass.

although an accurate geometrical figure was constructed to clearly illustrate the example, all results were determined using standard trigonometry.

The problem is to determine a good AR coating for SK-16 glass ($n = 1.62$) at 2.0 μm. A single layer of MgF_2 would still have over 0.6% reflectance at a minimum. Therefore, we choose to go to a two-layer structure with the inner layer being ZrO_2 ($n = 2.05$). The Fresnel coefficients at each boundary are computed, and the vector diagram is constructed. Note that we have a bonus: There are two solutions to this problem. Both will be evaluated. Compute the angles in the vector diagram using the law of cosines:

$$\cos A = \frac{r_b^2 + r_c^2 - r_a^2}{2r_b r_c}$$

$$\cos B = \frac{r_a^2 + r_c^2 - r_b^2}{2r_a r_c}$$

$$C = 180° - A - B$$

Next, rearrange the vectors around the origin with all vectors for reflectance from a higher index material pointing toward the center (r_a, r_b) and with all vectors for reflectance from a lower index material pointing away from the center (r_c). The angles between the vectors progressing from r_a to r_b to r_c are equal to $2\times$ the phase thickness for each layer. These angles can be determined from the angles calculated for the vector solution to the problem. Divide by 2 to obtain the film phase thicknesses, and then divide by 90 to obtain the film QWOT thicknesses. Both are shown in the table included in Fig. 8.17. A computed plot of the performance of each of these designs as well as a plot of a single layer of MgF_2 are shown in Fig. 8.18.

To clarify this procedure and to illustrate a situation where all wave amplitudes have a 180° phase change at the boundary, a second situation is illustrated in Fig. 8.19. In this case, we are designing an AR coating for ZnS ($n = 2.25$) using SiO ($n = 1.90$) and SiO_2 ($n = 1.46$) as the coating materials. The sequence of steps is as indicated before, and both computed results for an AR coating centered at 4 μm are included in Fig. 8.20.

The plots of the two examples illustrating the use of this technique show that one solution in each example is a simple V-coat shape, and the other solution, involving thicker film layers, has a second minimum at another wavelength. This thicker film structure has an asymmetric double minimum shape similar in appearance to another type of two-layer coating, the quarter-wave/half-wave structure. If a quarter-wave-thick low-index layer is deposited over half-wave-thick higher-index layers, a broader AR region with a double minimum can result. Figure 8.21 contains the computed plots for a quarter-wave-thick layer of magnesium fluoride deposited over a half-wave-thick layer of films of various indi-

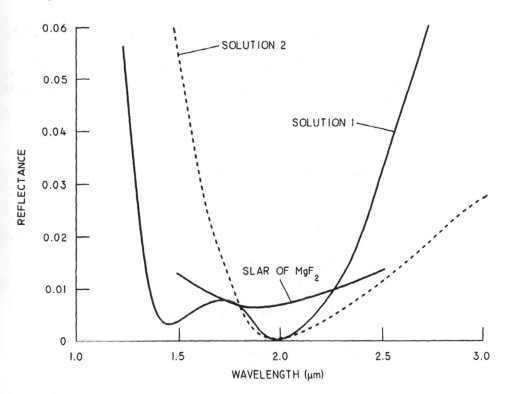

Figure 8.18 Plots of the solutions derived in Fig. 8.17.

ces ranging from 1.50 to 2.50 index on SK-16 glass (n = 1.62), where the coatings are centered at 2.0 μm. Since the half-wave-thick layers are absentee, the reflectance for each plot at the central wavelength is 0.65%, which is determined by the 1.38 QWOT on the 1.62 index glass. As the index of the half-wave-thick inner layers is increased, the double-lobe minimum shifts away from the central wavelength on either side, and the minima decrease toward a lower reflectance value. At one index for the half-wave material, the minima achieve a zero value. As the index of the inner half-wave material is increased even further, the minima increase and shift back toward the central wavelength. At a sufficiently higher inner-layer index, the shape becomes a narrower U-shape with a single minimum.

The solutions shown in Fig. 8.21 are just a special case of the solutions arrived at using the vector approach discussed previously and as represented in Fig. 8.18, which is also on SK-16 glass. The 2.05 index for the inner layer in Fig. 8.18 was chosen because ZrO_2 has this index and is a good material to use

MEDIUM	MATERIAL	INDEX		SOLUTION 1 Δx (QWOT)	SOLUTION 2 Δx (QWOT)
3	AIR	1.00		–	–
2	SiO_2	1.46	$0.1870 = r_a$	78.39° (0.871)	101.61° (1.12?)
1	SiO	1.90	$0.1310 = r_b$	30.61° (0.340)	156.78° (1.74?)
0	ZnS	2.25	$0.0843 = r_c$	–	–

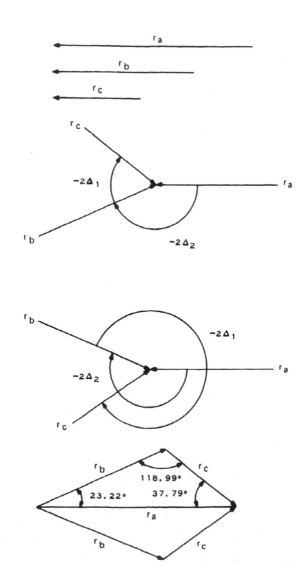

Figure 8.19 Vector solutions for two-layer AR coating of SiO_2 and SiO on ZnS.

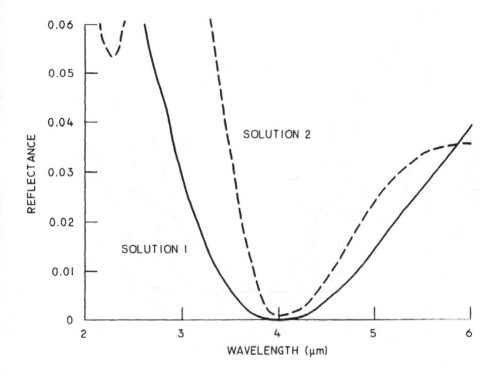

Figure 8.20   Plots of the solutions derived in Fig. 8.19 (λ_0 = 4 µm).

for the situation being represented. It is close to the 2.10 inner index example
from Fig. 8.21, which is approaching a 0 value for the reflectance at 2.35 µm.
At 2.35 µm, the relative thicknesses for the outer and inner layer, respectively,
of a quarter-wave/half-wave-thick structure are 0.816 and 1.63. In Fig. 8.18, the
first solution with QWOTs of 0.7967 and 1.7091 came close to matching this
situation. In fact, if the added constraint of the inner layer being optically twice
as thick as the outer layer had been enforced, the solution would have resulted
in something close to that arrived at in Fig. 8.21. Further consideration of the
vector approach indicates that there are an infinite number of additional layers
where each or both layers are increased in thickness by an integral number of
HWOT layers (which corresponds to going around the unit circle an integral
number of turns). Figure 8.22 shows three examples where the inner thickness
for solution 1 from Fig. 8.18 is made 1 and 2 HWOTs greater than the original
solution. Indeed, the reflectance of each is zero at the central wavelength.
However, the region of low reflectance becomes progressively narrower as the
thickness of the layers increases. The most efficient and broadest AR structure is
made of the thinnest layers possible to achieve the desired result.

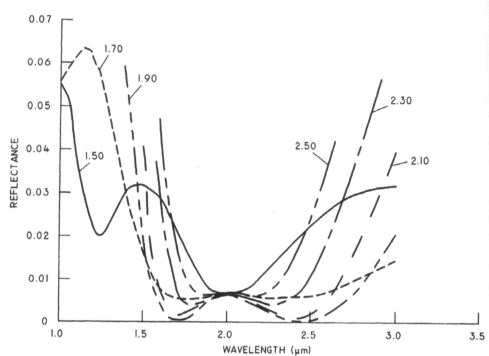

Figure 8.21 λ/4 − λ/2 antireflection structures where the QWOT outer layer is MgF$_2$ (N = 1.38) and the HWOT layer index is as shown.

8.2.2. Three-Layer Antireflection Structures

Broader and flatter AR bands can be achieved by increasing the number of layers in the AR structure. Cox et al. (1962) show examples of this for low-index glasses in the visible and the near-infrared. In this effort, they discuss three types of three-layer structures: quarter–quarter–quarter, quarter–half-quarter, and three-quarter–half–quarter. Of these options, they discounted the first as being too narrow and concluded that there was no clear choice between the latter two. They concentrated on studying examples of the second type. No particular design technique was employed, but many examples were calculated for different film combinations and at various angles of incidence.

Telford (1969) devises a quantitative procedure to solve for various three-layer solutions not limited to QWOT or HWOT layers. Mouchart (1977) presents an analytical approach to the three-layer MLAR design that defined the regions of refractive index for two of the layers where the index of one layer is fixed or two of the layers are made of the same material. He also discusses the stability of the various designs by taking the partial derivatives of the reflectances and the

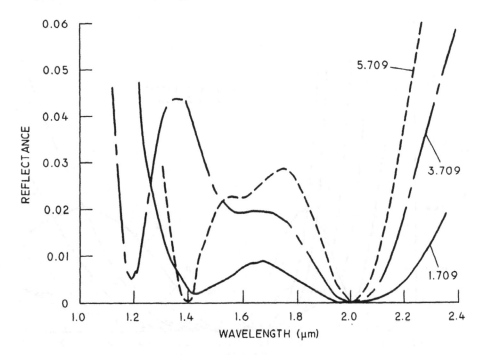

Figure 8.22 Two-layer AR structures where the outer layer is 0.797 QWOT of MgF_2 and the 1.709 solution from Fig. 8.17 is increased by 1 HWOT (3.709) and then by 2 HWOTs (5.709).

layer thicknesses used in the designs. Nagendra and Thutupalli (1983) discuss a numerical technique for design and optimization that is probably representative of the approach being used by most thin-film designers now that computing power is so widely available and easy to use. However, most optimization programs work best, or at least more efficiently, when starting with a good design. A good starting design can be obtained by using Eqs. (23) and (24) and solving for situations where the admittance is 1.

Thus the quarter–quarter–quarter structure gives

$$Y = n_3{}^2 n_1^2 / n_2^2 n_0 \qquad (31)$$

and the quarter–half–quarter structure (where the half-wave is broken into two quarter-waves) gives

$$Y = n_3{}^2 n_2^2 n_0 / n_2^2 n_1^2 = n_3{}^2 n_0 / n_1^2 \qquad (32)$$

Any materials with refractive indices that satisfy these relationships for $Y = 1$ will give a zero reflectance at the central wavelength. To achieve a broad AR

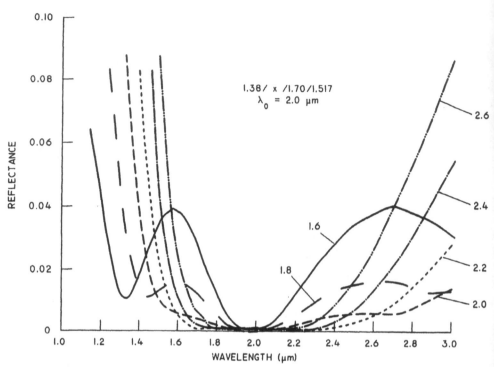

Figure 8.23 Reflectance versus wavelength for QWOT-HWOT-QWOT thick AR coatings on BK-7 ($n_3 = 1.38$, $n_2 =$ variable, $n_1 = 1.70$, and $n_0 = 1.517$).

region, however, it is necessary to apply them in the proper sequence. A descending order is most logical for high-index substrates, whereas a step up followed by a step down should be employed for low-index substrates.

Consider a low-index situation such as BK-7 with an index of about 1.517. The two common materials with refractive indices lower than BK-7 are quartz ($n = 1.46$) and magnesium fluoride ($n = 1.38$). As a single-layer AR coating they result in minima of 2.8 and 1.3%, respectively. Yet, in conjunction with higher index materials in a quarter–half–quarter-wave structure, they can produce fairly wide regions of low reflectance. Using Eq. (32) and letting $n_3 = 1.38$ and $Y = 1$, n_1 must be 1.70 to give zero reflectance at the central wavelength. Then n_2 can take on any value and will determine the reflectance on either side of the central wavelength. Figure 8.23 contains plots for this situation when n_2 is varied from 1.6 to 2.6. As this plot shows, the broadband AR characteristics are low in reflectance over a fairly broad range for a half-wave of about 2.05–2.10 refractive index. The higher refractive indices for n_2 result in narrow U-shaped curves like those obtained in the two-layer structure of Fig. 8.20.

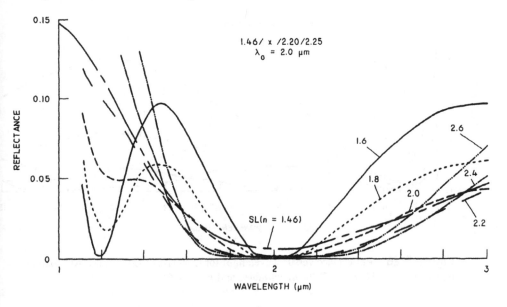

Figure 8.24 Reflectance versus wavelength for QWOT-HWOT-QWOT AR coatings on ZnS ($n_3 = 1.46$, $n_2 =$ variable, $n_1 = 1.70$, and $n_0 = 2.25$).

When the same structure and approach are used on ZnS ($n_0 = 2.25$), they are not quite so effective. Consider the situation when quartz ($n_3 = 1.46$) is chosen for the outer layer. This layer is not effective as a SLAR coating. For the three-layer case, solving for the ideal inner layer results in $n_1 = 2.20$, which is very close to the index of the substrate. Again plotting the reflectance versus wavelength with the refractive index for n_2 varying from 1.6 to 2.6 results in curves such as those plotted in Fig. 8.24. The lower reflectances for the different indices for the HWOT layers do not produce a broader layer. The reflectance of the three-layer combination is, however, significantly lower than that for a single layer of quartz. The intermediate thin-film HWOT material (n_2) can have any index in the 2.1–2.4 range. The critical element in this design is for the first QWOT layer to have an index lower than that of the substrate to match it better to the quartz.

On high-index substrates, fairly wide AR regions can be obtained using the quarter–quarter–quarter-wave thick structures. Consider germanium with a refractive index of 4.0. Choosing $n_3 = 1.34$ and $n_2 = 2.20$ and plotting the reflectance versus wavelength for values of n_1 ranging from 2.9 to 3.7 in increments of 0.2 gives the results plotted in Fig. 8.25. A refractive index of 3.30 gives a reflectance of 0 at 10 μm and a broad AR region on either side. The minimum at 10 μm is not surprising, because, using Eq. (31) with $Y = 1$, $n_3 =$

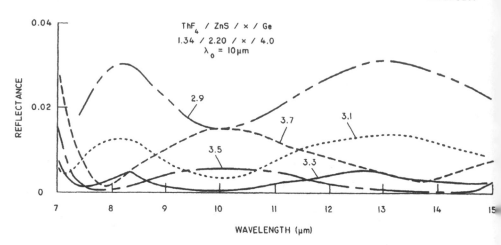

Figure 8.25 Reflectance versus wavelength for QWOT-QWOT-QWOT AR coatings on Ge ($n_3 = 1.34$, $n_2 = 2.20$, n_1 variable, and $n_0 = 4.0$).

1.34, $n_2 = 2.20$ and solving for n_1 gives 3.28 as a solution. The indices of 1.34 and 2.20 were chosen because they are the indices of ThF_4 and ZnS at 10 μm, both good coating materials. Although the 3.28 index cannot be obtained from common coating materials, it can be synthesized easily using Herpin periods of ZnS and Ge. This situation is an ideal opportunity to apply this concept a second time and to demonstrate a practical application. Consider the symmetrical period xH yL xH (L and H are QWOT layers of low- and high-index materials) when the high-index material is Ge in which $n = 4.30$ (note that the thin-film index for Ge deposited by physical vapor deposition at medium temperatures is typically higher than for the bulk material), the low-index material is ZnS, and $2x + y = 1$. Figure 8.26 contains a plot of the equivalent index versus wavelength for various x,y combinations. Also shown on this plot is the wavelength for which the phase thickness of the period is a QWOT. This plot shows that x,y values of 0.38, 0.24 give an index of 3.28 for a QWOT at approximately 10.5 μm. Scaling these thicknesses to give a QWOT at 10 μm and combining them gives a design of ThF_4/ZnS/0.36 Ge/0.229 ZnS/0.36 Ge on a Ge substrate. A plot of this design is shown in Fig. 8.27. Even though the Ge film has a higher index than the bulk value of Ge, it can be ignored when the coating is put down. This fact is also shown in Fig. 8.27. The central wavelength for the first two curves was centered at 8.3 μm to shift the AR region shorter and center it on either side of the 10 μm. Also shown in Fig. 8.27 is the alternative solution arrived at by solving for the xL yH xL structure that gives 3.28 index period. The equivalent index for this combination is plotted in Fig. 8.28 for various x,y pairs. This plot shows that x,y values of 0.33, 0.33 give an index of 3.28 at a QWOT of approximately 10.6

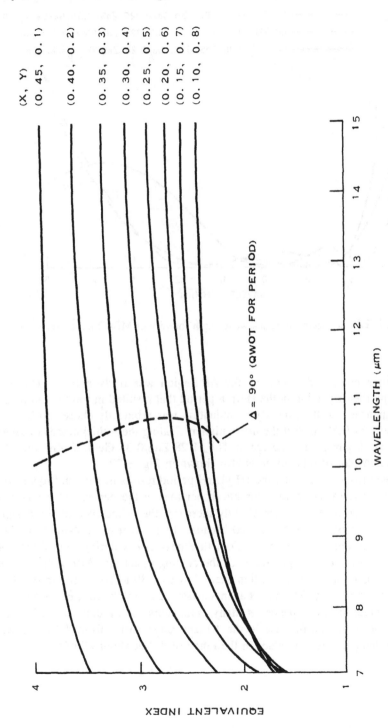

Figure 8.26 Equivalent index versus wavelength for xH yL xH period, where $\lambda_0 = 10$ μm, QWOT, $2x + y = 1$, $n_H = 4.3$, and $n_L = 2.2$.

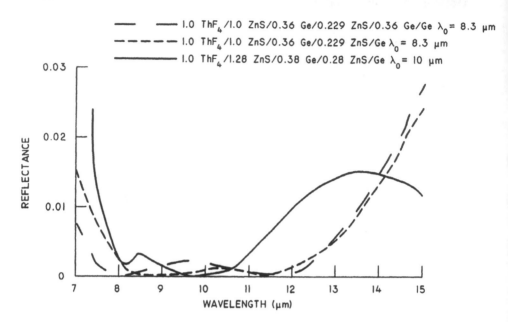

Figure 8.27   Reflectance versus wavelength for various MLAR structures on germanium.

μm. When computed, however, the AR region was fairly narrow. This was caused by the dispersion in the Herpin period that resulted in too low an index. To compensate for this, the x,y combination was arbitrarily chosen to be $x = 0.28$ and $y = 0.38$ to shift the index higher. Scaling these thicknesses to give a QWOT at 10 μm gives a design of $ThF_4/1.28$ ZnS/0.38 Ge/0.28 ZnS on a Ge substrate. A plot of this design is also shown in Fig. 8.27.

As seen from these plots, the xH yL xH period results in a much broader AR region because of the more favorable dispersion for the period. Also note that leaving out the inner Ge layer slightly increases the reflectance of the design. This would probably not be noticed because the exact result obtained in fabricating this design depends on random thickness errors during deposition and dispersion in the optical properties of the coating materials. Also note that the refractive index used for Ge in this example was 4.30 as compared to the 4.15 value used previously. This is not as inconsistent as it might appear, because the absolute index of Ge depends on many parameters during deposition. The refractive index of thin-film Ge in the 10 μm range varies from 4.15 at lower substrate temperatures to values in the range of 4.3 at about 250°C.

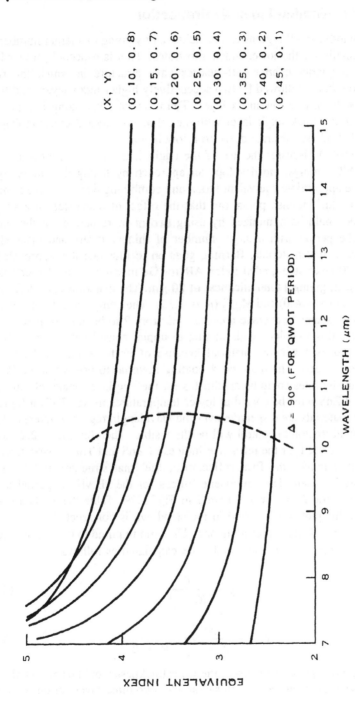

Figure 8.28 Equivalent index versus wavelength for xL yH xL period, where $\lambda_0 = 10$ μm, QWOT, $2x + y = 1$, $n_H = 4.3$, and $n_L = 2.2$.

8.2.3. Step-Graded Index Antireflection

Additional width for AR regions can be achieved by going to a larger number of layers, especially on the lower index substrates. This is especially true when considering a quarter–quarter–half–quarter-wave structure in which the first layer is of low index followed by two successively higher index layers and then coming back down to a low-index layer. The ultimate AR coating is a graded index coating of a thick single layer with the refractive index decreasing from a value near that of the substrate down to a low index.

Berning (1962) discusses the use of the equivalent index concept to design broadband AR coatings. He develops an approach by taking thin three-layer periods of the appropriate equivalent index and combining them to make a thick graded-index film. He also points out that the effect of dispersion in a single Herpin period could be minimized by using two or more periods of the same thickness. The period with a larger number of thinner layers had dispersion closer to that of real materials. Berning goes on to illustrate this approach by designing a 20-layer step-graded index AR for Ge in which each "layer" is a QWOT at 1 μm, giving a total thickness of 20 μm. The materials chosen for the example are Ge ($n = 4.0$) and MgF$_2$ ($n = 1.35$). The computed reflectance for the final design seems to average about 3–4% from 2 to beyond 20 μm.

To illustrate this technique with another example, consider the case of ZnS, commonly used as a window material because of its transmittance from the visible to the infrared. The high-index coating material to use would be ZnS. Several of the fluorides could be candidates for the low-index material, but the limitation that ZnS must be coated at lower temperatures makes ThF$_4$ a logical choice. Both materials are dispersive over the wavelength region of interest, but the most difficult region to control will be the visible. Therefore, $n = 2.20$ and $n = 1.50$ will be used for the refractive indices of ZnS and ThF$_4$, respectively. Also, to further reduce the final reflectance, the final structure will have a QWOT of MgF$_2$ added. The equivalent indices for the xL yH xL period as a function of the ratio $R = y/x$ are plotted in Fig. 8.29. Table 8.1 is then constructed from this plot. Each period in the stepdown is constructed so that the index is 0.05 lower than the preceding one. The ratio required to give each index is determined from the plot, and x and y are calculated as follows:

$$y = \frac{r}{R + 2} \tag{33}$$

$$x = \frac{1 - y}{2} \quad or \quad \frac{1}{R + 2} \tag{34}$$

If the three-layer period were used, the layer thickness would be xL yH xL as indicated previously. However, for this example, two three-layer periods half as

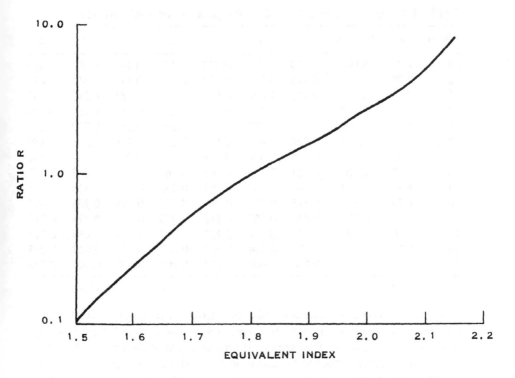

Figure 8.29 *R* versus equivalent index for *x*L *y*H *x*L period where $R = y/x$.

thick are used, which result in the same index but less dispersion. Therefore, each layer is made up of five layers of thicknesses $(x/2)$L, $(y/2)$H, xL, $(y/2)$H, and $(x/2)$L. Table 8.1 also lists the thicknesses for each layer needed to give a five-layer period with the required indices. To put the final design together, start with the first layer of the first period and progress through the fourth layer. The fifth layer of the design has the thickness of the fifth layer of the first period plus the first layer of the second period, because they are made of the same material $(0.123 = 0.050 + 0.073)$. The 14th period is really just a QWOT of ThF_4 because there is no ZnS in it. Thus, the last layer of the design becomes 1.237 QWOT of ThF_4 when coalesced with the fifth layer of the 13th period. The total number of layers in the design is now 52 and becomes 53 when the structure finished with a QWOT of MgF_2. The computed performance of this design is shown in Fig. 8.30, curve a. As previously indicated, ThF_4 has a significant decrease in index at longer wavelengths. The index of ZnS also decreases, although not so much. Therefore, the reflectance of this design as shown in Fig. 8.30, curve A, should be different when deposited. To estimate this effect, the

Table 8.1 Five-Layer Period Step-Graded Equivalent Index AR Structure

Period	n	$R = y/x$	x	y	1	2	3	4	5
1	2.15	8.000	0.100	0.800	0.050	0.400	0.100	0.400	0.050
2	2.00	5.000	0.145	0.710	0.073	0.355	0.145	0.355	0.073
3	2.05	3.400	0.185	0.630	0.093	0.315	0.185	0.315	0.093
4	2.00	2.600	0.217	0.565	0.108	0.282	0.217	0.282	0.108
5	1.95	2.000	0.250	0.500	0.125	0.250	0.250	0.250	0.125
6	1.90	1.600	0.278	0.444	0.139	0.222	0.278	0.222	0.139
7	1.85	1.280	0.305	0.390	0.152	0.145	0.305	0.145	0.152
8	1.80	1.000	0.333	0.333	0.167	0.167	0.333	0.167	0.167
9	1.75	0.730	0.366	0.267	0.183	0.133	0.366	0.133	0.183
10	1.70	0.530	0.395	0.209	0.197	0.105	0.395	0.105	0.197
11	1.65	0.360	0.424	0.152	0.212	0.076	0.424	0.076	0.212
12	1.60	0.240	0.446	0.108	0.223	0.054	0.446	0.054	0.223
13	1.55	0.155	0.473	0.054	0.237	0.027	0.473	0.027	0.273
14	1.50	0.000	0.500	0.000	0.250	0.000	0.500	0.000	0.250

design was recomputed using dispersive properties for both materials. A plot of this calculation is shown as curve b of Figure 8.30. As expected, there is significant change in the longer wavelength infrared where the reflectance is decreased because the indices of both materials are lower, and there is some change in the visible spectrum where the refractive index of the ZnS is even higher than 2.20.

The example developed herein is for a linear graded index AR coating. Yeh and Sari (1983) report on the use of exponentially graded refractive indices. Southwell (1983) concentrates on examples of profiles applicable specifically to the visible region. Sankur and Southwell (1984) present a timely application of this technique for ZnSe covering 0.6–12.6 μm, a band encompassing 4.5 octaves and comparable to the development reported here. They report a reflectance of 3% per surface using a ZnSe/CaF$_2$ system employing laser-assisted deposition techniques. Southwell (1985) expands on this technique by showing an alternative design that employs a digital structure by using a flip-flop optimization routine with very thin high- and low-index film layers.

Considerable amounts of theoretical data and design techniques are now available to the thin-film designer. The factors limiting future advances are the ability to achieve sufficient control in depositing specific layer thicknesses and exact knowledge of the optical properties of films being deposited. The latter factor has always been a problem. Divergent properties for various materials have been widely reported in the literature, and justly so. The properties of many materials depend closely on the conditions under which they are deposited. Variations as great as 10% are not unusual and will be reported later. Any coating facility

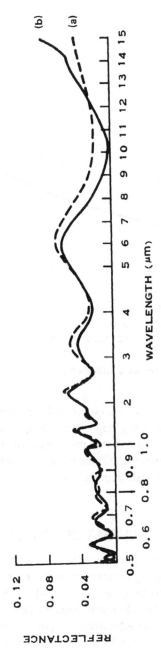

Figure 8.30 Computed reflectance versus wavelength for a step-graded equivalent index AR coating for ZnS where λ_0 = 0.95 μm. (a) n_L = 2.20, n_H = 1.50; (b) using dispersive refractive indexes for ZnS and ThF$_4$.

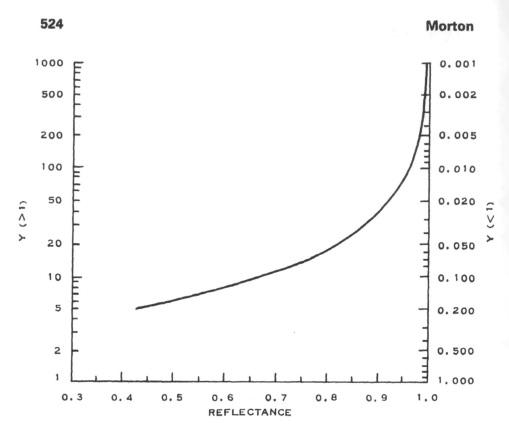

Figure 8.31   Reflectance versus admittance Y, for $Y >> 1$ and $Y << 1$.

expecting to do good work must understand and control parameters that affect the properties of the materials being deposited. Temperature, pressure, electrical biasing, reactive gases, and ion-assisted processes can aid in controlling material properties of thin films. Also, techniques must be developed to characterize the optical properties of the materials deposited so designers will know what material parameters to use.

8.3.   ALL-DIELECTRIC MIRROR COATINGS

Previously, it was shown that thin-film structures with admittances close to that of air resulted in regions of high transmission. Conversely, thin-film structures with either very low (close to zero) or very high admittances will result in regions of high reflectance. Equation (17) was used to develop a graph plotting the reflectance as a function of the admittance shown in Fig. 8.31. Note that equal reflectance is obtained for $Y > 1$, which is the reciprocal of $Y < 1$. That alternate QWOT layers of high- and low-index materials yield either high or low admit-

Table 8.2 Reflectance for Alternate QWOT Layers of TiO_2 and SiO_2

	Example 1			Example 2		
No.	Type	Y	R	Type	Y	R
1	L	1.405	2.84	H	3.487	30.72
2	H	3.765	33.67	L	0.611	5.82
3	L	0.566	13.07	H	8.654	62.85
4	H	9.343	65.07	L	0.243	36.57
5	L	0.228	32.50	H	21.477	83.00
6	H	23.187	84.15	L	0.099	67.14
7	L	0.092	69.16	H	53.299	92.77
8	H	57.542	93.28	L	0.040	85.21
9	L	0.037	86.22	H	132.273	97.02
10	H	142.802	97.24	L	0.016	93.76
11	L	0.015	94.20	H	328.261	98.79
12	H	354.393	99.44	L	0.006	97.44
13	L	0.006	97.62	H	814.648	99.51
14	H	879.500	99.55	L	0.003	98.96
15	L	0.002	99.52	H	2021.714	99.80

tances can be shown using Eqs. (23) and (24). An example of this is shown in Table 8.2 for alternate layers of TiO_2 ($n = 2.30$) and SiO_2 ($n = 1.46$) on BK-7 glass. The first example is for 15 QWOT layers starting with a low-index material. For odd-numbered layers the admittance becomes progressively smaller, and for even-numbered layers the admittance becomes progressively larger (in fact, the factor is the square of the ratio of the refractive indices, 0.40295 or 2.4817).

The second example starts with the high-index material. Here, the admittance becomes progressively smaller for the even-numbered layers and progressively larger for the odd-numbered layers by the same factor as in the first example. The addition of a high-index layer increases the reflectance, and the addition of a low-index layer decreases the reflectance. Thus, for a given number of layers, film structures ending in high-index layers give higher reflectance. The increase in reflectance from adding an LH pair becomes smaller as the reflectance of the film stack increases. In fact, from a practical standpoint, there is an upper limit to the reflectance that can be achieved with this type of structure because of the absorption in the film layers and because added layers cause increased scattering, which reduces the obtainable reflectance. Early work done by Perry (1965) gives an upper limit of 99.5% reflectance for a 27-layer all-dielectric mirror (ADM) composed of cryolite and ZnS and used in the visible region. Perry added 23 more layers to the 27-layer structure without improving the reflectance. In a similar study by Behrndt and Doughty (1967), the reflectance of ZnS/thorium oxyfluoride ADM structures is found to increase and the transmission decrease up to 21 layers. Then both the reflectance and the transmission decrease with additional layers. The reduction in reflectance is attributed to an increase in the number of scattering centers. This work initially achieved a maximum of 99.5% reflectance. Improved reflectance (99.8–99.9%) was finally achieved by carefully cleaning the substrates and minimizing spitting from the sources during deposition.

All-dielectric mirrors consisting of QWOT film stacks have a high reflectance at and near the central wavelength. At wavelengths farther from the central wavelength, the reflectance decreases significantly. The shape of the reflector depends on the number of layers in the filter stack and the ratio of indices. Figure 8.32 contains plots of the second example from Table 8.2 for a mirror made of up 3, 5, 7, and 15 layers. Each situation has about the same 50% reflectance width. However, added layers steepen the slope for the transition from low reflectance to high reflectance and increase the high-reflectance region between the 50% reflectance points. As indicated previously, the width of the high-reflectance band depends on the materials used. That is, the higher the index ratio, the broader the reflectance band. Epstein (1952) has shown that the half-width of the high-rejection region (Δg) is given by

$$\Delta g = \frac{2}{\pi} \text{ arc sin } \frac{n_H - n_L}{n_H + n_L} \tag{35}$$

Figure 8.33 shows the width Δg plotted as a function of the ratios of the refractive index of film materials that could be used in fabrication of ADM structures. Also shown is a plot of the 50% points versus the index ratio for thin-film structures with enough layers to have the high-reflectance region as broad as it will get. This latter plot will aid the designer who may use this type of structure as a beamsplitter. The 50% width is slightly broader than the high-reflectance bandwidth as calculated by Epstein.

On either side of the region of high rejection, the light is transmitted and reflected in an oscillatory pattern depending on the total optical thickness of the ADM stack. Thicker stacks will consist of lobes spaced closer than those in thinner stacks, as shown in Fig. 8.32. As the wavelength decreases and the layer becomes HWOT, the entire stack is absentee, and the reflectance is just that of the uncoated substrate except for absorption and scattering losses. As shown in Fig. 8.34, continued progression to shorter wavelengths results in a situation where the stack layers become three-quarter wavelength thick. At that point there is another region of high rejection. This pattern continues ad infinitum as the wavelength continues to decrease. There is high reflectance at wavelengths that are an odd-multiple QWOT thickness, and there is high transmission at wave-

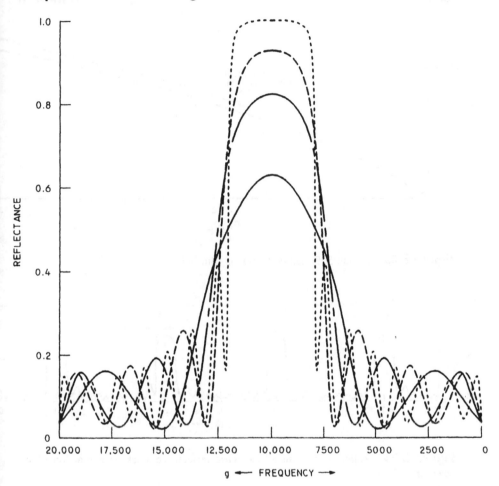

Figure 8.32   Reflectance versus frequency (cm^{-1}) for 3-, 5-, 7-, and 15-layer reflector stacks. Design = H(LH)^x where x = 1,2, 3, and 7; n_H = 2.30; and n_L = 1.46.

lengths that are an even-multiple QWOT thick. Random thickness errors in depositing the film layers will not greatly affect the central reflectance region but will tend to round off the transition (decrease the slope) from high reflectance to low reflectance. These errors can significantly alter the shape and distribution of the lobes in the region of high transmission. Often, ADM structures are used as beamsplitters where it is desirable to reduce or minimize the lobes on one side or the other of the rejection region. This is relatively easy to do by choosing the appropriate film structure and starting and ending it with an eighth-wave-thick layer instead of a QWOT. An example of a short-wave-pass (SWP) structure is

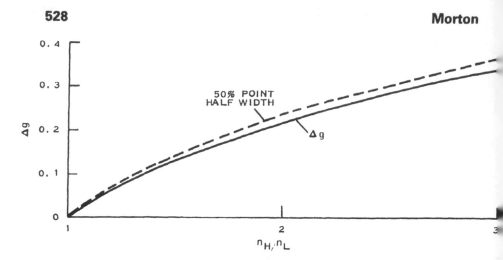

Figure 8.33 Δ_g and 50% halfwidth versus film index ratio.

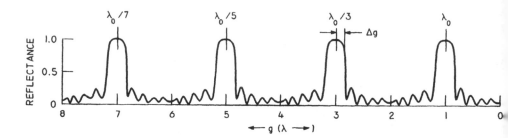

Figure 8.34 Reflectance of an all-dielectric reflectance stack as a function of g, where $g = \lambda_0/\lambda$.

shown in Fig. 8.35. In this example, the film sequence is L/2 H L/2 repeated many times, giving a L/2 H L H L . . . H L H L/2 design. The lobes in the transmission region on the shorter wavelength side of the rejection bands have been significantly reduced. In a similar manner, an H/2 L H/2 structure repeated many times will result in a long-wave-pass (LWP) filter with lobes suppressed on the long-wavelength side of the rejection band.

It is often desirable to use an ADM structure to notch out transmission over a narrow wavelength band and yet retain high transmission on both sides of the rejection band. There are many approaches to solving this problem. Thelen (1971) uses the dispersive refractive index of the A/2 B_1 A/2 structure to devise an AR structure of the type A/2 B_2 A/2 in which B_2 is a different index than B_1 and would match the index of an odd number of ADM periods into a substrate.

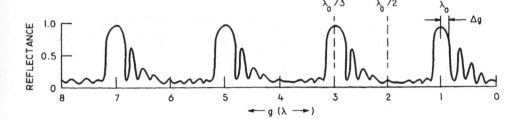

Figure 8.35   All-dielectric stack smoothed for SWP performance.

The dispersive nature of the period structure is effective on either side of the stack. For improved performance, additional periods with other B_x indices could be used to step-grade from the rejection stack to the substrate. Thelen then makes the concept producible by replacing the many different indices, B_x, with Herpin equivalent structures. Young (1967) describes an approach that resulted in equal ripple on either side of the rejection band. His approach is based on the similar concept for designing stepped-impedance filters for waveguides. The resulting filter is a structure in which all of the periods (xH, yL, xH) are of equal thickness but x and y change symmetrically from the center of the structure to the ends in a graded-index manner, resulting in high rejection in the center and relatively smooth high transmission on either side.

If a QWOT stack is tilted, the reflectance band center shifts toward the shorter wavelengths and the shape is distorted because of the differences between the spectral properties of the two polarizations. For large H/L ratios and higher angles of incidence, the difference can be significant and distort the edge or transition between transmission and reflection (Fig. 8.36). The difference in spectral performance between the two polarizations can be used to split polarizations over short-wavelength bands. Figure 8.36 also shows the dramatic spread in the polarizations if the coating is immersed in a cube. For this example, the split between the edges of the two polarizations is more than three times as great for the immersed case as it is for a plate. Immersion in higher index glass will exaggerate this even further. Since the indices for the two materials for the p polarization have the lowest ratio, it has the narrowest rejection band and a lower rejection level.

8.4.   BANDPASS FILTERS

A bandpass filter can be constructed by taking two symmetrical mirror structures and putting a multiple HWOT layer between them. An example of such a structure would be $(HL)^3(LH)^3$, which when written out becomes HLHLHL

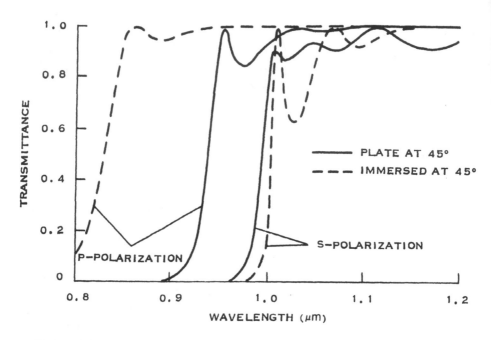

Figure 8.36 Transmittance edge of reflector stack at an angle showing polarization splitting.

LHLHLH. Upon rearranging layers, it becomes HLHLH LL HLHLH. Using Eq. (23), we see that the admittance is

$$Y = \frac{L^2L^2\ L^2\ H^2H^2H^2}{H^2H^2H^2\ L^2\ L^2L^2\ n_0} \tag{36}$$

All of the layers become absentee at the central wavelength, which means that the transmittance is that of the uncoated substrate. On either side of the central wavelength, the transmittance will fall off in a manner dependent on the relative indices of the materials being used and the number of layers in the mirror structures. The half-power bandwidth (HPBW) and the level of rejection in the side lobes of this structure are determined by the relative index of the two materials and the number of periods making up the mirror structure. A higher index ratio results in a narrower bandwidth and greater rejection in the side lobes. Increasing the number of layers that compose the mirror part of the structure gives a narrower bandwidth. An example of this for the above structure using H = 4.15, L = 2.15, M = 2.15, and a central wavelength of 10,000 nm (10 μm) is illustrated in Fig. 8.37 for three different mirror structures. In this example, the substrate was Ge (n_0 = 4.0), and the initial M layer was included as a SLAR

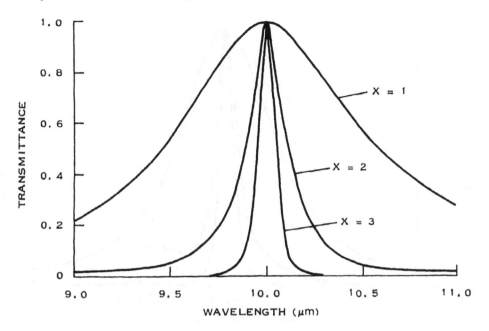

Figure 8.37 Single-cavity bandpass filters with low-index cavity layers, M (HL)$_x$ (LH)$_x$ design ($n_L = n_M = 2.15$, $n_{H} = 4.15$, and $n_0 = 4.0$).

coating. This leaves the low-index material as the inner cavity layer. If the materials are reversed, the high-index material becomes the inner cavity layer, and the filter performance for different numbers of periods is as illustrated in Fig. 8.38.

Single-cavity structures of the type just discussed have a triangular shape and a gradual transition from high transmittance to high reflection. (*Note*: If the film materials are nonabsorbing, then all energy not transmitted is reflected. If the film materials are absorbing, then some energy is also absorbed, with most of the absorption being concentrated in the layers closest to or adjacent to the cavity layers.) It is often desirable to sharply isolate a wavelength or band of wavelengths. This can be done by using multiple-cavity filters. Figure 8.39 illustrates this for one, two, and three cavities. Again, the HPBW and rejection levels are determined by the ratio of the indices of the materials chosen. In this case, however, the addition of layers increased the number of cavity layers but did not change the HPBW, only the sharpness of the transition between high transmittance and high reflectance. The bandwidth of this type of filter can be changed by changing the index ratio of the materials being used. It can also be changed by changing the design to include multiple HWOT cavity layers. Although only

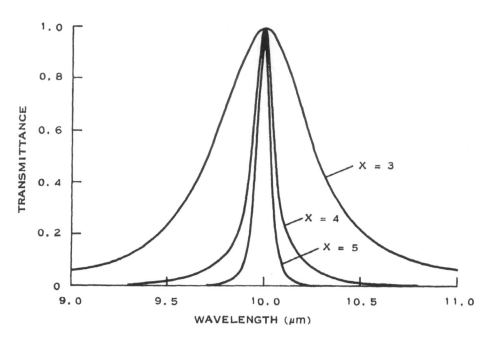

Figure 8.38   Single-cavity bandpass filters with high-index cavity layers, M $(LH)_x$ $(HL)_x$ design ($n_L = n_M = 2.15$, $n_H = 4.15$, and $n_0 = 4.0$).

limited natural bandwidths can be achieved by using reasonable designs and the few real materials available, almost any bandwidth can be achieved by using reasonable designs and the few real materials available, almost any bandwidth can be achieved by using symmetrical periods consisting of non-QWOT layers repeated many times with multiple HWOT cavity layers between them. It is also possible to put together SWP/LWP structures of the previous section to define a passband. For the multiple-cavity structures, the optical thickness of the cavity layers must be identical to achieve optimum performance. This becomes more critical when very narrow passbands are being made. Again, when tilted, the passband shifts toward the shorter wavelengths and distorts. The overall effects on the transmission region caused by tilting a filter is greatest for very narrow passbands.

8.5. METAL MIRROR COATINGS

Although highly absorbing, many metals have high reflectance and are used in mirror coatings, particularly aluminum, silver, and gold. Figure 8.40 shows the reflectance of common mirror coatings. In the longer wavelength infrared, they

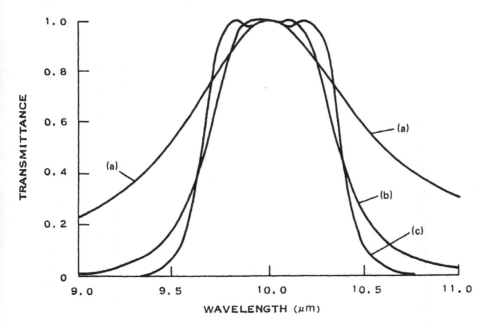

Figure 8.39 Multiple-cavity bandpass filters. (a) M H MM H; (b) M H MM HM-HMH MM H; (c) M H MM HMHMH MM HMHMH MM H.

basically reflect 97–98%, with aluminum typically averaging about 1% less than gold or silver. The metal films suffer from polarization splitting when tilted through high angles of incidence. The reflectance, however, of the s polarization increases to 100% as a limit, and the p polarization reflectance decreases only slightly, resulting in a small decrease in average reflectance over high tilt angles (0.1% for Al at 45°, and 0.4% for Al at 60° at 10-μm wavelength).

At all angles of incidence, the performance of metal mirrors maintains a similar color appearance, which is not the case for all-dielectric mirrors. Toward the shorter wavelengths, the spectral performance of the mirrors deviates more from each other. The reflectance of the aluminum starts decreasing below 2 μm and reaches a minimum of approximately 86% at 0.85 μm then increases to about 92% through most of the visible and near-UV. The reflectance of gold (copper is similar to gold) remains above 97% down to about 0.9 μm and then falls off through the visible and is highly absorbing in the green and blue part of the visible region and into the near-UV, which is the reason for its characteristic color.

Silver is the most neutral, maintaining a high reflectance (>97%) down through most of the visible. The reflectance of silver falls off rapidly below 0.4 μm and reaches a minimum at approximately 0.315 μm. The optical properties

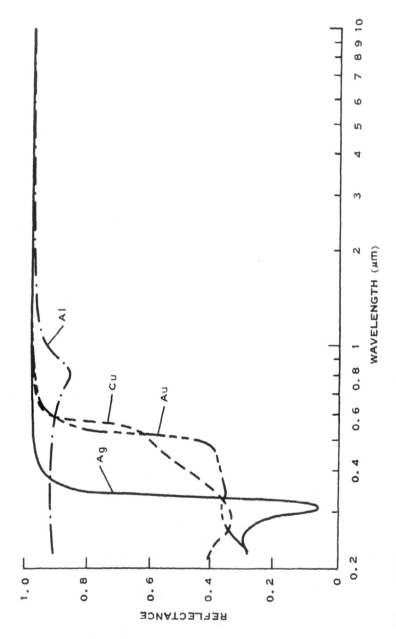

Figure 8.40   Reflectance versus wavelength for evaporated films of Ag, Al, Au, and Cu.

of silver in the UV are such that the decrease in reflectance is not caused by absorption alone. Because energy is transmitted into the film, thin films of silver can be made fairly transparent to UV (and even visible) energy while maintaining higher IR reflectance.

Although these mirror coatings have high reflectance when freshly deposited, all but gold react with oxygen and other constituents when exposed to the air. This is not as noticeable for aluminum, which oxidizes without changing appearance, as it is for silver and copper, which tarnish. Also, the latter metals are relatively soft, which makes them susceptible to scratching or sleeking when cleaned or handled. Protective layers can be deposited on these metals to protect them from reactants and make the surfaces more durable. Single-layer overcoats are commonly used. These overcoats will decrease the reflectance at wavelengths other than those at which they are a multiple HWOT. Thin overcoats that are HWOT in the visible or near-IR do not significantly affect the far-IR reflectance of meter mirrors at normal incidence and will have only a small effect at higher angles of incidence unless the overcoat is highly absorbing at that wavelength.

A good example is SiO-overcoated aluminum, which is a common inexpensive mirror reflecting better than 97% in the 8–12 μm region. When tilted, the reflectance at 8.8 μm drops to 86.3% at 45° and to 76.6% at 60° (Cox et al., 1975). They show that magnesium fluoride gives reflectances greater than 97.4% for angles to 60° over the same spectral region. In a later report, Cox and Hass (1978) report on the possibility of using yttrium oxide and hafnium oxide for similar applications. These oxides are almost as good spectrally as magnesium fluoride, with a slight degradation in reflectance toward the longer wavelength side of the region caused by a Reststrahl absorption band just beyond the region of interest.

The reflectance of a metal mirror can be enhanced by overcoating it with an all-dielectric stack consisting of QWOT periods similar to those used in the ADM structures discussed previously. Typically, only two or three periods are needed to increase the reflectance to 99% or more, because the reflectance of the bare metal is already high. Of course, the enhanced performance is achieved only over a narrow wavelength band. The enhanced performance would also be theoretically effective at all wavelengths where the period was an odd QWOT for the layers making up the enhancement period. Unfortunately, errors in depositing the layers and dispersion in the optical properties reduce the effectiveness of the enhancement at the higher order wavelengths. Outside the band of enhanced reflectance, the reflectance can be degraded significantly from that of the uncoated metal.

Thin layers of metal films are partially reflective and partially transmissive and have some absorption. Although not effective mirrors, they do occupy a unique niche in optical coating as neutral beamsplitters over fairly broad wave-

Table 8.3 Transmittance of Various Thin Metals (%)

Thickness (Å)	Ag[a]	Al[a,b]	Cu[a]	Pd[a]	Pt[a]	Ti[c]
10	75	85			80	
30	50	70				
50				50		
75		30				
100	20	18	50	25	40	27
140		40				19
170		7				15
200		4	30	15		
230					10	9.5
300				7		
300						5

Data are approximately those of the listed sources.
[a]Moses (1971), pp. 6, 28, 52, 54, 74.
[b]Hass and Waylonis (1961).
[c]Hass and Bradford (1957).

length bands. Table 8.3 consists of a summary of the approximate transmittance of different thicknesses of several metals for visible wavelengths. These values are also applicable to the near-IR. Extension to the mid- and far-IR is not possible, because the performance characteristics for those regions depends on the specific size and density of the nucleation sites and the growth characteristics of the films. For the most part, these data are empirical, depending on the conditions that existed when the films were deposited, and cannot be calculated from the thin-film properties reported in the next section for thick films of the metals. References for the various sources have also been included. Some liberty has been taken with the data from the various sources to fit them into the table. That is, thicknesses or transmission values have been adjusted to some extent. This table is being presented only as an aid to identifying approximate transmissions for the indicated thicknesses in the visible spectral region. As usual, the reader is cautioned to record data for his or her own procedures; the data here are intended to be used only as indicators of what might be achieved. Because of gold's unique dispersive nature, it has been excluded from this table.

8.6. THIN-FILM MATERIAL PROPERTIES

The properties of bulk materials were discussed in Chapter 1. Thin films of many such materials are used in coatings to enhance the optical performance of a system, to improve durability, or both. There has been considerable information reported over the last 40 or 50 years as the optical thin-film coating industry has developed a mature technology involving the use of highly sophisticated equipment. Cartwright (1940) notes that in 1892 Dennis Taylor reported that the tarnishing of camera lenses led to increases in their effective speed. Although not necessarily understood at that time, it was eventually deduced that the tarnish had a lower refractive index than the relatively high glass index. Cartwright further states that separate efforts by Taylor, F. Kollmorgan, and E.F. Wright in attempting to artificially tarnish glass were apparently not sufficiently effective to justify adoption by manufacturers of camera lenses. Cartwright then reports on his own efforts at the George Eastman Research Laboratory of Physics at the Massachusetts Institute of Technology on the deposition of metal fluorites (calcium fluoride) in vacuum to study the effectiveness of the films in increasing the speed of photographic lenses. The basic result of this effort was to increase the transmission of a 10-element system from 60 to 94% at the central wavelength. It is also interesting to note that several elements of his system are uncoated but obviously tarnished and transmit more than expected (e.g., a barium crown with $n = 1.603$ that transmits 92%). In the same time frame, Nicoll and Williams (1943) report a procedure to produce a low reflecting film of skeletonized silica. Their process involved the controlled exposure of silica-based glasses to dilute hydrofluoric acid vapor. After treatment, the glass surface was covered with a white deposit that was removed with water, leaving a hard, insoluble, low-reflecting film on the glass. Studies of these films showed that the chemical action selectively attacked the glass, producing soluble compounds of the metal constituents and leaving most of the silica unattacked. This conclusion was supported by chemical analysis, electron micrographs, and calculations of the refractive index requirements on the film.

It is not possible for this chapter to provide a comprehensive history of optical thin-film coatings. However, this discussion presents enough information with related references to provide both novices and experienced technologists with an understanding of key contributions from the last 50 years, concentrating on the last two decades.

Early work, from the 1930s to the 1950s, was concerned with developing the techniques for evaporating materials in a vacuum and depositing coatings uniformly over large areas. Primary efforts concentrated on antireflection coatings, metal mirrors, and more-or-less simple quarter-wavelength-thick stacks. These coatings suffered from poor durability and less than ideal spectral performance. Considerable effort in the 1950s and 1960s resulted in a rapidly expanding technology based on techniques to produce more durable coatings, techniques to use in characterizing materials, and an understanding of the physical characteristics needed to gain enhanced durability and optical performance. Efforts over the last decade have concentrated on improved deposition control and monitoring techniques resulting in near theoretical performance for many designs, an understanding of the importance of the microstructure of thin films in achieving both enhanced optical performance and durability, and novel techniques to achieve desired physical properties.

Much diverse and, to some extent, contradictory data for thin films have been published. This does not make the data suspect; it only points out that there is considerable latitude of control available to the optical thin-film manufacturer, which allows thin-film properties to be tailored to desired characteristics. Thus, anyone manufacturing or experimenting with thin films must be able to study and characterize the films deposited to know what optical and other physical properties are produced. Again, much has been written about the techniques used to characterize thin film, and these techniques are also reported here.

8.6.1. Film Preparation Techniques

For the most part, thin-filmcoatings are prepared by getting a source material into a gaseous state and then having the molecules condense on a relatively cool surface to form a thin film. Physical vapor deposition (PVD) and sputtering (SPT) are the two primary methods used to achieve this. Chemical vapor deposition (CVD) and reactive plasma deposition (RPD) use a vapor transport mechanism but have gaseous or liquid sources containing the desired atomic or molecular species. The gaseous reactants decompose and recombine to form the desired thin film. The decomposition and reaction is assisted by elevated substrate temperature in the case of CVD and by an rf plasma in the case of RPD. Solutions of titania and silica can also be used to form thin films by a spin or dip process. A technique applicable to a glass with the appropriate composition that could be heated and then leached to a porous structure near the surface was reported by Minot (1976). Although not a deposited film, this structure acts as a broadband gradient-index AR structure with 0.1–0.2% reflectance from the visible out to 2 μm. Structures of this type are limited to the appropriate glass type but are very effective and were reported to have damage thresholds of 12 J/cm^2 for 1-ns, 1.06-μm laser pulses. Lowdermilk and Mukerjee (1982) report on a process to deposit single-layer gradient-index AR film on any type of glass. Their sol–gel process involves deposition of a multicomponent noncrystalline metal oxide gel film and then subsequent controlled leaching with an appropriate solvent to leave a gradient refractive index on the parts. This work results in AR structures with laser damage threshold levels in the 20 ± 4 J/cm^2 range. The leach–etch and sol–gel processes are mentioned for completeness; however, they will not be developed any further, because to date they have been used only in the visible and near-IR spectra.

Each procedure for depositing optical coatings has its advantages and disadvantages. Physical vapor deposition from a resistive source, one of the earliest techniques used, involves relatively simple equipment and is probably the least expensive to set up. Essentially, a controlled low-voltage source capable of supplying a high current is needed. Early systems consisted of a variac and a stepdown transformer. Even in this day of computers, the only addition needed is a silicon-controlled rectifier (SCR).

However, not all materials can be evaporated from a resistive source. Some require a much higher temperature than can be achieved without burning out the resistive element; others might react with the hot element and contaminate the films. Electron guns (E-guns) can be used to evaporate materials requiring higher temperatures than can be achieved by using resistive sources. E-guns use a controlled beam of electrons striking the surface of a source to elevate a local area to evaporation temperature. The power supply, sweep controls, and the E-guns themselves are significantly more expensive than resistive source equipment. The advantage of PVD over other deposition processes is the relative ease of uniformly coating large areas and the relative ease of controlling and obtaining the desired film thicknesses. The disadvantage of PVD is that the films are somewhat porous with a packing density less than bulk. For this reason, the refractive indices are typically lower than bulk, and some of the films have a tendency to adsorb moisture into pores, which causes the coating to shift. Recent efforts in using IAD have resulted in higher packing densities for some materials being deposited by PVD. Ion-assisted deposition processing is relatively new, and improvements can be expected as it becomes more commonly used.

Sputtering is a process in which typically an inert gas, such as argon, is ionized and the anion is accelerated to collide with a target material that is deposited nearby as a thin film. Direct current sputtering involves a simple diode arrangement with a high negative potential on the target material. In a variation of this technique, a hot filament and secondary anode accelerate electrons through the space in front of the target, thus ionizing the gas and either increasing the sputtering rates or reducing the pressure of the sputtering process.

Radio-frequency sputtering (RFS) involves the use of radio-frequency signals. It is needed to sputter dielectric materials except in the case of ion beam sputtering. Ion beam sputtering involves the use of an ion source and lens system to form a plasma in one chamber. The ion beam is then extracted by the accelerator and lens system and focused on the target material. This system can be used to sputter materials in the low 10^{-6} torr range. The advantage of sputtering is that it is a high-energy process that produces denser and potentially more durable coatings than PVD and has the possibility of doing this at low substrate temperatures. However, early sputtering work had a significant temperature rise during the process because substrates were bombarded with free electrons.

The development of planar magnetron sputtering resulted in a system that controlled the electrons by collecting them before they could bombard the substrates and gave improved control over uniformity. The disadvantages here are low deposition rates and difficulty in controlling thicknesses and uniformity.

All thin-film coating processes have the potential of causing molecular species to decompose during processing. This is particularly true of sputter deposition of fluorides and PVD and sputtering of certain oxides. Partially decomposed films are nonstoichiometric, typically have higher than normal refractive indices, and are absorbing. Therefore, many thin-film deposition processes involve reactive deposition where the cation is introduced to make up the deficiency caused by the initial decomposition, thereby returning the film to stoichiometry. Thus, reactive deposition of an oxide typically has an oxygen ambient background, which becomes a control parameter in achieving the desired optical properties. This is also true of sputtering, where recombination with oxygen is aided by the ionizing plasma environment. Recent developments involving the use of ion beam sources in evaporators have brought this advantage to PVD.

8.6.2. Characterization of Optical Properties of Thin Films

A variety of methods can be used to determine the optical properties of thin films. Essentially, any characteristic that depends on n and k can be determined. Consider the periodicity for the reflectance and/or transmittance of a thin film as a function of film thickness, wavelength, and refractive index as calculated from Eqs. (17)–(21) and as shown in Fig. 8.41. These relationships suggest several approaches that might be used. First, the maxima and minima represent wavelengths for which the films are an integral number of QWOT's thick (assuming normal incidence). The index is

$$n = m\lambda/4d \qquad (37)$$

where $m = 1$ for 1 QWOT, 2 for 2 QWOT, 3 for 3 QWOT, . . . ; λ is the wavelength of a maximum or a minimum; and d is the physical thickness in the same units as the wavelength. Assuming that the film is homogeneous, the accuracy in determining n is limited only by the accuracy in determining the wavelength and film thickness. Wavelength data taken from a spectrophotometer scan are typically accurate. The limiting factor in using the relationship is an independent measurement of film thickness.

A noncontact technique for measuring film thickness if multiple-beam interferometry, which was developed to a high degree of accuracy by Tolansky as reported by Heavens (1955). This technique involves the interference between reflections from two surfaces in the region at the edge of a film on a flat substrate, which produces a fringe pattern displaced at the film edge (Fig. 8.42). The reference flat is a partially silvered reflector, and the sample surface is a highly silvered reflector. The reference surface is adjusted to be almost parallel to the sample surface, with a slight tilt along the film step to give the desired fringe pattern. Ideal conditions for the reflectance from the reference surface and

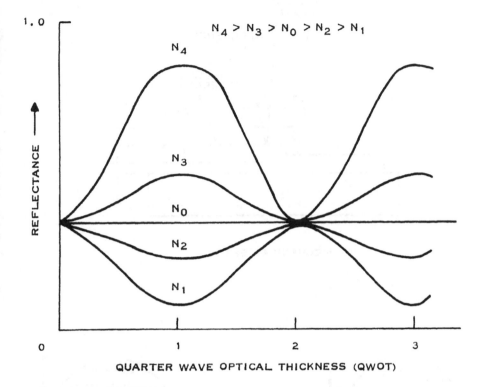

Figure 8.41 Reflectance versus wavelength for various refractive index film–substrate combinations.

the surface of the film being measured include the relative reflectance from each surface, the flatness of the sample surface, and the sharpness of the film edge. Under ideal conditions, steps as small as 20 Å can be detected, and film thicknesses can typically be determined to an accuracy of ±50 Å. This accuracy is independent of film thickness as long as the interference fringes can be followed or counted at the film edge. Both exceedingly sharp thick-film steps and shallow-film steps extending over the surface of the sample can be problems. Multiple-beam interferometry requires careful preparation of samples so that the measurement can be made. Also, the most accurate measurements are made by the aid of photography, which involves additional effort.

Surface profilometry, a contact method that became equally accurate in the late 1960s, requires no special preparation techniques unless the film material is so soft that a hard protective overcoat is needed. The technique involves a stylus

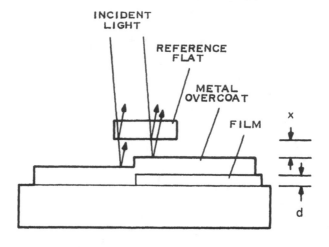

(a) MECHANICAL SETUP

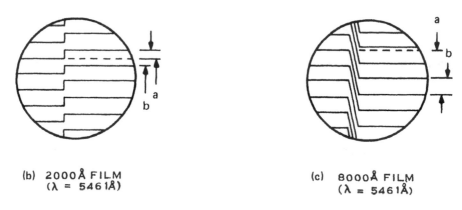

(b) 2000Å FILM
     (λ = 5461Å)

(c) 8000Å FILM
     (λ = 5461Å)

Figure 8.42   Multiple-beam interferometry for Fizeau fringes.

traversing the surface of a sample over the film-step edge. The output of the device is a strip-chart recording of the surface level at a fixed magnification. These devices typically have a range as low as 500 Å full scale and, under ideal conditions, can detect sharp steps as low as 10 Å on the most sensitive setting. However, thick films can be measured only to an accuracy of a fraction of the full-scale setting (typically, $\pm 2\%$). The two advantages of the profilometer over multiple-beam interferometry are the relative ease in making the measurement without special preparations and the ability to measure thick-film steps even when they are sharp.

To compute the index using Eq. (37), you must know the order in the interference pattern. This is relatively easy to determine for thin films. For thick films, it can be determined as follows. If m is the order at λ_1, $m + x$ is the order at a shorter wavelength (λ_2), and x is the number of inflections (maximum and minimum), then the second wavelength is displaced from the first where

$$ m = \frac{x \lambda_2}{\lambda_1 - \lambda_2} \tag{38} $$

Some judgment is still needed in choosing m, since the value computed from eq. (38) is unlikely to be an integer because λ_1 and λ_2 cannot be determined precisely and the refractive index will be dispersive, which also affects the wavelength for the maxima and the minima. It is best to use Eq. (38) only for small values of x or over wavelength bands where the material is relatively nondispersive.

At wavelengths where the film is an odd-numbered multiple of a QWOT, the reflectance is at a maximum displacement from the reflectance of the uncoated sample. Solving Eqs. (14) or (17) for the reflectance at a QWOT position gives

$$ R = \left(\frac{n_0 n_{j-1} - n_1^2}{n_0 n_{j-1} + n_1^2} \right)^2 \tag{39} $$

where n_- is the substrate refractive index, n_1 is the film refractive index, and n_{j+1} is the massive median refractive index (1 in air). Solving this equation for film index n from the measured reflectance gives

$$ n_1 = \left(n_0 n_{j+1} \frac{1 \pm \sqrt{R}}{1 \mp \sqrt{R}} \right)^{1/2} \tag{40} $$

Equation (40) has two solutions. Using the top sign gives a refractive index greater than $(n_0 n_{j+1})^{1/2}$, and using the lower sign gives a lower refractive index. The index values obtained using Eqs. (38) and (40) will agree within experimental accuracy assuming the films are ideal (that is, nonabsorbing, smooth, and homogeneous). As has been indicated previously, no film meets these conditions. However, the differences between the index values tell something about the film. The interferometric value obtained from Eq. (38) is an average value through the film thickness, and the specular value obtained using Eq. (40) depends on the Fresnel reflectances at the film boundaries.

In general, film deficiencies can be deduced from inspection of the interferometric spectral scans of reflectance and transmittance as a function of wavelength. Transparent films with little absorption will have an amplitude between maxima and minima that depends on the relative refractive index of the film–substrate combination. At HWOT points, the film will be absentee, and the sample should transmit and reflect as if it were uncoated. Inhomogeneous films will not be entirely absentee at HWOT points, and the reflectance compared with the reflectance of the uncoated sample will vary, depending on the nature of the inhomogeneity. Feldman et al. (1976) showed that if the film index is increasing with thickness, the reflectance will be greater than for the uncoated substrate, and if the film index is decreasing with thickness, the reflectance will be less than that of the uncoated substrate (Fig. 8.43).

The evidence of absorption and scattering (roughness) can also be deduced. Southwell (1986) showed that for low absorption and roughness there is little effect on the reflectance of a film, whereas there is significant effect on the transmission (Fig. 8.44) at the HWOT points (peaks). He accounts for this decrease by noting that the standing wave pattern or field strength is greatest at this point. There is little decrease in transmission at QWOT wavelengths that are close to a null in the standing wave pattern. For the absorbing situation, there is a general decrease in the transmission at all wavelengths, with a greater decrease at the HWOT points (peaks). If the absorption is significantly high, the amplitude of the interference patterns for both transmittance and reflectance will be significantly attenuated and there will be a general decrease in transmittance toward the shorter wavelengths.

Ellipsometry

The interferometric technique requires a knowledge of film thickness. There are many ways to obtain thin-film optical properties without first making an independent thickness measurement. Most of these result in obtaining the thickness as well as the optical properties, and some will be discussed later. An ellipsometer can be used to measure the ellipsometric parameters ψ and Δ, which are the differential amplitude and phase change of the two polarizations of light reflected at some angle of incidence. They are defined by the relationships

$$\tan \psi = \left| \frac{r_p}{r_s} \right| \qquad \text{(reflectance ratio)} \qquad (41)$$

$$\Delta = \delta_p - \delta_s \qquad \text{(differential phase shift)} \qquad (42)$$

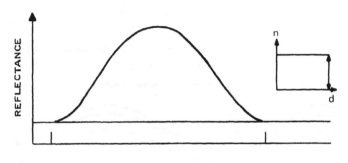

HOMOGENEOUS FILM, REFRACTIVE INDEX CONSTANT

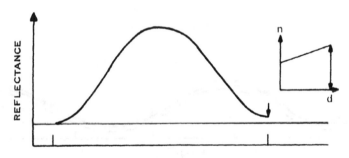

POSITIVE INHOMOGENEOUS FILM, REFRACTIVE INDEX INCREASING WITH FILM THICKNESS

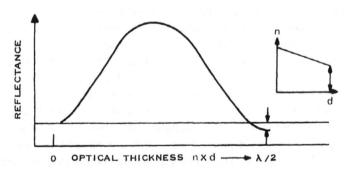

NEGATIVE INHOMOGENEOUS FILM, REFRACTIVE INDEX DECREASING WITH FILM THICKNESS

Figure 8.43 Homogeneity of thin films. (Used with permission of EM Chemicals.)

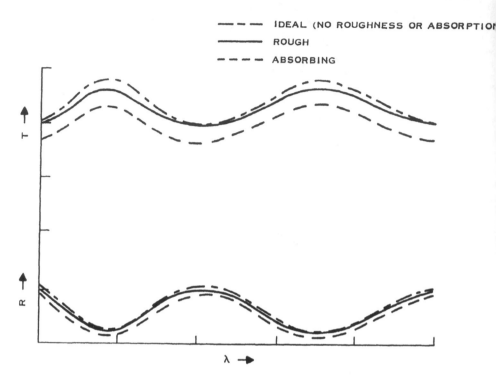

Figure 8.44   Effects of roughness and absorption on reflectance and transmission.

Consider the arrangement shown in Fig. 8.45, where the incident path consists of a light source (and a collimator unless the source is a laser), polarizer,
and compensator directed at an angle ϕ onto a surface. The reflected light passes
through an analyzer and a telescope and onto a detector. For a given sample, the
compensator is adjusted until the phase shift between the s and p polarizations is
zero, thus determining Δ. Then the analyzer is adjusted to extinguish the reflected radiation to determine ψ. For an uncoated sample, n and k can then be
determined for the substrate explicitly from ψ and Δ by using the reduced case
for a zero-thickness film. If the sample has a film on it where $k \ll 1$ (or k can be
assumed to be zero), the solution is in the real plane but still fairly complicated.
If k is not small, the expressions are complex and, when separated into real and
imaginary parts, yield two equations that can be solved simultaneously for any
combination of n, d, and k. These expressions are complicated and are usually
solved by computer computation, from graphic representation, or from previously computed tables. If the film thickness is known, the limitation on having
$k \ll 1$ can be removed and n and k can be determined. Archer (date unknown)

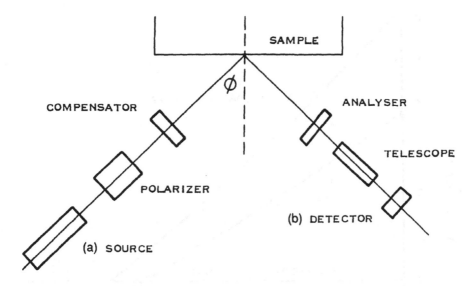

Figure 8.45 Ellipsometer arrangement with (a) laser or broadband source, monochromator, and collimator and (b) eye, photomultiplier, or IR detector.

has prepared a fairly detailed manual for Gaertner Scientific Corporation on ellipsometry. The manual is a good tutorial for this technique. Archer (1962) also published an article explaining the technique.

Ellipsometry has been used for many years in the visible and near-IR spectral regions. Leonard et al. (1983) at the University of Dayton Research Institute developed three IR ellipsometers. One instrument uses a tunable diode laser source or an HeNe laser and covers the 3.39–4.00-μm spectral band. A second instrument with only an HeNe laser (3.39 μm) was developed specifically for use as an in situ monitor for film growth studies at Kirtland Air Force Base. The third instrument had a Nernst glower and a monochromator as a source and covered the 1–12 μm spectral band. Because the computations required to obtain information on optical properties from an ellipsometer are fairly complicated, most instruments (including these) are connected to a computer and are interfaced to run automatically. The cited reference refers to the many optimization procedures that could be incorporated into the control software.

Ellipsometry is not a good technique to use on absorbing films but is well suited for the nonabsorbing or low-absorbing films because it does not require an independent thickness measurement. One drawback of ellipsometry (and of many of the techniques used to characterize optical films) is the need for relatively complicated computations that can be done only with a computer.

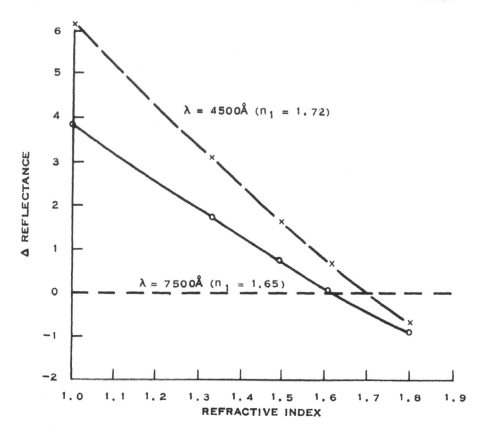

Figure 8.46   Reflectance difference versus immersion material refractive index.

Immersion Spectroscopy

Ellis (1963) and Ellis and Megarity (1964) reported on an interesting approach using immersion spectroscopy, which also does not require an independent measurement of thickness. Their approach involves measuring the reflectance interference pattern for a film–substrate combination when the sample is immersed in various index solutions. The resulting maxima in reflectance are then defined by Eq. (14). As the refractive index of the inversion fluids approaches that of the film material, the interference amplitudes approach zero. It is not necessary to exactly match the film index, because a plot of the reflectance amplitudes versus the immersion fluid index can be extrapolated to the index for zero amplitude. A computer-generated example illustrating this approach is shown in Fig. 8.46. The example was computed for a dispersive film for which the index was 1.65 at 7500 Å and 1.72 at 4500 Å. The Δ reflectance plot is the difference between the QWOT or higher order QWOT maxima and the value for uncoated substrate. If the index of the immersion material exceeds that of the film, the QWOT thickness values become a minimum and the pattern is inverted. This is shown as a negative Δ in the example. Although illustrated with reflectance, it is obvious that this technique readily lends itself to transmission measurements for films on nonabsorbing substrates as long as care is exercised in arranging the measurement setup. In either case, errors develop if the film or the immersion medium is absorbing, because even though the real part of the refractive indices match, the imaginary part of the reflectance amplitude, $k/(4n^2 + k^2)$, does not.

Ellis's original work was based on measuring the reflectance in the optical density mode of a spectrophotometer. Thus, his Δ_{max} was

$$\Delta_{max} = \log R_{QWOT} - \log R_{uncoated} \tag{43}$$

$$= 2 \log \left(\frac{1 + R_{21}/r_{10}}{1 - r_{21}/r_{10}} \right) - 2 \log \left(\frac{1 + R_{21} r_{10}}{1 - r_{21} r_{10}} \right) \tag{44}$$

where r_{10} refers to the film–substrate boundary and r_{21} refers to the immersive medium–film boundary. As before, Δ_{max} becomes zero where the r_{21} term becomes zero or the refractive index of the film matches that of the immersion medium. Ellis notes that experimentally the optical density extrapolated nearly linearly to zero. It is also true that the maximum reflectance amplitudes, although not linear, can be extrapolated to zero with a fair degree of accuracy with a sufficient number of data points. In addition to the effects of absorption already noted, there is a problem in using this technique on films with refractive indices sufficiently high that $r_{21} > r_{10}$ in air. Under this condition, the optical density is increasing rather than decreasing as the refractive index of the immersion medium increases. The optical density continues to increase, becoming infinite where $r_{21} = r_{10}$. Further increases in the refractive index of the immersion medium will then result in the optical density decreasing as described previously. This is not a problem in using the reflectance difference technique, because it just inverts the slope of the line. The best accuracy in any case is obtained with the use of immersion fluids whose refractive index is close to that of the film. It is not easy to find higher index oils or fluids for this technique, because many of them are absorbing. Nevertheless, this is not much of a problem for the lower index films being characterized in the visible and near-IR.

Reflectance and Transmittance Inversion

A variety of numerical techniques developed and reported over the last 20 years typically involve a spectrophotometer measurement of reflectance and transmittance of a coated sample at normal incidence and for the two polarizations at various angles of incidence. To derive n, k, and possibly d, an iterative search routine is used that involves the minimization of a merit function derived from the measured data and the computed performance from closely spaced variations of the parameters being determined. An effort of this type was carried out by Baldini and Rigaldi (1970), who used the transmittance and reflectance for energy incident on both sides of a sample to give them three measured parameters for determining n, k, and d. Sulzbach and Morton (1972) used the work of Baldini and Rigaldi to develop a similar procedure. They note that the greatest inaccuracy reported by Baldini and Rigaldi was the computed thickness. Since thickness measurements of a fairly high degree of accuracy were possible on Sulzbach and Morton's films, this factor was removed as a variable in their work. They include a computed example of the accuracy of this approach, showing accuracy out to the fourth decimal place in recovering the parameters used to set up the test. The merit function being optimized is the sum of the squares of the difference between the measured and computed values for reflectance and transmittance. Optimization is possible using any angle of incidence and polarization (random, s, or p). This technique is absolute for precisely known values of R and T. Unfortunately, R and T are usually not known to better than a few tenths of a percent. However, this technique and others like it are usually accurate enough for determining indices (typically 1–2%), especially if the materials are being characterized at closely spaced wavelength intervals and then plotted to average out variations dispersively. Typically, k cannot be determined accurately, especially if it is small (less than 0.001). This program has also been structured to handle the situation where the material being characterized can be embedded as a single layer or as many layers in a multilayer structure for which the optical properties of the other materials are well known. However, the thickness (optical or physical) of all the layers being characterized must be known fairly accurately.

Demner and Shamir (1978) developed an approach using interferometric measurements with improved accuracy in determining k. In their setup, a beamsplitter is used to split an incident beam into two paths incident simultaneously at the sample front surface I_1 and back surface I_4 at the same angle of incidence. Each reflected beam combines with a transmitted beam coming from the opposite side and is measured by a detector I_1 or I_4 (Fig. 8.47). The outputs of the detectors are then the intensities of the interference patterns between the coherent waves. A phase shift or difference between waves is generated by having one incident beam reflected off a mirror mounted to a transducer so the mirror can be moved back and forth in a controlled manner. The detector outputs are then applied to a strip-chart recorder or an oscilloscope. As the phase between the two beams is varied, the detector outputs sweep out an ellipse on the X-Y recorder or on the oscilloscope. Three terms are obtained: an amplitude ratio (A_y/A_x), a center coordinate ratio (Y_o/X_o), and a phase angle (Δ) that can be determined from the ellipse. These terms are related as follows (the medium subscripts developed in this text are the opposite to those used by Demner and Shamir):

$$\Delta = \arcsin (y^1/A_4) \qquad (45)$$

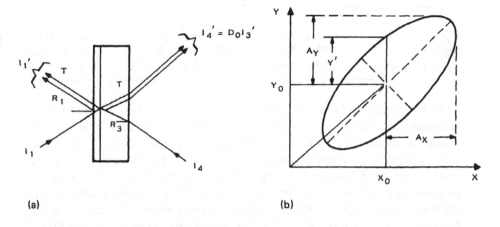

Figure 8.47 Ellipsometer setup; (a) wavepath through sample, (b) measured parameters from ellipse.

$$\frac{A_y}{A_x} = D_o \frac{n_1 \cos\theta_1 \, |t_3| \, |r_1|}{n_3 \cos\theta_3 \, |t_1| \, |r_3|} = D_o \left(\frac{R_3}{R_1}\right)^{1/2} \tag{46}$$

$$\frac{Y_o}{X_o} = D_o \frac{n_1 \cos\theta_1 \, |t_3|^2 + \eta_2 |r_1|^2}{n_3 \cos\theta_3 \, |r_3|^2 + \eta_2 |t_1|^2} = D_o \frac{n_3 \cos\theta_3}{n_1 \cos\theta_1} \left(\frac{1 + \eta_2^2 \, p}{\eta_2^2 + q}\right) \tag{47}$$

$$\Delta_1 = \Delta\pi$$

where

$$p \triangleq \frac{|r_1|^2}{|t_3|^2} = \frac{n_1 \cos\theta_1}{n_3 \cos\theta_3} \left(\frac{R_1}{T}\right) \tag{48}$$

$$q \triangleq \frac{|r_3|^2}{|t_3|^2} = \frac{n_3 \cos\theta_3}{n_1 \cos\theta_1} \left(\frac{R_3}{T}\right) \tag{49}$$

R_1 is the reflectance for film-side incidence; R_3, the reflectance for back-surface incidence; and T is the transmittance.

By using Eqs. (48) and (49), p and q can be measured directly by successively blocking each incident wave and ratioing the energies of the two detectors (I_1' and I_4') assuming that n_1, n_3, θ_1, and θ_3 are known. These measurements allow Y_o/X_o and A_y/A_x to be determined. Values for Δ or Δ_1 can be determined dynamically by applying a sinusoidal signal to the transducer and measuring it directly from the oscilloscope screen. Finally, values for n, k, and/or d are obtained from the derived relationships by numerical inversion or plots computed for the values of the measured parameters of A_y/A_x, p, q, or Δ_1. Several examples are presented in the original work as well as a complete derivation of the method sketched here. Final results for three sample materials are reported as follows:

	n	k	d
In_2O_3	2.02 ± 0.06	$(19 \pm 4) \times 10^{-4}$	605 ± 16nm
MgF_2	1.42 ± 0.02	$(45 \pm 5) \times 10^{-4}$	202 ± 12nm
Thin plastic sheet	1.53 ± 0.001	$(4 \pm 0.06) \times 10^{-4}$	499 ± 3 nm

In the preceding examples, the accuracy of the k values has been more accurately determined to an order of magnitude from the prior example. Although an improvement, it is still not sufficiently accurate for obtaining extinction coefficients for materials that are sufficiently nonabsorbing to be usable in coatings for higher energy laser applications. As already indicated, there are problems in using the numeric inversion of R and T spectrophotometric data to determine n and k. One is that the solutions are not unique. The second is that the accuracy for any given R, T, d, or λ situation depends on the intersection of the R, T plots in the nk plane. For a given film–substrate combination, these contours depend on specific values of d and λ. The most accurate inversion occurs when the R and T contours intersect at normal incidence, and the least accurate occurs when the plots intersect at small angles to each other. This limitation can be alleviated to some extent by taking data for films of several different thicknesses (assuming that n and k are independent of thickness, which is not always the case, especially for thin films) and using a curve-plotting technique to minimize errors.

Another way to avoid this limitation is to use multiple-wavelength techniques rather than limiting the inversion to a single wavelength. This could be done assuming nondispersive characteristics, but generally when multiple wavelengths are used some type of dispersion equation is used. Several dispersion relationships are available. Emiliani et al. (1986) used the Sellmeier formulas as a basis for their work:

$$n^2(\lambda) = A + B/(\lambda^2 - \lambda_o^2) \tag{50}$$

$$k(\lambda) = (1/nQ)(D\lambda + C/\lambda + E/\lambda^3) \tag{51}$$

Dobrowolski et al. (1983) used several different dispersion relationships including a Lorentzian model (A), a Drude model (B), a combination of the Lorentzian and Drude models (C), and a simplified version of the Sellmeier equation (D):

(A)
$$n^2(\lambda) = A + k^2(\lambda) + \sum_{i=1}^{j} \frac{B_i \, \lambda^2 \, (\lambda^2 - C_i^2)}{(\lambda^2 - C_i^2)^2 + D_i^2 \, \lambda^2} \tag{52}$$

$$k(\lambda) = 2n \, \lambda^{-1} \sum_{i=1}^{j} \frac{B_i \, D_i \, \lambda^3}{(\lambda^2 - C_i^2)^2 + D_i^2 \, \lambda^2} \tag{53}$$

(B)
$$n^2(\lambda) = A + k^2(\lambda) + B\lambda^2/(1 + C^2\lambda^2) \tag{54}$$

$$k(\lambda) = BC\lambda^3/[2n(\lambda)(1 + C^2\lambda^2)] \tag{55}$$

(C) $$n^2(\lambda) = A + k^2(\lambda) + \frac{B\,\lambda^2}{1 + C^2\,\lambda^2} + \sum_{i=1}^{j} \frac{B_i\,\lambda^2\,(\lambda^2 - C_i^2)}{(\lambda^2 - C_i^2)^2 + D_i^2\,\lambda^2} \quad (56)$$

$$k(\lambda) = 2n(\lambda)^{-1}\frac{BC\,\lambda^3}{1 + C^2\,\lambda^2} + \sum_{i=1}^{j} \frac{B_i\,D_i\,\lambda^3}{(\lambda^2 - C_i^2)^2 + D_i^2\,\lambda^2} \quad (57)$$

(D) $$n^2(\lambda) = 1 + A/[1 + B\lambda^{-1}]^2 \quad (58)$$

$$k(\lambda) = C/[n(\lambda)\,D\lambda + E\lambda^{-1} + \lambda^{-3}] \quad (59)$$

As before, the technique involves entering R and T data from spectrophotometer scans. This time, however, the data are taken over a wavelength bandwidth with enough data points to give a single solution to determine the coefficients for the selected relationships. A random search is then made, varying the coefficients for the selected relationship to give n and k dispersively over the selected band. The merit function is again a combination of the computed and measured differences. The coefficients thus determined are not intended to be applicable over extended bands, just over the selected band. They are used as a vehicle to characterize materials over the selected band and to minimize the uncertainty in values of n and k determined from a single-wavelength calculation. Application over wide wavelength regions would require additional terms in the dispersion relationships.

Laser Calorimetry

Laser calorimetry offers the opportunity to measure absorption directly. Loomis (1973a) used a technique whereby a sample is suspended from a thermally insulating frame between four thermocouple junctions (two front and two back). The reflected power, transmitted power, and temperature of the sample are taken before, during, and after exposure to a laser beam (Fig. 8.48). The data are then digitized by a computer and reduced to the absorptance. The analysis consisted first of determining a second-order polynomial fit to the thermopile temperatures before and after exposure and then extrapolating both curves back to the center of the exposure to give the temperature change:

$$\Delta T = T_2 - T_1 \quad (60)$$

An alternative procedure for obtaining the temperature change is to use a third-order polynomial fit to calculate the slopes for each step in the test cycle and extrapolate them to the center of the exposure by

$$\frac{\Delta T}{\Delta t} = \frac{m_b + m_c}{2} + m_h \quad (61)$$

A comparison of the rate of change in temperature by these two techniques is a measure of how good the data are. Regardless of the results of this comparison, the first result was used by Loomis in further data reduction because he believed

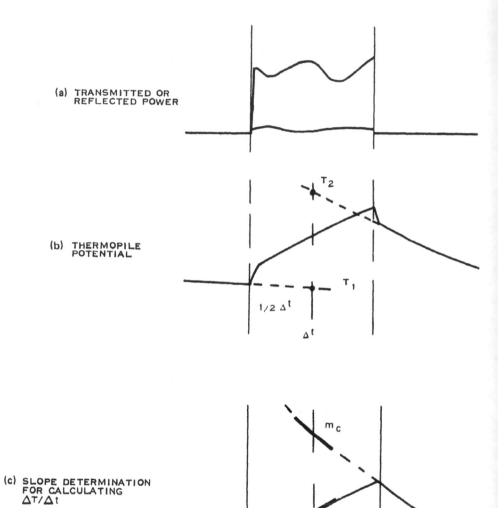

(a) TRANSMITTED OR
    REFLECTED POWER

(b) THERMOPILE
    POTENTIAL

(c) SLOPE DETERMINATION
    FOR CALCULATING
    $\Delta T/\Delta t$

Figure 8.48   Display of data collected during laser calorimetry measurement.

that it was a more accurate calculation. The absorbed power is then (where H_c is the heat capacity of the sample)

$$PA = H_c = \Delta T / \Delta t \tag{62}$$

To complete the calculation, the incident power (PI) was computed by numerically integrating the reflected power (PR) and transmitted power (PT) curves over the exposure time, adding them to the absorbed power (PA), and computing the absorption coefficient as follows:

$$\alpha = \frac{A}{d} = \frac{PA}{(PA + PT + PR)\, d} \tag{63}$$

Because of the rather limited accuracy of the power meters ($\sim 3\%$ each) and other factors, the accuracy in determining the absorption was thought to be, in the worst case, $\sim 10\%$. However, this direct measurement of the absorptance uses a very sensitive means and makes it possible to measure absorption many orders of magnitude below that which can be determined from spectrophotometric data.

The preceding derivation is for an uncoated sample of bulk material. Determining the absorption coefficient for a film on a coated sample is much more complicated because of the interactions of radiation at the various interfaces, the multiple reflections taking place, the phase problem dictated by consideration of coherency, and the transcendental functions that relate all of the above. Loomis (1973b) resolved this problem for the general case where either one or both surfaces of a sample might be coated. His derivation assumed incoherency between waves reflected from opposite faces of the sample. Derivation of the general case is more complicated than space allows. However, for a single film on a substrate, the absorption coefficient becomes

$$\alpha = \frac{A_1 n_1 \{ (n_2 + n_o)^2 \cos \delta^2 + [n_i + (n_2 n_o / n_1)]^2 \sin^2 \delta \}}{2\, n_2 \, (n_1^2 + n_o^2)\, d} \tag{64}$$

where A_1 is the absorption coefficient of the coated sample minus the absorption coefficient of the uncoated sample, and all other terms are as defined previously.

Temple (1979) shows the absorption from a single film as having three parts: (1) within the bulk of the film, (2) at the air–film interface, and (3) at the film–substrate interface. The technique involves the preparation of a wedge-shaped film on a sample during a single deposition sequence and the collection of data for specific thickness (0 and multiple $\lambda/4$ QWOT values). The data are then reduced to show the absorption contributed at each source. It is interesting to note that the example of the technique (As_2Se film on CaF_2) referenced in this citation had little bulk absorption in the film and had most of the absorption at the film interfaces, especially at the film–substrate interface. Temple used a scanning adiabatic calorimeter for this effort (Decker and Temple 1977). However the method is applicable to any laser calorimeter setup.

A variety of techniques are applicable to the measurement of small absorption coefficients. Hordvik (1977) wrote a review that summarizes many of these. They include three basic approaches: transmission, calorimetry, and emittance. The transmission techniques he refers to are similar to those previously discussed for the numerical inversion of spectrophotometric data, ellipsometry, and attenuated total reflection (Harrick 1976). The sensitivity of the best of these techniques is probably about 10^{-3} cm^{-1}. Laser calorimetry is judged to have an order-of-magnitude greater sensitivity. Most laser calorimetric measurements use techniques similar to the two examples cited. However, photoacoustic laser calorimetry offers even greater sensitivity. In this technique, a periodically interrupted laser beam is incident on the sample, generating a photoacoustic wave whose amplitude is proportional to the absorbed energy. Various approaches with this technique allow one to distinguish bulk and surface effects and to measure absorption in thin films. Measurements have been made as low as 10^{-5} cm^{-1}. Hordvik also reports on some preliminary efforts to use interferometric techniques to measure the change in path length resulting from heating and thus give the temperature rise in the sample. The advantage of this technique would be to eliminate the scatter problem inherent in the other approaches.

Emittance Spectroscopy

Emittance measurements probably have the most potential for realizing high sensitivity in absorption measurements. Stierwalt (1973) used a setup consisting of a Beckman IR3 spectrophotometer with the source replaced by an evacuated source chamber that could be held at $25 = 0.05°C$ (Fig. 8.49). Stierwalt and Potter (1962, 1963) and Stierwalt et al. (1963) have published extensively on this technique. The sample is mounted in a Dewar assembly. The system is calibrated by first making a scan with a blackbody in place of the sample. During the calibration, a servo system controls the slit width to maintain a constant output from the detector. The slit program is used in subsequent measurements. When measuring samples, the output of the detector is the emittance of the sample and is thus a direct measurement of the absorptance.

Absorption as low as 10^{-5} has been achieved, and measurements as low as 10^{-6}–10^{-7} cm^{-1} are theoretically possible if stray radiation and fluctuations in the blackbody can be controlled. Measurements can be made at various temperatures and over the wavelength range of the spectrophotometer. Measurements made at ambient room temperature are better in the 10-μm region but fall off drastically toward the near-IR because of the spectral emittance distribution. Emittance spectroscopy is a highly versatile technique, but the amount of time, the controls, and the care required to take good measurements probably limit its use to the laboratory. For routine work, laser calorimetry is probably the best approach.

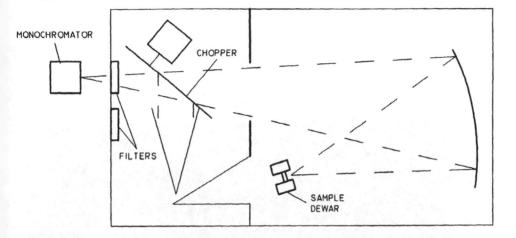

Figure 8.49 Source arrangement for emittance–absorbtance measurement. Detector, monochromator, and source are enclosed in an evacuated isothermal held at 25°C.

Thin-Film Microstructure

The performance of any single coating does not depend on current conditions or exposure but rather on its entire history and, to some extent, the history of the surface upon which it is deposited. This is becoming more evident as the factors necessary to produce coatings with ''good'' characteristics become better understood. Obviously, the microstructures of the coatings play a key role in both the spectral performance and the environmental stability of coatings. MacCleod (1982) has written an excellent review of the work done in studying columnar growth and stress in thin films. Because most coatings have a packing density of less than 1, there are voids in the coating. A typical morphology evident from scanning electron microscopic (SEM) studies of various films is that many have a columnar structure (Fig. 8.50). The voids or capillaries around the columns leave room for moisture to be absorbed into the structure, which locally changes the refractive index of the film and ''shifts'' the spectral performance. If the absorption is uneven, the shifting will be uneven, which results in a blotchy appearance to the coating. Sometimes it is possible to bake the coatings to drive out the moisture. However, this fix is usually neither complete nor permanent.

Harris et al. (1979) discuss various models for columnar structure and relate them to studies for ZnS, cryolite (Na_3AlF_6), and TiO_2, three materials known to have columnar structures. In this effort, they devise columnar models to demonstrate that by varying the widths of the columns (thus, the packing density of the film) it is possible to match measured film profiles. Although their results were not decisive, their measurements were not inconsistent with the columnar model and laid the grounds for additional effort. This effort was an alternative to

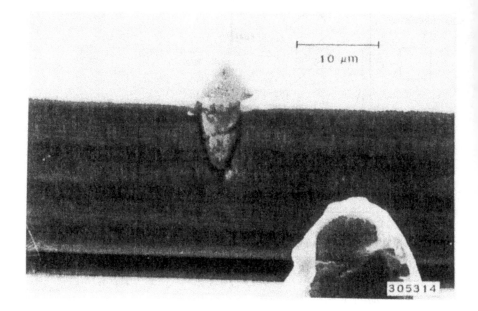

Figure 8.50   SEM micrograph of Ge/ZnS stack showing columnar structure in the physically thicker ZnS films. (Photo courtesy of Texas Instruments, Inc.)

earlier efforts to model thin-film inhomogeneity as a series of horizontal isotropic layers. Although easy to calculate, this latter approach can never be independently supported by any observable physical characteristic and therefore is of no real significance.

The microstructure also explains the stress that develops in thin-film coatings. Ennos (1966) did one of the more extensive sets of measurements on stress. There are two types of stress in thin films: extrinsic and intrinsic. A difference in the expansion coefficients of thin-film materials and the substrate on which they are deposited results in extrinsic stress. Intrinsic stress is caused by forces internal in the film. Attraction between columns across voids leads to tensile stress. Most thin films exhibit initial tensile stress. As the growth continues, the stress can become more tensile or can change to compressive. A compressive stress develops if the columns become larger in diameter as the film grows. The effects of stress may be small in many cases, but enough stress can cause the films to crack, delaminate, or pull completely off the surface of the substrate. Stress in a film and the cracking that develops from it leaves the film vulnerable to further degradation when it is exposed to harsh environments such as humidity, salt spray, and high-energy lasers. Therefore, it is necessary to control stress by whatever means are available. Although not much can be done about extrinsic stress, intrinsic stress can be influenced by controlling the microstructure of the film during deposition.

A variety of factors, including evaporation rate, substrate temperature, bias potentials, angle of impingement, sticking coefficients, and ion milling during deposition, affect the microstructure of films. In PVD, the growth of the material is essentially toward the arriving material. Thus, under usual circumstances, the columns are essentially vertical to the surface. When the deposition is oblique, an anistropy can develop, with the films being biaxially birefringent and no longer orthogonal to the surface. In cases where the impingement varies over large angles of incidence, severe inhomogeneity can develop, degrading the optical performance of the film. The other factors can also have a significant effect on the durability of films. A prime example is magnesium fluoride, which is porous and soft when deposited at low substrate temperatures but becomes very dense with a packing density close to 1 when deposited at 300°C substrate temperatures. There are few voids in a good MgF_2 film, and thus there is strong bonding between columns, which makes it impervious to attack by moisture. Films with lower packing densities take moisture into the voids, which weakens bonds between the columns. These weak bonds result in less durable films, even to the extent of causing or allowing cracking and delamination to occur.

Unfortunately, it is not possible to just increase substrate temperatures to get the benefits of increased packaging density. Higher substrate temperatures have many other effects, such as increasing the crystallinity of the film. Amorphous films are less dense, they are homogeneous, and they offer few opportunities for scattering. Crystalline films, on the other hand, can have grain boundaries and crystallites that scatter light and degrade the optical performance. Also, the sticking coefficient (the ratio of molecules that nucleate and stay in the film to the number of impinging molecules) for all materials depends on the substrate temperature. Thus, for some materials, when the substrate temperature is high, the possibility for an arriving molecule to migrate over the surface and find a nucleation site before it is ejected from the surface is low. ZnS is a prime example of this. Typically, it is not deposited above 160°C, because its highly temperature-dependent sticking coefficient makes it impossible to sufficiently control the thickness of the deposit.

Just as the voids between columns weaken the durability of thin films, so does any other inclusion or discontinuity. Particulate matter left on the surface before deposition and spit particles impinging on the surface during deposition are particularly bad. Not only do they leave an opening for moisture, but this opening can be severe if the particles shadow the surface from deposition. In addition to particulate matter, SEM studies to understand the microstructure have revealed that thin films can also have nodules that grow in the film. Typically, these nodules grow from a small seed and end up having a conical shape. One such nodule is present in the micrograph shown in Fig. 8.50. Another nodule is

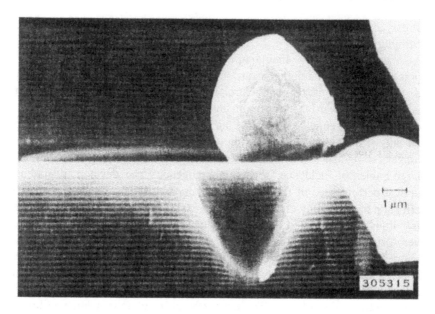

Figure 8.51 SEM micrograph of TiO_2/SiO_2 stack showing a nodule. (Photo courtesy of Texas Instruments, Inc.)

shown in Fig. 8.51. The latter micrograph is interesting in that the material from the nodule has flipped out and is setting above it. It is not conclusively known what causes a nodule to start growing; however, once started, they usually continue to grow and form a convex cap at the surface of the film. In addition to creating areas that allow moisture to penetrate the film, particulate matter and nodules cause scattering, which degrades the optical performance; provide potential areas with lower laser damage threshold; and lower the abrasion resistance of the films. Because the particles and nodules are not as tightly bound as the film material, they can be dragged across the surface, causing sleeks and scratches.

Identification of contaminants, impurities, and inclusions in thin films is not straightforward. However, spectrographic techniques have undergone a dramatic change over the last 15–20 years. These techniques use electron beams, ion beams, and X-rays as sources to probe the samples being studied. The following list includes some of these techniques:

1. Auger electron spectroscopy (AES)
2. Electron-induced X-ray fluorescence
3. Electron microprobe analysis (EMP), which includes energy-dispersive spectroscopy (EDS) or X-ray analysis (EDX) and wavelength-dispersive spectroscopy (WDS) or X-ray analysis (WDX)

4. Rutherford backscattering spectroscopy (RBS)
5. Secondary ion-mass spectrometry (SIMS), which includes ion-microprobe mass analysis (IMMA)
6. X-ray-induced X-ray fluorescence (XRF)
7. X-ray photoelectron microscopy (XPS), also known as electron microscopy for chemical analysis (ESCA)

Each of these techniques results in electrons, ions, or X-rays as the species being detected.

Bowling (1984) has prepared an excellent summary review and comparison of the above techniques. He notes that electrons can be focused sufficiently to give spatial resolution as low as 300 Å, whereas ions can resolve only about 1000Å at best and typically about 10,000 Å. X-rays are the least focusable and are limited to about 1–5 mm, although some suppliers of XPS instruments expected to get down to 150 μm. Sampling depth of electron techniques are limited to a few tens of angstroms, whereas ions can penetrate 100 Å. X-rays are the least surface-sensitive; they can penetrate 0.5–10 μm. Detection limits vary from about 1% of the sample for electron techniques to 0.01% for X-ray techniques. Electron techniques are least sensitive to mass of the nucleus being detected, whereas X-ray techniques have a mass detection sensitivity range as large 10^4–10^5. Ion detection techniques can achieve detectivity levels as good as 1 ppb under ideal circumstances but suffer from the greatest mass sensitivity range, which can be as great as 10^6. Table 8.4 presents Bowling's list of data comparing the various techniques. Note that some parameters in the table are qualified in the footnotes.

Laser Damage Testing

Another important attribute of optical thin-film coatings is the laser damage threshold (LDT) level. Because coated surfaces typically have LTDs lower than those of uncoated surfaces, the coatings are a major factor, limiting fluence levels for laser optics.

As the use of lasers becomes more widespread, the need for a system to protect personnel and optical equipment against accidental or deliberate exposure to high-energy laser threats becomes mandatory. Many factors are involved in determining the cause of damage. Generally, damage occurs when energy is absorbed and cannot be dissipated fast enough, resulting in either melting or rapid expansion, which develops stress and causes delamination in the coating or fracturing of the substrate itself. Obviously, absorption is a major contributor to this type of failure and is a factor in the failure of both AR and ADR coating structures. Although coated optics have lower LDTs than uncoated optics, the damage in coated optics is still influenced by the surface morphology and structure upon which it is deposited. Bloembergen (1973) showed that scratches, grooves, and defects in the surface result in a significant reduction in LDT.

Table 8.4   Summary of Technique Detection Parameters

Parameter	Technique					
	AES	EXD	RBS	SIMS	XPS	XRF
Excitation source	e^-	e^-	Ion	Ion	X-ray	X-ray
Detected species	e^-	X-ray	Ion	Ion	e^-	X-ray
Elements detected	$Z > 2$	$Z > 10^a$	$Z > 1$	All[b]	All[c]	$Z > 12$
Spatial resolution	300 Å	1 μm	1 mm	2–20 μm	1–5 mm	1 cm
Depth probed	5–50 Å	0.5–5 μm	1 μm	0–100 Å	5–50 Å	1–10 μm
Detection limits	0.1%	0.01%	0.01%	<1 ppb	0.1%	0.01%
Sensitivity variations	50	10^3–10^5	10^3	10^4–10^6	50	10–10^{5d}
Chemical information	Some[e]	No	No	No	Yes	No
Standards required	Yes	Yes	No	Yes	Yes	Yes
Sample charging	Yes	Yes[f]	No	Yes	No	No
Beam damage	Yes	Yes	Yes	Yes	No	No
Vacuum required	UHV	Low	High	High	UHV	None

[a]$Z > 7$ for windowless detectors; $Z > 4$ for WDX.
[b]Depends on source ion (e.g., oxygen not detectable with an oxygen ion source).
[c]Not sensitive for H or He.
[d]The factor of 10 can be achieved with a system using multiple interchangeable X-ray sources.
[e]AES chemical information is not yet well characterized.
[f]Samples can be coated to overcome charging.
Source: Texas Instruments, Dallas, Texas, by permission of R.A. Bowling.

Polishing processes introduce subsurface damage and stress, which weaken the surface. The surface can also be weakened by the presence of voids, grain boundaries, and dislocations. Surface damage can be minimized by appropriate surface preparation techniques.

Sharma et al. (1977) used optical microscopy and X-ray topography to study the surface and subsurface of commercially prepared Ge samples. Some of the surfaces were then ion-etched to remove the damaged surface. Subsequent testing showed a marked improvement for the ion-etched surface. A similar result for coated optics is reported by Carniglia (1981). In his work, LDT testing was performed on four different AR structures on surfaces prepared by two different methods. One method is a conventional polish; the second method is a "super polish" using a bowl-feed process. For all coating types and for two different substrate types, the coatings on the super-polished surface had higher LDTs, some almost twice as high. Figure 8.52 shows an example of damage caused by polishing sleeks on the surface of a Ge sample coated with a 10.6-μm bandpass filter.

The nature of the test setup strongly influences results. Factors that must be considered include wavelength, pulsewidth or duration, repetition rate, and spot size. For the most part, these are determined by the type of laser used; however,

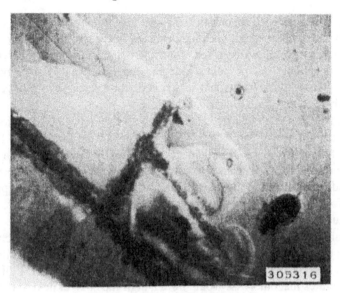

Figure 8.52 Laser damage in 10.6-μm bandpass filter coating. (Damage is promoted by polishing sleeks.)

the spot size and fluence levels often are not controlled independently. That is, for many test setups, the power density is determined by locating the test sample in a converging or diverging beam, with the higher exposure levels being closer to the focal point. Exposing different size areas of the sample is not much of a problem if the failure mechanism is the result of intrinsic characteristics of the film(s). However, it can be a serious problem if the failures are caused by defects, because the probability of exposing those defects is determined by the area being irradiated and the density of the defects.

Surface defects are more of a problem for AR coatings, because the energy is being transmitted through the film and the sample. On the other hand, surface defects can be less of a problem for ADR mirror coatings, because the energy is being reflected more from the outer layers than from the inner layers. Since PVD coatings essentially mirror the morphology of the substrate surface, defects grow through to the surface. Sputtered coatings have a higher energy of arrival, which results in more surface mobility and a tendency for the coating to smooth out local defects and to produce denser and more durable films. This advantage for sputtered coatings can be offset by the greater difficulty in controlling thickness and a tendency for the films to be less stoichiometric and possibly more absorbing.

There are three basic modes of operation for laser damage testing: single-pulse, multiple-pulse, and continuous-mode. LDT levels depend on which type of test is being run. Obviously, the LDT level is lower for continuous operation because there is less opportunity for the heat to be dissipated. Conversely, single-pulse testing yields the highest LDT, because there is more time for absorbed energy to be dissipated. Multiple-pulse LDT falls somewhere in between, depending on the pulsewidth and repetition rate. For a given energy density, longer pulsewidths and higher repetition rates have a lower LDT because there is less time to dissipate the energy.

Other factors that influence any test and all data repeated by different experiments are the spatial and temporal distributions of the laser beam. Discontinuities or hot spots can result in damage at apparently lower levels than would be experienced with a well-controlled Gaussian distribution. All of these problems have led to a prolific amount of conflicting data being published on specific LDT levels for thin-film materials and/or material combinations. However, for a given test setup, trends have been well established.

A variety of techniques have been used to improve LDT. One such technique isolates the coatings from the substrate by using a barrier layer. Carniglia (1981) used an HWOT barrier of evaporated quartz to improve the LDT of four-layer AR structures that otherwise would have a high-index layer next to the substrate. He notes that there is 50% improvement when alumina was used as the high-index material, a 35% improvement for zirconia, a 30% improvement for tantalum pentoxide, and next to no improvement for titanium dioxide. To explain the reason for the improvement, Carniglia decided that the major effect of the barrier layer is due to the mechanical properties of the silica layer itself. He speculates that this is because the compressive stress of the silica layer offsets the tensile stress from other materials. Also, the silica is amorphous and thus able to withstand the thermal and mechanical shock of the laser pulse better than the more crystalline high-index materials. It is also true that silica is similar in composition to the substrates and thus adheres to them more strongly.

In a similar manner, Carniglia et al. (1979) demonstrated that HWOT overcoats on TiO_2/SiO_2 stacks improved damage thresholds from just less than 9 J/cm^2 to more than 14 J/cm^2. This is not easily explained, because mapping of the electric field vectors show that the field strengths in the reflectors are the same with and without the overcoat. Again, the improvement is attributed to the mechanical strength and durability of the silica itself and the stress compensation resulting from the compressive stress of the silica.

Ristau et al. (1985) show that damage levels for single films decrease as the film thickness increases. It was shown previously that stress in coatings also increases as thickness increases. These two facts support Carniglia's contention that the improvement in damage resistance of the undercoats and overcoats is related to stress and the microstructure of the materials. The increase in stress during film growth is more rapid for thin films, falls off as the film becomes thicker, and reaches a maximum after 0.2–0.3 µm has been deposited. Another conclusion from the work of Ristau et al. is that stacks of alternate high- and low-index layers have about half the LDT of a block of the same thickness of the two materials deposited one on the other. This result also supports the results of Temple where the absorption in thin-film structures is shown to be concentrated in the film boundaries. Because the stacks have more film boundaries, they should have lower LDT levels.

Another factor that enters into the technique of laser damage testing is a self-annealing effect noted when multiple pulses are incident on a site. Frink et al. (1987) note that this is the cause of a difference in LDT levels measured on the samples from the same runs at different test facilities. This is especially true when a single site is irradiated at successively higher energy levels. Initial evidence of damage (a small scattering of light from initial pulses) disappears, and on closer inspection there appears to be no damage. This effect is thought to be caused by a healing or elimination of superficial defects in the film, leaving a denser and more durable coating than was there before. Thus, the LDT resulting for a given structure can appear to be higher than it actually is if a test uses single pulses of increasing energy levels that are incident at the same location on the surface. Frink et al. also determined that the improvement remained for about 4 days on some samples and was completely gone after 12 days. Further tests were run on samples that were exposed to high humidity and then tested after being dried. Frink et al. reached the conclusion that absorbed water in the films is the primary factor in different LDTs of the conditioned and unconditioned films. Irradiating different locations also allows one to eliminate defect-oriented failures and to determine threshold levels possible for defect-free films on defect-free surfaces.

The commercial and captive coating houses tend to be secretive about the materials and process used to produce their films. Thus, many of the data reported in the literature are generic (i.e., AR, MLAR, ADR, etc.) rather than specific as to material types. Fortunately, there are sufficient data published to determine damage thresholds of the more commonly used thin-film materials, which are also probably the better materials available. Test data from various sources have been gathered together and are reported in Table 8.5 for Nd:YAG, Table 8.6 for iodine, Table 8.7 for HF, and Table 8.8 for CO_2. The order of the data is more or less by material type and then by pulse length. Ambiguities between experimenters still exist, which makes comparisons difficult. One uncertainty is how to decide whether damage has occurred and what number to assign as the damage threshold. Some report the highest exposure level without damage, others report the lowest level at which damage occurs, and still others report an average of these two numbers. Some take data based on a few shots, while others (such as Seitel et al., 1981) used approximately 50 sites per test and a maximum likelihood analysis [see Porteous et al. (1977)] to arrive at the LDT.

8.6.3. Material Properties

Any material that can be put into a gaseous or fluid state is a candidate for use in making optical thin films. Those materials known to have the best optical and

Table 8.5   1.064-μm Pulsed Laser Damage Thresholds

Material	Type	LDT (J/cm²)	FWHM (nsec)	Reference
TiO$_2$/SiO$_2$	ADR	5–9	1	Carniglia (1981)
TiO$_2$/SiO$_2$	ADR	7–13	1	Wirtenson (1986)
TiO$_2$/SiO$_2$	ADR	35	15	Ristau et al. (1985)
TiO$_2$/SiO$_2$ (coevap)	ADR	75	15	Ristau et al. (1985)
ZrO$_2$/SiO$_2$	ADR	8–10	1	Wirtenson (1986)
ZrO$_2$/SiO$_2$	ADR	4–8	1	Carniglia (1981)
Ta$_2$O$_5$/SiO$_2$	AR	4–18	1	Milam et al. (1981)
Ta$_2$O$_5$/SiO$_2$	AR	4–9	1	Carniglia (1981)
Ta$_2$O$_5$/SiO$_2$	ADR	72	15	Ristau et al. (1985)
Leach etch[a]	AR	4–13	1	Deaton et al. (1980)
Sol-gel	AR	8–15	1	Milam et al. (1981)
Sol-gel	AR	15–50	9	Deaton et al. (1980)
HfO$_2$/SiO$_2$	ADR	166	15	Ristau et al. (1985)
Al$_2$O$_3$/SiO$_2$	AR	4–10	1	Carniglia (1981)
Al$_2$O$_3$/SiO$_2$	ADR	78	15	Ristau et al. (1985)
Al$_2$O$_3$	AR	12	9.5	Seitel et al. (1981)
Al$_2$O$_3$	SL	13	5	Walker et al. (1979)
Al$_2$O$_3$	SL	20	15	Walker et al. (1979)
HfO$_2$	SL	5	1	Pawlewicz et al. (1979)
HfO$_2$	SL	8	5	Walker et al. (1979)
HfO$_2$	SL	13	15	Walker et al. (1979)
MgF$_2$	AR	13	5	Walker et al. (1979)
SiO$_2$	AR	12	9.5	Seitel et al. (1981)
SiO$_2$	SL	47	15	Walker et al. (1979)
ThF$_4$	AR	2.5–5	9.5	Seitel et al. (1981)
ThF$_4$	SL	41	15	Walker et al. (1979)
ZrO$_2$	SL	10	5	Walker et al. (1979)
ZrO$_2$	SL	10	15	Walker et al. (1979)

[a]The leach-etch data was taken at 0.532 μm, but Deaton et al. note that the results are comparable to data taken another time at 1.064 μm.

physical properties have been studied and used extensively. Other, less likely, materials have also been characterized, and some are used in situations where the advantage of one good property outweighs the disadvantage of some other less desirable property (e.g., using a relatively soft radioactive material such as ThF$_4$ on infrared optics because of its low index at 10 μm, the relative ease of depositing it, and the fact that the films are generally smooth and adherent). One can usually infer the optical and phsyical properties of thin films from the bulk properties; however, techniques in depositing the films can cause variations.

Table 8.6   1.315-μm Pulsed Laser Damage Thresholds[a]

Material	Type	LDT (J/cm^2)	FWHM (nsec)	Supplier[b]
PbF$_2$/ZnS	ADR	17–27	8	Laser Power Optics
ThF$_4$/ZnS	ADR	74–85	8	Laser Power Optics
ThF$_4$/ZnS	ADR	54–111	8	OCLI
ThF$_4$/ZnS	ADR	70–92	8	Northrop
PbF$_2$/ZnSe	ADR	5–11	8	Laser Power Optics
ThF$_4$/ZnSe	ADR	9–13	8	Laser Power Optics
TiO$_2$/SiO$_2$	ADR	18–24	8	OCLI
TiO$_2$/SiO$_2$	ADR	32–117	8	Spectra Physics
ZrO$_2$/SiO$_2$	ADR	47–55	8	Spectra Physics
Al$_2$O$_3$/SiO$_3$	ADR	35	8	Northrop

[a]Data for reflector stacks deposited on silicon and molybdenum mirror surfaces and reported by Deaton and Seiler (1984). Data for coating types for which materials are not identified are not included.
[b]Companies supplying the test samples.

Table 8.7   2.7-μm Pulsed Laser Image Thresholds

Material	Type	LDT (J/cm^2)	FWHM (nsec)
Si (evaporated)	SL	20	100
Si (sputtered)	SL	55	100
SiH	SL	40	100
SiO$_2$	SL	120	100
Si/SiO$_x$ (evaporated)	ADR	70	100
Si/SiO$_2$ (sputtered)	AR	98	100
SiH/SiO$_2$ (sputtered)	AR	114	100

Source: Donovan et al. (1981).

Unfortunately, thin films generally are less durable than bulk materials because of the presence of voids, inclusions, and sometimes impurities. Good processing techniques are needed to achieve optimum optical performance and durability.

In general, the best coatings are made using the best starting material. Certainly, high purity is desirable. Purity is typically stated as a percent (e.g., 99.99% or "four 9s"). Designations of this type rarely indicate the absolute purity of the material but rather the difference from 100% and the total of a limited number of other possible impurities for which the lot has been tested. Four 9s pure ThF$_4$ probably contains no more than 0.01% rare earth metals but

Table 8.8   10.6-μm Pulsed Laser Damage Thresholds

Material	Type	LDT (J/cm^2)	FWHM (nsec)	Reference
TlI/Kcl/TlI	ML	8	0.01	Detrio and Dempsey (1980)
DLC on Ge	SL	3	20	Gibson and Wilson (1985)
DLC on Ge	SL	4	50	Gibson and Wilson (1985)
DLC on Ge	SL	28	100	Wood (1986)
ThF$_4$	SL	67	100	Wood (1986)
ZnS	SL	40	20	Gibson and Wilson (1985)
ZnS	SL	16	33	Lewis et al. (1985)
ZnS	SL	60	50	Lewis et al. (1985)
ZnSe	SL	40	20	Gibson and Wilson (1985)
ZnSe	SL	60–73	33	Lewis et al. (1985)
ZnSe	SL	60	50	Gibson and Wilson (1985)
ZnSe	SL	56	100	Wood (1986)
ZnSe	SL	31–38	340	Lewis et al. (1985)
ZnSe/ThF$_4$	ADR	90	230	Deng et al. (1985)
ZnSe/ThF$_4$	AR	30–40	230	Deng et al. (1985)

might contain a higher percentage of oxygen. A specific analysis is required where absolute high purity is required or where there are known contaminants that cannot be tolerated (i.e., soluble salts such as NaCl and KCl and water). Thin-film coating materials for most routine work are specified as vacuum-annealed or vacuum-deposition grade materials. Materials can also be purchased in a wide range of forms such as powder, granular, pressed pellets, crystalline, chunks, wire, and even vacuum-pressed specialty shapes for sputtering targets and E-gun pockets.

The remainder of this section presents data on the optical and physical properties of various thin-film materials. Every attempt has been made to ensure the quality of these data, which come from a large number of sources. For the most part, the discussion of each material will refer to more than one data source. Also, the range of properties for certain materials shown by the reviewed data is usually listed in tables. The range of data uncovered reinforces the contention that individuals must take responsibility for characterizing their own work and/or capability. Where differing properties can be obtained by processing control, the range of control parameters and the results are indicated.

Each subsection starts with a brief summary of the class of materials in that section and the range of properties expected in general. This is then followed with expanded details on selected materials for which considerable data are available or for which emphasis is deemed important. Finally, there is a table of abbreviated data for as many materials as information could be found for. References for the data in the table are numbered, and these references are listed in appendix. A letter M in the reference column indicates that some of the data for that material are the result of the author's personal experience but not previously published. The comments are used to qualify or define parameters or processing techniques where known. The following abbreviations are used extensively to compress the data:

Abbreviation	Definition
AB	Air bake
EB	E-beam source deposition
hygsc	Hygroscopic
RE	Reactive deposition
RPD	Reactive plasma deposition
RS	Resistive source deposition
spt	Sputter-deposited film
ST	Substrate temperature in °C

The material has been broken into subsections to make it more manageable. The organization is generally from lower to higher index groups and alphabetically within a section. The subsections are halides, oxides, sulfur/selenium/tellurium compounds, other compounds, and metals.

Halides

The halides are relatively low-index films transparent from the UV region out to the FLIR window of about 10 μm. This class of materials consists of the fluorides, chlorides, bromides, and iodides. All but the fluorides are generally water-soluble and soft and are therefore usable only in specialty, protected, or space environments. The fluorides are basically insoluble and are moderately durable when properly prepared.

Typically, more durable films are prepared by depositing at higher temperatures or when processing includes a postdeposition vacuum anneal cycle before removal from the chamber. After removal from the chamber, air baking may also improve durability, but some caution should be exercised because an air bake at sufficiently high temperature might convert some materials to an oxide of a considerably different index and absorption characteristics. This is particularly true of 300°C or higher air bake of CeF_3 and ThF_4. Many other fluorides exhibit a slight increase in index when air baked. This may be caused by a densification of the film but is more likely due to a slight or partial oxidation.

The evaporation characteristics are such that fluorides usually can be evaporated from a resistive source or an E-gun with no problem. When removed from the vacuum chamber, films have been known to increase in index because of the absorption of moisture. This effect can be fairly rapid or take a long time. Thus, some investigators have reported an aging effect for films. Sputtered films have a tendency to decompose and can be somewhat absorbing. However, those facilities that specialize in sputtered coatings seemed to have control over this problem for many materials.

Ion-assisted deposition studies have not been as extensive for the halides as for the oxides; however, what has been done indicates that durability can be improved and denser, moisture-free films can be prepared under the proper conditions. Martin (1986) reported that oxygen-assisted depositions improve the durability of MgF_2 and Na_3AlF_6 films. The Na_3AlF_6 film also exhibited an increase in refractive index from 1.34 for the unassisted ambient temperature deposition to 1.37 for the oxygen-assisted deposition. No data were reported for the refractive index of the MgF.

Ion-assisted deposition of MgF_2 using argon ions was reported by Gaskill et al. (1986). They reported a change of the UV properties with a 75-Å shift in the short-wavelength absorption edge to longer wavelengths and the development of an absorption band at 260 nm. The films had excellent mechanical properties for deposition at ambient room temperature, with some surpassing those of conventionally deposited films. These workers also observed a depletion of fluorine and the inclusion of oxygen. It was surmised that argon bombardment left dangling bonds that were filled with oxygen from the vacuum background. The denser films with oxygen had a slightly higher index, just below 1.40.

Table 8.9 lists the properties of halides.

Oxides

The oxides are generally higher in index than the halides and have a narrower transmission range. The transmission range extends from the longer UV wavelengths or the shorter visible wavelengths out to the FLIR band, with some cutting off in the mid-IR, and other wavelengths deeper in the IR. Oxides are generally insoluble and more durable than the fluorides. Some exhibit a tendency to disassociate on evaporation and thus require a reactive environment of oxygen to maintain stoichiometry.

Instabilities in stoichiometry and film density can result in a wider range of refractive indices for films and the possibility of absorption, both of which are controllable by process parameters. Slightly absorbing films can often be cleared up by baking at higher temperatures in air. Some oxides evaporate at fairly high temperatures, requiring the use of E-guns. The oxides are usually deposited at higher substrate temperatures with a partial background of oxygen introduced to achieve a stoichiometric nonabsorbing film.

The partial pressure of oxygen, the deposition rate, and the substrate temperature are all parameters that determine optical and physical properties of the films (Table 8.10). More complete oxidation occurs for higher substrate temperatures, higher oxygen background pressure, and slower deposition rates.

a. SiO_2. Of the most commonly deposited oxides, quartz shows the least tendency to dissociate. Good films can be formed with minimal variation in optical properties for a wide range of deposition parameters. Fluck et al. (1986) studied SiO_2 deposited at 8–18 Å/s, from 0.6 to 1.65×10^{-4} torr and at substrate temperatures ranging from 20 to 350°C. They obtained refractive indices ranging from 1.43 to 1.47 in the visible range, with the higher value resulting from the higher temperature, the higher rate, and an intermediate partial pressure of oxygen. The lower index value was obtained from a condition with the lower substrate temperature, higher rate, and higher partial pressure of oxygen.

Quartz has a tendency to take in moisture. This is shown by a spectral absorption band present at about 3 μm in the less dense films. The more durable films obtained from depositions at higher substrate temperatures are more moisture-stable (Pliskin 1968). More recent efforts depositing SiO_2 by IAD (McNeil et al. 1984; Saxe et al. 1984) have shown that durable films of quartz can also be obtained at substrate temperatures as low as 50°C. Although the water band at 3 μm was not measured in these papers, the films were deposited with other high-index materials, and the band shifts normally present in the non-IAD depositions were shown to be reduced for the IAD deposition. Also, the hydrogen content of the films was shown to be reduced. These results imply a reduced sensitivity to absorbed moisture. There is also an absorption band caused by Si—O stretching vibrations in the 9.4 μm region. The exact location of this band and subsequent absorption bands at 11.0 and 12.5 μm depend on the degree of oxidation (mixture of SiO, SiO_2O_3, SiO, SiO_2O_3, and SiO_2) and the presence of silanol (SiOH) [see Pliskin (1968)].

Quartz is a very stable material, more so than many of the other oxides. Some other oxides have a tendency to dissociate and become absorbing when deposited directly to glass surfaces. For this reason, quartz is often used as a sealing layer against glass surfaces before the deposition of less stable oxides.

b. Al_2O_3. Aluminum oxide is a fairly stable material but has a slight tendency to dissociate. Therefore, it is usually deposited with a slight oxygen background of about 5×10^{-5} torr. The index is much lower than bulk sapphire, generally about 1.62–1.63 in the visible and less than 1.60 in the near-IR when deposited at 300°C substrate temperatures. The refractive index drops by about 0.06 when deposited at 40°C substrate temperatures.

Its intermediate index makes it an ideal material for use in antireflection coatings and narrowband QWOT rejection stacks. Conventional deposition techniques result in fairly durable films. IAD films have an improved durability and stability to absorbing moisture.

c. TiO_2. Titanium dioxide is possibly the most widely studied oxide because of its high indices and the ability to produce films with indices from around 2.1 to 2.7 depending on processing parameters. The exact index obtained depends

Table 8.9 Properties of Halide Thin Films

Material (comments)	References[a]	Transmission range	Visible		Near-IR		IR		Stress
			n	k	n	k	n	k	
AlF$_3$	1,2	0.2–11	1.385						
BaF$_2$ (R S, EB, 300ST) (RS, EB, 200ST)	M, 3	0.25–15	1.40		1.385		1.37 1.27	3×10^{-3}	T
BiF$_1$ (RE, spt)	4–6	0.26–20	1.74		1.30 1.70		1.65		C T
CaF$_2$ (RS, EB, 300ST, hygsc, soft, scattering in thick films)	M,1,7–9	0.15–12	1.25		1.20–1.24				
CeF$_3$ (RS, EB, 160ST, 300ST, AB300 converts to oxide n = 2.0–2.1)	M,7,10–12		1.55	1.4×10^{-4}	1.45–1.52		1.36–0.45	2×10^{-3}	T
CsBr (hygsc)	13,14	0.3–55	1.709		1.667		1.66		
CsI	13,14	0.25–80	1.806		1.742		1.70		
ErF$_4$ (EB, 300ST, AB300 improves durability, increases n by 0.05)	M		1.49		1.43		1.36	2×10^{-3}	
HfF$_4$	15		1.55		1.55		1.40		
KBr (soft, hygsc)	13,14	0.25–40	1.59		1.536		1.52	5×10^{-3}	
KCl	14	0.20–25	1.48		1.47		1.45		
KI (soft, hygsc)	13	0.25–45	1.68		1.65		1.62		
LaF$_3$ (RS, EB, 160–300ST; ST < 160 stress cracks; ST > 300 improves durability)	M,1,9–11	0.25	1.51—1.55		1.53		1.45–1.50		
PbCl$_2$ (RS low evaporation temperature)	M,7,16	0.3–14	2.20				2.20		T
PbF$_2$ (RS, 160ST, not durable)	M,1,7	0.24–20	1.74		1.63		1.62		T
	17		1.78		1.71		1.64	1×10^{-2}	

Material									
LiF (RS, hygsc)	1,8,13,14,18	0.11–9.0	1.40		1.38		1.109	1×10^{-3}	T
MgF_2 (RS, EB, 300ST, thk < 4 μm; QWOT are very durable; low STs are not durable)	1,8,19–21	0.11–7.5	1.385	9×10^{5}					T
Na_3AlF_3 (cryolite) (RS. hygsc, soft)	13		1.33						T
NaCl (water-soluble)	1,7,22	0.2–14	1.30–1.33		1.52		1.50	1×10^{-3}	
NaF	9,14,17,23	0.2–25	1.54		1.30		1.23	5×10^{-5}	
NdF_3 (EB, 160ST, 300ST. AB300 increases visible n by 0.02)	1,14,18	0.2–14	1.30–1.33	2×10^{-3}	1.50		1.40	6×10^{-3}	
	M	0.17	1.57	1×10^{-5}		3×10^{-3}			
PrF_3 (EB, 160ST—fairly durable; 300ST—stressed; AB improves durability)	12		1.56–1.58		1.49		1.47	$\sim\!10^{-2}$	
	M								
SrF_2 (RS, 160ST, soft)	M, 24	0.13–11	1.30–1.37		1.26–1.34		1.22–1.30	$\sim\!10^{-2}$	
TlCl (soft, water-soluble)	7,17,25	? to 25	2.25		2.26				
TlBr (soft, water-soluble)	14,17	0.6–40	2.45–2.70		2.39		2.33–2.38		
ThF_4 (RS, EB, 300ST—durable, AB300 oxidizes to $n = 1.86$)	M,7,19,22,24	0.2–15	1.48–1.54	5×10^{-6}	1.49	1.5×10^{-3}	1.35	1×10^{-3}	T
YF_3 (EB, 160ST low stress; 300ST stressed; AB300 increases visible n by 0.1)	M		1.43–1.46		1.40		1.37		
ZnF_2 (EB, 200ST. films smooth, not durable; AB300 improves durability)	M		2.3		1.75	5×10^{-3}	1.60		

Abbreviations: AB, air bake; C, compressive; EB, E-beam source deposition; hygsc, hygroscopic; RE, reactive deposition; RPD, reactive plasma deposition; RS, resistive source deposition; spt, sputter-deposited film; ST, substrate temperature in °C; T, tensile.

[a]References are listed in the appendix. M = Morton, previously unpublished data.

Table 8.10 Properties of Oxide Thin Films

Material (comments)	References[1]	Transmission range	Visible n	Visible k	Near-IR n	Near-IR k	IR n	IR k	Stress
AlO_2 (EB, 275 ST) (40 ST)	M,22,26–28	0.3–5.5	1.62–1.66		1.55–1.59		1.0	0.1	T
	1		1.54	2.3×10^{-5}		8×10^{-6}			
Al_2O_3:MgO (EB stoichiometry varies, spinel)	70		1.60–0.70						
Al_2O_3:Ta_2O_5 (EB, see reference)	29		1.58–2.05						
BcO (RE)	30	0.2–6	1.72						
Bi_2O_3	4,16,24	0.2–2.6	2.45						
CeO_2 (laser-assisted deposition)	M,10,16,22	0.4–5.0	2.18–2.42		2.2				T
	31		2.0						
GeO	23		1.65						
HfO_2 (laser-assisted deposition)	10,32	0.2–12.0	2.00		1.95				
	31		2.20						
InO_2 (cond trans film)	M	0.4–1.0	2.1	3×10^{-3}	>3.0	$>1 \times 10^{-1}$			
$InSnO_2$ (spt cond trans film)	M	0.4–1.0	2.0		2.0	6.0	5.5	1.0	
La_2O_3 (EB, RS)	10	0.35 to ?	1.85		1.83				
(laser-assisted deposition)	31		1.70						
MgO (EB, hazy, abs. moisture)	M,1,18,31	0.2–8.0	1.73		1.63	$\sim 10^{-6}$		$\sim 10^{-4}$	C
Nb_2O_3	33	—	2.16						
Nd_2O_3 (EB, 300 ST; RS, RE)	M,11	0.4 to ?	2.0		1.93				
Pr_6O_{11} (300 ST)	11	0.4 to ?	1.92		1.83				

Material	Ref	Range (μm)	n	k	n	k	
Sc₃O₃ (RS-howitzer)	1,31		1.86				C
SiO (EB, RE of SiO)	M,22,26,34–36	0.5–8.0	1.76–1.90	1.5×10^{-5}	1.40	2×10^{-6}	
SiO₂ (EB, RE of SiO)	M,1,18,26,36,37	0.2–8.0	1.46–1.47		1.2	1×10^{-1}	
(ion plating)	28		1.485				
SnO₂ (laser-assisted deposition)	M	0.4–1.0	2.0	$2–9 \times 10^{-3}$			
	31		1.96–2.0				
Ta₂O₅ (EB, RE at low deposition rates)	M,1,10,17,31	0.3–5.0	2.15				
(spt)	33		2.02				
(ion plating)	28		2.24				
TiO₂ (EB, RE)	M,1,7,10,37	0.35–12.0	2.10–2.30	5.5×10^{-3}		8×10^{5}	C
(ion plate or anodized Ti)	5,28,38,39		2.40–2.70	4×10^{-3}	2.3	1×10^{-3}	
(laser-aided deposition)	31		2.57				
ThO₂	1,7,22,40,41	0.25–15.0	1.80		1.75		
Y₂O₃ (EB, RE)	M,1,10,37	0.25–12.0	1.82		1.76		
(laser-assisted deposition)	31		1.70				
ZnO	31,33	0.35–12.0	1.85–2.0	0.4			C
ZrO₂	M,1,10,22		2.03	2×10^{-4}	1.97	2×10^{-5}	
(EB-inhomogeneous) (laser-assisted deposition)	31		2.10	2×10^{-4}		2×10^{-5}	
(ion plating)	28		2.18				
ZrO₂-TiO₂ (n dept on ST; subst 1)	42	0.4–7	1.9–2.1				

Abbreviations: Cond trans, condition transparent; C, compressive; EB, E-beam source deposition; RE, reactive deposition; spt, sputter-deposited film; ST, substrate temperature, °C; T, tensile.

[a]References are listed in Appendix A; M = Morton, previously unpublished data.

highly on substrate temperature, deposition rate, and oxygen background pressure. Figure 8.53 shows the refractive index for TiO_2 over the 0.4–2.0 μm range for films prepared by various methods. Data reported by Ritter (1965) for reactive evaporated films of 1×10^{-4} torr. His data are consistent with data reported by others for conventional reactive evaporations. Partial pressures below 1×10^{-4} torr usually result in a decrease in refractive index, and the films become absorbing.

Slight absorption in films can often be eliminated by postdeposition baking in air around 300°C. Baking at much higher temperatures could result in absorbing films caused by crystallization and a partial conversion to the rutile phase as reported by Rujkorakarn et al. (1986). Usually, films prepared at pressures of 1.5 $\times 10^{-4}$ torr or higher are nonabsorbing. There is less absorption in films deposited by conventional reactive evaporation if the deposition rate is kept fairly slow, in the 4 min per visible QWOT range. Pulker et al. (1986) report on films prepared by conventional reactive evaporation at 300°C substrate temperatures and ion plating at 50–100°C substrate temperatures. The ion-plated films had a higher refractive index, were denser, and were less susceptible to water vapor absorption. Rujkorakarn et al. (1986) study postdeposition annealing effects on films that are ion beam sputter-deposited at low substrate temperatures. The films were basically amorphous and of the anatase phase when annealed from 200°C to about 350°C. Annealing above 350°C causes crystallization to begin, and the films convert to the rutile phase, which is much more absorbing in the visible wavelengths. At 2 μm, the films remain nonabsorbing in either state.

d. Ta_2O_5. Tantulum pentoxide shows characteristics similar to those of titanium dioxide; however, the refractive index is lower. Conventional refractive evaporated films have visible refractive indices ranging from 1.95 to 2.10 depending on substrate temperature and the partial pressure of oxygen present during deposition. Films prepared by ion plating have been reported with refractive indices as high as 2.24 at 0.550 μm. Tantalum oxide deposited by conventional reactive evaporation has a tendency to become absorbing if deposited too fast. Typical deposition rates are in the range of 8–10 minutes per visible QWOT.

e. ZrO_3. Zirconium dioxide is another good film material. It has a refractive index slightly lower than that of tantalum pentoxide. Conventional reactive evaporation at 300°C can produce films with a refractive index from 1.90 to about 2.10, depending on the oxygen partial pressure. Deposition by reactive ion plating at low substrate temperatures has resulted in films with a refractive index of 2.18 at 0.550 μm. Low deposition rates are not critical to the production of good films of zirconium dioxide.

II–VI Compounds

This class of materials consists of the sulfides, selenides, and tellurides. Their refractive indices are higher than those of the oxides, and the useful transmission

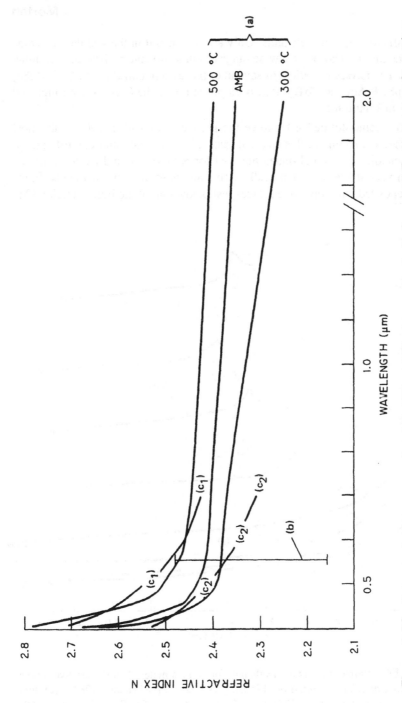

Figure 8.53 Refractive index versus wavelength for TiO_2. (a) Titanium sputtered with O_2 and Ar. Temperatures are for postdeposition anneal. [Rujkorakarn et al. (1986)]. (b) E-gun reactive evaporation, 300°C substrate temperature ppO_2 from 10^{-4} to 8×10^{-4} [Ritter (1966)]. (c) c_1, ion plating; c_2, reactive E-gun evaporation [Pulker et al. (1986)].

range shifts to longer wavelengths. Only a few transmit in the visible, and some only in the far-IR. They have a wide range of durability and resistance to staining and acid. The refractive indices of several of those compounds (ZnS, CdS, ZnSe, CdTe, PbS, PbSe, and PbTe) have been plotted over the 0.4–14 μm range and recorded in Figure 8.54.

a. ZnS. Zinc sulfide has the lowest refractive index and is probably the most widely used compound in this class. Its ease of evaporation and control made it an early candidate as a high-index material for both visible and IR applications. Although very stable, it is chemically active and is soluble in most acids. Films can be deposited with refractive indices just below that of the bulk material. One

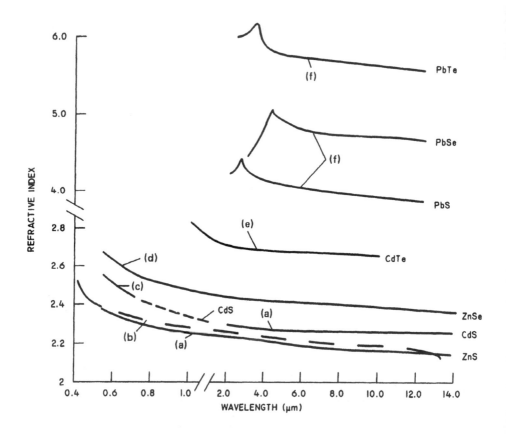

Figure 8.54 Refractive versus wavelength for various materials. (a) Resistive source ZnS and bulk CdS [Hall and Ferguson (1955)]. (b) CVD ZnS [Debenham (1984)]. (c) Bulk CdS [Gottesman and Ferguson (1954)]. (d) CVD Raytran ZmSe (Raytheon data sheet). (e) Hot-pressed CdTe [Ladd (1966)]. (f) PbS, ppSe, and PbTe [Zemel et al. (1965)].

problem in depositing ZnS is that the sticking coefficient (the ratio of molecules that nucleate and remain as part of the film to the number of arriving molecules) for evaporated films is highly temperature-sensitive. Thus, depositions using indirect monitoring for rate and thickness control are usually limited to substrate temperatures of 160°C or less.

Al-Douri (1986) reports on the losses in ZnS films as a function of the deposition rate and the substrate temperature. His study shows that the smallest losses are present for low deposition rates (<25 Å/min) and low substrate temperatures (50°C). As the substrate temperature increases, the differences diminish until reaching 150°C. Above 150°C, the differences appear minimal. Although data were taken for films deposited on substrates at temperatures up to 300°C, Al-Douri alludes to the difficulty in maintaining control and has data only for the lower deposition rates at the higher substrate temperatures. The higher losses noted for low substrate temperatures and for high deposition rates are probably due to more particulate matter being emitted from the source and subsequently buried in the film. This matter would represent a discontinuity that would cause light scattering. As the substrate temperature increases, the increase in losses at all deposition rates is probably caused by increased crystallinity and surface roughness.

Lewis et al. (1985) prepared ZnS films on both sides of Ge substrates by molecular beam epitaxy (MBE) under ultrahigh-vacuum (UHV) conditions that were very smooth, free of voids, free of OH bands, and transmitted close to 100%. Samples prepared at higher pressures (4×10^{-6} torr) simulating conventional optical coater environments transmitted about 1% less and showed the presence of OH species.

b. ZnSe.   Zinc selenide has a slightly higher refractive index than ZnS and does not exhibit as sensitive a sticking coefficient dependence on substrate temperature. Thus, ZnSe can be deposited at higher temperatures. Other than that, it is similar to ZnS and is a very useful material where a higher refractive index is needed. Table 8.11 lists the properties of II–IV and some III–V compound thin films.

Other Materials

A variety of other materials are used to coat IR optics that do not fall into the previous classifications. These are not necessarily related by a common property but are grouped here for convenience. One of the most exciting materials to come into use recently is a form of carbon with properties similar to those of diamond; thus, the phrase "diamond-like carbon" (DLC) is being used. Many also refer to it as alpha-carbon or α-C. The films are prepared by a CVD transport mechanism with some type of electronic activation, usually rf power. They can also be prepared by sputtering from a graphite or carbon target and by ion beam techniques. Because of wide interest in diamond-like and true diamond films, a considerable amount of material has been published on a variety of techniques, resulting in considerable variation in published physicochemical properties for the films. The many potential applications for diamond films include their use as doped semiconductors, as insulated gates in field-effect transistors (FETs), as protective coatings in electronics, and as protective coatings on optics. Although somewhat absorbing, the hardness and durability of diamond coupled with a natural refractive index of 2.0–2.4 makes it an ideal coating material in the IR region and as a single-layer AR coating for Ge.

a. DLC/Diamond.   Early work on preparing carbon films resulted in depositions consisting of mixed phases including graphite and diamond. Although the Russians discovered in the mid-1970s [see Messier et al. (1987)] that active hydrogen was a selective etchant of graphite and essential in producing purer diamond films, this was first reported in English by Spitsyn et al. (1981). Bubenzer et al. (1981) used an rf-excited plasma of various hydrocarbon gases to produce α-C:H films. In this effort, they studied the refractive index, C–H stretching absorption band intensity at 3.4 μm, and the visible absorption of the material. They note that the properties of the films depend primarily on the kinetic energy with which the ions hit the substrates. They obtained films with refractive indices ranging from 1.86 to 2.24, with the higher index resulting from ions arriving with the higher energies. The C–H stretching band is present because of the 25% hydrogen present in their process. However, the 3.4-μm absorption is also significantly reduced for the higher arrival energies. The short-wavelength absorption edge is moved toward the longer wavelengths, changing the visible appearance of the films from yellowish to black. An SLAR coating of this material on Ge in which the index is controlled at 2.9 μm and the QWOT thickness adjusted to 10.6 μm results in a sample with a peak transmission of about 92%. A 4% loss per surface appears to be the lower limit achievable from the better films.

Mirtich et al. (1985) report on single and dual ion-beam depositions to produce amorphous films of DLC. The single-beam setup uses a primary 30-cm ion gun masked to give a 10-cm beam using argon gas in the main chamber and as a neutralizer. Once a discharge is established, they introduce 28% CH_4 in the discharge chamber as the carbon source. This arrangement allows for two things to take place simultaneously: the deposition of carbon and the sputtering of carbon as a result of argon bombardment. At higher levels of CH_4, the discharge is extinguished, and at low levels there is no net deposition of film. The resulting films are a mixture of graphite and diamond structure, and it is suspected that the argon sputtering preferentially removes more graphite than diamond and results in improved film quality. Because of this, they added a second 8-cm ion gun. Auger analysis of these films showed primarily carbon with traces of oxygen and argon incorporated. Argon ion sputtering of the films removed the oxygen signal

Table 8.11 Properties of II–VI and III–V Compound Thin Films

Material (comments)	References[a]	Transmission range	Visible n	Visible k	Near-IR n	Near-IR k	IR n	IR k
AgS	43	0.6–12.0	3.0	0.8	2.4			
As$_2$S$_3$	44,45	0.6–12.0	2.66		2.41		2.38	
As$_2$Se$_3$	24,45	0.8–?			2.79			
CdS	1,17,46,46	0.6–7.0	2.5–2.6		2.30		2.25	
CdSe	48	0.7–25.0			2.45		2.42	
CdTe	5,7,17,31	0.86–28.0			2.7–3.05		2.6	
GaAs	47	0.9–11.0	3.8		3.3–3.5		3.1–3.3	
GaP	34,S&B		3.65	$\sim 1 \times 10^{-3}$	2.94	4×10^{-3}	2.9	2.7×10^{-2}
GaSb	S&B; 17				3.82; 2.76	2.5×10^{-3}; 50 cm^{-1}	3.84; 2.75	8×10^{-3}; 85 cm^{-1}
GeS	45	0.9–12.0			2.3		3.7	
InAs	1,47	3.8–7.0	4.5		3.7		3.42	
InSb	17,S&B; 1,47	7–16	1.15	$\sim 1 \times 10^{-3}$	3.46; 4.3		4.0	
PbS	17,S&B; 49–52	3.3–?		0.19	4.12–4.2	0.091; 15 cm^{-1}; 40 cm^{-1}; 80 cm^{-1}	3.95; 3.9–4.06	2×10^{-3}; 2 cm^{-1}; 50 cm^{-1}; 90 cm^{-1}
PbSe	17,47,52				4.6–4.8		4.77	
PbTe	1,52				5.6–5.74		5.71	
Sb$_2$S$_3$	1,16,25	0.5–10.0	3.0		2.7		2.5	
Se	53		3.1		2.48		2.47	
Te (soft)	M,25,52; 17; 53	3.5–7.8			5.1; 4.86; 4.40		4.80; 4.30	
ZnS (Irtran II) (EB, RS, spt)	1,31,33,46,54	0.4–14.0	2.25–2.4		2.26		2.17–2.2	
ZnSe (Irtran IV) (EB, RS, spt)	1,31,55	0.5–20.0	2.45–2.6		2.42		2.40	
ZnTe	10,31	0.65–?	2.8		2.7			

Abbreviations: EB, E-beam source deposition; RS, resistive source deposition; spt, sputter-deposited film.

[a]References are listed in Appendix A; M = Morton, previously unpublished data; S&B, Seraphin and Bennett, publication unknown.

and enhanced the argon signal. The electron diffraction patterns of films produced by both methods are characteristic of an amorphous solid. Optically, the films produced using the dual-beam source have lower absorption and higher refractive indices than those produced using the single beam only. In fact, the dual-beam-produced films have a refractive index of 2.46, close to that of dia mond, and show similar chemical resistance to immersion in 3:1 mixtures of concentrated H_2SO_4 and HNO_3. The films magnified 30,000 times in an electron microscope were very smooth and did not have the long-range order of the diamond crystal structure; thus, they were only "diamond-like."

Although carbon films are relatively easy to make, they do not adhere well to many materials. Therefore, before using carbon films, one must know the materials to which carbon will adhere and the techniques that improve adherence. It is difficult to compile a definitive list of these materials, because sticking depends on the process itself. However, there have been studies on techniques to improve adherence. Stein et al. (1981) showed that implanting carbon by accelerating the arriving carbon ions up to 10 keV resulted in embedding carbon to depths of 600 Å in caldium fluoride, a material to which carbon does not adhere well. The resulting films have a tenacious bond to the substrate and exhibit superior chemical resistance, allowing the carbon to be used as a barrier layer over other coatings. Mirtich et al. (1986) studied the problem of making DLC adhere to ZnS and ZnSe. The study shows that ion beam cleaning, ion implantation of nitrogen ions, and ion implantation of neon and helium ions do not help. However, an intermediate layer of sputtered Ge or Si works well.

b. Si_xN_y. Silicon nitride is another hard, durable material that can be used as a protective or barrier layer. When prepared by reactive sputtering, the composition, and thus the refractive index, of the material can be varied. Serenyi and Habermeier (1987) studied this process as a technique to control the deposition of the material for use as an SLAR for semiconductor lasers. They were able to obtain a range of indices from 1.55 to 2.03 by varying the sputter gas pressure ratio of Ar/N_2 from 1 to 150. The resultant films had low absorption for low and high refractive index films and relatively high absorption for the intermediate index values.

Table 8.12 lists the properties of the various thin-film materials.

Mixtures

All materials discussed up to now consist primarily of a singular molecular or elemental species. Attempts have been made to produce intermediate index films by mixing two or more film materials by one of two techniques. One method is to deposit films from two separate sources where the physical setup allows individual control of each source; thus, the ratio of the number of molecules of each source arriving at the substrate results in films with a refractive index

Table 8.12 Properties of Some Thin Films

Material (comments)	Ref.[a]	Transmission range	Visible		Near-IR		IR		Stress
			n	k	n	k	n	k	
α-Carbon (single-beam source)	56		2.0				2.0		C
(DLC) (dual-beam source)	56						2.46		
Diamond	57		2.46		1.95	1.2×10^{-10}	1.93		
	17		2.42					120	
GaAs (spt)	25		3.8		3.4		3.3		
Ge (EB, spt)	22,43	1.8–20.0	4.72	1.86	<4–4.5	$<10^{-2}$			
(>1000-Å films, 160 ST)	M		5.0		4.2		4.17	$<10^{-3}$	T
(>100-Å films, 80–100 ST)	M		6.0–6.5	4.0	4.5–5.0	$\sim10^{-2}$			
Si (EB, spt)	1,22	0.9–8.0	4.4	3×10^{-2}	3.4				
SiC	17		2.68		1.95				
Si$_3$N$_4$ (CVD)	33,58		2.0						
Si$_x$N$_y$ (reactive spt)	59		1.6–2.1	5.2×10^{-4}	1.5–1.7		1.75		
BN (RPD)	60								

Abbreviations: C, compressive; CVD, chemical vapor deposition; DLC, diamondlike carbon; EB, E-beam source deposition; RPD, reactive plasma deposition; spt, sputter-deposited film; ST, substrate temperature; T, tensile

[a]References are listed in the appendix; M = Morton, previously unpublished data.

determined by the final mixture. The second method, which is discussed in this section, is to premix the materials and deposit them from a single source.

a. Al_2O_3:MgO.   Heaney et al. (1981) used sapphire (n = 1.60–1.63) and magnesium oxide (n = 1.70–1.74) to deposit on substrates at 50 and 300°C, respectively. This effort attempted to combine the good characteristics of Al_2O_3 to overcome the haziness usually obtained for MgO and give a film with a good intermediate index (1.74). E-beam evaporation of synthetic single-crystal spinel Al_2O_3:MgO on 300°C substrates was used. Sequential deposition from a charge to which no fresh material was added between runs resulted in an initial film with a 1.72 refractive index decreasing to 1.66 on the fourth and fifth use. Although this particular effort was unsuccessful in achieving the desired results for this situation, similar attempts using other materials have had better success.

b. Al_2O_3:Ta_2O_5.   Ganner (1986) used various weighted mixtures of Al_2O_3 and Ta_2O_5 in an E-beam gun to produce films with refractive index from 1.58 to about 2.05 on substrates heated to 290°C and with a 4×10^{-4} torr backfill of oxygen. He also studied the dispersion of the films produced over the visible range and reports no anomalies. That is, the dispersion is well within the range expected considering the dispersion characteristics of the separate materials. Subsequent depositions produced consistent results. The greater success of this effort probably can be attributed to the fact that the components have nearly the same melting points and vapor pressures.

c. CeF_3:CeO_2.   Fujiwara (1963a) mixed CeF_3 and CeO_2 powders in various combinations and used resistive source evaporation from molybdenum boats to deposit films on substrates heated to 100°C. He obtained refractive indices ranging from 1.60 to 2.13. He also states that the films had reasonable durability in chemical and mechanical tests.

d. CeF_3:ZnS.   Fujiwara (1963b) also mixed CeF_3 and ZnS powders in various combinations to produce films ranging in index from 1.58 to 2.40. However, in this case, the films were evaporated by a radiation-heated vapor source.

e. MgF_2:ZnS.   Yadava et al. (1973) compared films prepared by depositing from a single source using various mixtures of MgF_2 and ZnS with films prepared from dual sources of each material in which an equivalent mixture was obtained discretely by alternately depositing very thin layers of each material. The mixed sources were prepared in pellet form and evaporated from a molybdenum boat onto ambient room temperature substrates. The refractive indices ranged from 1.38 to 2.35 as expected. Spectrophotometer scans of the films verify that they were basically homogeneous.

Metals

The optical properties of thin metal films have been determined with varying results. Studies have been conducted since the time of Drude to improve measurements and the performance of thin metal films for use as mirrors. The reflectance of thick (optically opaque) films commonly used for mirrors is reported in Section 8.5. A compilation of optical properties for these metals from various sources is included in Table 8.13. Because data were collected from

Table 8.13 Optical Properties of Selected Metal Thin Films

Wavelength (μm)	Ag		Al		Au		Cu	
	n	k	n	k	n	k	n	k
0.25	1.31	1.39	0.17	2.60	1.33	1.87	1.37	1.78
0.30	1.34	0.96	0.25	3.35	1.53	1.89	1.40	1.34
0.35	0.10	1.42	0.33	3.90	1.50	1.87	1.37	1.92
0.40	0.075	1.93	0.40	3.92	1.45	1.47	0.85	2.11
0.45	0.055	2.24	0.49	4.32	1.40	1.88	0.87	2.20
0.50	0.050	2.87	0.66	4.80	0.84	1.84	0.88	2.42
0.55	0.055	3.32	0.76	5.30	0.34	2.37	0.72	2.42
0.60	0.060	3.75	0.97	6.00	0.23	2.97	0.17	3.07
0.65	0.070	4.20	1.24	6.60	0.19	3.50	0.13	3.65
0.70	0.075	4.62	1.55	7.00	0.17	3.97	0.12	4.17
0.75	0.080	5.05	1.80	7.12	0.16	4.42	0.12	4.62
0.80	0.090	5.45	1.99	7.05	0.16	4.84	0.12	5.07
0.85	0.100	5.85	2.08	7.15	0.17	5.30	0.12	5.47
0.90	0.105	6.22	1.96	7.70	0.18	5.72	0.13	5.86
0.95	0.110	6.56	1.75	8.50	0.19	6.10	0.13	6.22
1.0	0.28	6.7	0.98	7.65	0.22	6.71	0.20	6.27
2.0	0.78	13.2	1.75	16.1	0.55	13.9		
3.0			3.2	23.5	1.17	21.0	1.22	17.1
4.0	1.89	28.7	4.8	30.0	2.04	27.9		
5.0			6.7	37.6	3.27	35.2		
6.0	4.15	42.6	9.5	44.4	4.70	41.7		
7.0			12.6	51.0	6.3	48.0	5.25	40.7
8.0	7.14	56.1	15.6	58.1	7.82	54.6		
10.0	10.69	69.0	26.0	67.3	11.5	67.5	12.6	64.3
12.0			33.1	78.0	15.4	80.5		

Note: The following references, listed in full in the appendix, provide data as noted:
[61] Hass and Waylonis (1961), Al 0.25–0.35
[62] Schulz and Tangherlini (1954), n of Ag, Al, Au, Cu 0.4–0.95
[63] Schulz (1954), k of Ag, Al 0.4–0.95; k of Au, Cu 0.45–0.95
[64] Valkonen et al. (1986), Ag 1.0–2.0
[65] Johnson and Christy (1972), n of Ag, Au, Cu 0.25–0.35; k of Ag, Au, Cu 0.25–0.40
[66] Motulevich (1973), Al 1.0–8.0; Au 1.0–12.0
[67] Cox et al. (1975), Al 10.0–12.0
[68] Beattie (1957), Ag 4.0–10.0
[69] Shklyarevskii and Padalka (1959), Cu 3.0–10.0

many sources, it was necessary, in some cases, to interpolate between points to match the wavelengths chosen for the table. Interpolation was also needed when using data from a graph. For the latter situation, data are present to two significant digits only. Tabular data are presented to the accuracy of the original source, if possible, or have been rounded off to fit the table.

APPENDIX: REFERENCES FOR TABLES 8.9–8.13

1. H.K. Pulker, *Appl. Opt. 18*:1969 (1979).
2. W. Heitmann, *Thin Solid Films 5*:61 (1970).
3. D.E. McCarty, *Appl. Opt. 2*:591 (1963).
4. T.J. Moravec, R.A. Skogman, and G.E. Bernal, *Appl. Opt. 18*:105 (1979).
5. H.A. MacLeod, *Thin Film Optical Filters*, 2nd ed. Macmillan, New York, 1986, p. 504.
6. T.J. Moravec and G. Bernal, *Appl. Opt. 17*:1938 (1978).
7. A.E. Enos, *Appl. Opt. 5*:51 (1966).
8. O.S. Heavens and S.D. Smith, *J. Opt. Soc. Am. 47*:469 (1957).
9. R.H. Hopkins, R.A. Hoffman, and W.E. Kramer, *Appl. Opt. 14*:2631 (1975).
10. D. Smith and P. Baumeister, *Appl. Opt. 18*:111 (1979).
11. G. Hass, J.B. Ramsay, and R. Thun, *J. Opt. Soc. Am. 49*:116 (1959).
12. R.T. Kersten, H.F. Mahlein, and W. Rauscher, *Thin Solid Films 28*:369 (1975).
13. W.J. Smith, *Modern Optical Engineering*, McGraw-Hill, New York, 1966, p. 158.
14. A.J. Moses, *Handbook of Electronic Materials*, Vol. 1, IFI/Plenum, New York, 1971, p. 12.
15. A.M. Ledger and R.C. Bastein, Nonradioactive antireflection coatings for infrared optics, FTR, Contract No. DAAG53-75-0168, 1975, p. 28.
16. L. Holland, *Vacuum Deposition of Thin Films*, Wiley, New York, 1961, p. 510.
17. W.L. Wolf, Properties of Optical Materials, in *Handbook of Optics*, W.G. Driscoll, W. Vaughan, Eds., McGraw-Hill, New York, 1978, p. 7–93.
18. M. Hertzberger and C.D. Salzberg, *J. Opt. Soc. Am. 52*:420 (1962).
19. H.K. Paulker, *Thin Solid Films 34*:343 (1976).
20. D. Hacman, *Opt. Acta 17*:661 (1970).
21. G. Emiliani, E. Masetti, and A. Piegari, Thin film refractive index determination by different techniques, *Thin Film Technologies II, Proc. SPIE 652*:153 (1986).
22. G. Hass and E. Ritter, *J. Vac. Sci. Technol. 4*:71 (1967).
23. Chemical Rubber Company, *Handbook of Chemistry and Physics*, 39th ed., 1959.
24. J. Feldman, M. Friz, and F. Stetter, *Res./Develop. 27*:49 (1976).
25. P.W. Black and J. Wales, *Infrared Physics*, Pergamon, New York, 1968, Vol. 8, p. 200.
26. J.T. Cox, G. Hass, and J.B. Ramsay, *J. Phys. 25*:250 (1964).
27. M.L. Lang and W.L. Wolfe, *Appl. Opt. 22*:1267 (1983).
28. H.K. Pülker, M. Bühler, and R. Hora, paper presented at SPIE's 30th Annual International Technical Symposium on Optical and Opto-Electronic Applied Sciences and Engineering, 1986.

29. P. Ganner, Medium index mixed-oxide layers for use in AR-coatings, *Thin Film Technologies II, Proc. SPIE 652*:69 (1986).
30. J. Ebert, Activated reactive evaporation, *Optical Thin Films, Proc. SPIE 325*:29 (1982).
31. H. Sanker and R. Hall, *Appl. Opt. 24*:3344 (1985).
32. P.W. Baumeister, and O. Arnon, *Appl. Opt. 16*:439 (1977).
33. Burgeil, J.C., V.S. Chen, F. Vrantny, and G. Smolinsky, *J. Elec. Chem. Soc. 115*:229 (1968).
34. A.P. Bradford and G. Hass, *J. Opt. Soc. Am. 53*:1096 (1963).
35. G. Hass and C.D. Salzberg, *J. Opt. Soc. Am. 44*:181 (1954).
36. W.A. Pliskin, *Thin Solid Films 2*:1 (1968).
37. K. Fluck, G. Szalai, J. Kojnok, and A. Szasz, Optimization of technology for the systems of hard-oxide layers by spectroscopic methods, *Thin Film Technologies II, Proc. SPIE 652*:307 (1986).
38. R.C. Menard, *J. Opt. Soc. Am. 52*:427 (1962).
39. R. Rujkorakarn, L.S. Hsu, and C.Y. She, Crystallization of titania films by thermal heating, *NBS Spec. Publ. 727*:253 (1986).
40. W. Heitman and E. Ritter, *Appl. Opt. 7*:307 (1968).
41. K.H. Behrndt and D.W. Doughty, *J. Vac. Sci. Technol. 3*:264 (1966).
42. F. Stetter, R. Esselborn, N. Harder, and P. Tolles, presented at Topical Meeting on Interference Coatings, Asilomar, Calif., 1976.
43. W.J. Anderson and W.N. Hansen, *J. Opt. Soc. Am. 67*:1053 (1977).
44. W.S. Rodney, I.H. Malitson, and T.A. King, *J. Opt. Soc. Am. 48*:633 (1958).
45. A.R. Hilton, *Appl. Opt. 5*:1877 (1966).
46. J.F. Hall and W.F.C. Ferguson, *J. Opt. Soc. Am. 45*:714 (1955).
47. J.A. Jamison, R.H. McFee, G.N. Plass, R.H. Grube,and R.G. Richards, *Infrared Physics and Engineering*, McGraw-Hill, New York, 1963, p. 267.
48. D.E. McCarthy, *Appl. Opt. 4*:507 (1965).
49. R.B. Schoolar and J.R. Dixon, *Phys. Rev. 137*:A667 (1965).
50. H.R. Reidel and R.B. Schoolar, *Phys. Rev. 131*:2082 (1959).
51. R.B. Schoolar and J.N. Zemel, *J. Appl. Phy. 35*:1848 (1964).
52. J.N. Zemel, J.D. Jensen, and R.B. Schoolar, *Phys. Rev. 140*:A330 (1968).
53. A.J. Moses, Optical material properties, in *Handbook of Electronic Materials*, Vol. 1, compiled by EPIC, (1971), talurium, p. 86; selenium, p. 62.
54. M. Debenhan, *Appl. Opt. 23*:2238 (1984).
55. A. Feldman, S. Malitson, D. Horwitz, D. Waxler, and M.J. Dodge, *Proc. 4th Annual Conf. on Infrared Laser Window Material*, 1984, p. 118.
56. J.M. Mirtich, D.M. Swee, and J.C. Angus, *Thin Solid Films 131*:245 (1985).
57. A.A. Déon, J.M. Mackowski, D.L. Balageas, and P. Robert (1986), Rain erosion damage of diamond-like coated germanium, in *Infrared Technology and Applications*, L.R. Baker and A. Masson, Eds., *Proc. SPIE 590*:106.
58. K.E. Bean, P.S. Gleim, R.L. Yeakley, and W.R. Runyan, *J. Electrochem. Soc.: Solid State Sci. 114*:733 (1967).
59. M. Serenyi and H.-U. Habermeier, *Appl. Opt. 26*:845 (1987).
60. F.C. Sulzbach and D.E. Morton, Development of infrared coating techniques for reconnaissance and weapons delivery, Tech. Report AFAL-TR-73-272 (1972).

61. G. Hass and J.E. Waylonia, *J. Opt. Soc. Am. 51*:719 (1961).
62. L.G. Schulz and F.R. Tangherlini, *J. Opt. Soc. Am. 44*:362 (1954).
63. L.G. Schulz, *J. Opt. Soc. Am. 44*:357 (1954).
64. E. Valkonen, C.-G. Ribbing, and J.-E. Sundgren (1986), Optical constants of thin silver and titanium nitride films, *Thin Film Technologies II, Proc. SPIE 652*:235.
65. P.B. Johnson and R.W. Christy, *Phys. Rev. B6*:4370 (1972).
66. B. Motulevich, Optical properties of metals, translated from *Proceedings of the P.N. Lebedev Physics Institute,* Vol. 55, Plenum, New York, 1973, Al, p. 65; Au, p. 80.
67. J.T. Cox, G. Hass, and W.R. Hunter, *Appl. Opt. 14*:1247 (1975).
68. J.R. Beattie, *Physica 23*:898 (1957).
69. I.N. Shklyarevskii and V.G. Padalka, *Opt. Spektrosk. 6*:78 (1959).
70. J.B. Heaney, G. Hass, and M. McFarland, *Appl. Opt. 20*:2335 (1981).

SYMBOLS

A	absorptance
B	numerator of the matrix product of a thin-film structure on a substrate
C	denominator of the matrix product of a thin-film structure on a substrate
H	quarter-wave optical thickness of a high-index layer
L	quarter-wave optical thickness of a low-index layer
N_x	complex refractive index for waves traveling in a medium and obliquely incident on an interface (x is polarization)
R	reflectance, or Fresnel coefficient
T	transmittance
Y	admittance of a thin-film structure
d	film thickness, or coefficient used in calculating reflectance from Fresnel coefficient
k	extinction coefficient: imaginary part of refractive index
n	real part of refractive index
r	reflected wave amplitude
α	absorption coefficient (cm^{-1})
Δg	half-bandwidth of high-reflectance region for an all-dielectric reflector stack
δ	phase thickness of film
η	$n - ik$ (complex refractive index)
λ	wavelength

REFERENCES

Al-Douri, A.A.J. (1986), *J. Vac. Sci. Technol. A4*:2477.
Archer, R.S. (1962), *J. Opt. Soc. Am. 52*:970.

Archer, R.J. (date unknown). *Manual on Ellipsometry*, Gaertner Scientific Corporation.

Baldini, G., and L. Rigaldi (1970). *J. Opt. Soc. Am. 60*:495.

Behrndt, K.H., and D.W. Doughty (1967). *J. Vac. Sci. Technol. 4*:199.

Berning, P.H. (1962). *J. Opt. Soc. Am. 52*:431.

Bloembergen, N. (1973). *Appl. Opt. 12*:661.

Bowling, R.A. (1984). Laboratory Methods To Study Thin-Film Phenomena, presented at the 3rd Topical Meeting on Optical Interference Coatings, Monterey, Calif., April 17–19, 1984.

Bubenzer, A., B. Dischler, and A. Nyaiesh (1981), Optical Properties of Hydrogenated Amorphous Carbon (α-C:H)—a Hard Coating for IR-Optical Elements. NBS Spec. Publ. 638, p. 477.

Carniglia, C.K. (1981). *Thin Solids Films 77*:225.

Carniglia, C.K., J.H. Apfel, T.H. Allen, T.A. Tuttle, W.H. Loudermilk, and D. Milam (1979). NBS Spec. Publ. 568, p. 377.

Cartwright, C.H. (1940). *J. Opt. Soc. Am. 30*:110.

Catalan, L.A. (1962). *J. Opt. Soc. Am. 52*:437.

Cox, J.T., G. Hass, and R.F. Roundtree (1954). *Vacuum 4*:445.

Cox, J.T., G. Hass, and A. Thelen (1962). *J. Opt. Soc. Am. 52*:965.

Cox, J.T., G. Hass, and W.R. Hunter (1975). *Appl. Opt. 14*:1247.

Deaton, T.F., F. Rainer, D. Milam, and W.L. Smith (1980). Survey of Damage Thresholds at 532 nm for Production-Run Optical Components. NBS Spec. Publ. 620, p. 297.

Debenham, M. (1984). *Appl. Opt. 23*:2238.

Decker, D.L., and P.A. Temple (1977). In *Laser Induced Damage in Optical Materials*, A.J. Glass and A.H. Guenther, Eds., NBS Spec. Publ. 509, p. 281.

Demner, Y., and J. Shamir (1978). *Appl. Opt. 17*:3738.

Deng, H., M. Bass, and N. Koumvakalis (1985). Single Pulse Laser Induced Damage in IR Coatings at 10.6 Micrometers, NBS Spec. Publ. 727, p. 371.

Detrio, J.A., and D.A. Dempsey (1980). Pulsed CO_2 Damage Threshold Measurements of Rb:KCl and NaCl, NBS Spec. Publ. 620, p. 88.

Dobrowolski, J.A., F.C. Ho, and A. Waldorf (1983). *Appl. Opt. 22*:3191.

Donovan, T.M., E.J. Ashley, J.B. Franck, and J.O. Porteus (1981). Hydrogenated Amorphous Silicon Films: Preparation, Characterization, Absorption, and Laser-Damage Resistance, NBS Spec. Publ. 638, p. 472.

Ellis, W.P. (1963). *J. Opt. Soc. Am. 53*:613.

Ellis, W.P., and E.D. Megarity (1964). *J. Opt. Soc. Am. 54*:225.

Emiliani, G., E. Masetti, and A. Piegari (1986). Thin Film Refractive Index Determination by Different Techniques, *Thin Film Technologies II, Proc. SPIE 652*:153.

Ennos, A.E. (1966). *Appl. Opt. 5*:51.

Epstein, L.I. (1952). *J. Opt. Soc. Am. 42*:806.

Feldman, J., M. Fritz, and F. Stetter (1976). *Res./Develop. 27*:49.

Fluck, K., G. Szalai, J. Kojnok, and A. Szasz (1986). Optimization of Technology for the Systems of Hard-Oxide Layers by Spectroscopic Methods, *Thin Film Technologies II, Proc. SPIE 652*:307.

Frink, M.M., J.W. Arenberg, D.W. Mordaunt, S.C. Seitel, M.T. Babb, and E.A. Teppo (1987). *Appl. Phys. Lett. 51*:415.

Fujiwara, S. (1963a). *J. Opt. Soc. Am. 53*:880.

Fujiwara, S. (1963b). *J. Opt. Soc. Am. 53*:1317.

Ganner, P. (1986). Medium Index Mixed-Oxide Layers for Use in AR-Coatings. *Thin Film Technologies II, Proc. SPIE 652*:69.

Gaskill, J.D., U.J. Gibson, C.M. Kennemore, H.A. MacLeod, R.S. Moshrefzadeh, S.G. Saxe, and G.I. Stegeman (1986). Graduate Fellowships in Modern Optics, Final Report 830301-860828 for Contract DAAG29-83-G-0019.

Gibson, D.R., and A.D. Wilson (1985). Studies of CO_2 Laser Induced Damage to Infrared Optical Materials and Coatings, NBS Spec. Publ. 727, p. 100.

Gottesman, J., and W.F.C. Ferguson (1984). *J. Opt. Soc. Am. 44*:368.

Hall, J.P., Jr., and W.F.C. Ferguson (1985). *J. Opt. Soc. Am. 45*:714.

Harrick (1967). *Internal Reflection Spectroscopy*, Interscience, New York.

Harris, M., H.A. MacLeod, S. Ogura, E. Pelletier, and B. Vidal (1979). *Thin Solid Films 57*:173.

Hass, G., and A.P. Bradford (1957). *J. Opt. Soc. Am. 46*:125.

Hass, G., and J.E. Waylomis (1961). *J. Opt. Soc. Am. 51*:719.

Hass, G., H.H. Schroeder, and A.F. Turner (1956). *J. Opt. Soc. Am. 46*:125.

Heaney, J.B., G. Hass, and W. McFarland (1981). *Appl. Opt. 20*:2335.

Heavens, O.S. (1955). *Optical Properties of Thin Solid Films*, Butterworth, London.

Herpin, A.C.R. (1947). *Acad. Sci. Paris 225*:182.

Hordvik, A. (1977). *Appl. Opt. 16*:2827.

Ladd, L.R.S. (1966). *Infrared Phys. 6*:145.

Leonard, T.A., J. Loomis, K.G. Harding, and M. Scott (1983). Design and Construction of Three Infrared Ellipsometers for Thin-Film Research, NBS Spec. Publ. 620, p. 345.

Lewis, K.L., J.A. Savage, A.G. Cullis, N.G. Chew, L. Charlwood, and D.W. Craig (1985). Assessment of Optical Coatings Prepared by Molecular Beam Techniques, NBS Spec. Publ. 727, p. 162.

Loomis, J. (1973a). Private communication, April 9–11, 1973.

Loomis, J. (1973b). *Appl. Opt. 12*:877.

Loudermilk, W.H., and S.P. Mukherjee (1982). Graded-Index Antireflection Coatings for High Power Lasers Deposited by the Sol-Gel Process, NBS Spec. Publ. 638, p. 432.

MacLeod, H.A. (1969). *Thin Film Optical Filters*, American Elsevier, New York, p. 19.

MacLeod, H.A. (1982). *Optical Thin Films, Publ. 325*:21.

McNeil, J.R., A.C. Barron, S.R. Wilson, and W.C. Herrmann, Jr. (1984). *Appl. Opt. 23*:552.

Martin, P.J. (1986). *Vacuum 36*:585.

Messier, R., A.R. Badziam, K.E. Badzian, P. Bachmann, and R. Roy (1987). From Diamond-Like to Diamond Coating, presented at International Conference on Metallurgical Coatings, San Diego, Calif., Mar. 23–27, 1987.

Milam, D., F. Rainer, W.H. Loudermilk, J. Swain, C.K. Carniglia, and T.T. Hart (1981). A Review of the 1064-nm Damage Tests of Electron-Beam Deposited Ta_2O_5/ SiO_2 Antireflection Coatings, NBS Spec. Publ. 638:446.

Minot, M.J. (1976). *J. Opt. Soc. Am. 66*:515.

Mirtick, M.J., D.M. Swee, and J.C. Angus (1985). *Thin Solid Films 131*:245.

Mirtich, M.J., D. Nir, D. Swee, and B. Banks (1986). *J. Vac. Sci. Technol. A 4*:2680.

Moses, A.J. (1971). Optical Material Properties. Handbook of Electronic Materials (EPIC) 1:28.

Mouchart, J. (1977). *Appl. Opt. 16*:2722.

Nagendra, C.L., and G.K.M. Thutupalli (1983). *Appl. Opt. 22*:4118.

Nicoll, F.H., and F.E. Williams (1943). *J. Opt. Soc. Am. 33*:434.

Pawlewicz, W.T., R. Busch, D.D. Hays, P.M. Martin, and N. Laegreid (1979). NBS Spec. Publ. 568:359.

Perry, D.L. (1965). *Appl. Opt. 4*:987.

Pliskin, W.A. (1968). The Evaluation of Thin Film Insulators, *Thin Solid Films 2*:1.

Porteous, J.O., J.L. Jernigan, and W.N. Faith (1977), Multithreshold Measurements and Analysis of Pulsed Laser Damage on Optical Surfaces, A.J. Glass and A.H. Guenther, Eds., Proc. 9th Ann. Symp. on Optical Materials for High Power Lasers. Boulder, Colo., (Oct. 4–6, 1977), NBS Spec. Publ. 509, p. 507.

Pulker, H.K., M. Buhler, and R. Hora (1986). SPIE 30th Annual Int. Tech. Symp. on Optical and Optoelectronic Applied Sciences and Engineering, Aug. 17–22, 1986, San Diego, Calif.

Ristau, D., X.C. Dang, and J. Ebert (1985). Interface and Bulk Absorption of Oxide Layers and Correlation to Damage Threshold at 10.6 Micrometers, NBS Spec. Publ. 727, 298.

Ritter, E. (1965). Deposition of Oxide Films by Reactive Evaporation, presented at 12th National Vacuum Symposium, New York (Sept. 30, 1965).

Rujkorakarn, R., L.S. Hsu, and C.Y. She (1986). Crystallization of Titania Films by Thermal Heating, NBS Spec. Publ. 727:253.

Sankur, H., and W.H. Southwell (1984). *Appl. Opt. 23*:2770.

Saxe, S.G., M.J. Messerly, B. Bovard, L. DeSandre, F.J. Van Milligen, and H.A. MacLeod (1984). *Appl. Opt. 23*:3633.

Seitel, S.C., J.B. Franck, C.D. Marrs, J.H. Dancy, W.N. Faith, and G.D. Williams (1981). Selective and Uniform Laser-Induced Failure of Antireflection Coated LiNbO$_3$ Surfaces, NBS Spec. Publ. 638:439.

Sharma, S.K., R.M. Wood, and R.C. Wood (1977). *NBS Spec. Publ. 509*:183.

Southwell, W.H. (1983). *Opt. Lett. 8*:584.

Southwell, W.H. (1985). *Appl. Opt. 24*:457.

Southwell, W.H. (1986). Coherence Loss Due to Thin Film Interface Roughness, *Thin Film Technologies II, Proc. SPIE 652*:300.

Spitsyn, B.V., L.L. Bouilov, and B.V. Derjaguin (1981). *J. Cryst. Growth 52*:219.

Stein, M.L., S. Aisenberg, and B. Bendov (1981). Studies of Diamond-Like Carbon Coatings for Protection of Optical Components, NBS Spec. Publ. 638:482.

Stierwalt, D.L. (1973). *Private communication (Apr. 12, 1973)*.

Stierwalt, D.L., and R.F. Potter (1962). *Proc. International Conference on the Physics of Semiconductors*, Exeter, p. 513.

Stierwalt, D.L., and R.F. Potter (1963). Emittance Studies of III-V Compound Semiconductors, NOLC Report 630.

Stierwalt, D.L., J.B. Bernstein, and D.D. Dirk (1963). *J. Appl. Opt. 2*:1169.

Sulzbach, F.C., and D.E. Morton (1972). Development of Infrared Coating Techniques for Reconnaissance and Weapon Delivery, Tech. Report AFAL-TR-73-272.

Temple, P.A. (1979). *Appl. Phys. Lett. 34*:677.

Thelen, A. (1966). *J. Opt. Soc. Am. 56*:1533.

Thelen, A. (1971). *J. Opt. Soc. Am. 61*:365.

Thetford, A. (1969). *Opt. Acta 16*:37.

Walker, T.W., A.H. Guenther, C.G. Fry, and P. Nielson (1979). NBS Spec. Publ. 568:405.

Wirtenson, G.R. (1986). Coatings for High Energy Applications: The Nova Laser, *Proc. SPIE 607*:75.

Wood, R.M. (1986). *Laser Damage in Optical Materials*, IOP Publishing Ltd., p. 112.

Yadava, V.N., S.K. Jharma, and K.L. Chopra (1973). *Thin Solid Films 27*:243.

Yeh, P., and S. Sari (1983). *Appl. Opt. 22*:4142.

Young, L. (1967). *Appl. Opt. 6*:297.

Zemel, J.N., J.D. Jensen, and R.B. Schoolar (1968). *Phys. Rev. 140*:A330.

Appendix

REFERENCES FOR TABLES 5.4–5.30

1. *The Infrared Handbook*, Wolfe, W. L. and Zissis, G. J., eds, IRIA Center, Environmental Research Institute of Michigan, Office of Naval Research, U.S. Government Printing Office, Washington, D.C., 1985, p. 7–22.
2. *Handbook of Chemistry and Physics*, Weast, R. C. and Astle, M. J., Eds., CRC Press, Boca Raton, Fla., 1982.
3. *Handbook of Laser Science and Technology, Volume IV, Part 2, Optical Materials*, Weber, M.J., ed., CRC Press, Boca Raton, Fla., 1986, p. 52.
4. Ibid., p. 51.
5. Ibid., p. 53.
6. Breckenridge, R. G., et al., *Phys. Rev. 96*: 571 (1954).
7. Welker, H., *Physica 20*: 893 (1954).
8. *The Infrared Handbook*, op.cit., p. 7–24.
9. Klocek, P., Stone, L., Boucher, M., and DeMilo, C., *SPIE Proc. 929*: 69 (1988).
10. II–VI Incorporated brochure 1979.
11. *The Infrared Handbook*, op.cit., p. 7–23.
12. Musikant, S., *Optical Materials, An Introduction to Selection and Application*, Marcel Dekker, Inc., New York, 1985, p. 203.
13. Raytheon data.
14. Rhodes, W. H., Trickett, E. A., Wei, G. C., "Transparent Polycrystalline La_2O_3-Doped Y_2O_3," *SPIE Proc. 505*: 9–14 (1984).

15. Slack, G. A., "Advanced Materials for Optical Windows," General Electric Co., Report #79CRD071, 1979, p. 13.
16. Advanced Composite Materials Corp. brochure.
17. Slack, G. A., op.cit., p. 12.
18. Abrikosov, N., Kh., Bankina, V. F., Poretskaya, L. V., Shelimova, L. E., and Skudnova, E. V., *Semiconducting II–VI, IV–VI, and V–VI Compounds*, Plenum Press, New York, 1969, p. 104.
19. Ibid, p. 110.
20. *Thermophysical Properties of Matter*, Touloukina, Y. S., and Ho, C. Y., eds., Plenum Press, New York, Vol. 5, 1977, p. 305.
21. Ibid., p. 722.
22. Blakemore, J. S., *J. Appl. Phys. 53*: 10, R175 (1982).
23. Connolly, J., diBenedetto, B., and Donadio, R., *SPIE Proc. 181*: 141–144 (1971).
24. Koenig, J. R., "Thermal and Mechanical; Properties of Calcium Lanthanum Sulfide," SoRI Rpt. No. SoRI-EAS-85-401-5267-I-F, April, 1985.
25. *Thermophysical Properties of Matter*, op.cit., p. 920.
26. Taylor, R.E., Groot, H., and Larimore, J. "Thermophysical Properties of La_2O_3-Doped Y_2O_3, A Report to GTE Labs," Purdue University Report No. TPRL581, 1987.
27. "Thermophysical Properties of Y_2O_3 Based Materials," TPRL Report No. 581, Purdue University, February 1987.
28. Kodak Publication No. U-72, Eastman Kodak Company, NY, 24, 1971.
29. *Thermophysical Properties of Matter*, op.cit., p. 910.
30. Harris, R., Graves, G., Dempsey, D., Greason, P., Gangl, M., O'Quinn, D., and Lefebvre, M., "Optical and Mechanical Properties of Water-Clear ZnS," in *Laser-Induced Damage in Optical Materials: 1983*, NBS Special Production 688, Washington, D.C., 1985, pp. 52–58.
31. *Thermophysical Properties of Matter*, op.cit., p. 1005.
32. *Thermophysical Properties of Matter*, op.cit., p. 1077.
33. "Thermal Conductivity of Spinel, ALON, and Yttria," SoRI Rpt. No. SoRI-EAS-86-695-6081-I-F, July, 1986.
34. *Thermophysical Properties of Matter*, op.cit., p. 1089.
35. *Handbook of Optics*, Driscoll, W. G., ed., McGraw-Hill, New York, 1978, p. 7–130.
36. *American Institute of Physics Handbook*, Gray, D. E., Ed., McGraw-Hill, New York, 1972, p. 4–76.
37. *Thermophysical Properties of Matter*, op.cit., p. 791.
38. Neuberger, M., *Handbook of Electronic Materials Vol. 2, III–V Semiconducting Compounds*, Plenum Press, New York, 1971, p. 66.
39. Musikant, op.cit., p. 113.
40. Moses, A. J., *Handbook of Electronic Materials, Volume 1: Optical Materials Properties*, Plenum Press, New York, 1971, p. 16.
41. Feldman, A., Horowitz, D., Waxler, R., and Dodge, M., "Optical Materials

Characterization: Final Technical Report," NBS Technical Note 993, U.S. Government Printing Office, Washington, D.C., 1979, p. 63.

42. Kodak, op.cit., p. 17.
43. Sahagian, C. S., and Pitha, C. A., "Compendium on High-Power Infrared Laser Window Materials," LQ-10 Program, Air Force Cambridge Research Laboratories, AFCRL-72-0170, 1972, p. 132.
44. Rhodes, W.H., "Transparent Polycrystalline Yttria for IR Applications," Proc. 16th Symp. on Electromagnetic Windows, Atlanta, June, 1982.
45. Bailey and Yates, *Proc. Phys. Soc.* (London) *91*: 390 (1967).
46. Batchelder, D. N., and Simmons, R. O., *J.C.P. 41*: 2324 (1964).
47. Buchanan, R. C., *Ceramic Materials for Electronics Processing, Properties, and Applications*, Buchanan, R. C., ed., Marcel Dekker Inc., New York, 1986, p. 7.
48. *Thermophysical Properties of Matter*, op.cit., p. 816.
49. *Handbook of Optics*, op.cit., p. 7–132.
50. *Handbook of Chemistry and Physics*, op.cit., p. B-132.
51. Ibid., p. B-147.
52. Ibid., p. B-84.
53. Ibid., p. B-143.
54. *Handbook of Optics*, op.cit., p. 7–138.
55. *Handbook of Chemistry and Physics*, op.cit., p. B-148.
56. Ibid, p. B-113.
57. *Handbook of Laser Science and Technology*, op.cit., p. 9.
58. *Handbook of Chemistry and Physics*, op.cit., p. B-131.
59. *Handbook of Laser Science and Technology*, op.cit., p. 10.
60. *Handbook of Chemistry and Physics*, op.cit., p. B-139.
61. Ibid, p. B-164.
62. Ibid, p. B-134.
63. Ibid, p. B-87.
64. Kodak, op.cit., p. 18.
65. *Handbook of Laser Science and Technology*, op.cit., p. 11.
66. *Handbook of Chemistry and Physics*, op.cit., p. B-74.
67. Ibid, p. B-139.
68. Ibid, p. B-204.
69. Ibid, p. B-140.
70. Ibid, p. B-206.
71. Ibid, p. B-116.
72. Willingham, C., Klein, C., and Pappis, J., "Multispectral Chemically Vapor-Deposited ZnS: An Initial Characterization," in *Laser-Induced Damage in Optical Materials: 1981*, NBS Special Publication 638, Washington, D.C., 1983, p. 53.
73. *Handbook of Chemistry and Physics*, op.cit., p. B-98.
74. Ibid, p. B-160.
75. Ibid, p. B-92.
76. Ibid, p. B-80.
77. Ibid, p. B-173.

78. Rhodes, W. H., "Controlled Transient Solid Second-Phase Sintering of Yttria," *J. Am. Cer. Soc. 64*: 1, 12 (1981).
79. *Handbook of Chemistry and Physics*, op.cit., p. B-102.
80. Ibid, p. B-165.
81. Ibid, p. B-144.
82. Ibid, p. B-86.
83. DeVaux, L. H., and Pizzarello, F. A., *Phys. Rev. 102*: 85 (1956).
84. *Handbook of Chemistry and Physics*, op.cit., p. B-78.
85. *Thermophysical Properties of Matter*, op.cit., p. 540.
86. *Handbook of Chemistry and Physics*, op.cit., p. B-155.
87. Ibid, p. B-112.
88. *Handbook of Laser Science and Technology*, op.cit., p. 12.
89. *Handbook of Chemistry and Physics*, op.cit., p. B-111.
90. Sahagian, op.cit., p. 74.
91. *The Infrared Handbook*, op.cit., p. 7–35.
92. Ibid, p. 7–34.
93. Fernelius, N., Graves, G, and Knecht, W., *SPIE Proc. 297*: 188–195 (1982).
94. *Handbook of Laser Science and Technology*, op.cit., p. 13.
95. Optovac, Inc. brochure, 1969.
96. Kodak, op.cit., p. 16.
97. Ballard, S., McCarthy, K., and Wolfe, W. "Optical Materials For Infrared Instrumentation," IRIA Center, Report #2389-11-S, The University of Michigan, Willow Run Laboratories, 1959.
98. Sahagian, op.cit., p. 78.
99. Moses, op.cit., p. 44.
100. *The Infrared Handbook*, op.cit., p. 7–36.
101. Ibid, p. 7–37.
102. Texas Instruments unpublished data.
103. *Handbook of Optics*, op.cit., p. 7–137.
104. *The Infrared Handbook*, op.cit., p. 7–20.
105. Duncanson, A., and Stevenson, R. W., *Proc. Phys. Soc. 72*: 1001–1006 (1958).
106. *Handbook of Optical Constants of Solids*, Palik, E. D., ed., Academic Press, Orlando, 1985.
107. *The Infrared Handbook*, op.cit., p. 7–21.
108. *Landolt-Bornstein, Numerical Data and Functional Relationships in Science and Technology*, Hellwege, K.-H., ed., Springer-Verlag, Berlin, New Series, Volume 2, Part 6, 1959, p. 453.
109. Private communication.
110. *Landolt-Bornstein*, op.cit., Group III, Volume 2, 73, 1969.
111. Ibid, p. 71.
112. Klein, C., *Applied Optics 25*: 1873–1875 (1986).
113. Neuberger, op.cit., p. 67.
114. Ibid, p. 109.
115. *Landolt-Bornstein*, op.cit., Volume 2, Part 6, 1959, p. 489.
116. Neuberger, op.cit., p. 94.
117. Buchanan, op.cit., p. 4.

118. Trickett, E. A., Rhodes, W. H., Wei, G. C., and Sordelet, D., "Infrared Transmitting La_2O_3-Doped Y_2O_3," *Proc. 18th Symp. on Electromagnetic Windows*, Atlanta, 1986.
119. *Landolt-Bornstein*, op.cit., p. 482.
120. Ibid, p. 490.
121. Neuberger, op.cit., p. 79.
122. *Dielectric Materials and Applications*, von Hipple, A. R., ed., John Wiley and Son, Inc., New York, 1954, p. 487.
123. *Handbook of Laser Science and Technology*, op.cit., p. 350.
124. *American Institute of Physics Handbook*, op.cit., p. 4–116.
125. Sahagian, op.cit., p. 68.
126. *American Institute of Physics Handbook*, op.cit., p. 4–115.
127. Sahagian, op.cit., p. 64.
128. Ibid, p. 100.
129. Ibid, p. 124.
130. Ibid, p. 92.
131. Ibid, p. 88.
132. Cardona, M., *Semiconductors and Semimetals*, Beer, A. C., and Willardson, R. K., eds., Vol. 3, Academic Press, New York, 1967, p. 140.
133. *The Infrared Handbook*, op.cit., p. 7–31.
134. Ibid, p. 7–33.
135. Moses, op.cit., p. 76.
136. Ibid, p. 18.
137. Ibid, p. 100.
138. Klein, C. and Willingham, C., "Elastic Properties of Chemically Vapor-Deposited ZnS and ZnSe," in *Basic Properties of Optical Materials*, NBS Special Publication 697, Washington, D.C., 1985, pp. 137–140.
139. Blakemore, op.cit., p. R128.
140. Private communication.
141. Rhodes, W. H., Baldoni, J. G., and Wei, G. C., "The Mechanical Properties of La_2O_3-Doped Y_2O_3," GTE Labs Technical Report TR 86-818-1, July, 1986.
142. *Silicon Carbide—1968*, Henisch, H. K., and Roy, R., eds., Pergamon Press, New York, 1969, p. S367.
143. *Landolt-Bornstein*, op.cit., Volume 2, Part 1, 1971, p. 583.
144. Ibid, p. 580.
145. Ibid, p. 526.
146. Ibid, p. 574.
147. Baldwin, C. M. and Mackenzie, J. D., *J. Non-Crystalline Solids 31*: 441 (1979).
148. Wray, J. H. and Neu, J. T., *J. Opt. Soc. Am. 59*: 774 (1969).
149. Corning brochure—7940.
150. Corning brochure—7957.
151. Corning brochure—7971.
152. Schott brochure—IRG 9.
153. Drexhage, M. G., El-Bayoumi, O. H., and Moynihan, C. T., SPIE Proc. *320*: 27 (1982).

154. Loretz, T. J., Mansfield, J. L., Drexhage, M. G., Moynihan, C. T., Abstract in *Am. Ceram. Soc. Bull. 60*(3): 413 (1981).
155. Ohara brochure—HTF-1.
156. Drexhage, M. G., El-Bayoumi, O. H., Moynihan, C. T., Bruce, A. J., Chung, K. H., Gavin, D. L., and Loretz, T. J., *J. Am. Ceram. Soc. 65*, (10): C-168–C-171 (1982).
157. Schott brochure—IRG 7.
158. Schott brochure—IRG N6.
159. Corning brochure—9753.
160. Corning brochure—9754.
161. Schott brochure—IRG 11.
162. Schott brochure—IRG 3.
163. Schott brochure—IRG 2.
164. Borrelli, N. F. and Bumbaugh, W. H., *SPIE Proc. 843*: 6, (1987).
165. Klocek, P., and Colombo, L., *J. Non-Cryst. Sol. 93*: 1, 1987.
166. Malitson, I. H., Rodney, W. S., and King, T. A., *J. Opt. Soc. Am. 48*: 633 (1958).
167. Amorphous Materials Inc. brochure—AMTIR-1.
168. Schott brochure—IRG 100.
169. Savage, J. A., *Infrared Optical Materials and Their Antireflection Coatings*, Adam Hilger Ltd., Bristol, 1985, p. 84.
170. Gavin, D. L., Chung, K. H., Bruce, A. J., Moynihan, C. T., Drexhage, M. G., and El-Bayoumi, O. H., *Comm. Am. Ceram. Soc.*, C-182, 1982.
171. *Properties of Glasses and Glass-Ceramics*, Corning Glass Works, Corning, 1973.
172. Texas Instruments unpublished data.
173. Sahagian, op.cit., p. 140.
174. *Handbook of Laser Science and Technology*, op.cit., p. 73.
175. *American Institute of Physics Handbook*, op.cit., p. 4–158.
176. *The Infrared Handbook*, op.cit., p. 7–26.
177. Ibid, p. 7–37.
178. *Handbook of Chemistry and Physics*. Weast, R. C., ed., CRC Press, Cleveland, 1972, p. B-70.
179. Mecholsky, J. J., abstract #31 in Proc. of Second Int'l. Symposium on Halide Glasses, 1983, Rensselaer Polytechnic Institute, Troy, New York.
180. *Handbook of Optics*, op.cit., p. 7–136.
181. Ballard, op.cit., p. 32.
182. Perazzo, N. L., Gavin, D. L., Bruce, A. J., Loehr, S. R., and Moynihan, C. T., Abstract P6 in Proc. of the Second Int'l. Symp. on Halide Glasses, 1983, Rensselaer Polytechnic Institute, Troy, New York.
183. *The Infrared Handbook*, op.cit., p. 7–33.
184. Tesar, A., Bradt, R. C., and Pantano, C. G., Abstract #49 in Proc. of Second Int'l. Symp. on Halide Glasses, 1983, Rensselaer Polytechnic Institute, Troy, New York.
185. Ballard, op.cit., p. 20.

Index

Milton Keynes UK
Ingram Content Group UK Ltd.
UKHW020004071024
449327UK00031B/2647